OXFORD IB STUDY GUIDES

Andrew Allott

Biology

FOR THE IB DIPLOMA

2014 edition

OXFORD
UNIVERSITY PRESS

OXFORD
UNIVERSITY PRESS

Great Clarendon Street, Oxford, OX2 6DP, United Kingdom

Oxford University Press is a department of the University of Oxford. It furthers the University's objective of excellence in research, scholarship, and education by publishing worldwide. Oxford is a registered trade mark of Oxford University Press in the UK and in certain other countries

First published in 2014

British Library Cataloguing in Publication Data
Data available

978-0-19-839351-1

10 9 8 7

Paper used in the production of this book is a natural, recyclable product made from wood grown in sustainable forests. The manufacturing process conforms to the environmental regulations of the country of origin.

Printed in China by Golden Cup

Acknowledgements

The publishers would like to thank the following for permissions to use their photographs:

Artwork by OUP and Six Red Marbles

Cover image: © Martin Harvey / Alamy **p1:** http://www.ncbi.nlm.nih.gov; **p1:** OUP; **p1:** CreativeNature.nl/Shutterstock; **p1:** Public Domain **p1:** Public Domain; **p1:** © Nigel Cattlin/Visuals Unlimited/Corbis; **p1:** OUP; **p1:** OUP; **p1:** ZEPHYR/SCIENCE PHOTO LIBRARY; **p2:** OUP; **p2:** Public Domain; **p2:** PR. PHILIPPE VAGO, ISM/SCIENCE PHOTO LIBRARY; **p2:** Public Domain; **p2:** Public Domain; **p2:** DR KEITH WHEELER/SCIENCE PHOTO LIBRARY; **p2:** OUP; **p2:** OUP; **p2:** Public Domain; **p3:** SCI-COMM STUDIOS/SCIENCE PHOTO LIBRARY; **p3:** DR JEREMY BURGESS/SCIENCE PHOTO LIBRARY; **p3:** OUP; **p3:** DR KEITH WHEELER/SCIENCE PHOTO LIBRARY; **p3:** BIOPHOTO ASSOCIATES/SCIENCE PHOTO LIBRARY; **p3:** DR KEITH WHEELER/SCIENCE PHOTO LIBRARY; **p3:** NATIONAL LIBRARY OF MEDICINE/SCIENCE PHOTO LIBRARY; **p3:** © Bettmann/CORBIS; **p3:** OUP; **p3:** BIOPHOTO ASSOCIATES/SCIENCE PHOTO LIBRARY; **p3:** MICROSCAPE/SCIENCE PHOTO LIBRARY; **p4:** http://myibsource.com; **p4:** Dr Graham Beards/Wikipedia; **p4:** DR DAVID FURNESS, KEELE UNIVERSITY/SCIENCE PHOTO LIBRARY; **p4:** ASTRID & HANNS-FRIEDER MICHLER/SCIENCE PHOTO LIBRARY; **p4:** Michael Abbey/SCIENCE PHOTO LIBRARY; **p5:** OUP; **p5:** The American Association for the Advancement of Science; **p5:** OUP; **p5:** OUP; **p5:** THOMAS DEERINCK, NCMIR/SCIENCE PHOTO LIBRARY; **p6:** MOREDUN ANIMAL HEALTH LTD/SCIENCE PHOTO LIBRARY; **p6:** OUP; **p6:** Jmol; **p6:** Jmol; **p6:** Jmol; **p6:** Public Domain; **p6:** OUP; **p6:** http://www3.nd.edu; **p6:** Dr. Michaël Laurent, KULeuven, Belgium; **p7:** OUP; **p8:** OUP; **p8:** OUP; **p8:** Public Domain; **p8:** Dr. Gladden Willis/Getty Images; **p8:** Image Source/Getty Images; **p9:** OUP; **p9:** OUP; **p9:** OUP; **p9:** Public Domain; **p10:** MEDICAL SCHOOL, UNIVERSITY OF NEWCASTLE UPON TYNE/SIMON FRASER/SCIENCE PHOTO LIBRARY; **p11:** OUP; **p11:** Public Domain; **p11:** © Dennis Degnan/CORBIS; **p12:** Public Domain; **p12:** STEVE GSCHMEISSNER/SCIENCE PHOTO LIBRARY; **p13:** DAVID PARKER/SCIENCE PHOTO LIBRARY; **p15:** OUP; **p17:** staticd/Wikipedia; **p16:** OUP; **p16:** M.I. WALKER/SCIENCE PHOTO LIBRARY; **p16:** SCIENCE PICTURES LTD/SCIENCE PHOTO LIBRARY; **p71:** OUP; **p71:** OUP; **p71:** OUP; **p71:** OUP; **p71:** OUP; **p71:** OUP; **p87:** DR KEITH WHEELER/SCIENCE PHOTO LIBRARY; **p100:** OUP; **p121:** DR KEITH WHEELER/SCIENCE PHOTO LIBRARY; **p128:** OUP; **p128:** OUP; **p128:** OUP; **p145:** OUP; **p145:** DR P. MARAZZI/SCIENCE PHOTO LIBRARY; **p145:** Sam Droege/Flickr; **p165:** Public Domain; **p183:** SUSUMU NISHINAGA/SCIENCE PHOTO LIBRARY.

Although we have made every effort to trace and contact all copyright holders before publication this has not been possible in all cases. If notified, the publisher will rectify any errors or omissions at the earliest opportunity.

Introduction and acknowledgements

The IB Biology Programme has been comprehensively reviewed for teaching from September 2014 onwards. This book has been written in response to the curriculum review and is intended to help students find the information that they need quickly and easily when studying the new programme.

All topics in Higher Level (HL) and Standard Level (SL) Biology are covered, including all four options. The topics covered are in the same sequence as in the syllabus, but within some topics the sequence of sub-topics has been slightly altered, to give a more coherent progression of ideas.

- Topics 1–6 are core topics studied at both HL and SL.
- Topics 7–11 are additional topics studied only at HL.
- Options A–D can be studied at HL or SL, with extra material needed at HL, separated on clearly marked pages at the end of the option. Please note that on the new programme, only one option is studied.

Practice questions are included at the end of topics and options. Answers to each question are given, though students and teachers may be able to find other valid answers!

Guidance is given for students working on internal assessment or preparing for final exams.

There has never been a more important and exciting time to study biology. There are unprecedented opportunities for using recently developed techniques in beneficial ways, but there are also greater threats to the natural world than for millions of years. A thorough understanding of the principles of biology is essential if we are to counter the threats and make the most of the opportunities. Biology teachers worldwide should continue to be commended for the work they do in promoting this understanding. Teachers of IB Biology often take on an additional challenge – to promote international understanding. There are many opportunities for this in Biology. Apart from humans, living organisms do not recognize national frontiers. Living organisms throughout the biosphere, including humans, are interdependent. Human activities have international impacts, so international cooperation is essential to protect the biosphere and its treasure-house of biodiversity.

I am very grateful for the help that fellow teachers have given me during the writing of this book. Eleanor Walter at Oxford University Press was also a great help, as was Julian Thomas in his role as copy editor. I am indebted to my wife Alison and son William for their support and forbearance during the many hours that I have spent on it. I would like to dedicate the work that I did on the book to all biologists around the world, who are striving to conserve living organisms and their habitats.

Contents

Cell theory

INTRODUCING THE CELL THEORY

One the most important theories in biology is that cells are the smallest possible units of life and that living organisms are made of cells. The ancient Greeks had debated whether living organisms were composed of an endlessly divisible fluid or of indivisible subunits, but the invention of the microscope settled this debate. Cells consist of cytoplasm, enclosed in a plasma membrane. In plant and animal cells there is usually a nucleus that contains genes.

Human cheek cell

plasma membrane
cytoplasm
nucleus
mitochondria

Moss leaf cell

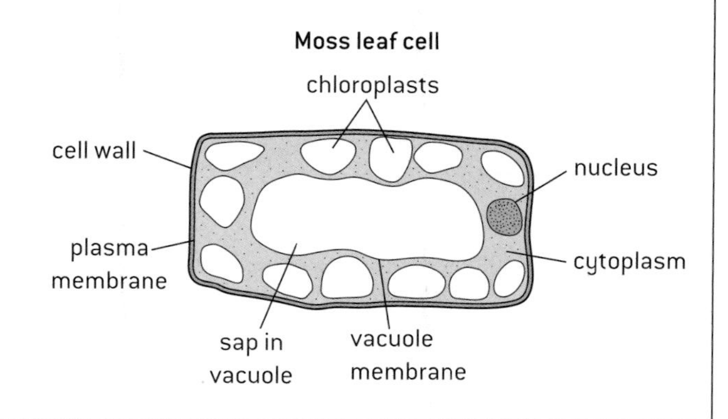

EXCEPTIONS TO THE CELL THEORY

The cell theory was developed because biologists observed a trend for living organisms to be composed of cells. Scientific theories can be tested by looking for discrepancies – cases that do not fit the theory. There are some tissues and organisms that are not made of typical cells:

1. **Skeletal muscle** is made up of muscle fibres. Like cells these fibres are enclosed inside a membrane, but they are much larger than most cells (300 or more mm long) and contain hundreds of nuclei.
2. **Giant algae** such as *Acetabularia* (below) can grow to a length of as much as 100mm so we would expect them to consist of many small cells but they only contain a single nucleus so are not multicellular.

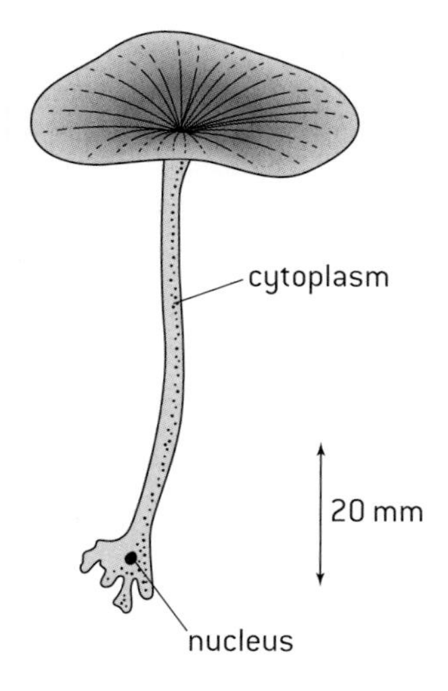

3. **Aseptate fungi** consist of thread-like structures called hyphae. These hyphae are not divided up into sub-units containing a single nucleus. Instead there are long undivided sections of hypha which contain many nuclei.

Despite these and some other discrepancies, there is still a strong overall trend for living organisms to be composed of cells, so the cell theory has not been abandoned.

DRAWINGS IN BIOLOGY

The command term 'draw' is defined by IB as: *'Represent by means of a labelled, accurate diagram or graph, using a pencil. A ruler (straight edge) should be used for straight lines. Diagrams should be drawn to scale.'*

A sharp pencil with a hard lead (2H) should be used. This allows clear, sharp single lines to be drawn. In exams, diagrams should not be drawn faintly as they will not show clearly in scans.

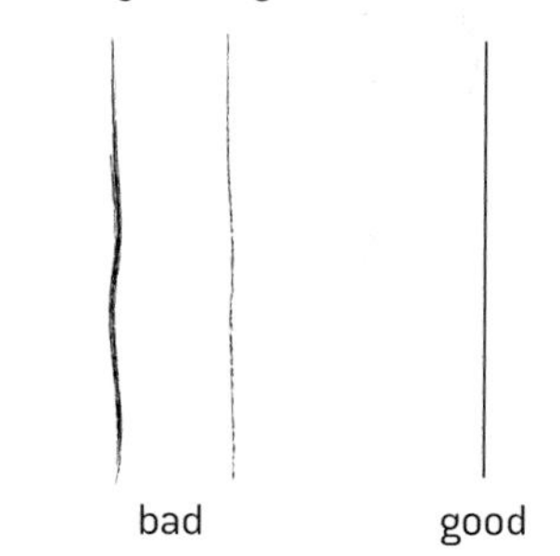

There should be no gaps, overlaps or multiple lines.

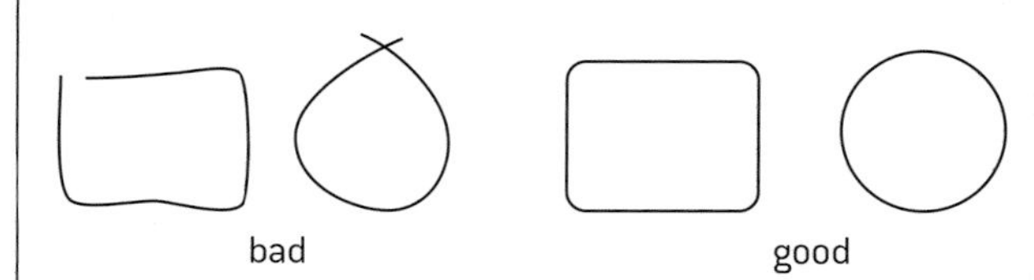

Labelling can be in ink or pencil, with labelling lines rather than arrows. Labelling lines should be drawn using a ruler and they should point precisely to the structure being labelled.

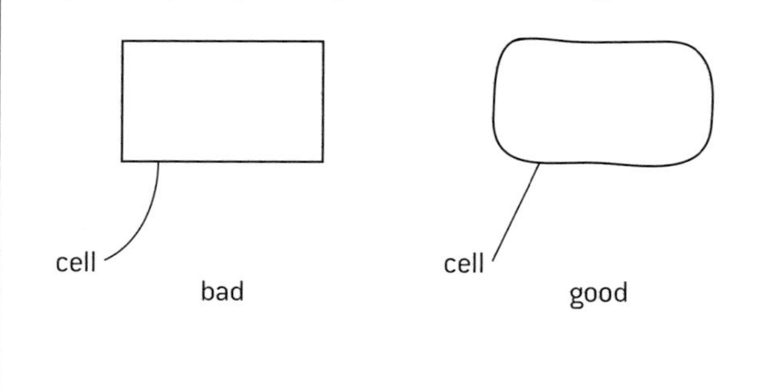

Unicellular and multicellular organisms

FUNCTIONS OF LIFE IN UNICELLULAR ORGANISMS

Unicellular organisms consist of only one cell. They carry out all **functions of life** in that cell. Two examples are given here: ***Paramecium*** lives in ponds and is between a twentieth and a third of a millimetre long. ***Chlamydomonas*** lives in freshwater habitats and is between 0.002 and 0.010 millimetres in diameter. They are similar in how they carry out some functions of life and different in others.

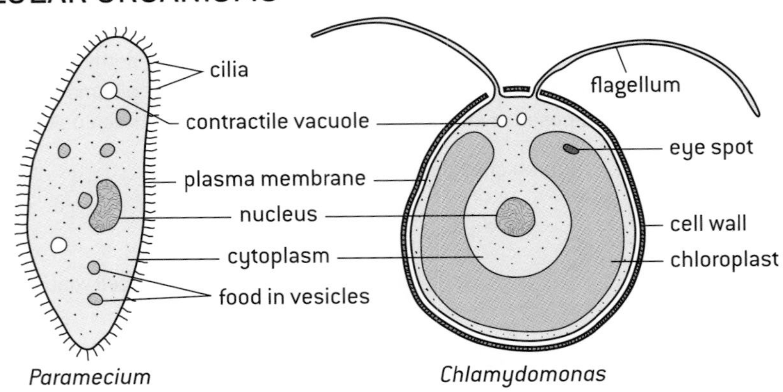

Function	*Paramecium*	*Chlamydomonas*
Nutrition	Feeds on smaller organisms by ingesting and digesting them in vesicles (*endocytosis*)	Produces its own food by photosynthesis using a chloroplast that occupies much of the cell
Growth	Increases in size and dry mass by accumulating organic matter and minerals from its food	Increases in size and dry mass due to photosynthesis and absorption of minerals
Response	Reacts to stimuli, e.g. reverses its direction of movement when it touches a solid object	Reacts to stimuli, e.g. senses where the brightest light is with its eyespot and swims towards it
Excretion	Expels waste products of metabolism, e.g. CO_2 from respiration diffuses out of the cell	Expels waste products of metabolism, e.g. oxygen from photosynthesis diffuses out of the cell
Metabolism	Both: produces enzymes which catalyse many different chemical reactions in the cytoplasm	
Homeostasis	Both: keeps internal conditions within limits, e.g. expels excess water using contractile vacuoles	
Reproduction	Both: reproduces asexually using mitosis or sexually using meiosis and gametes	

MULTICELLULAR ORGANISMS

As a cell grows larger its **surface area to volume ratio** becomes smaller. The rate at which materials enter or leave a cell depends on the surface area of the cell. However, the rate at which materials are used or produced depends on the volume. A cell that becomes too large may not be able to take in essential materials or excrete waste substances quickly enough. Large organisms are therefore multicellular – they consist of many cells.

Being multicellular has another advantage. It allows division of labour – different groups of cells (tissues) become specialized for different functions by the process of **differentiation**. The drawings (right) show two of the hundreds of types of differentiated cell in humans.

EMERGENT PROPERTIES

Emergent properties arise from the interaction of the component parts of a complex structure. We sometimes sum this up with the phrase: 'the whole is greater than the sum of its parts'. Multicellular organisms have properties that emerge from the interaction of their cellular components.

For example, each cell in a tiger is a unit of life that has distinctive properties such as sensitivity to light in retina cells, but all of a tiger's cells combined give additional emergent properties – for example the tiger can hunt and kill and have a profound ecological effect on its ecosystem.

DIFFERENTIATION

An organism's entire set of genes is its **genome**. In a multicellular organism each cell has the full genome, so it has the instructions to develop into any type of cell. During differentiation a cell uses only the genes that it needs to follow its pathway of development. Other genes are unused. For example, the genes for making hemoglobin are only expressed in developing red blood cells. Once a pathway of development has begun in a cell, it is usually fixed and the cell cannot change to a different pathway. The cell is said to be 'committed'.

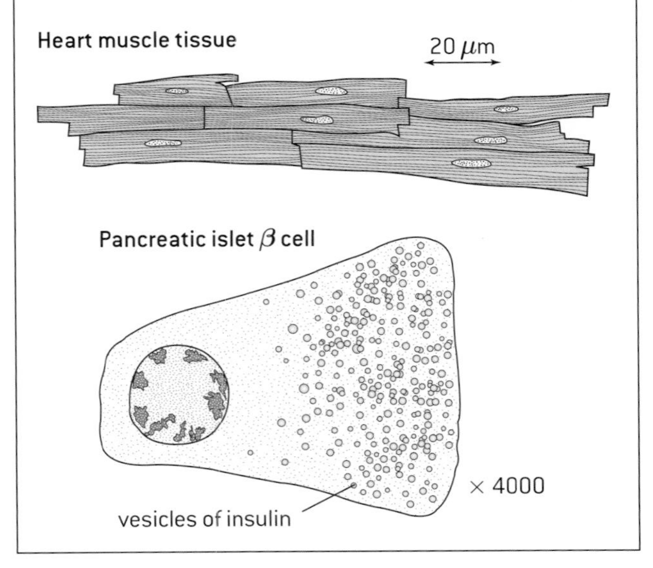

Stem cells

STEM CELLS

Stem cells are defined as cells that have the capacity to divide and to differentiate along different pathways. Human embryos consist entirely of stem cells in their early stages, but gradually the cells in the embryo commit themselves to a pattern of differentiation. Once committed, a cell may still divide several more times, but all of the cells formed will differentiate in the same way and so they are no longer stem cells.

Small numbers of cells persist as stem cells and are still present in the adult body. They are found in most human tissues, including bone marrow, skin and liver. They give some human tissues considerable powers of regeneration and repair, though they do not have as great a capacity to differentiate in different ways as embryonic stem cells.

Other tissues lack the stem cells needed for effective repair – brain, kidney and heart, for example. There has been great interest in the therapeutic use of embryonic stem cells with organs such as these. There is great potential for the use of embryonic stem cells for tissue repair and for treating a variety of degenerative conditions, for example Parkinson's disease.

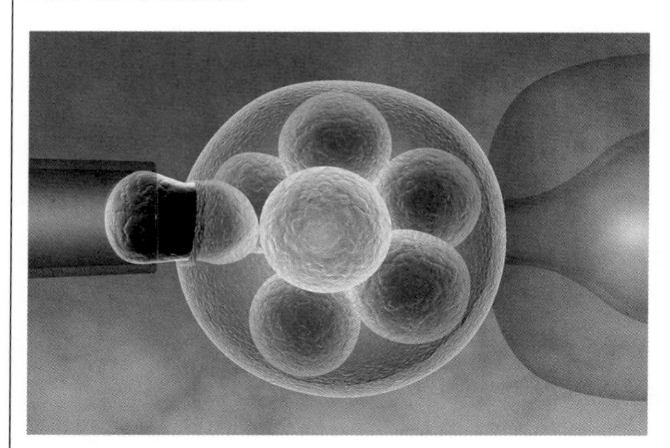

Removing a stem cell from an embryo

ETHICS OF THERAPEUTIC USE OF STEM CELLS

Ethics are moral principles that allow us to decide whether something is morally right or wrong. Scientists should always consider the ethics of research and its consequences before doing it.

The main argument in favour of therapeutic use of stem cells is that the health and quality of life of patients suffering from otherwise incurable conditions may be greatly improved. Ethical arguments against stem cell therapies depend on the source of the stem cells. There are few objections to the use of an adult's own stem cells or cells from an adult volunteer. Newborn babies cannot give informed consent for stem cells to be harvested from their umbilical cord, but parental consent is given and the cells are stored in case they are needed during the baby's subsequent life, which seems unobjectionable.

However, the ethical issues concerning stem cells taken from specially created embryos are more controversial. Some argue that an embryo is a human life even at the earliest stage and if the embryo dies as a result of the procedure it is immoral, because a life has been ended and benefits from therapies using embryonic stem cells do not justify the taking of a life.

There are a several counter-arguments:

- early stage embryos are little more than balls of cells that have yet to develop the essential features of a human life
- early stage embryos lack a nervous system so do not feel pain or suffer in other ways during stem cell procedures
- if embryos are produced deliberately, no individual that would otherwise have had the chance of living is denied the chance of life
- large numbers of embryos produced by IVF are never implanted and do not get the chance of life; rather than kill these embryos it is better to use stem cells from them to treat diseases and save lives.

EXAMPLES OF THERAPEUTIC STEM CELL USE

1. **Stargardt's macular dystrophy** is a genetic disease that develops in children between the ages of 6 and 12. Most cases are due to a recessive mutation of a gene called ABCA4. This causes a membrane protein used for active transport in retina cells to malfunction, so photoreceptive cells degenerate and vision becomes progressively worse. The loss of vision can be severe enough for the person to be registered as blind.

   Researchers have developed methods for making embryonic stem cells develop into retina cells. This was done initially with mouse cells but, in 2010, a woman in her 50s with Stargardt's disease was treated by having 50,000 retina cells derived from embryonic stem cells injected into her eyes. The cells attached to the retina and remained there during the four-month trial. There was an improvement in the woman's vision, and no harmful side effects. Further trials with larger numbers of patients are needed, but after these initial trials at least, we can be optimistic about the development of treatments for Stargardt's disease using embryonic stem cells.

2. **Leukemia** is a type of cancer in which abnormally large numbers of white blood cells are produced in the bone marrow. A normal adult white blood cell count is 4,000–11,000 per mm^3 of blood. With leukemia the count rises above 30,000 and with acute leukemia above 100,000 per mm^3.

Adult stem cells are used in the treatment of leukemia:

- A large needle is inserted into a large bone, usually the pelvis and fluid is removed from the bone marrow.
- Stem cells are extracted from this fluid and are stored by freezing them. They are adult stem cells and only have the potential for producing blood cells.
- A high dose of chemotherapy drugs is given to the patient, to kill all the cancer cells in the bone marrow. The bone marrow loses its ability to produce blood cells.
- The stem cells are then returned to the patient's body. They re-establish themselves in the bone marrow, multiply and start to produce red and white blood cells. In many cases this procedure cures the leukemia completely.

Light microscopes and drawing skills

USING LIGHT MICROSCOPES

1. Treat the specimen with a stain that makes parts of the cells of the specimen visible.
2. Mount the specimen on a microscope slide with a cover slip to make it flat and protect the microscope.
3. Put the microscope slide on the stage so the specimen is below the objective lens.
4. Plug in the microscope and switch on the power so that light passes through the specimen.
5. Focus with the low power objective lens first.
6. Use the focusing knobs to bring the slide and objective lens as close as possible without touching.
7. Look through the eyepiece lens and move the slide and objective lens apart with the coarse focusing knob until the specimen comes into focus.
8. Use the fine focusing knob to focus on particular parts of the specimen.
9. Move the slide to bring the most interesting part of the specimen into the centre of the field of view.
10. Turn the revolving nose piece to select the high power objective, then refocus using steps 5–7 again.
11. Adjust the illumination using the diaphragm.

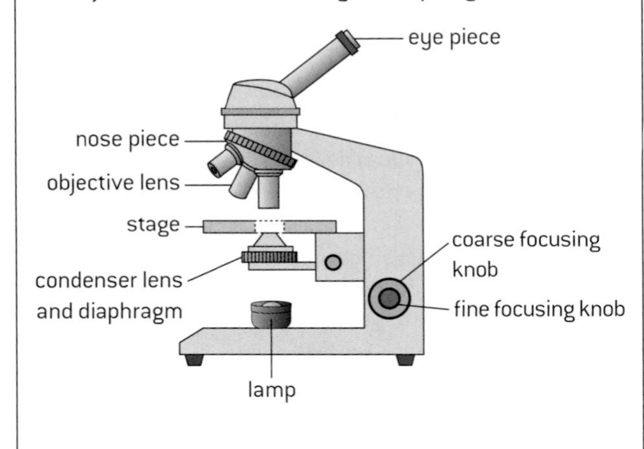

MAGNIFICATION CALCULATIONS

Microscopes are used to investigate the structure of cells and tissues. Most microscopes use light to form an image and can make structures appear up to 400 times larger than their actual size. Electron microscopes give much higher magnifications.

The structures seen with a microscope can be recorded with a neat drawing or a photograph can be taken down the microscope – called a micrograph. An important skill in biology is calculating the magnification of a drawing or micrograph. Use these instructions:

1. Choose an obvious length, for example the maximum diameter of a cell. Measure it on the drawing.
2. Measure the same length on the actual specimen.
3. If the units used for the two measurements are different, convert them into the same units.

   One millimetre (mm) = 1,000 micrometres (μm)
4. Divide the length on the drawing by the length on the actual specimen. The result is the **magnification**.

$$\text{Magnification} = \frac{\text{size of image}}{\text{size of specimen}}$$

Example

The scale bar on the drawing of heart muscle tissue on page 2 represents a length on the specimen of 20 μm and is 10 mm long, which is 10,000 μm.

$$\text{Magnification} = \frac{10{,}000}{20} = 500$$

The magnification equation can be rearranged and used to calculate the actual size of a specimen if the magnification and size of the image are known.

$$\text{Size of specimen} = \frac{\text{size of image}}{\text{magnification}}$$

Example

The length of the beta cell in the pancreatic islet on page 2 is 48mm, which is 48,000 μm, and the magnification of the drawing is × 4000.

$$\text{Actual length of the cell} = \frac{48{,}000\ \mu m}{4000} = 12\ \mu m$$

SCALE BARS

A scale bar is a line added to a micrograph or a drawing to help to show the actual size of the structures. For example, a 10 μm bar shows how large a 10 μm object would appear. The figure below is a scanning electron micrograph (SEM) of a leaf with the magnification and a scale bar both shown.

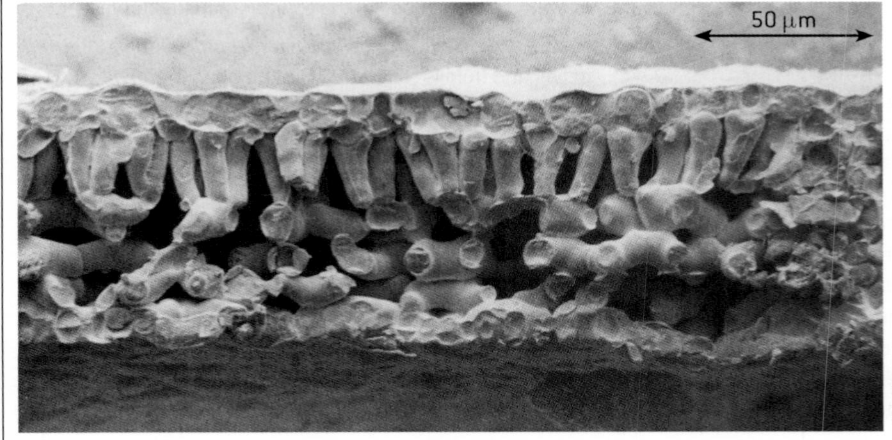

S.I. size units

1000mm = 1 m

1000μm = 1mm

1000nm = 1μm

Scanning electron micrograph of leaf (x480)

Electron microscopes and ultrastructure

RESOLUTION AND MAGNIFICATION

In every type of microscope magnification can be increased until a point above which the image can no longer be focused sharply. This is because the resolution of the microscope has been exceeded.

Resolution is the ability of the microscope to show two close objects separately in the image.

The resolution of a microscope depends on the wavelength of the rays used to form the image – the shorter the wavelength the higher the resolution. Electrons have a much shorter wavelength than light, so electron microscopes have a higher resolution than light microscopes. They can therefore produce a sharp image at much higher magnifications.

	Light microscope	Electron microscope
Resolution	0.25 μm	0.25 nm
Magnification	× 500	× 500,000

Transmission electron microscopes (TEM) are used to view ultra-thin sections. (Names of parts of this microscope do not have to be memorized.)

Scanning electron microscopes (SEM) produce an image of the surfaces of structures.

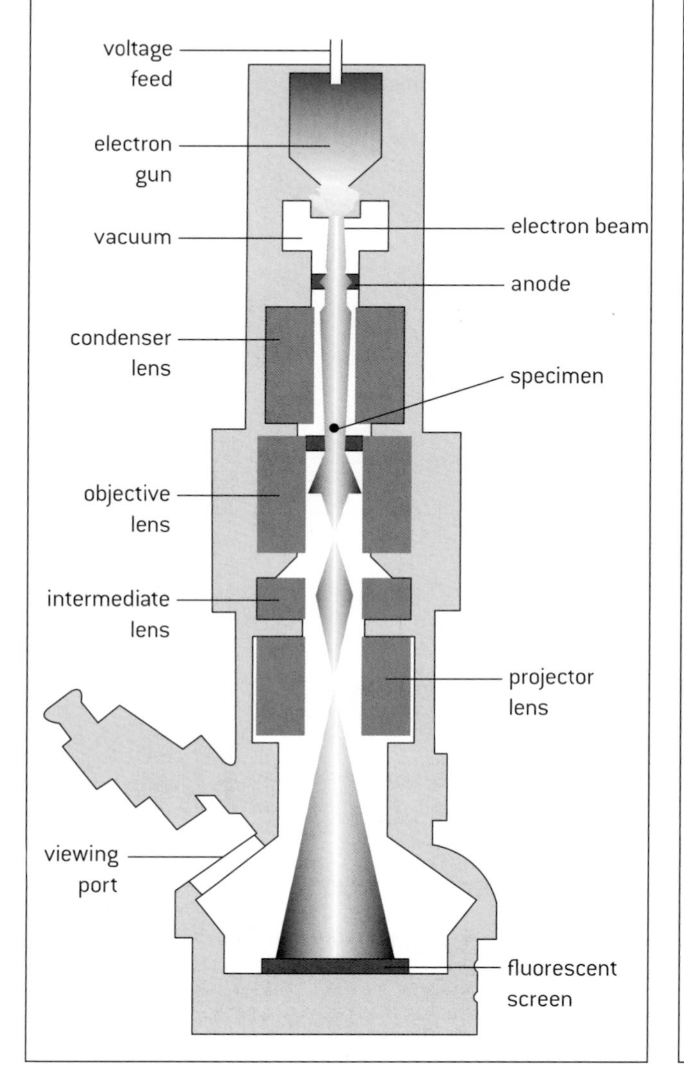

TECHNOLOGY AND SCIENCE

The diagram (below left) shows a simplified version of the technology of an electron microscope. The electron microscope is a good example of an important trend in science – improvements in technology or apparatus lead to developments in scientific research.

The invention of the electron microscope led to a much greater understanding of the structure of cells and the discovery of many structures within living organisms.
The detailed structure of the cell that was revealed by the electron microscope is known as **ultrastructure**.

ULTRASTRUCTURE OF PALISADE CELLS

The electron micrograph below is an example of the detailed ultrastructure that the electron microscope reveals.

Chloroplast – carries out photosynthesis.

Cell wall – supports and protects the cell.

Plasma membrance – controls entry and exit of substances.

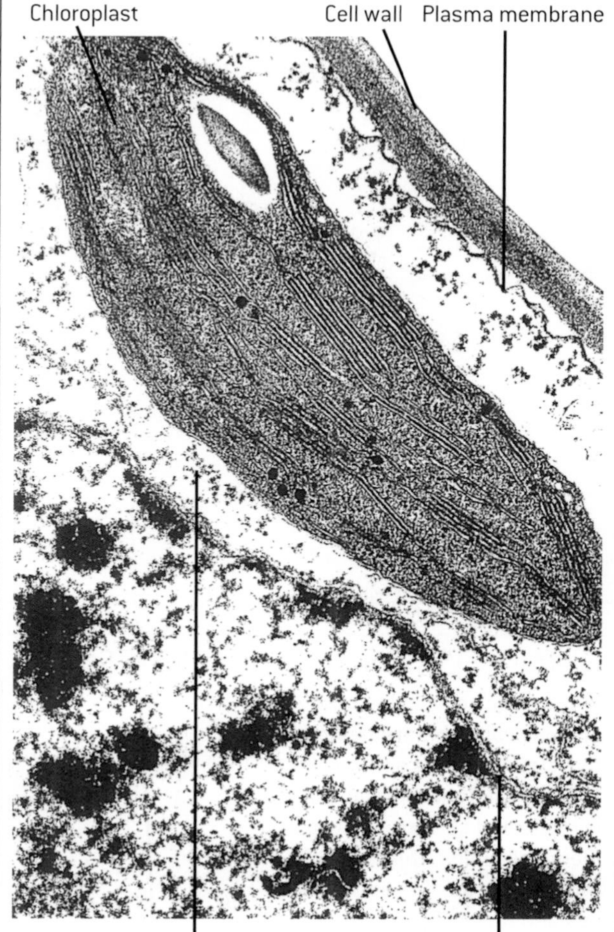

Free ribosomes – synthesize cytoplasmic proteins.
Nuclear membrane – protects chromosomes.

In the other parts of this cell there were many more chloroplasts and a large vacuole, indicating that the function of this cell was photosynthesis. It is a palisade mesophyll cell from the leaf of a plant.

Prokaryotic cells

STRUCTURE OF PROKARYOTIC CELLS

Cells are divided into two types according to their structure, prokaryotic and eukaryotic. The first cells to evolve were prokaryotic and many organisms still have prokaryotic cells, including all bacteria.

Prokaryotic cells have a relatively simple cell structure. Eukaryotic cells are divided up by membranes into separate compartments such as the nucleus and mitochondria, whereas prokaryotic cells are not compartmentalized. They do not have a nucleus, mitochondria or any other membrane-bound organelles within their cytoplasm.

SURFACE AREA TO VOLUME RATIOS

As the size of any object is increased, the ratio between the surface area and the volume decreases. Consider the surface area to volume ratio of cubes of varying size as an example. The rate at which materials enter or leave a cell depends on the surface area of the cell. However, the rate at which materials are used or produced depends on the volume.

A cell that becomes too large may not be able to take in essential materials or excrete waste substances quickly enough. Surface area to volume ratio is important in biology. It helps to explain many phenomena apart from maximum cell sizes.

DRAWING PROKARYOTIC CELLS

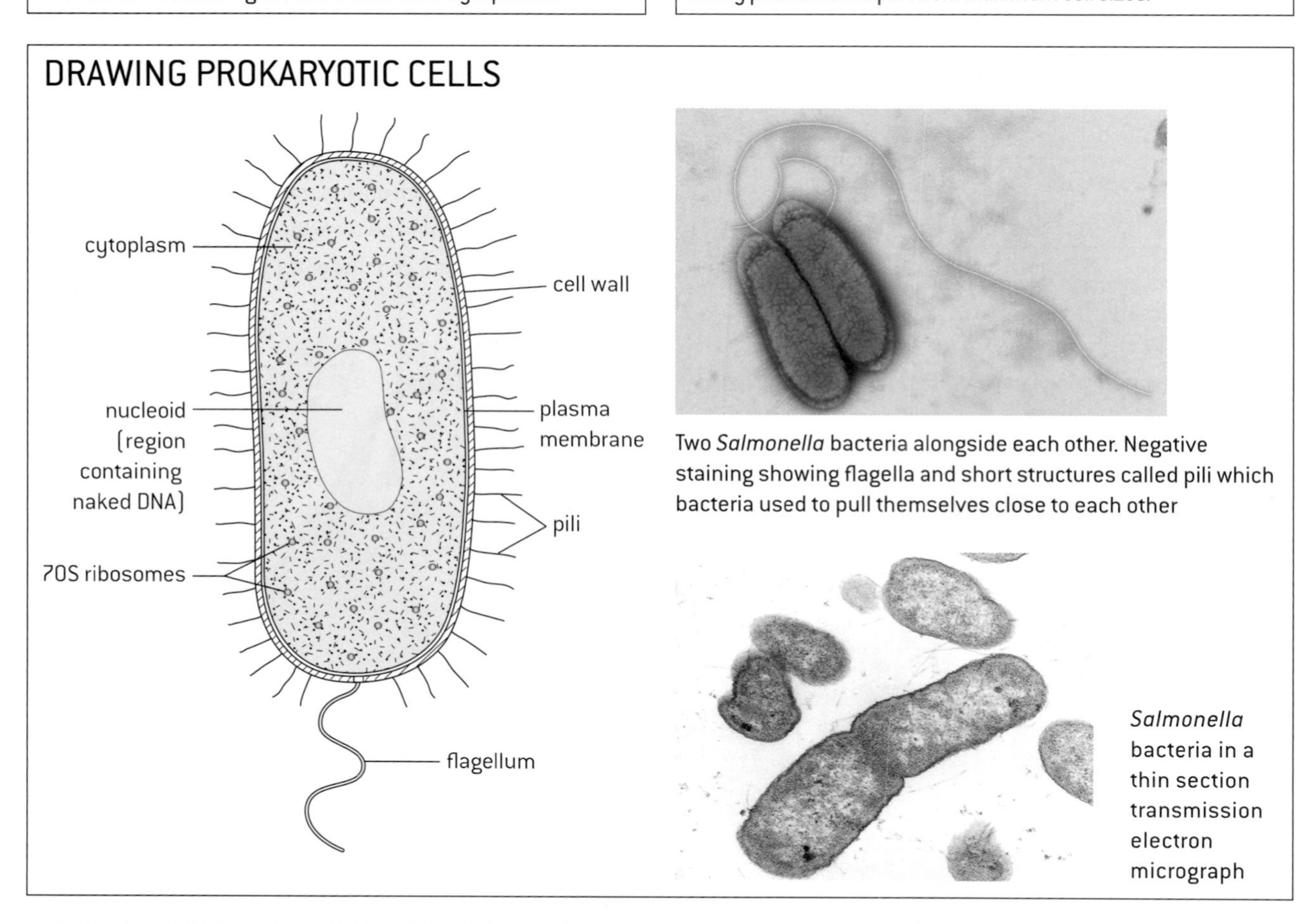

Two *Salmonella* bacteria alongside each other. Negative staining showing flagella and short structures called pili which bacteria used to pull themselves close to each other

Salmonella bacteria in a thin section transmission electron micrograph

BINARY FISSION IN PROKARYOTES

Prokaryotic cells divide by a process called **binary fission** – this simply means splitting in two. The bacterial chromosome is replicated so there are two identical copies. These are moved to opposite ends of the cell and the wall and plasma membrane are then pulled inwards so the cell pinches apart to form two identical cells. Some prokaryotes can double in volume and divide by binary fission every 30 minutes.

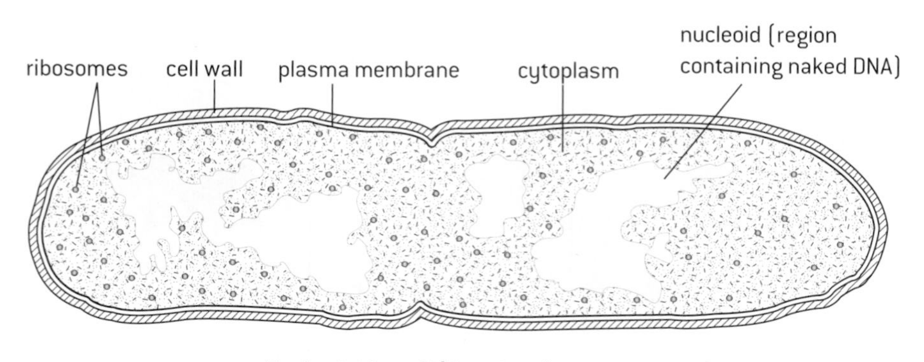

***Escherichia coli* (2μm long) starting to divide**

Eukaryotic cells

STRUCTURE OF EUKARYOTIC CELLS

Using a light microscope it is possible to see that eukaryotic cells have cytoplasm enclosed in a plasma membrane, like prokaryotic cells. However, unlike prokaryotic cells, they usually contain a nucleus.

Under the electron microscope details of much smaller structures within the cell are visible. This is called the **ultrastructure** of a cell. There are a number of different types of organelle that form compartments in eukaryotic cells, each bounded by either one or two membranes:

Organelles with a single membrane:

- Rough endoplasmic reticulum
- Smooth endoplasmic reticulum
- Golgi apparatus
- Lysosomes
- Vesicles and vacuoles

Organelles with a double membrane:

- Nucleus
- Mitochondrion
- Chloroplast

Advantage of compartmentalization:

Enzymes and substrates used in a process can be concentrated in a small area, with pH and other conditions at optimum levels and with no other enzymes that might disrupt the process.

DRAWING EUKARYOTIC CELLS

The drawing shows the types of organelle that occur in eukaryotic cells. Chloroplasts and cell walls are part of plant cells but not animal cells.

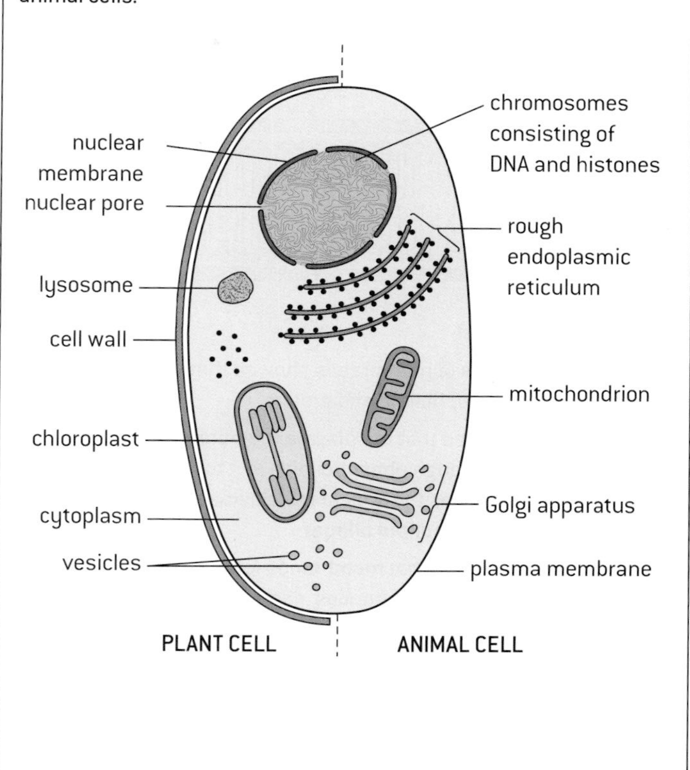

IDENTIFYING ORGANELLES AND DEDUCING FUNCTIONS

The electron micrograph shows the structure of a cell in the pancreas.

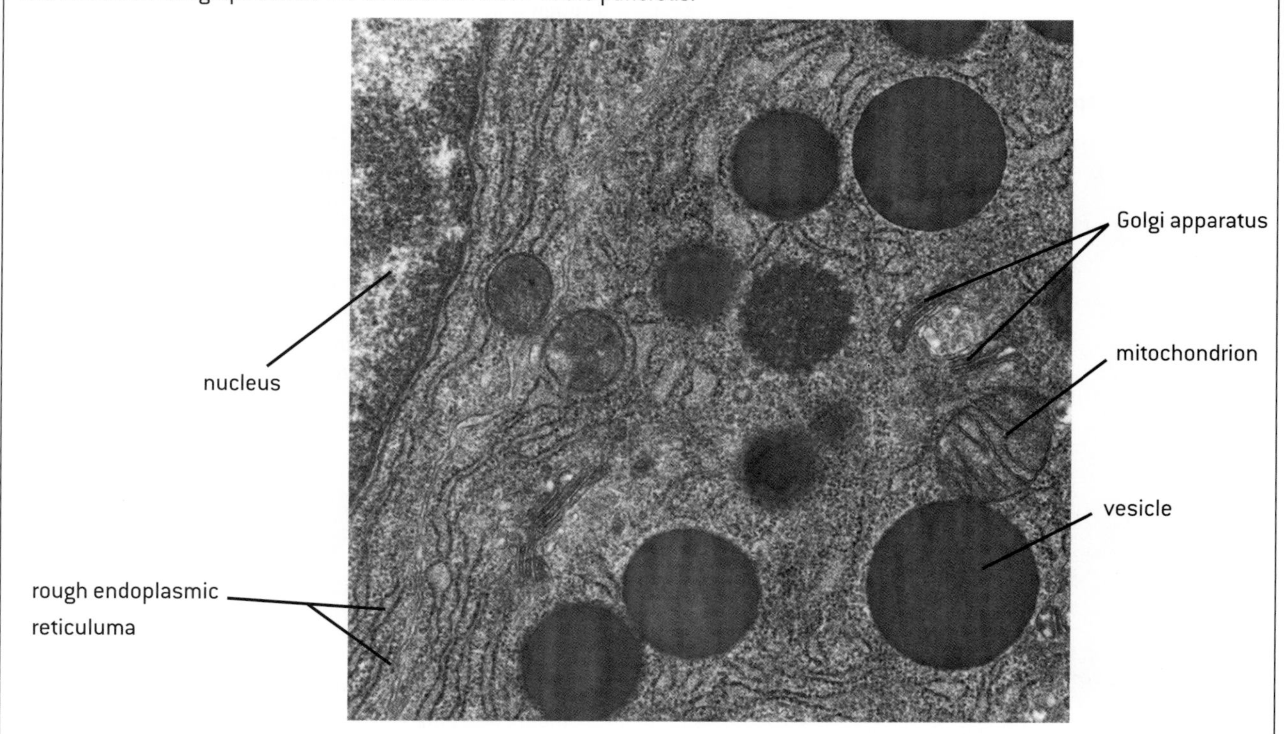

The presence of large amounts of rough endoplasmic reticulum and many Golgi apparatuses shows that the main function of this cell is to synthesize and secrete proteins, presumably the enzymes in pancreatic juice.

Models of membrane structure

THE DAVSON–DANIELLI MODEL

In this model of membrane structure there is a bilayer of phospholipids in the centre of the membrane with layers of protein on either side. It was developed by Davson and Danielli in the 1930s.

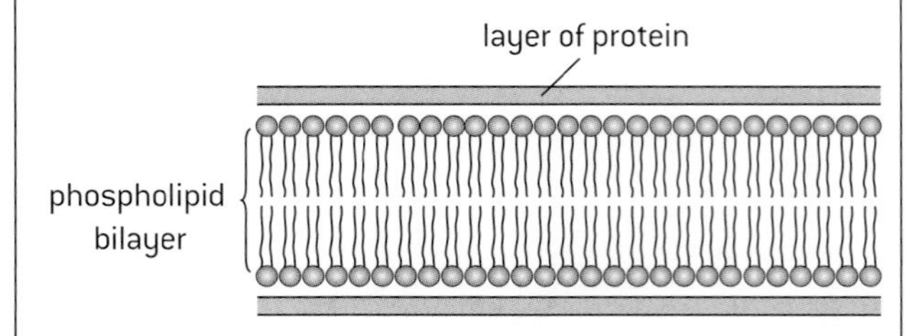

Reasons for the model:

1. Chemical analysis of membranes showed that they were composed of phospholipid and protein.
2. Evidence suggested that the plasma membrane of red blood cells has enough phospholipids in it to form an area twice as large as the area of the plasma membrane, suggesting a **phospholipid bilayer**.
3. Experiments showed that membranes form a barrier to the passage of some substances, despite being very thin, and layers of protein could act as the barrier.

Testing the model:

High magnification electron micrographs were first produced in the 1950s. In these micrographs membranes appeared as two dark lines separated by a lighter band.

This seemed to fit the Davson–Danielli model, as proteins usually appear darker than phospholipids in electron micrographs. The electron micrograph below shows membranes both at the surfaces of cells and around vesicles with the appearance that seemed to back up the Davson–Danielli model.

Electron micrograph of biological membranes

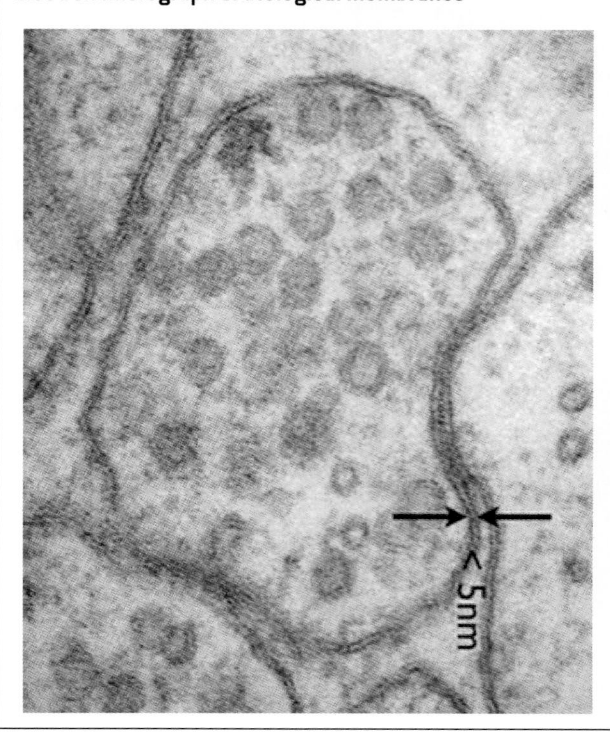

THE SINGER–NICOLSON MODEL

In the 1950s and 60s evidence accumulated that did not fit the Davson–Danielli model:

1. Freeze-fracture electron micrographs showed that globular proteins were present in the centre of the phospholipid bilayer (below).

2. Analysis of membrane proteins showed that parts of their surfaces were hydrophobic, so they would be positioned in the bilayer and in some cases would extend from one side to the other.

Non-polar amino acids in the centre of water-soluble proteins stabilize their structure.

Polar amino acids on the surface of proteins make them water soluble.

Non-polar amino acids cause proteins to remain embedded in membranes.

Polar amino acids create channels through which hydrophilic substances can diffuse. Positively charged R groups allow negatively charged ions through and vice versa.

Polar amino acids cause parts of membrane proteins to protrude from the membrane. Transmembrane proteins have two such regions.

3. Fusion of cells with membrane proteins tagged with different coloured fluorescent markers showed that these proteins can move within the membrane as the colours became mixed within a few minutes of cell fusion.

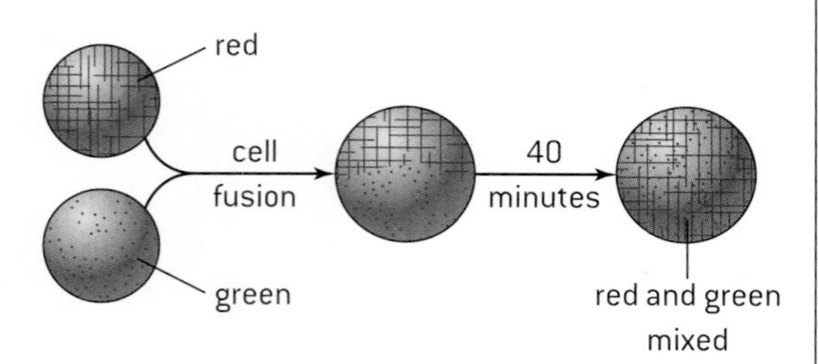

This evidence falsified the Davson–Danielli model. A new model was proposed in 1966 by Singer and Nicolson. This model is still used today. It is called either the Singer–Nicolson model or fluid mosaic model.

Membrane structure

FLUID MOSAIC MODEL OF MEMBRANE STRUCTURE

Phospholipid molecules are shown as an oval with two parallel lines because they have a phosphate head with two fatty acid tails attached. Proteins occupy a range of different positions in the membrane. **Integral** proteins are embedded in the phospholipid bilayer. **Peripheral** proteins are attached to an outer surface of the membrane. **Glycoproteins** have sugar units attached on the outer surface of the membrane.

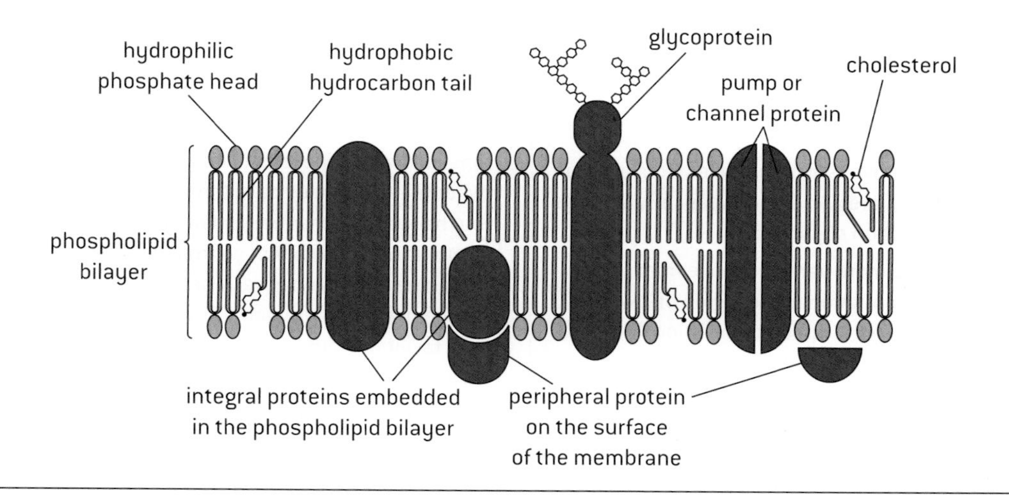

PHOSPHOLIPIDS

Phospholipids are the basic component of all biological membranes. Phospholipid molecules are **amphipathic**. This means that part of the molecule is attracted to water (**hydrophilic**) and part is not attracted to water (**hydrophobic**).

The phosphate head is hydrophilic and the two fatty acid tails, which are composed of hydrocarbon chains, are hydrophobic. When phospholipids are mixed with water they naturally become arranged into bilayers, with the hydrophilic heads facing outwards and making contact with the water and the hydrocarbon tails facing inwards away from the water. The attraction between the hydrophobic tails in the centre of the phospholipid bilayer and between the hydrophilic heads and the surrounding water makes membranes very stable.

CHOLESTEROL

Cholesterol is a component of animal cell membranes. Most of the cholesterol molecule is hydrophobic but, like phospholipids, there is one hydrophilic end; so cholesterol fits between phospholipids in the membrane.

Cholesterol restricts the movement of phospholipid molecules. It therefore **reduces the fluidity** of the membrane. It also **reduces the permeability of the membrane** to hydrophilic particles such as sodium ions and hydrogen ions. This is important, as animal cells need to maintain concentration differences of these ions across their membranes, so diffusion through the membrane must be restricted.

MEMBRANE PROTEINS

Membrane proteins are diverse in structure, function and position in the membrane. The diagram above shows a glycoprotein, used for cell-to-cell communication. The diagram below shows examples of other membrane proteins.

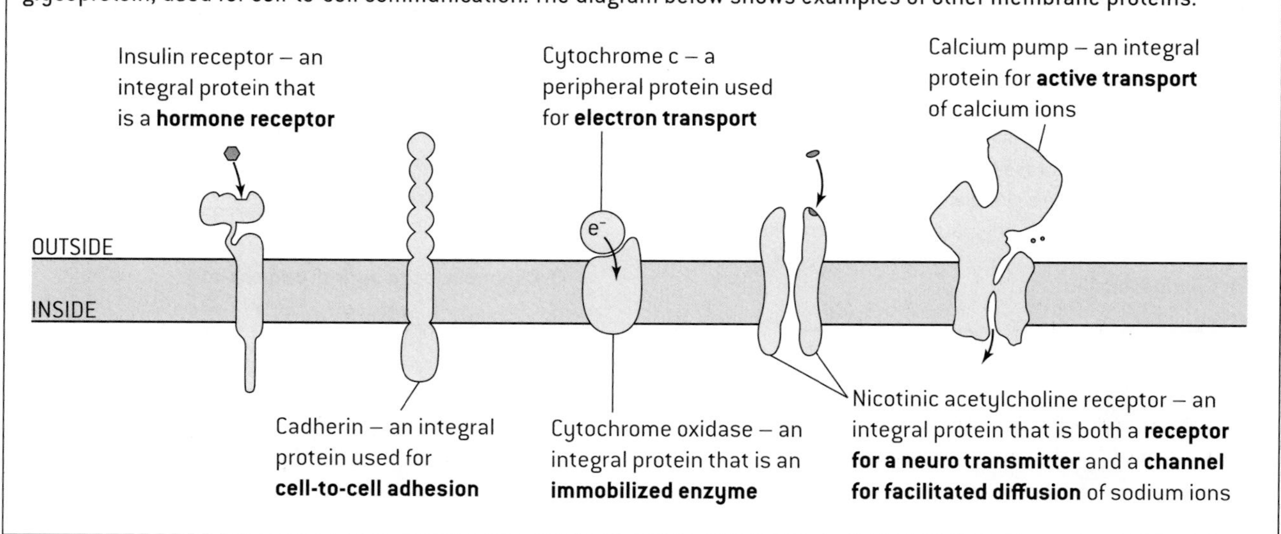

Diffusion and facilitated diffusion

DIFFUSION

Solids, liquids and gases consist of particles – atoms, ions and molecules. In liquids and gases, these particles are in continual motion. The direction of movement is random. If particles are evenly spread then their movement in all directions is even and there is no net movement – they remain evenly spread despite continually moving. Sometimes particles are unevenly spread – there is a higher concentration in one region than another. This causes diffusion.

Diffusion is the passive movement of particles from a region of higher concentration to a region of lower concentration, as a result of the random motion of particles.

Diffusion occurs because more particles move from the region of higher concentration to the region of lower concentration than move in the opposite direction. Diffusion can occur across membranes if there is a concentration gradient and the membrane is permeable to the particle. For example, membranes are freely permeable to oxygen, so if there is a lower concentration of oxygen inside a cell than outside, it will diffuse into the cell. Membranes are not permeable to cellulose, so it does not diffuse across.

SIMPLE AND FACILITATED DIFFUSION

Membranes allow some substances to diffuse through but not others – they are **partially permeable**. Some of these substances move between the phospholipid molecules in the membrane – this is **simple diffusion**. Other substances are unable to pass between the phospholipids. To allow these substances to diffuse through membranes, channel proteins are needed. This is called **facilitated diffusion**. Channel proteins are specific – they only allow one type of substance to pass through. For example, chloride channels only allow chloride ions to pass through. Cells can control whether substances pass through their plasma membranes, by the types of channel protein that are inserted into the membrane. Cells cannot control the direction of movement. Facilitated diffusion always occurs from a region of higher concentration to a region of lower concentration. Both simple and facilitated diffusion are passive processes – no energy has to be used by the cell to make them occur.

There are sodium and potassium channel proteins in the membranes of neurons that open and close, depending on the voltage across the membrane. They are voltage-gated channels and are used to transmit nerve impulses.

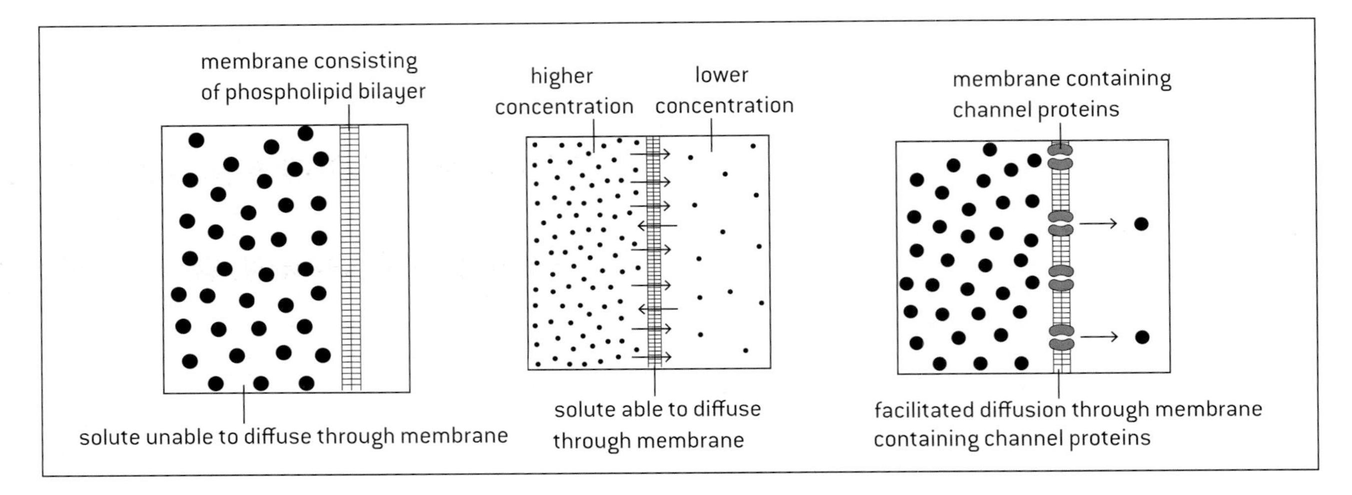

STRUCTURE AND FUNCTION OF POTASSIUM CHANNELS IN AXONS

The axons of neurons contain potassium channels that are used during an action potential. They are closed when the axon is polarized but open in response to depolarization of the axon membrane, allowing K^+ ions to exit by facilitated diffusion, which repolarizes the axon. Potassium channels only remain open for a very short time before a globular sub-unit blocks the pore. The channel then returns to its original closed conformation.

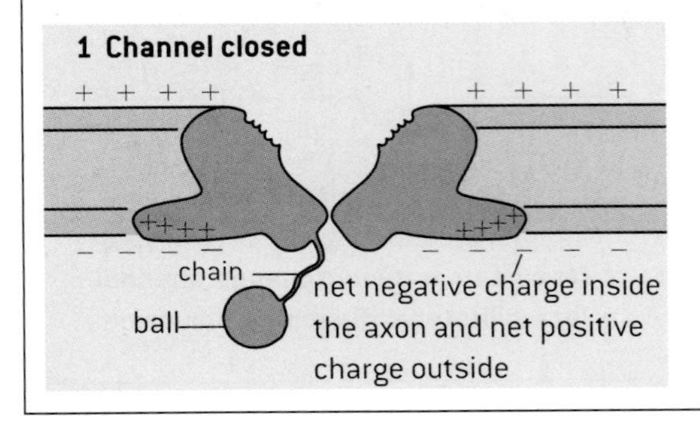

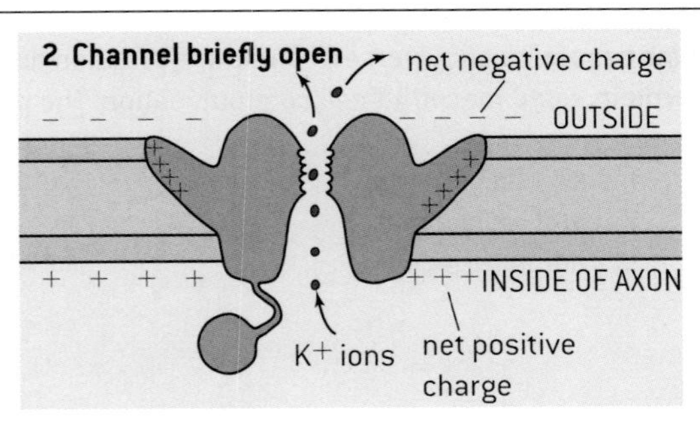

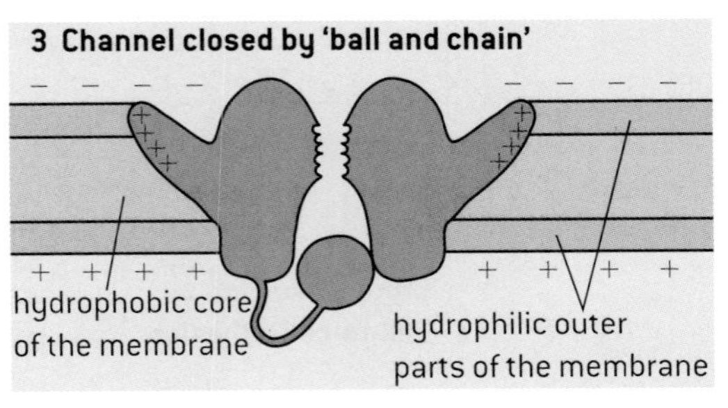

Osmosis

WATER MOVEMENT BY OSMOSIS

Plasma membranes are usually freely permeable to water. The passive movement of water across membranes is different from diffusion across membranes, because water is the solvent. A solvent is a liquid in which particles dissolve. Dissolved particles are called solutes. The direction in which water moves is due to the concentration of **solutes**, rather than the concentration of water molecules, so it is called osmosis, rather than diffusion.

Osmosis is the passive movement of water molecules from a region of lower solute concentration to a region of higher solute concentration, across a partially permeable membrane.

Attractions between solute particles and water molecules are the reason for water moving to regions with a higher solute concentration.

Solute molecules cannot diffuse out as the membrane is impermeable to them

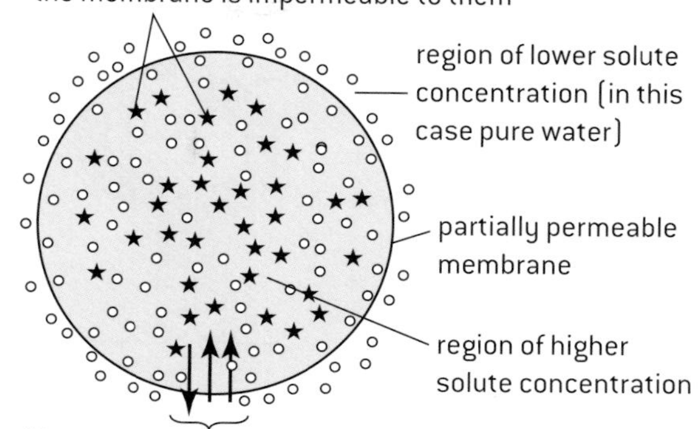

Water molecules move in and out through the membrane but more move in than out. There is a net movement from the region of lower solute concentration to the region of higher solute concentration

ESTIMATING OSMOLARITY

The osmolarity of a solution is the number of moles of solute particles per unit volume of solution. Pure water has an osmolarity of zero. The greater the concentration of solutes, the higher the osmolarity.

If two solutions at equal pressure but with different osmolarity are separated by a partially permeable membrane, water will move by osmosis from the solution with the lower osmolarity to the solution with the higher osmolarity. Plant cells absorb water from a surrounding solution if their osmolarity is higher than that of the solution (i.e. the surrounding solution is **hypotonic**) or lose water if their osmolarity is lower (i.e. the solution is **hypertonic**). This principle can be used to estimate the osmolarity of a type of plant tissue, such as potato.

Method:

1. Prepare a series of solutions with a suitable range of solute concentrations, such as 0.0, 0.1, 0.2, 0.3, 0.4 and 0.5 moles/litre.
2. Cut the tissue into samples of equal size and shape.
3. Find the mass of each sample, using an electronic balance.
4. Bathe tissue samples in each of the range of solutions for long enough to get measurable mass changes, usually between 10 and 60 minutes.
5. Remove the tissue samples from the bathing solutions, dry them and find their mass again.
6. Calculate percentage mass change using this formula:
$$\% \text{ change} = \frac{(\text{final mass} - \text{initial mass})}{\text{initial mass}} \times 100$$
7. Plot the results on a graph.
8. Read off the solute concentration which would give no mass change. It has the same osmolarity as the tissue.
NB The osmolarity of a glucose solution is equal to its molarity because glucose remains as single molecules when it dissolves.

   The osmolarity of a sodium chloride solution is double its molarity because one mole of NaCl gives two moles of ions when it dissolves – one mole of Na^+ and one mole of Cl^-.

SAMPLE OSMOLARITY RESULTS

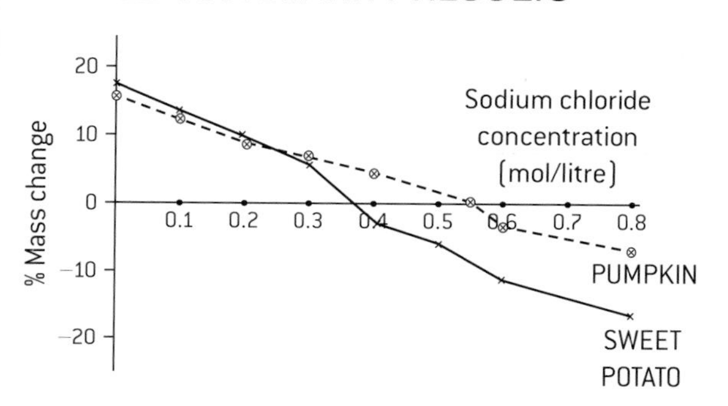

The estimated osmolarity of the pumpkin is equal to 0.55 moles / litre NaCl solution, which is 1.1 osmoles / litre.

ACCURACY IN OSMOSIS EXPERIMENTS

Estimates of osmolarity from this experiment will only be as accurate as the quantitative measurements, so it is essential for these to be as accurate as possible:

- the volume of water used for making solutions should be measured with a volumetric flask
- the initial and final mass of tissue samples should be measured with the same electronic balance that is accurate to 0.01 grams (10 mg).

AVOIDING OSMOSIS IN DONOR ORGANS

Osmosis can cause cells in human tissues or organs to swell up and burst, or to shrink due to gain or loss of water by osmosis. To prevent this, tissues or organs used in medical procedures such as kidney transplants must be bathed in a solution with the same osmolarity as human cytoplasm.

- A solution of salts called **isotonic saline** is used for some procedures.
- Donor organs are surrounded by isotonic slush when they are being transported, with the low temperatures helping to keep them in a healthy state.

Active transport

PUMP PROTEINS AND ACTIVE TRANSPORT

Active transport is the movement of substances across membranes using energy from ATP. Active transport can move substances against the concentration gradient – from a region of lower to a region of higher concentration.

Protein pumps in the membrane are used for active transport. Each pump only transports particular substances, so cells can control what is absorbed and what is expelled. Pumps work in a specific direction – the substance can only enter the pump on one side and can only exit on the other side.

1. Particle enters the pump from the side with a lower concentration
2. Particle binds to a specific site. Other types of particle cannot bind
3. Energy from ATP is used to change the shape of the pump
4. Particle is released on the side with a higher concentration and the pump then returns to its original shape

STRUCTURE AND FUNCTION OF SODIUM–POTASSIUM PUMPS IN AXONS

The axons of neurons contain a pump protein that moves sodium ions and potassium ions across the membrane. As sodium and potassium are pumped in opposite directions it is an **antiporter**. The energy that is required for the pumping is obtained by converting ATP to ADP and phosphate, so it is an ATPase. It is known to biochemists as **Na^+/K^+-ATPase**. One ATP provides enough energy to pump two potassium ions in and three sodium ions out of the cell. The concentration gradients generated by this active transport are needed for the transmission of nerve impulses in axons.

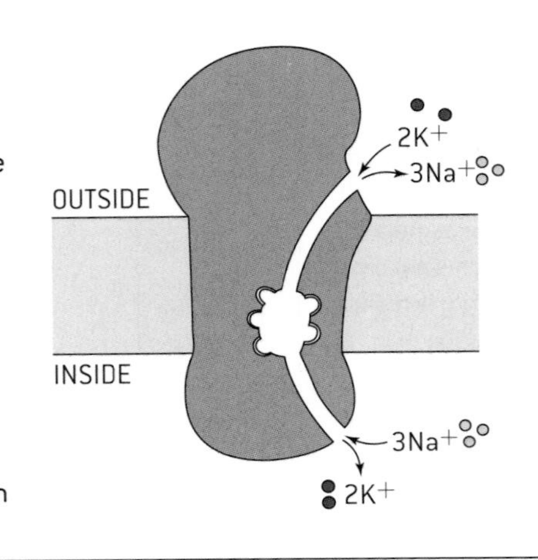

In the centre of the pump there are two binding sites for K^+ ions and three for Na^+ ions. The pump has two alternate states. In one, there is access to the binding sites from the outer side of the membrane and there is a stronger attraction to K^+ ions, so Na^+ are discharged from the cell and K^+ bind. In the other state there is access to the binding sites from inside and there is a stronger attraction for Na^+ ions, so K^+ ions are discharged into the cell and Na^+ bind. Energy from ATP causes the switch from one state to the other and then back again.

TRANSPORT USING VESICLES

The fluidity of membranes allows them to move and change shape. Small pieces of membrane can be pinched off the plasma membrane to create a vesicle containing some material from outside the cell. This is **endocytosis**. Vesicles can also move to the plasma membrane and fuse with it, releasing the contents of the vesicle outside the cell This is **exocytosis**. Vesicles are used to move materials from one part of the cell to another. For example, vesicles move proteins from the rough ER to the Golgi apparatus.

1. Proteins are synthesized by ribosomes and then enter the rough endoplasmic reticulum
2. Vesicles bud off from the rER and carry the proteins to the Golgi apparatus
3. The Golgi apparatus modifies the proteins
4. Vesicles bud off from the Golgi apparatus and carry the modified proteins to the plasma membrane

ENDOCYTOSIS

1. Part of the plasma membrane is pulled inwards
2. A droplet of fluid becomes enclosed when a vesicle is pinched off
3. Vesicles can then move through the cytoplasm carrying their contents

EXOCYTOSIS

1. Vesicles fuse with the plasma membrane
2. The contents of the vesicle are expelled
3. The membrane then flattens out again

Origins of cells

CELL DIVISION AND CELL ORIGINS

Until the 19th century some biologists believed that life could appear in non-living material. This was called 'spontaneous generation'. There is no evidence that living cells can be formed on Earth today except by division of pre-existing cells. Spontaneous generation of cells is not currently possible.

ORIGINS OF THE FIRST CELLS

The general principle that cells are only formed by division of pre-existing cells can be used to trace life back to its origins. All of the billions of cells in a human or other multicellular organism are formed by repeated cell division, starting with a single cell formed by reproduction. We can trace the origins of cells back through the generations and through hundreds of millions of years of evolution. Eventually we must reach the first cells, as life has not always existed on Earth. Before these cells existed there was only non-living material on Earth. One of the great challenges in biology is to understand how the first living cells evolved from non-living matter and why spontaneous generation could take place then but not now.

It is a remarkable fact that the sixty-four codons of the genetic code have the same meanings in the cells of all organisms, apart from minor variations. The universality of the genetic code suggests strongly that all life evolved from the same original cells. Minor differences in the genetic code will have accrued since the common origin of life on Earth.

PASTEUR'S EXPERIMENTS

The general principle that cells can only come from pre-existing ones was tested repeatedly by scientists in the 18th and 19th centuries. It was the experiments of Louis Pasteur that verified the principle beyond reasonable doubt.

The most famous of Pasteur's experiments involved the use of swan-necked flasks. He placed samples of broth in flasks with long necks and then melted the glass of the necks and bent it into a variety of shapes. Pasteur then boiled the broth in some of the flasks to kill any organisms present but left others unboiled as controls. Fungi and other organisms soon appeared in the unboiled flasks but not in the boiled ones, even after long periods of time. The broth in the flasks was in contact with air, which it had been suggested was needed for spontaneous generation, yet no spontaneous generation occurred. Pasteur snapped the necks of some of the flasks to leave a shorter vertical neck. Organisms were soon apparent in these flasks and decomposed the broth.

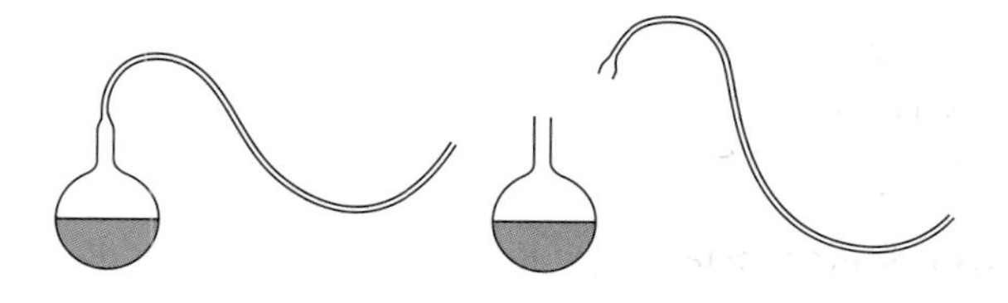

He concluded that the swan necks prevented organisms from the air getting into the flasks and that no organisms appeared spontaneously.

THE ENDOSYMBIOTIC THEORY

Symbiosis is two organisms living together. With endosymbiosis a larger cell takes in a smaller cell by endocytosis, so the smaller cell is inside a vesicle in the cytoplasm of the larger cell. Instead of the smaller cell being digested, it is kept alive and performs a useful function for the larger cell. The smaller cell divides at least as frequently as the larger cell so all cells produced by division of the larger cell contain one or more of the smaller cells inside its vesicle. According to the endosymbiotic theory, this process happened at least twice during the origin of eukaryotic cells.

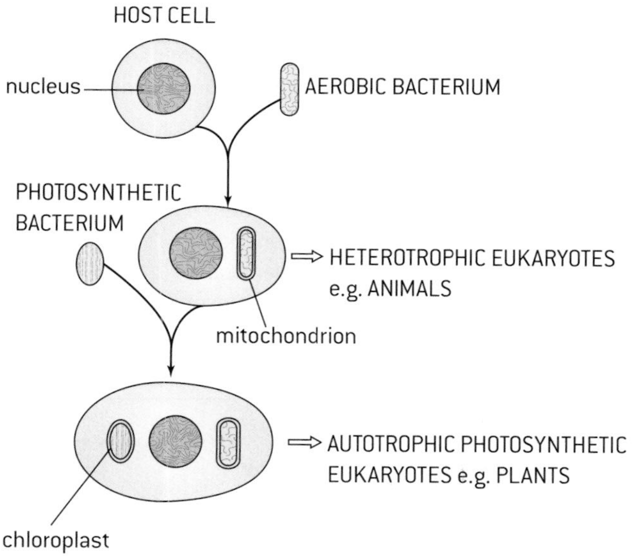

HOST CELL
nucleus
AEROBIC BACTERIUM
PHOTOSYNTHETIC BACTERIUM
⟹ HETEROTROPHIC EUKARYOTES e.g. ANIMALS
mitochondrion
⟹ AUTOTROPHIC PHOTOSYNTHETIC EUKARYOTES e.g. PLANTS
chloroplast

1. A cell that respired anaerobically took in a bacterium that respired aerobically, supplying both itself and the larger cell with energy in the form of ATP. This gave the larger cell a competitive advantage because aerobic respiration is more efficient than anaerobic. Gradually the aerobic bacterium evolved into mitochondria and the larger cell evolved into heterotrophic eukaryotes alive today such as animals.
2. A heterotrophic cell took in a smaller photosynthetic bacterium, which supplied it with organic compounds, thus making it an autotroph. The photosynthetic prokaryote evolved into chloroplasts and the larger cell evolved into photosynthetic eukaryotes alive today such as plants.

This theory explains the characteristics of mitochondria and chloroplasts:

- They grow and divide like cells.
- They have a naked loop of DNA, like prokaryotes.
- They synthesize some of their own proteins using 70S ribosomes, like prokaryotes.
- They have double membranes, as expected when cells are taken into a vesicle by endocytosis.

Mitosis

CHROMOSOMES AND CONDENSATION

In eukaryotes nearly all the DNA of a cell is stored in the nucleus. A human nucleus contains 2 metres of DNA and yet the nucleus is only about 5 μm in diameter. It fits in quite easily because the DNA molecule is so narrow – its width is 2 nm, which is 0.002 μm. A DNA molecule is far too small to be visible with a light microscope.

In eukaryotes the DNA molecules have proteins attached to them, forming structures called chromosomes. During mitosis the chromosomes become shorter and fatter. This is called **condensation** and occurs by a complex process of coiling, known as **supercoiling**.

CHROMATIDS AND CENTROMERES

The chromosomes become condensed enough during the early stages of mitosis to be visible with a light microscope. At this stage of mitosis each chromosome is a double structure. The two parts of the chromosome are called **sister chromatids**. They are held together at one point by a structure known as a **centromere**.

The term 'sister' indicates that the two chromatids contain an identical DNA molecule, produced by **DNA replication** before the start of mitosis. During mitosis the centromere divides and the sister chromatids separate. From then onwards they are referred to as **chromosomes** rather than chromatids.

THE PHASES OF MITOSIS

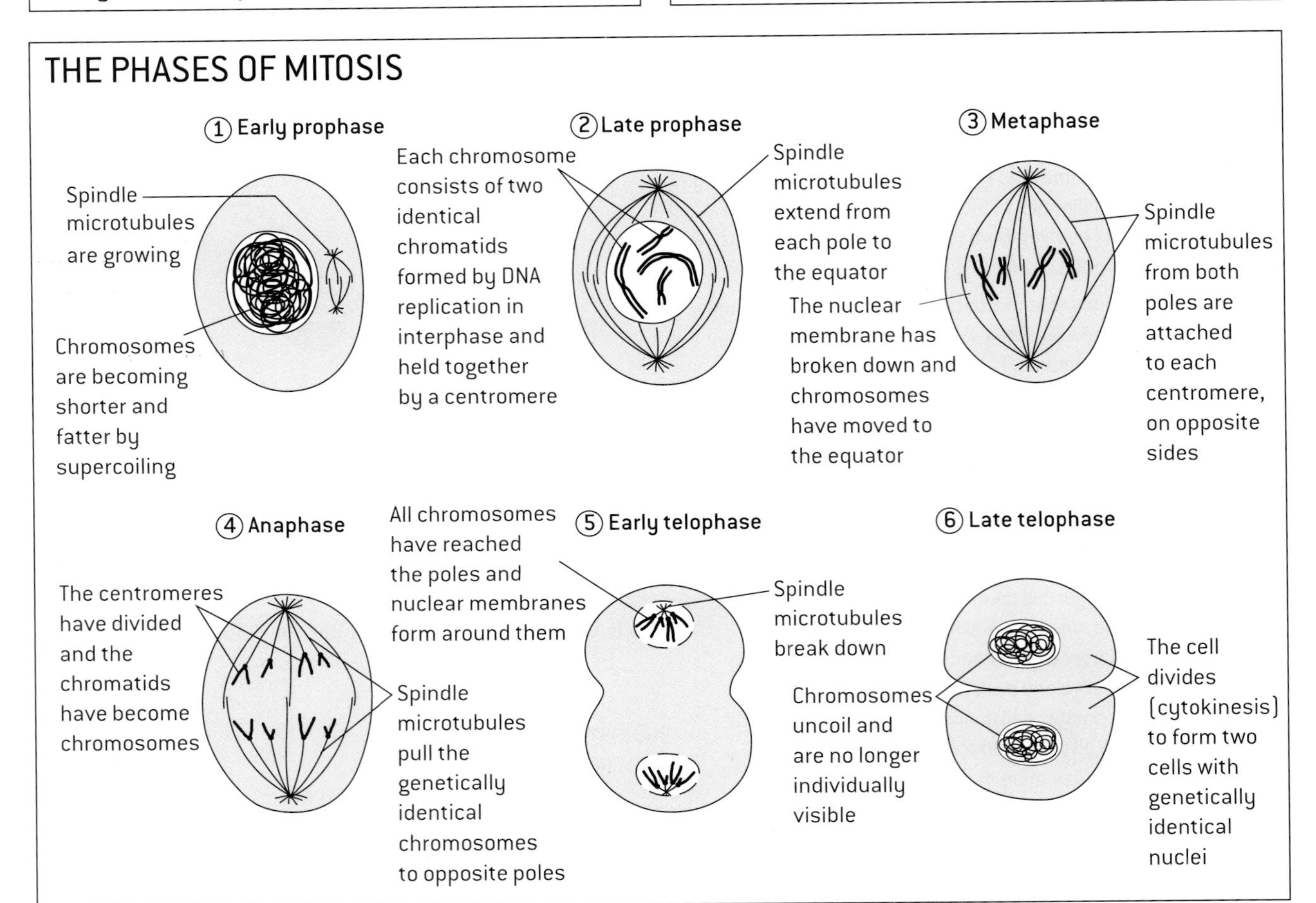

MITOTIC INDEX

The **mitotic index** is the ratio between the number of cells in mitosis in a tissue and the total number of observed cells.

$$\text{Mitotic index} = \frac{\text{number of cells in mitosis}}{\text{total number of cells}}$$

Count the total number of cells in the micrograph and then count the number of cells in any of the four phases of mitosis. The mitotic index can then be calculated.

The mitotic index is used by doctors to predict how rapidly a tumour will grow and therefore what treatment is needed. A high index indicates a fast-growing tumour. One cell in each of the four stages of mitosis is identified right.

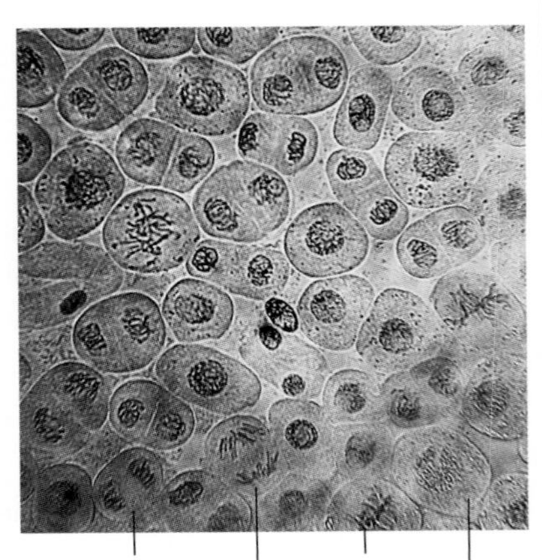

CYTOKINESIS

Cytokinesis is the division of the cytoplasm to form two cells. It occurs after mitosis and is different in plant and animal cells.

- In plant cells a new cell wall is formed across the equator of the cell, with plasma membrane on both sides. This divides the cell in two.
- In animal cells the plasma membrane at the equator is pulled inwards until it meets in the centre of the cell, dividing it in two.

Cell cycles and cancer

THE CELL CYCLE IN EUKARYOTES

The **cell cycle** is the sequence of events between one cell division and the next. It has two main phases: **interphase** and **cell division**. Interphase is a very active phase in the life of a cell when many metabolic reactions occur. Some of these, such as the reactions of cell respiration, also occur during cell division, but DNA replication in the nucleus and protein synthesis in the cytoplasm only happen during interphase.

During interphase the numbers of mitochondria in the cytoplasm increase, as they grow and divide. In plant cells the numbers of chloroplasts increase in the same way.

Interphase consists of three phases, the **G_1 phase, S phase** and **G_2 phase**. In S phase the cell replicates all the genetic material in its nucleus, so that after mitosis both the new cells have a complete set of genes. Some do not progress beyond G_1, because they are never going to divide so do not need to prepare for mitosis.

At the end of interphase, the cell begins **mitosis**, where the nucleus divides to form two genetically identical nuclei. At the end of mitosis, the cytoplasm of the cell starts to divide and two cells are formed, each containing one nucleus (**cytokinesis**).

CYCLINS AND CELL CYCLE CONTROL

Each of the phases of the cell cycle involves many important tasks. A group of proteins called cyclins is used to ensure that tasks are performed at the correct time and that the cell only moves on to the next stage of the cycle when it is appropriate.

Cyclins bind to enzymes called cyclin-dependent kinases. These kinases then become active and attach phosphate groups to other proteins in the cell. The attachment of phosphate triggers the other proteins to become active and carry out tasks specific to one of the phases of the cell cycle.

There are four main types of cyclin in human cells. The graph below shows how the levels of these cyclins rise and fall. Unless these cyclins reach a threshold concentration, the cell does not progress to the next stage of the cell cycle. Cyclins therefore control the cell cycle and ensure that cells divide when new cells are needed, but not at other times.

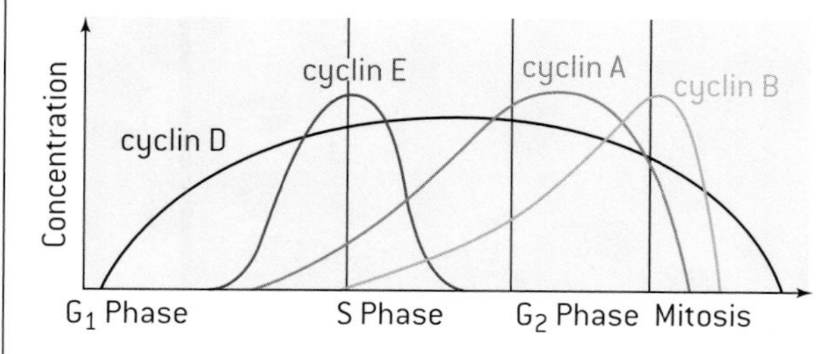

The discovery of cyclins is an example of what is known as **serendipity** – making happy and unexpected discoveries by accident. Tim Hunt was doing research into protein synthesis in sea urchin eggs. He noticed a protein that increased and decreased in concentration repeatedly and also that the increases and decreases corresponded with particular phases of the cell cycle. He named the protein cyclin. This and other cyclins were found to be key parts of the control of the cell cycle. Tim Hunt's discovery was partly due to luck but it was also due to being observant and realizing the significance of an unexpected observation.

TUMOUR FORMATION

Oncogenesis is the formation of tumours. The process starts with mutations in genes involved in the control of the cell cycle called **oncogenes**. Mutations have to occur in several oncogenes in the same cell for control to be lost. The chance of this is very small but the body contains billions of cells, any one of which could have mutations in its oncogenes, so the overall risk is significant.

Anything that increases the chance of mutations will increase the risk of tumour formation. Some chemical substances cause **mutations**. These chemicals are called mutagens. Ionizing radiation also causes mutations and therefore tumours.

When control of the cell cycle has been lost a cell undergoes repeated uncontrolled divisions that produce a mass of cells called a primary tumour. **Primary tumours** are often benign because they do not grow rapidly and do not spread, but others are malignant because cells become detached from them, are carried elsewhere in the body and there develop into a **secondary tumour**.

The spreading of cells to form tumours in a different part of the body is known as **metastasis**. Patients with secondary tumours are said to have cancer and unless the tumours are successfully treated they are likely to lead to a patient's death.

SMOKING AND CANCER

There is a positive correlation between cigarette smoking and the death rate due to cancer. The more cigarettes smoked per day the higher the chance of developing cancer of the lung and some other organs.

Although this correlation does not by itself prove that smoking causes cancer, there is also evidence that chemicals in tobacco smoke are mutagenic and therefore **carcinogenic** (cancer-causing). The best health advice that can be given to anyone is 'Don't smoke'.

Questions – cell biology

1. The micrograph below shows a transverse section of part of an animal cell.
   a) Identify the organelles labelled X and Y. [2]
   b) The maximum actual diameter of Y is 2 μm. Calculate the magnification of this organelle in the electron micrograph. [2]
   c) Determine, with two reasons, whether the cell is prokaryotic or eukaryotic. [2]
   d) From evidence in the electron micrograph, deduce two substances that were being synthesized in large quantities by this cell. [2]
   e) The dark granules in the cell are glycogen. Explain the conclusions that you draw from this information. [2]

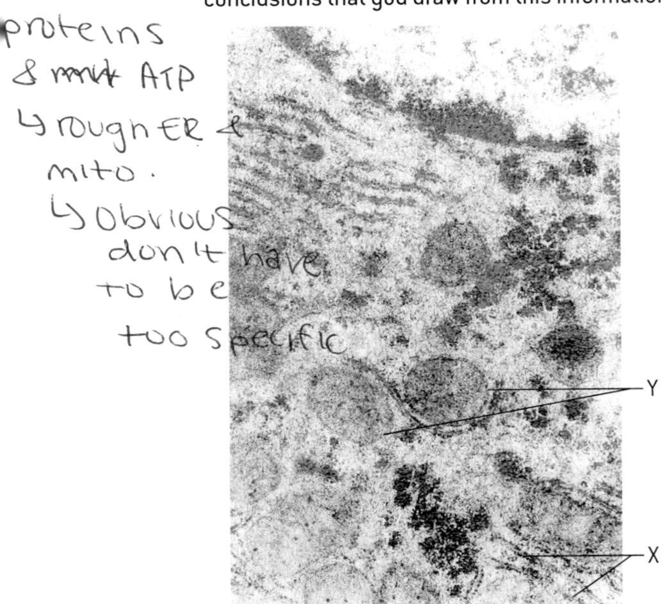

2. The table below gives the results of an experiment in which samples taken from a potato tuber were bathed in solutions with different concentrations of sucrose.

Concentration of sucrose (mol dm^{-3})	0.0	0.2	0.4	0.6
Initial mass (g)	22.5	21.8	23.7	22.2
Final mass (g)	25.9	22.9	22.8	18.7
Mass change (g)	+3.4			−3.5
% mass change				−16%

   a) Complete the table by calculating the missing mass changes and percentage mass changes. [3]
   b) Draw a graph to display the percentage mass changes. [4]
   c) (i) Estimate the osmolarity of the potato tissue. [1]
      (ii) Explain the reasons for your estimate. [2]

3. The freeze-etched scanning electron micrograph below shows part of a cell.

   a) Identify three organelles in the micrograph. [3]
   b) Outline the model of membrane structure proposed by Davson and Danielli. [2]
   c) Explain the evidence visible in the micrograph that falsified the Davson–Danielli model. [3]
   d) Outline one other type of evidence that could not be reconciled with Davson and Danielli's model. [2]

4. a) Identify the stage of mitosis in cells I to IV. [4]

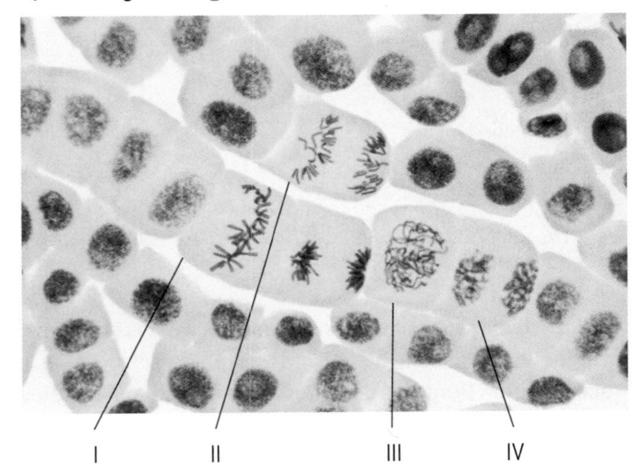

   b) Calculate the mitotic index of the root tissue in the micrograph. [4]
   c) State two processes that **must** occur in a plant cell before it starts mitosis. [2]

Molecules to metabolism

VITALISM AND MOLECULAR BIOLOGY

There are significant differences between living and non-living things, so at one time it was believed that they must be made of different materials. It was thought that living organisms were composed of organic chemicals that could only be produced in living organisms because a 'vital force' was needed. This was known as the **theory of vitalism** and it was falsified by a series of discoveries, including a method of synthesizing urea artificially.

Biologists now accept that living organisms are governed by the same chemical and physical forces as in non-living matter. The science of **molecular biology** aims to explain living processes in terms of the chemical substances involved. Since the discovery of the structure of DNA in the 1950s molecular biology has been tremendously successful and many processes have now been explained in molecular terms.

No 'vital force' has been discovered and a better answer to the question of what makes living organisms different from non-living matter is **natural selection**.

SYNTHESIS OF UREA

Urea was discovered in human urine in the 18th century. It is an organic compound with this structure:

```
       O
       ‖
       C
      / \
  H2N     NH2
```

According to the theory of vitalism it was predicted that urea could only be made in living organisms because it was an organic compound, so a vital force was needed.

In 1828 the German chemist Friedrich Wöhler synthesized urea artificially using silver isocyanate and ammonium chloride. This was the first time that an organic compound had been synthesized artificially. It helped to falsify the theory of vitalism but did not disprove it completely. Scientific theories are rarely abandoned until several pieces of evidence show that they are false.

ATOMS AND MOLECULES

An **atom** is a single particle of an element, consisting of a positively charged nucleus surrounded by a cloud of negatively charged electrons. A **molecule** is a group of two or more atoms held together by **covalent bonds**. These can be single, double or even occasionally triple covalent bonds.

In simple diagrams to show the structure of a molecule, the atom of an element is shown using the element's symbol and a covalent bond with a line.

Examples of molecular diagrams

```
    H   H
    |   |
H — C — C — O — H
    |   |
    H   H            O ═ C ═ O     N ≡ N     H — C ≡ N

   ethanol           carbon        nitrogen   hydrogen
                     dioxide                  cyanide
```

Nitrogen is an **element** but the other three molecules are **compounds** as two elements are bonded together.

The molecules used by living organisms are based on carbon. Each carbon atom forms four covalent bonds, allowing a great diversity of compounds to exist. Other elements used in molecules mostly form fewer covalent bonds:

Bonds	Element	Symbol
One	Hydrogen	H
Two	Oxygen	O
Three	Nitrogen	N
Four	Carbon	C

Covalent bonds are relatively strong, so molecules can be stable structures. Much weaker bonds form between molecules. They are called **intermolecular forces**.

The main types of molecule used by living organisms are carbohydrates, lipids, proteins and nucleic acids.

METABOLISM

Metabolism is the web of all the enzyme-catalysed reactions in a cell or organism. Most **metabolic pathways** consist of chains of reactions (below) but there are also some cycles of reactions (right).

Anabolism is the synthesis of complex molecules from simpler molecules. Living organisms produce **macromolecules** (very large molecules) from smaller single sub-units called **monomers**. Anabolic reactions are **condensation reactions** because water is produced.

Catabolism is the breakdown of complex molecules into simpler molecules including the hydrolysis of macromolecules into monomers. In **hydrolysis reactions** water molecules are split.

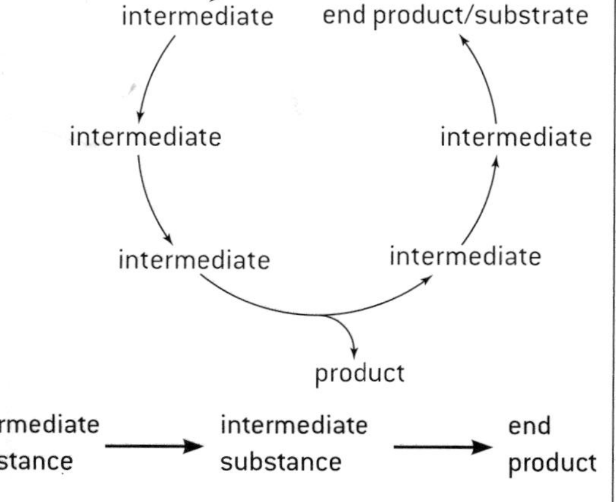

initial substrate → intermediate substance → intermediate substance → intermediate substance → intermediate substance → end product

Water

POLARITY OF WATER

Covalent bonds are formed when two atoms share a pair of electrons. In some cases the nucleus of one of the atoms is more attractive to the electrons than the other so the electrons are not shared equally. The consequence of this is that part of the molecule has a slight positive charge and another part has a slight negative charge. This feature of a molecule is called **polarity**.

Water molecules are polar. Hydrogen nuclei are less attractive to electrons than oxygen nuclei so the two hydrogen atoms have a slight positive charge and the oxygen atom has a slight negative charge. Water molecules have two poles and therefore are dipoles – they show **dipolarity**.

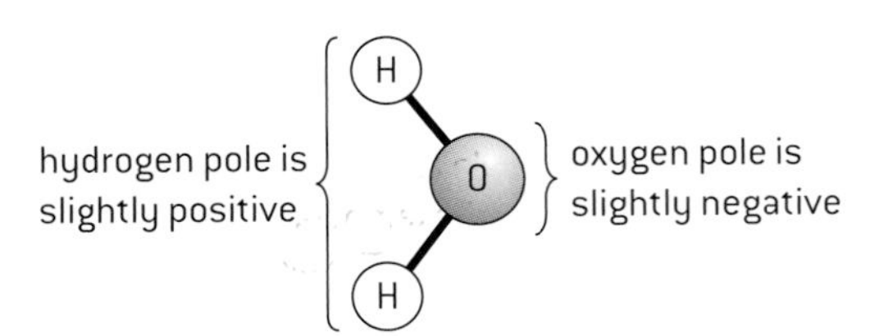

HYDROGEN BONDING IN WATER

An intermolecular bond can form between the positive pole of one water molecule and the negative pole of another. This is called a **hydrogen bond**. In liquid water many of these bonds form, giving water properties that make it a very useful substance for living organisms.

As with any chemical bond, energy is released when a hydrogen bond is made and used when a hydrogen bond is broken. For example, when a water molecule evaporates, hydrogen bonds between it and other water molecules must be broken. Heat energy is used to do this, explaining the use of sweat as a **coolant** – evaporation of water from sweat removes heat from the body.

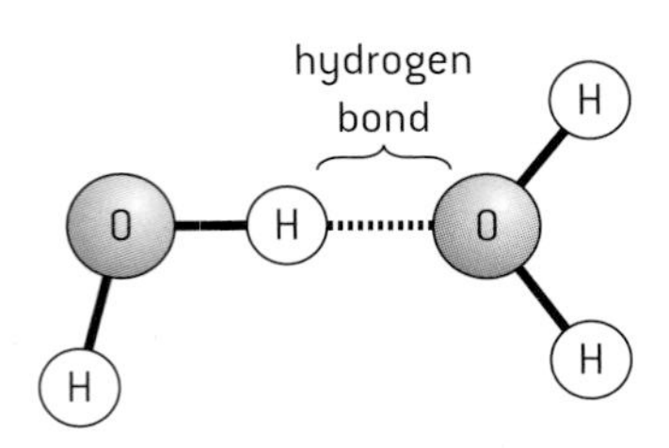

THERMAL PROPERTIES OF METHANE AND WATER COMPARED

The significance of hydrogen bonding in water can be illustrated by comparing the properties of water (H_2O) with those of methane (CH_4) – a substance with a similar molecular mass that has weaker intermolecular forces, not hydrogen bonds.

Property	Methane	Water	Explanation
Melting point	−182 °C	0 °C	Ice melts at a much higher temperature: hydrogen bonds restrict the movement of water molecules and heat is needed to overcome this.
Specific heat capacity	2.2 J per g per °C	4.2 J per g per °C	Water's heat capacity is higher: hydrogen bonds restrict movement so more energy is stored by moving molecules of water than methane.
Latent heat of vaporization	760 J/g	2257 J/g	Water has a much higher heat of vaporization: much heat energy is needed to break hydrogen bonds and allow a water molecule to evaporate.
Boiling point	−160 °C	100 °C	Water's boiling point is much higher: heat energy is needed to break hydrogen bonds and allow water to change from a liquid to a gas.

SOLUBILITY IN WATER

Some substances are attractive to water and form intermolecular bonds with water molecules. These substances are **hydrophilic**.

Ionic compounds and substances with polar molecules are hydrophilic. Many hydrophilic substances dissolve in water because their ions or molecules are more attracted to water than to each other.

If a substance is not hydrophilic it is said to be **hydrophobic**. This does not mean that it is repelled by water, but that water molecules are more strongly attracted to each other than to the non-polar molecules of hydrophobic substances. Hydrophobic substances are therefore insoluble in water.

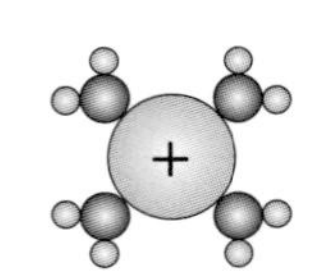

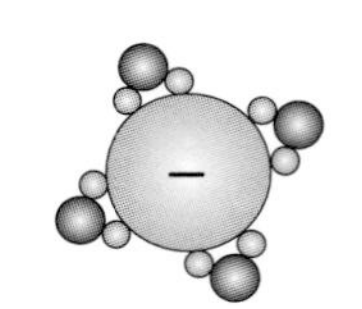

Ions with positive or negative charges dissolve as they are attracted to the negative or positive poles of water molecules.

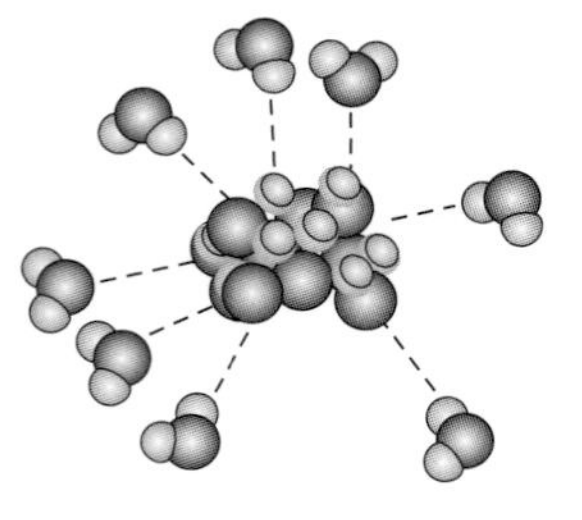

Many molecules are polar so are attracted to water molecules and dissolve.

TRANSPORT IN BLOOD

Blood transports a variety of substances. Most are transported in the blood plasma which contains many different solutes. The mode of transport of a substance depends on its solubility in water.

- Sodium chloride is soluble in water and is transported dissolved in the plasma as Na^+ and Cl^- ions.
- Glucose and amino acids are polar and so can be transported dissolved in the plasma.
- Oxygen is non-polar and the amount that dissolves in the plasma is insufficient so red blood cells are needed with hemoglobin to which oxygen binds.
- Cholesterol and fats are non-polar and insoluble in water so they are transported in small droplets called **lipoproteins**. The cholesterol and fats are inside, coated by phospholipids and proteins.

Water and life

PROPERTIES OF WATER

Water is very common on Earth but has some unusual properties. These properties can be explained using the theories of dipolarity and hydrogen bonding. This is a good example of one of the distinctive features of science – theories being used to explain natural phenomena. The remarkable properties of water make it so useful in many ways to living organisms that life could not exist without it. Water's uses as a coolant and as a transport medium in blood were described on the previous page.

Type of property	Explanation in terms of hydrogen bonding and dipolarity	Example of a benefit to living organisms
Cohesive	Water molecules **cohere** (stick to each other) because of the hydrogen bonds that form between them.	Strong pulling forces can be exerted to suck columns of water up to the tops of the tallest trees in tubes called xylem vessels. These columns of water rarely break despite the suction forces.
Adhesive	The dipolarity of water molecules makes them adhere to surfaces that are polar and therefore hydrophilic.	Adhesive forces between water and cellulose in cell walls in the leaf cause water to be drawn out of xylem vessels, keeping the cell walls moist and able to act as a gas exchange surface.
Thermal	Due to hydrogen bonding, water has high melting and boiling points, high latent heat of vaporization and high specific heat capacity.	These thermal properties cause water to be liquid in most habitats on Earth, making it suitable for living organisms. The high specific heat capacity makes its temperature change relatively slowly so it is a stable habitat. The high heat of vaporization makes it an effective coolant in leaves or in sweat.
Solvent	Many substances dissolve in water due to its polarity, including those composed of ions or polar molecules.	Most chemical reactions take place with all of the substances involved in the reaction dissolved in water, so water is the medium for metabolic reactions.

CONDENSATION REACTIONS

In a condensation reaction two molecules are joined together to form a larger molecule plus a molecule of water. Anabolic reactions are condensation reactions. A single sub-unit is a **monomer** and a pair of monomers bonded together is a **dimer**. A long chain of monomers is a **polymer**.

For example, two amino acids can be joined together to form a dipeptide by a condensation reaction. Further condensation reactions can link amino acids to either end of the dipeptide, eventually forming a chain of many amino acids. This is called a **polypeptide**. The new bond formed to link amino acids together is a **peptide bond**.

Condensation of two amino acids to form a dipeptide and water

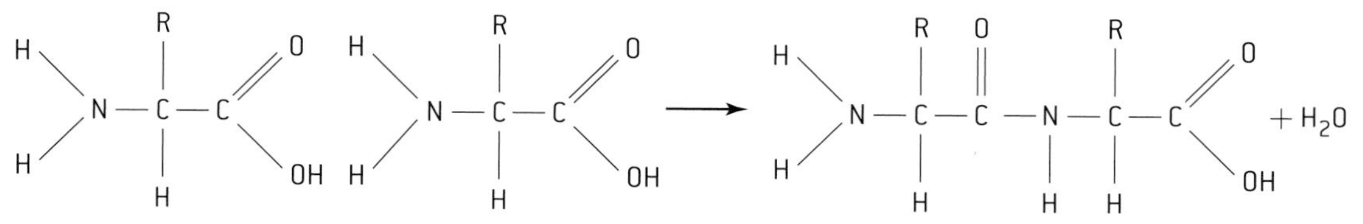

$$H_2N{-}CHR{-}C(=O)OH + H_2N{-}CHR{-}C(=O)OH \rightarrow H_2N{-}CHR{-}C(=O){-}NH{-}CHR{-}C(=O)OH + H_2O$$

Condensation reactions are used to build up carbohydrates and lipids.

- The basic sub-units of carbohydrates are **monosaccharides**. Two monosaccharides can be linked to form a **disaccharide** plus water and more monosaccharides can be linked to a disaccharide to form a large molecule called a **polysaccharide**.
- Fatty acids can be linked to glycerol by condensation reactions to produce lipids called **glycerides**. A maximum of three fatty acids can be linked to each glycerol, producing a **triglyceride** plus three water molecules.

HYDROLYSIS REACTIONS

Hydrolysis reactions are the reverse of condensation reactions. In a hydrolysis reaction a large molecule is broken down into smaller molecules. Water is used up in the process. Water molecules are split into –H and –OH groups, hence the name hydrolysis (*lysis* = splitting). The –H and –OH are needed to make new bonds after a bond in the large molecule has been broken. Catabolic reactions are hydrolysis reactions, including those used to digest food.

Examples:

polypeptides + water → dipeptides or amino acids

polysaccharides + water → disaccharides or monosaccharides

glycerides + water → fatty acids + glycerol

Carbohydrates

MONOSACCHARIDES

Monosaccharides are sugars that consist of a single sub-unit (monomer). They contain only atoms of carbon, hydrogen and oxygen in the ratio 1:2:1 so **ribose** for example is $C_5H_{10}O_5$ and **glucose** is $C_6H_{12}O_6$. Ribose and glucose are important molecules so it is useful to be able to draw and recognize their molecular structure. They share certain features:

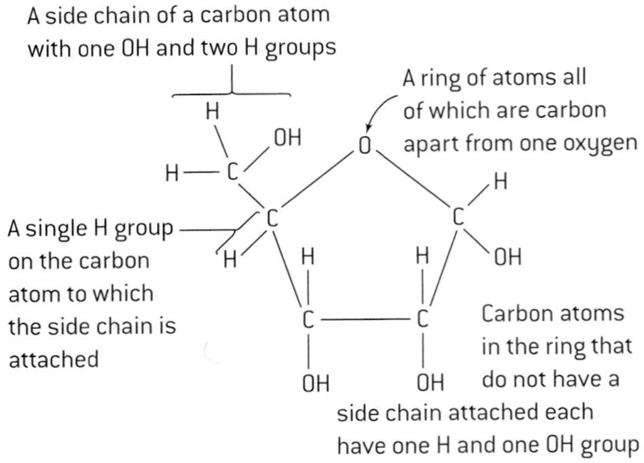

A side chain of a carbon atom with one OH and two H groups

A ring of atoms all of which are carbon apart from one oxygen

A single H group on the carbon atom to which the side chain is attached

Carbon atoms in the ring that do not have a side chain attached each have one H and one OH group

The molecule shown in the diagram above is D-ribose. The D indicates that this is the right-handed version of ribose. Left- and right-handed versions of ribose and glucose can exist but living organisms use only the right-handed versions (D-ribose and D-glucose).

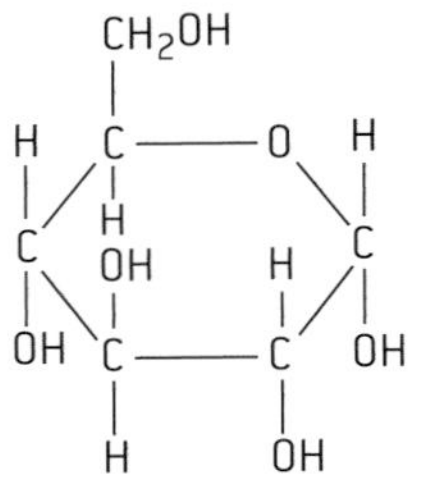

α-D-glucose

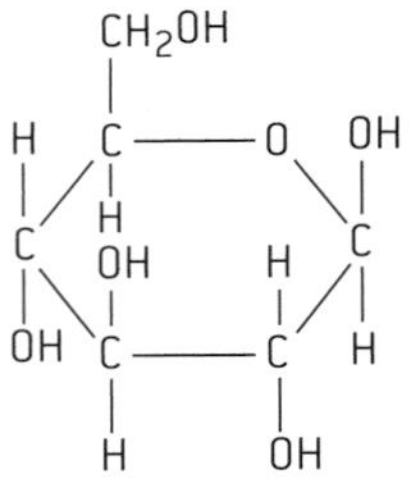

β-D-glucose

DISACCHARIDES

Pairs of monosaccharides are linked together by condensation to form disaccharides. Glucose, galactose and fructose are monosaccharides that are commonly used to make disaccharides:

glucose + glucose ⟶ maltose + H_2O

glucose + galactose ⟶ lactose + H_2O

glucose + fructose ⟶ sucrose + H_2O

Disaccharides are sugars. Their molecules can be recognized by the double ring structure.

POLYSACCHARIDES

The polysaccharides cellulose, glycogen and starch are all composed of glucose. To help describe their structure, a numbering system for the carbon atoms in glucose is used:

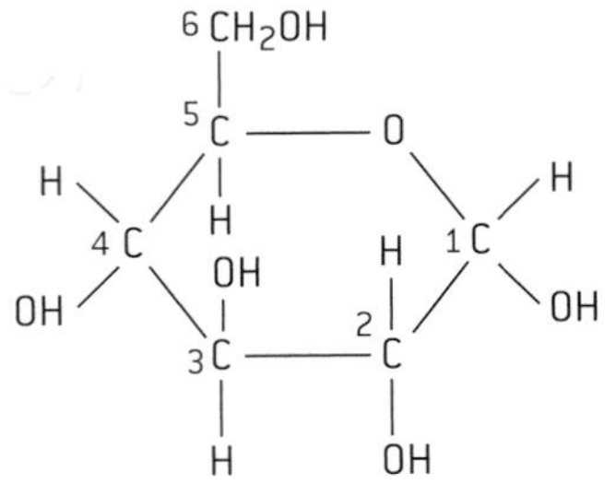

The basic linkage between the glucose subunits is a glycosidic bond from C_1 of a glucose to C_4 of the next, but some polysaccharides also have some 1,6 glycosidic bonds, giving them a branched structure.

1. **Cellulose** is an unbranched polymer of β-D-glucose. The orientation of the glucose units alternates (up-down-up and so on), which makes the polymer straight rather than curved, and allows groups of cellulose molecules to be arranged in parallel with hydrogen bonds forming cross links. These structures are cellulose microfibrils. They have enormous tensile strength and are the basis of plant cell walls.

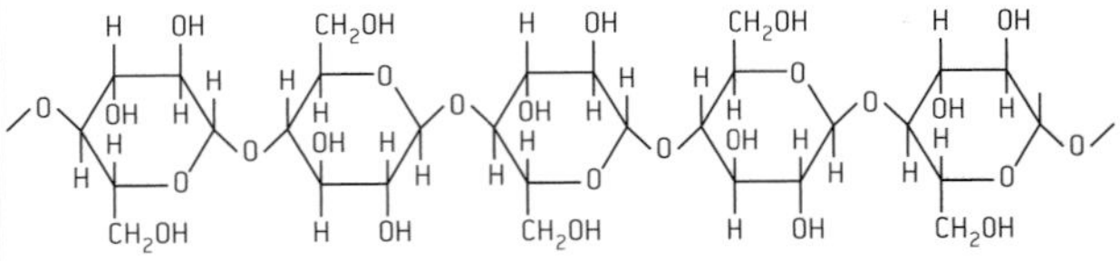

2. **Starch** is a polymer of α-D-glucose, with all of the glucose subunits in the same orientation, giving the polymer a helical shape. There are two forms of starch: **amylose** has only 1,4 linkages so is unbranched, whereas **amylopectin** has some 1,6 linkages so is a branched molecule (below).

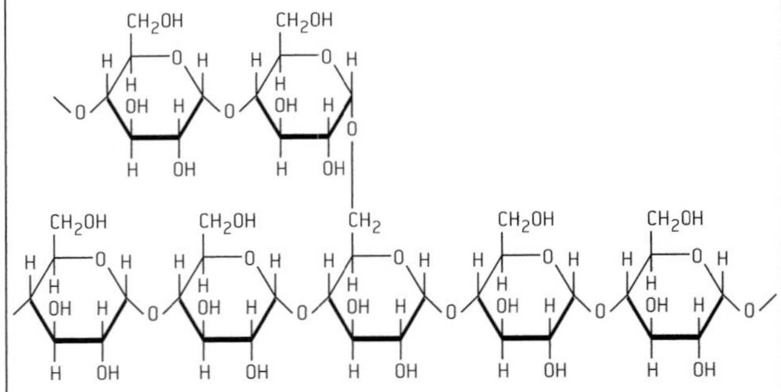

   Starch is used by plants to store glucose in an insoluble form that does not cause osmotic problems. By making the molecule branched it is possible to load or unload glucose more rapidly as there are more points on starch molecules to which glucose can be added or detached.

3. **Glycogen** is similar in structure to amylopectin – it is a branched polymer of α-D-glucose. There are more 1,6 linkages than in amylopectin so it is more branched. Glycogen is used by mammals to store glucose in liver and muscle cells. Because glycogen is insoluble, large amounts can be stored whereas if glucose was stored it would cause water to enter the cells by osmosis and there would be a danger of them bursting.

Molecular visualization of polysaccharides

MOLECULAR VISUALIZATION SOFTWARE

Computer programs are used to produce images of molecules. The most widely used molecular visualization software is JMol, which can be downloaded free of charge. There are also many websites that use JMol, which are easy to use.

You should be able to make these changes to the image of a molecule that you see on the screen:

- Use the scroll function on the mouse to make the image larger or smaller.
- Left click and move the mouse to rotate the image.
- Right click to display a menu that allows you to change the style of molecular model, label the atoms, make the molecule rotate continuously or change the background colour.

Examples of JMol images of polysaccharides:

Amylose (the unbranched form of starch)

Glycogen or amylopectin (the branched form of starch)

Cellulose

HYDROGEN BONDING IN CELLULOSE

Molecular visualization can be used to show interactions between molecules. This image shows how cellulose molecules consisting of chains of β-D-glucose can form a parallel array, with hydrogen bonding at regular intervals both within each molecule and between molecules. This structure occurs in the cellulose microfibrils of plant cell walls.

Because the chains of α-D-glucose in starch and glycogen are helical, they cannot become aligned in a parallel array so hydrogen bonds do not form.

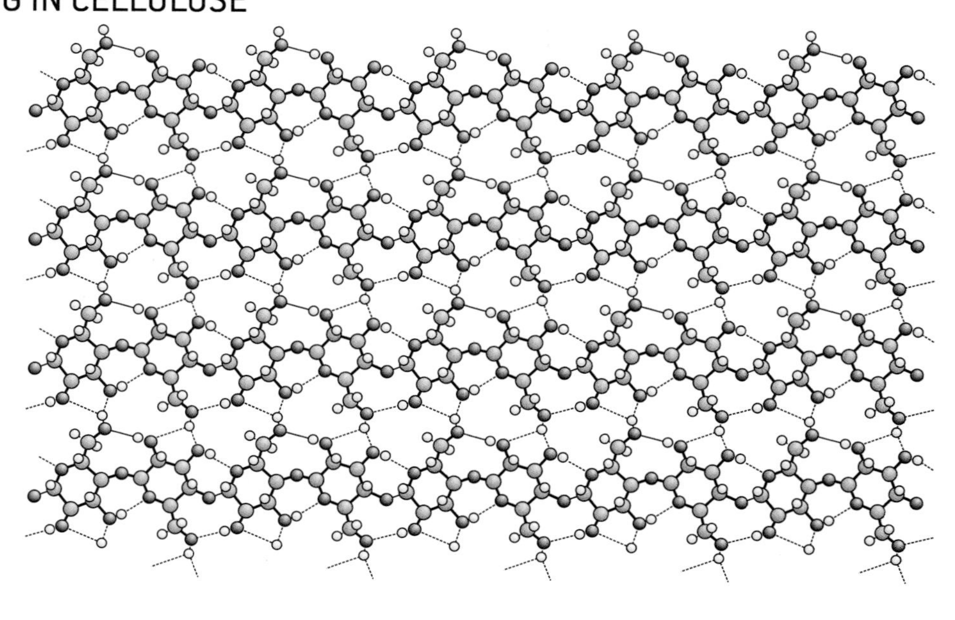

Lipids

LIPIDS

Lipids are carbon compounds made by living organisms that are mostly or entirely hydrophobic. There are three main types of lipid:

1. **Triglycerides** are made from three fatty acids and one glycerol by condensation reactions so they have three hydrocarbon tails. Fats and oils are triglycerides.
2. **Phospholipids** are similar to triglycerides but only have two fatty acids linked to glycerol, with a phosphate group instead of the third fatty acid. Phospholipids are only partly hydrophobic and form the basis of membranes.
3. **Steroids** all have a similar structure of four fused rings in their molecule. Cholesterol, progesterone, estrogen and testosterone are all steroids.

Molecules can be identified as lipids if they have two or three hydrocarbon chains or the quadruple ring structure of steroids. Hydrocarbon chains are often shown simply as a zigzag line in molecular diagrams of lipids and rings of carbon atoms are shown as hexagons or pentagons.

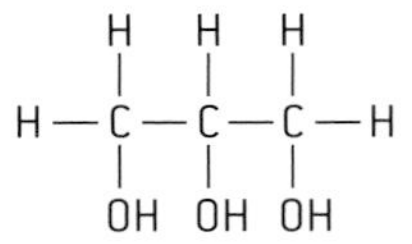

glycerol

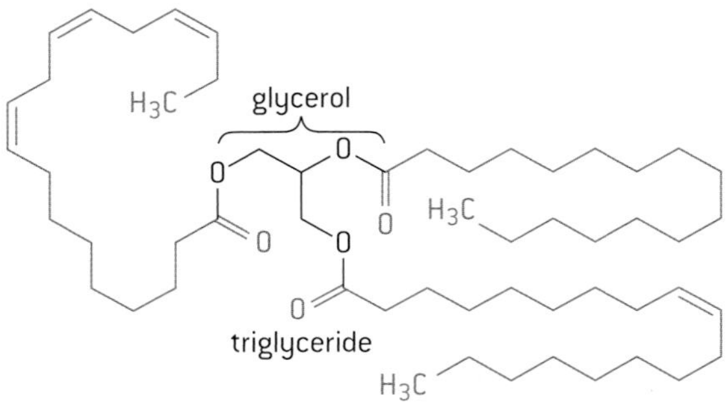

triglyceride

Triglyceride – the fatty acid tails are flexible and can change position

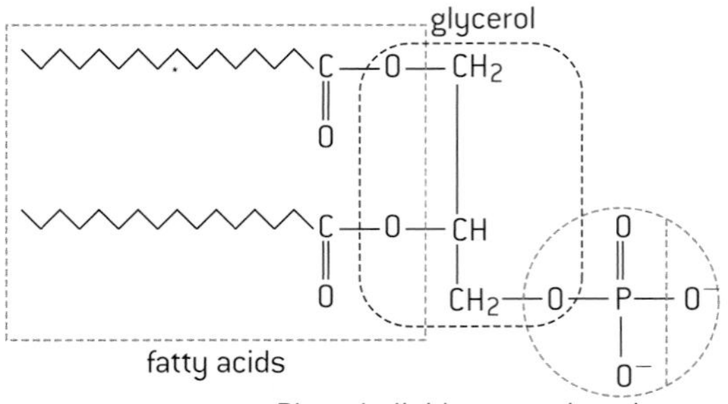

fatty acids

Phospholipid

phosphate

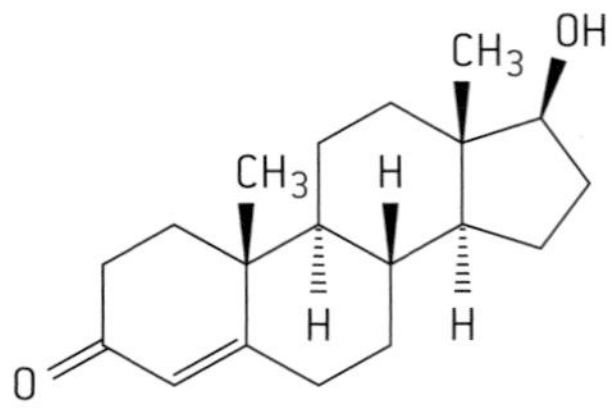

Testosterone – a steroid

DRAWING FATTY ACIDS

There are two parts to a fatty acid:

a carboxyl group that is acidic, which is shown as

$-COOH$ or:

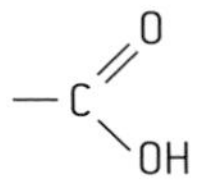

and an unbranched hydrocarbon chain, which can either be shown in full (right) or in a briefer form,

$-CH_2-(CH_2)_n-CH_3$

The fatty acid shown right is saturated as all the carbon atoms in the molecule are linked to each other by single covalent bonds, so the molecule holds as much hydrogen as possible. It is useful to be able to draw the structure of a saturated fatty acid such as this and also to recognize other types of fatty acid.

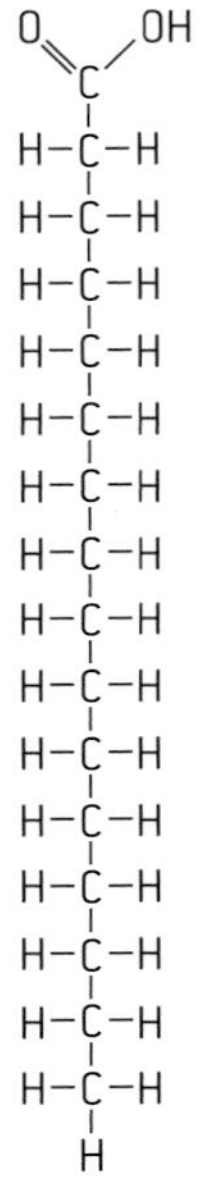

fatty acid (number of carbon atoms and bonding between carbon atoms varies)

TYPES OF FATTY ACID

Fatty acids vary in the number of carbon atoms in the hydrocarbon chain and in the bonding of the carbon atoms to each other and to hydrogens.

Saturated – all of the carbon atoms in the chain are connected by single covalent bonds so the number of hydrogen atoms bonded to the carbons cannot be increased.

Unsaturated – contain one or more double bonds between carbon atoms in the chain, so more hydrogen could be bonded to the carbons if a double bond was replaced by a single bond.

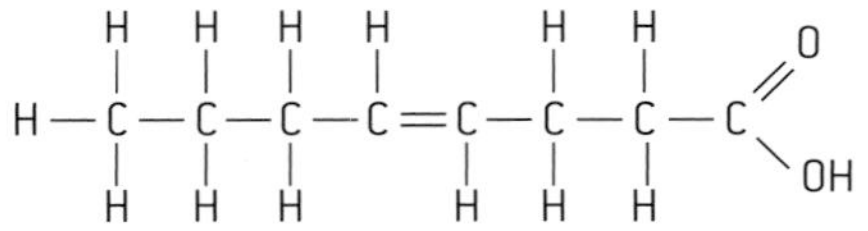

Unsaturated fatty acid
(naturally occurring ones have more carbon atoms)

Monounsaturated – only one double bond.

Polyunsaturated – two or more double bonds.

The position of the nearest double bond to the CH_3 terminal is significant. In omega-3 fatty acids, it is the third bond from CH_3, whereas in omega-6 fatty acids it is the sixth.

Cis unsaturated – hydrogen atoms are bonded to carbon atoms on the same side of a double bond.

Trans unsaturated – hydrogen atoms are bonded to carbon atoms on opposite sides of a double bond.

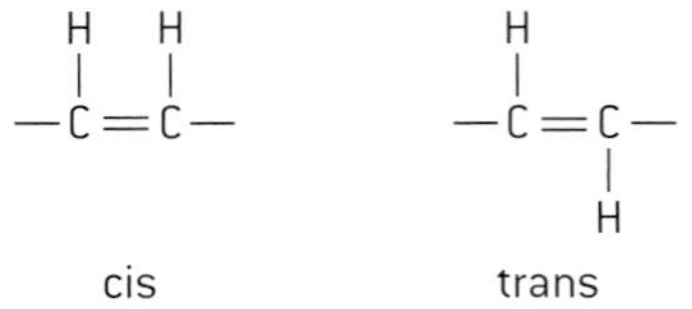

cis

trans

Lipids and health

ENERGY STORAGE

Fats or oils (lipids) and glycogen or starch (carbohydrates) are both used by living organisms as stores of energy. The seeds of plants contain starch or oil. In humans there are stores of glycogen in the liver and muscles of fat in adipose tissue.

There are advantages in using lipids rather than carbohydrate for long-term energy storage. The amount of energy released in cell respiration per gram of lipids is double the amount released from a gram of carbohydrates. The same amount of energy stored as lipid rather than carbohydrate therefore adds half as much to body mass. In fact the mass advantage of lipids is even greater because fats form pure droplets in cells with no water associated, whereas each gram of glycogen is associated with about two grams of water, so lipids are actually six times more efficient in the amount of energy that can be stored per gram of body mass. This is important, because we have to carry our energy stores around with us wherever we go. It is even more important for animals such as birds and bats that fly.

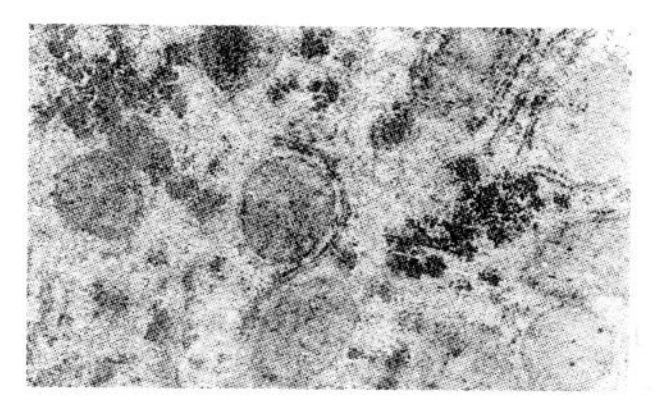

glycogen granules in liver

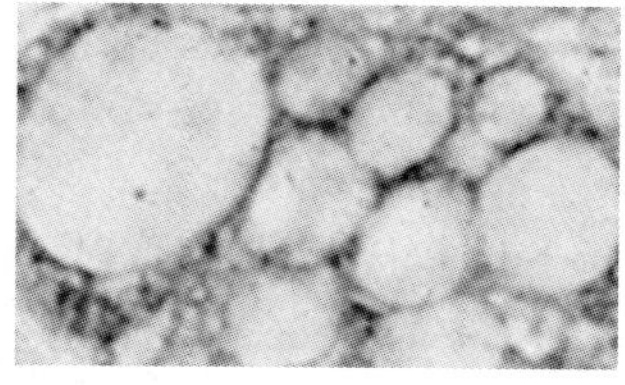

fat droplets in adipose cell

BODY MASS INDEX

It is not possible to assess whether a person's body mass is at a healthy level simply by weighing them, because of natural variation in size between adults. Instead, body mass index is calculated. The units for BMI are kg/m^2.

$$\text{BMI} = \frac{\text{mass in kilograms}}{(\text{height in metres})^2}$$

The table below can be used to draw conclusions from a person's BMI.

Body mass index	*Conclusion*
below 18.5	underweight
18.5–24.9	normal weight
25.0–29.9	overweight
30.0 or more	obese

BMI can also be worked out using a nomogram (right) A ruler is used to make a straight line linking mass and height and BMI can be read off from the central scale.

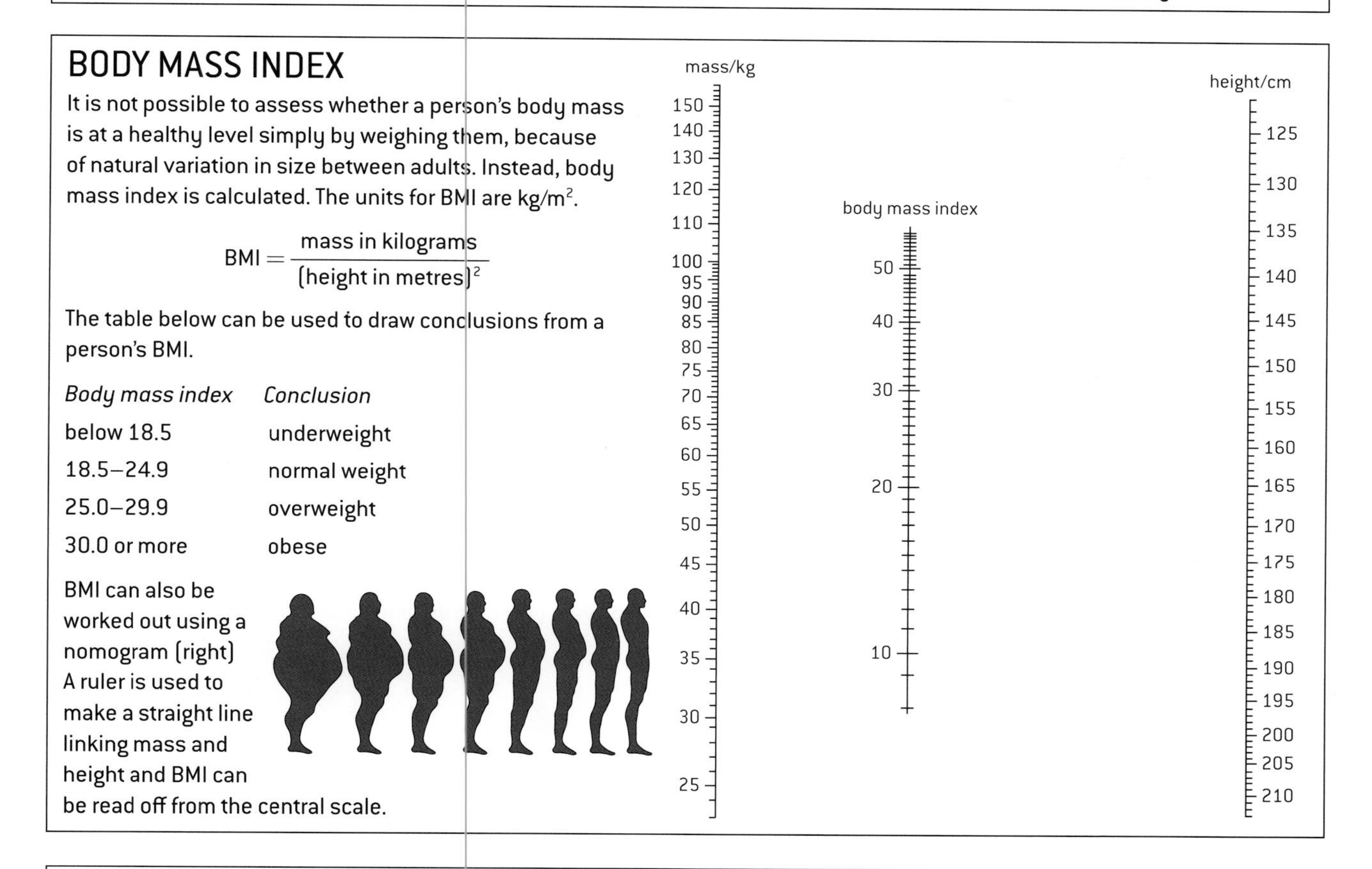

HEALTH RISKS OF TRANS-FATS AND SATURATED FATTY ACIDS

Trans-fats are mostly artificially produced but their use is now banned in some countries. There is a positive correlation between amounts of trans-fat consumed and rates of **coronary heart disease (CHD)**. Other risk factors have been tested, to see if they can account for the correlation, but they do not. Trans-fats therefore probably do cause CHD. In patients who died from CHD, fatty deposits in the diseased arteries have been found to contain high concentrations of trans-fats, which gives more evidence of a causal link.

Saturated fatty acids occur naturally in animal fats and some vegetable oils. A positive correlation has been found between saturated fatty acid intake and rates of CHD in many research programs, but there are populations that do not fit the correlation, such as the Maasai of Kenya that have a diet of foods rich in saturated fats yet CHD is very rare. It is possible that the actual cause of CHD is not saturated fat itself but another factor correlated with saturated fat intake, such as low amounts of dietary fibre.

Evaluation of evidence is an important process in science, especially with correlations between fat intake and health risks. Correlations are statistical links that may or may not be due to causation, so a positive correlation between saturated fat intake and rates of CHD does not prove that saturated fats cause CHD. Surveys based on large sample sizes are more trustworthy. Effects of factors other than the one being investigated should have been taken into account in the analysis. Results from a single survey should be treated with caution.

Amino acids and polypeptides

DRAWING AMINO ACIDS

Amino acids have a central carbon atom with four different atoms or groups linked to it:

- **hydrogen** atom
- **amine** group ($-NH_2$)
- **carboxyl** group ($-COOH$)
- **R group** or radical (R).

```
         R         O
H        |        //
  \      |       /
   N --- C --- C
  /      |       \
H        |        OH
         H
```

amino acids
(each of the twenty amino acids in proteins has a different R group)

AMINO ACID DIVERSITY

The R-group of amino acids is variable. Amino acids with hundreds of different R-groups could be produced in the laboratory, but most living organisms include only twenty of them in the polypeptides synthesized by their ribosomes. The same twenty amino acids are used. Trends such as this are often significant in science. In this case the use of the same repertoire of amino acids is one of the pieces of evidence supporting the theory that all living organisms share common ancestry.

There are a few discrepancies in the trend. Two other amino acids are included in a few polypeptides. These are selenocysteine and pyrrolysine. Only a minority of polypeptides in a minority of organisms contain either of these amino acids. In both cases special mechanisms are needed to incorporate them into polypeptides and it is more likely that these mechanisms evolved after the basic method of making polypeptides from the twenty amino acids. The two extra amino acids are therefore extra variations rather than a falsification of the theory that there are twenty basic amino acids in all organisms.

PEPTIDE BONDS AND POLYPEPTIDES

Amino acids are linked together by condensation reactions. This is shown on page 19. The new bond formed between the amine group of one amino acid and the carboxyl group of the next is a peptide bond.

```
    R1   O        R2
    |    ||       |
 -- C -- C -- N -- C --
    |         |    |
    H         H    H
```

peptide bond

A molecule consisting of two amino acids linked together is a dipeptide. Polypeptides consist of many amino acids linked by peptide bonds.

POLYPEPTIDES AND PROTEINS

- A **polypeptide** is an unbranched chain of amino acids.
- The number of amino acids is very variable and can be over 10,000, though most have between 50 and 2,000 amino acids.
- Chains of fewer than 40 amino acids are usually called **peptides** rather than polypeptides or proteins.
- Amino acids can be linked together in any sequence giving a huge range of possible polypeptides. If we consider a polypeptide with 100 amino acids, the number of possible sequences is 20^{100}, which is an almost unimaginably large number.
- Only a small proportion of the possible sequences of amino acids are ever made by living organisms.
- Particular sequences are made in very large quantities because they have useful properties.
- Over two million polypeptides have so far been discovered in living organisms.
- The amino acid sequence of a polypeptide is coded for by a **gene**. The sequence of bases in the DNA of the gene determines the sequence of amino acids in the polypeptide.
- A **protein** consists either of a single polypeptide or more than one polypeptide linked together.

PROTEINS AND PROTEOMES

A **proteome** is all of the proteins produced by a cell, a tissue or an organism. By contrast, the genome is all of its genes. Whereas the genome of an organism is fixed, the proteome is variable because different cells in an organism make different proteins. Even in a single cell the proteins that are made vary over time depending on the cell's activities.

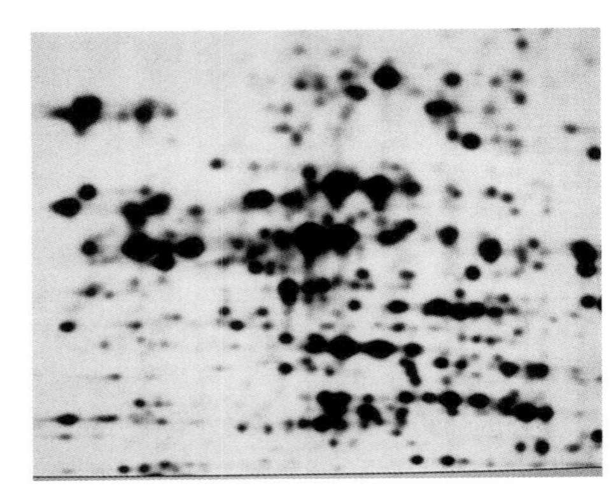

Proteins that are extracted from a tissue can be separated in a sheet of gel by electrophoresis and identified. This example shows proteins in the legume *Medicago trunculata*.

Within a species there are strong similarities in the proteome of all individuals but also differences. The proteome of each individual is unique, partly because of differences of activity but also because of small differences in the amino acid sequence of proteins. With the possible exception of identical twins, none of us have identical proteins, so each of us has a unique proteome. Even the proteomes of identical twins can become different with age.

Protein structure and function

PROTEIN CONFORMATIONS

- The conformation of a protein is its three-dimensional structure.
- The polypeptides of most proteins are folded up to produce a globular shape.
- The sequence of amino acids in a polypeptide determines how this folding is done and so determines the conformation of a protein.
- Each time a polypeptide with a particular sequence of amino acids is synthesized on a ribosome, the conformation will tend to be precisely the same.
- The structure is stabilized by intramolecular bonds between the amino acids in the polypeptides that are brought together by the folding process.

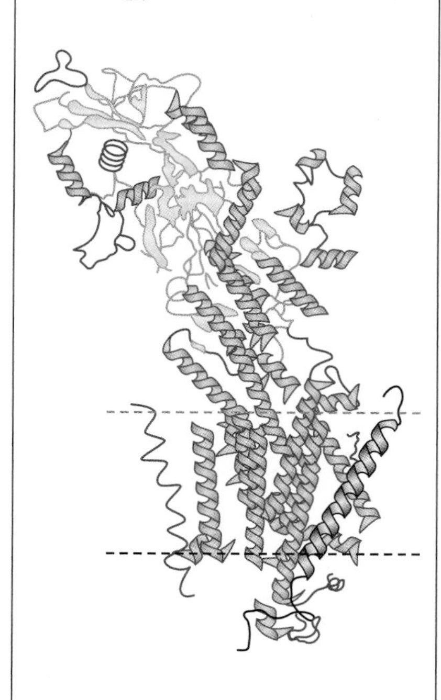

The image above represents the conformation of a protein. It shows the sodium–potassium pump and the position of the membrane where it is located.

This is an example of a protein that switches reversibly between alternative conformations. It allows the sodium–potassium pump to pick up ions from one side of the membrane and discharge them on the other side.

DENATURATION

The conformation of most proteins is delicate and it can be damaged by various substances and conditions. This is called **denaturation**.

1. **Heat** causes vibrations within protein molecules that break intramolecular bonds and cause the conformation to change. Heat denaturation is almost always irreversible.

   This can be demonstrated by heating egg white, which contains dissolved albumin proteins. The albumins are denatured by the heat and in their new conformation are insoluble. The causes the liquid egg white to turn into a white solid.

2. Every protein has an ideal or optimum **pH** at which its conformation is normal. If the pH is increased by adding alkali or decreased by adding acid, the conformation of the protein may initially stay the same but denaturation will eventually occur when the pH has deviated too far from the optimum. This is because the pH change causes intramolecular bonds to break within the protein molecule.

   The photograph shows egg white mixed with hydrochloric acid.

FUNCTIONS OF PROTEINS

Living organisms synthesize many different proteins with a wide range of functions. Six examples are given here.

1. Rubisco is the **enzyme** with an active site that catalyses the photosynthesis reaction that fixes carbon dioxide from the atmosphere, providing all the carbon needed by living organisms to make sugars and other carbon compounds.
2. Insulin is the **hormone** that is carried dissolved in the blood and binds specifically and reversibly to insulin receptors in the membranes of body cells, causing the cells to absorb glucose and lower the blood glucose concentration.
3. Immunoglobulins are **antibodies** that bind to antigens on pathogens. The immune system can produce a huge range of immunoglobulins, each with a different type of binding site, allowing specific immunity against many different diseases.
4. Rhodopsin is the **pigment** that makes the rod cells of the retina light-sensitive. It has a non-amino acid part called **retinal** that absorbs a photon of light and when this happens the rod cell sends a nerve impulse to the brain.
5. Collagen is a **structural** protein. It has three polypeptides wound together to form a rope-like conformation and is used in skin to prevent tearing, in bones to prevent fractures and in tendons and ligaments to give tensile strength.
6. Spider silk is a **structural** protein that is used to make webs for catching prey and lifelines on which spiders suspend themselves. It has very high tensile strength and becomes stronger when it is stretched, so resisting breakage.

Enzymes

SUBSTRATES AND ACTIVE SITES

Catalysts speed up chemical reactions without being changed themselves. Living organisms make biological catalysts called **enzymes** to speed up and control the rate of the reactions of metabolism. Enzymes are globular proteins. A reactant in an enzyme-catalysed reaction is known as a **substrate**.

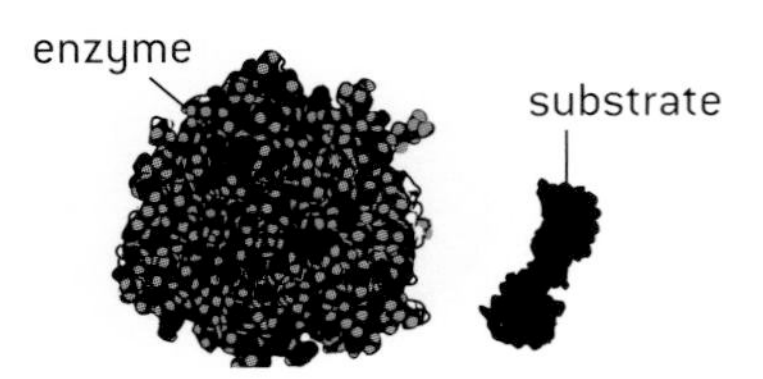

Enzymes catalyse reactions using a special region called the **active site**. Catalysis only occurs if the substrates are in a liquid so their molecules are in continual random motion and there is a chance of **collisions** between the substrates and the active site on the surface of the enzyme.

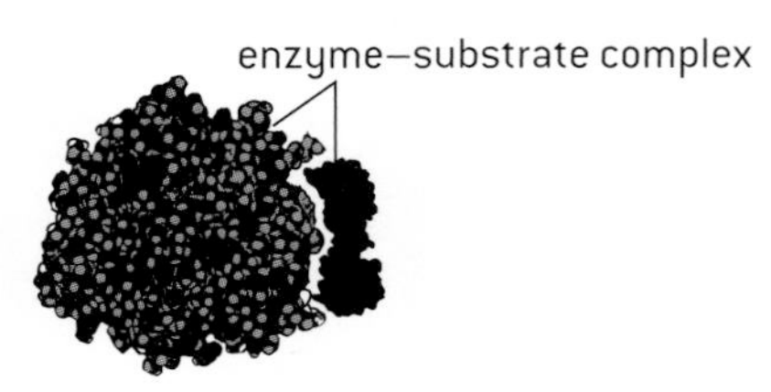

Substrate has collided with the active site on the enzyme and has become bound to it

Collisions can result in **binding** as the shape and chemical properties of the active site complement those of substrates. They are chemically attracted to each other and fit together. Molecules other than the substrate do not fit or are not attracted so do not bind, making enzymes **substrate-specific**.

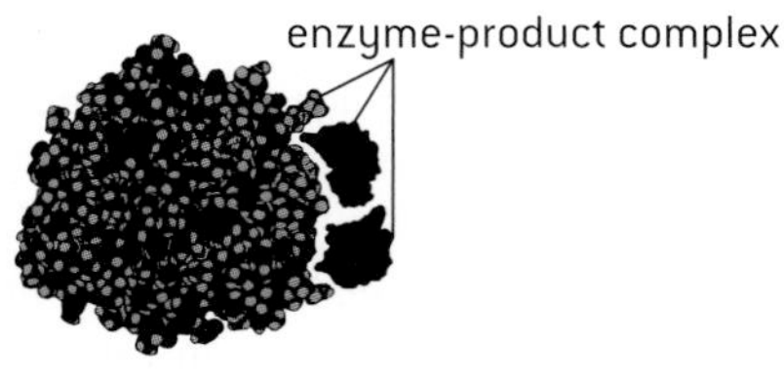

While bound to the active site the substrate has been converted into the products

The binding of substrates to the active site reduces the energy needed for them to be converted into products. The **products** are released from the active site, freeing it up to catalyse the reaction with more substrates. An enzyme can catalyse its reaction many times per second.

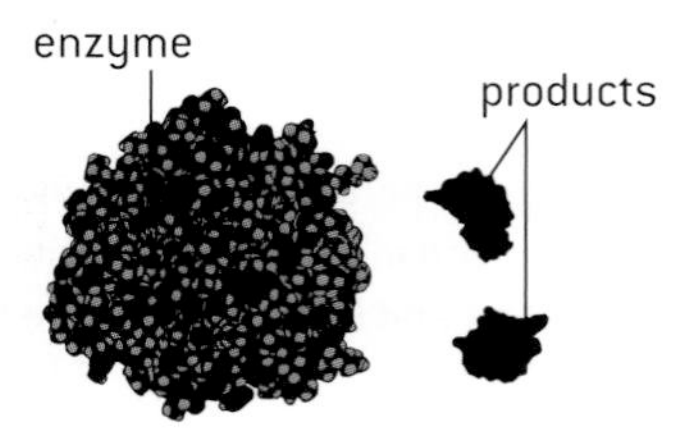

The products detach from the active site, leaving it free for more substrate to bind.

IMMOBILIZED ENZYMES

Enzymes are widely used in industry for catalysing specific reactions. The enzymes are usually **immobilized**, by attachment of enzymes to another material or into aggregations to restrict their movement. Enzyme immobilization has benefits:

1. Catalysis can be controlled by adding or removing enzymes promptly from the reaction mixture.
2. Enzyme concentrations can be higher.
3. Enzymes can be reused, saving money.
4. Enzymes are resistant to denaturation over greater ranges of pH and temperature.
5. Products are not contaminated with enzymes.

There are many methods of enzyme immobilization:

1. attachment to surfaces such as glass (adsorption)
2. entrapment in a membrane or a gel (e.g. alginate)
3. aggregation by bonding enzymes together into particles of up to 0.1 mm diameter.

PRODUCTION OF LACTOSE-FREE MILK

Lactose is the sugar in milk. It can be hydrolysed into glucose and galactose by the enzyme lactase.

$$\text{lactose} \xrightarrow{\text{lactase}} \text{glucose} + \text{galactose}$$

Lactose-free milk is produced either by adding free lactase to the milk or by using lactase that has been immobilized on a surface or in beads of a porous material. The enzyme is obtained from microorganisms such as *Kluveromyces lactis*, a yeast that grows in milk.

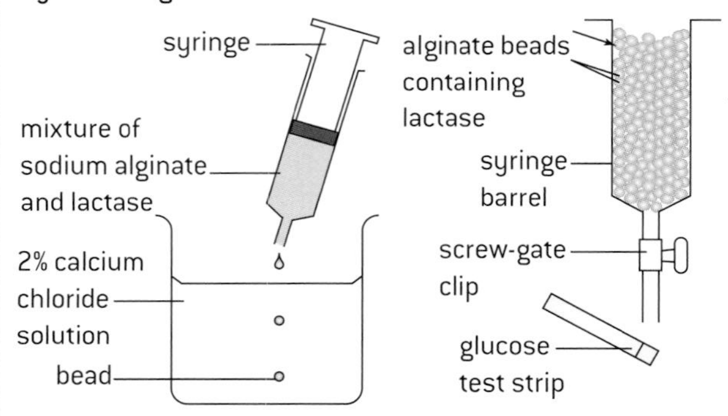

This process can be performed in the laboratory by making alginate beads containing lactase and putting them into milk. The lactose concentration of the milk drops and the glucose concentration rises.

Lactose-free milk has some advantages:

1. Many people are lactose intolerant and cannot drink more than about 250 ml of milk per day unless it is lactose-reduced.
2. Galactose and glucose are sweeter than lactose, so less sugar needs to be added to sweet foods containing milk, such as milk shakes or fruit yoghurt.
3. Lactose tends to crystallize during production of ice cream, giving a gritty texture. Because glucose and galactose are more soluble than lactose they remain dissolved, giving a smoother texture.
4. Bacteria ferment glucose and galactose more quickly than lactose, so the production of yoghurt and cottage cheese is faster.

Factors affecting enzyme activity

FACTORS AFFECTING ENZYME ACTIVITY

Wherever enzymes are used, it is important that they have the conditions that they need to work effectively. Temperature, pH and substrate concentration all affect the rate at which enzymes catalyse chemical reactions. The figures (below and right) show the relationships between enzyme activity and substrate concentration, temperature and pH.

EFFECT OF TEMPERATURE

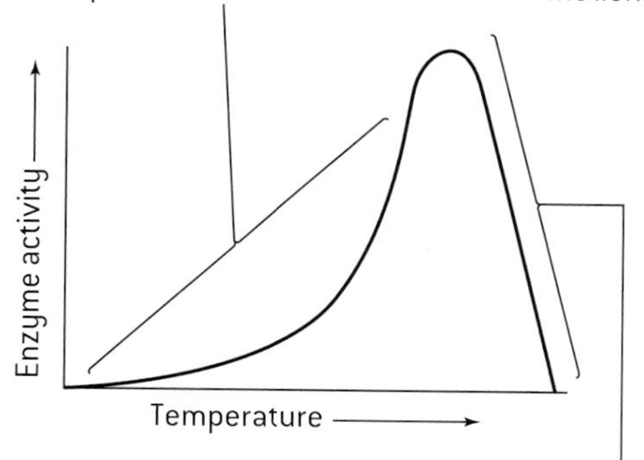

EFFECT OF pH

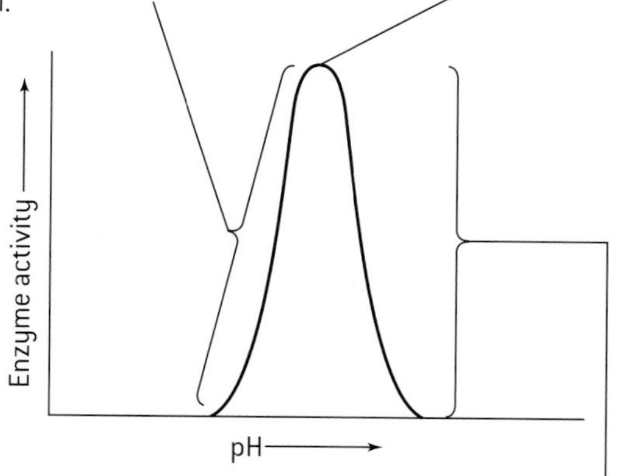

EFFECT OF SUBSTRATE CONCENTRATION

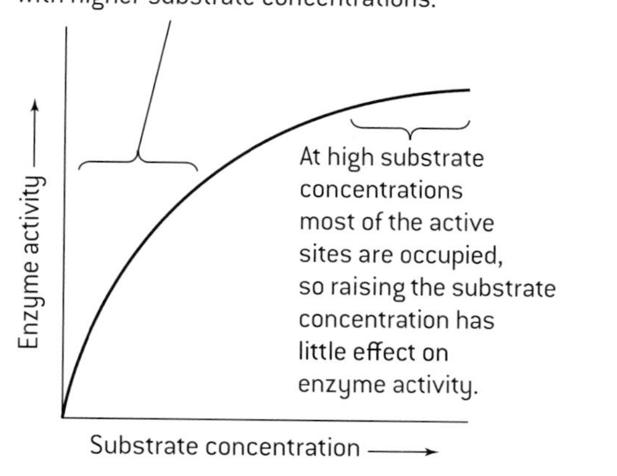

DESIGNING ENZYME EXPERIMENTS

The effect of temperature, pH or substrate concentration can be investigated experimentally. To design an experiment that will give reliable results these decisions must be made:

1. The **independent variable** – there should be just one independent variable and you choose the levels of it. You need a wide enough **range** to show all the trends, for example pH 1–14. Your method should make sure that each level of the variable is maintained as accurately as possible.
2. The **dependent variable** – this is the measurement you take to assess the rate of enzyme activity. S.I. units should be used. You might time how long it takes for the substrate to be used up or measure the quantity of a product formed after a certain time. The measurement should be **quantitative** and as **accurate** as possible, for example a time to the nearest second. The experiment should be repeated so there are **replicate results** that can be compared to evaluate whether they are **reliable**.
3. The **control variables** are other factors that could affect enzyme activity. They must be kept constant so that they do not cause differences in the results of the experiment. For example, if pH is the independent variable, temperature is a control variable and must be kept constant at 20 °C or some other suitable temperature, using a water bath.

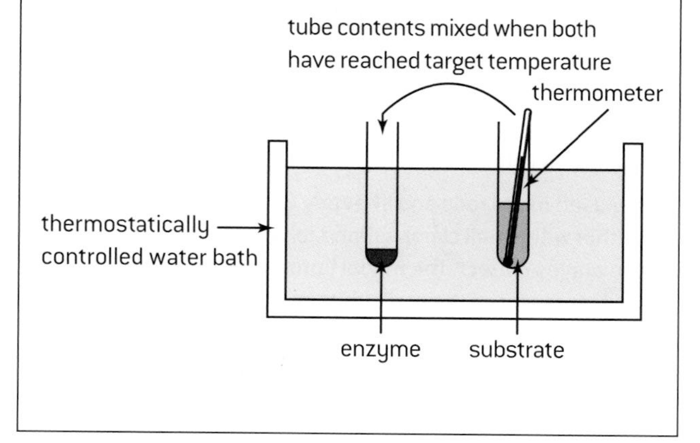

Structure of DNA and RNA

DRAWING NUCLEOTIDES

DNA and RNA are the two types of nucleic acid. They are both polymers of sub-units called nucleotides. Each **nucleotide** consists of three parts – a **pentose** sugar, a **phosphate** group and a **base**. In diagrams of nucleotides they are usually shown as pentagons, circles and rectangles, respectively. The figure (below) shows how the sugar, the phosphate and the base are linked up in a nucleotide.

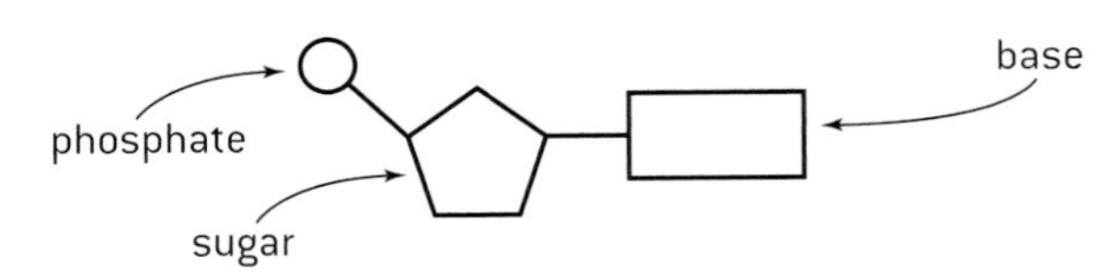

There are two differences between DNA and RNA nucleotides.

1. The type of pentose is **ribose** in RNA but **deoxyribose** in DNA.
2. In both DNA and RNA there are four possible bases. Three of these are the same: **adenine, cytosine** and **guanine**. The fourth base is **thymine** in DNA but is **uracil** in RNA.

STUCTURE OF RNA AND DNA

There is a third difference between DNA and RNA – the number of strands of nucleotides. RNA usually has one strand and DNA usually has two. The nucleotides in a strand of DNA or RNA are linked together by covalent bonds between the pentose sugar of one nucleotide and the phosphate of the next one. The diagram (right) shows a strand of RNA.

G
U
A
U
C
A

In DNA the two strands are **antiparallel** – they run alongside each other but in opposite directions. The two strands are linked by **hydrogen bonding** between their bases. Each base will only form hydrogen bonds with one other base, so two base pairs only are possible: adenine with thymine and cytosine with guanine (A–T and C–G). These are known as **complementary base pairs**. The diagram (far right) shows the structure of DNA. This diagram does not show how the two strands are wound to form a double helix, which is the overall shape of a DNA molecule.

STRUCTURE OF DNA

T A
A T
G C
T A
G C
T A
C G
G C
A T
C G
A T
T A
A T
T A

MODEL MAKING AND THE DISCOVERY OF THE STRUCTURE OF DNA

Model making played a critical part in Crick and Watson's discovery of the structure of DNA, but it took several attempts before they were successful. They used cardboard shapes to represent the bases in DNA and found that A–T and C–G base pairs could be formed, with hydrogen bonds linking the bases. The base pairs were equal in length so would fit into a molecule between two outer sugar–phosphate backbones.

Other scientists had produced X-ray diffraction data showing the DNA molecule to be helical. A flash of insight was needed to make the parts of the molecule fit together: the two strands in the helix had to run in opposite directions. Crick and Watson were then able to build their famous model of the structure of DNA.

They used metal rods and sheeting cut to shape and held together with small clamps. Bond lengths were all to scale and bond angles correct. The model immediately convinced others that it represented the real structure of DNA. Further testing of the model confirmed this.

DNA replication

SEMI-CONSERVATIVE REPLICATION

Crick and Watson's model of DNA structure immediately suggested a method of copying called **semi-conservative replication.** The two strands of the DNA molecule are separated by breaking the hydrogen bonds between their bases. New polymers of nucleotides are assembled on each of the two single strands. A strand of DNA on which a new strand is assembled is called a 'template strand'. Because of **complementary base pairing,** each of the new strands has the same base sequence as the old strand that was separated from the template strand. The two DNA molecules produced in this way are identical to each other and to the original parent DNA molecule. This is semi-conservative replication because each of the DNA molecules produced has one new strand and one strand conserved from the parent molecule.

Two alternative theories were rejected: **conservative replication** – both strands of the parent DNA remain together and another molecule is produced with 2 new strands and **dispersive replication** – every molecule produced by DNA replication has a mixture of old and new sections in both of its strands.

MESELSON AND STAHL AND DNA REPLICATION

Soon after the discovery of the structure of DNA, strong evidence for semi-conservative replication was published by Meselson and Stahl. They cultured *E. coli* bacteria for many generations in a medium where the only nitrogen source was ^{15}N, so the nitrogen in the bases of the bacterial DNA was ^{15}N. They then transferred the bacteria abruptly to a medium with the less dense ^{14}N isotope. A solution of caesium chloride was spun in an ultracentrifuge at 45,000 revolutions per minute for 24 hours. Caesium ions are heavy so tend to sink, establishing a gradient with the greatest caesium concentration and therefore density at the bottom. Any substance centrifuged with the caesium chloride solution becomes concentrated at the level of its density. Meselson and Stahl spun samples of DNA collected from their bacterial culture at different times after transfer to the ^{14}N medium. The DNA shows up as a dark band in UV light. After one generation the DNA was intermediate in density between ^{14}N and ^{15}N, as expected with one old and one new strand. After two generations there were two equal bands, one still $^{14}N/^{15}N$ and one at ^{14}N density. In the following generations the less dense ^{14}N band became stronger and the $^{14}N/^{15}N$ band weaker.

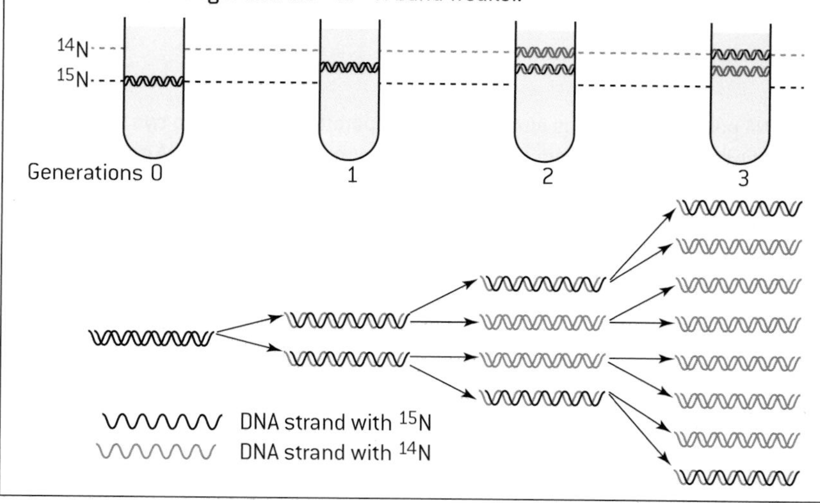

STAGES IN DNA REPLICATION

Stage 1
Helicase unwinds the double helix and separates the two strands by breaking hydrogen bonds.

free nucleotides

Stage 2
DNA polymerase links nucleotides together to form new strands, using the pre-existing strands as templates.

Stage 3
The daughter DNA molecules each rewind into a double helix.

The two daughter DNA molecules are identical in base sequence to each other and to the parent molecule, because of complementary base pairing. Adenine will only pair with thymine and cytosine will only pair with guanine. Each of the new strands is complementary to the template strand on which it was made and identical to the other template strand

Transcription and translation

RNA POLYMERASE AND TRANSCRIPTION

The sequence of bases in a polypeptide is stored in a coded form in the base sequence of a gene. The first stage in the synthesis of a polypeptide is to make a copy of the base sequence of the gene. The copy is made of RNA and is carried to the ribosomes in the cytoplasm to give them the information needed to synthesize a polypeptide, so is called mRNA (messenger RNA). The copying of the base sequence of a gene by making an RNA molecule is called **transcription**. The process begins when the enzyme RNA polymerase binds to a site on the DNA at the start of a gene. It then carries out all of the stages shown in the diagram below.

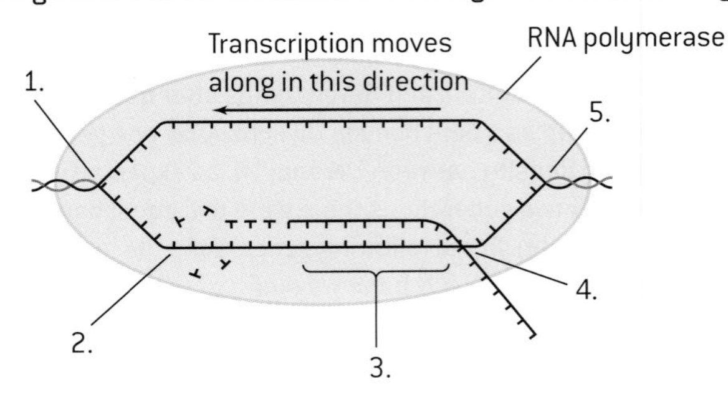

1. RNA polymerase moves along the gene separating the DNA into two single strands. (Stages 1, 2 and 3 are all carried out by the enzyme RNA polymerase).
2. RNA nucleotides are assembled along one of the two strands of DNA. The same rules of **complementary base pairing** are followed as in replication, except that uracil pairs with adenine, as RNA does not contain thymine.
3. The RNA nucleotides are linked together by covalent bonds between the pentose sugar of one nucleotide and the phosphate of the next.
4. The RNA strand separates from the DNA strand as it is produced and is released completely when the end of the gene is reached.
5. The DNA strands pair up again and twist back into a double helix.

DNA AND RNA SEQUENCES

If the base sequence of a strand of mRNA is known, the base sequence of the DNA strand from which it was transcribed can be deduced.

Example:

mRNA strand: A U C G C U

will have been transcribed by assembling RNA nucleotides on a DNA strand with this sequence:

transcribed DNA strand: T A G C G A

In a double-stranded DNA molecule the transcribed strand is paired with another strand that has complementary bases.

antisense strand: T A G C G A
sense strand: A T C G C T

The strand of DNA that is not transcribed has the same sequence as the mRNA, apart from having T in place of U, so it is called the **sense strand** and the transcribed strand is the **antisense strand**.

TRANSLATION

Translation is the synthesis of polypeptides on ribosomes, using mRNA and tRNA. The amino acid sequence of polypeptides is determined by mRNA according to the genetic code. The **genetic code** that is translated on the ribosome is a **triplet code** – three bases code for one amino acid. A group of three bases is called a **codon**. Translation depends on complementary base pairing between codons on mRNA and anticodons on tRNA.

anticodon

large sub-unit of ribosome

amino acid

small sub-unit of ribosome

direction of movement of ribosome

1. Messenger RNA binds to a site on the small sub-unit of the ribosome. The mRNA contains a series of **codons** consisting of three bases, each of which codes for one amino acid.

2. Transfer RNA molecules are present around the ribosome in large numbers. Each tRNA has a special triplet of bases called an **anticodon** and carries the amino acid corresponding to this anticodon.

3. There are three binding sites for tRNA molecules on the large sub-unit of the ribosome but only two ever bind at once. A tRNA can only bind if it has the anticodon that is complementary to the codon on the mRNA. The bases on the codon and anticodon link together by forming hydrogen bonds, following the same rules of complementary base pairing as in replication and transcription.

4. The amino acids carried by the tRNA molecules are bonded together by a peptide linkage. A dipeptide is formed, attached to the tRNA on the right. The tRNA on the left detaches. The ribosome moves along the mRNA to the next codon. Another tRNA carrying an amino acid binds. A chain of three amino acids is formed. These stages are repeated until a polypeptide is formed.

The genetic code

USING THE GENETIC CODE

The genetic code is a triplet code – three bases code for one amino acid. A group of three bases is called a **codon**. If codons consisted of two bases there would only be sixteen codons (4^2) – not enough for the twenty amino acids in polypeptides. With three bases in a codon there are 64 different codons (4^3). This gives more than enough codons to code for the twenty amino acids in proteins. None of the 64 codons are unused. There are two or more codons for most amino acids. The meaning of each codon is shown in the table.

First base of codon (5' end)	Second base of codon on messenger RNA: U	C	A	G	Third base of codon (3' end)
U	Phenylalanine Phenylalanine Leucine Leucine	Serine Serine Serine Serine	Tyrosine Tyrosine STOP STOP	Cysteine Cysteine STOP Tryptophan	U C A G
C	Leucine Leucine Leucine Leucine	Proline Proline Proline Proline	Histidine Histidine Glutamine Glutamine	Arginine Arginine Arginine Arginine	U C A G
A	Isoleucine Isoleucine Isoleucine Methionine / START	Threonine Threonine Threonine Threonine	Asparagine Asparagine Lysine Lysine	Serine Serine Arginine Arginine	U C A G
G	Valine Valine Valine Valine	Alanine Alanine Alanine Alanine	Aspartic acid Aspartic acid Glutamic acid Glutamic acid	Glycine Glycine Glycine Glycine	U C A G

There is no need to learn the meaning of each codon, but it is useful to be able to use the table to deduce the sequence of amino acids coded for by the base sequence of a length of mRNA. For example, the sequence CACAGAUGGGUC codes for histidine, arginine, tryptophan, valine.

The table can also be used to find the triplets of bases that code for an amino acid. For example, methionine is only coded for by the triplet AUG and this triplet is also used as the start codon.

PRODUCTION OF HUMAN INSULIN IN BACTERIA

Human insulin is a protein that contains just 51 amino acids. The gene that codes for insulin has been transferred from humans to the bacterium *E. coli* and to other organisms, to produce the insulin that is needed to treat diabetes. Details of the methods used for gene transfer are explained in Topic 3.

The amino acid sequence of the insulin that is produced in these organisms using the transferred gene is identical to the sequence produced in humans. This is because of the **universality of the genetic code** – *E. coli* and humans use the same genetic code so each codon in the mRNA is translated into the same amino acid when insulin is made.

Amino acid sequence of human insulin

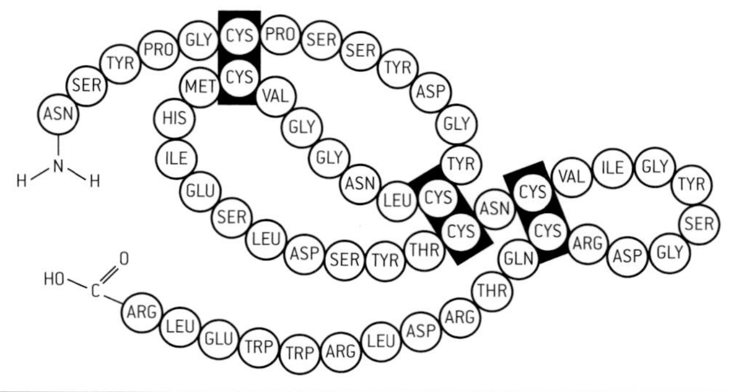

Although the genetic code is described as universal there are minor variations that occur in some organisms. For example, in some yeasts CUG codes for serine rather than leucine. In some organisms a stop codon is used for a non-standard amino acid.

POLYMERASE CHAIN REACTION

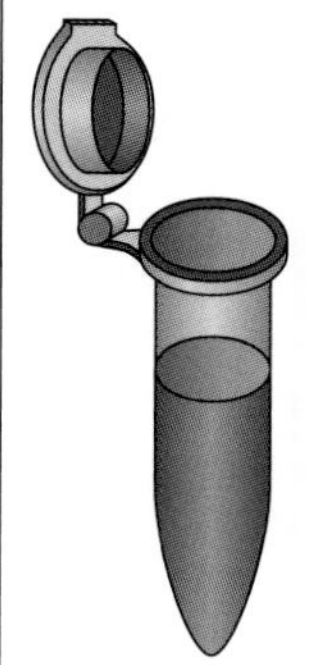

For gene transfer procedures, many copies of the desired gene are needed. It is also useful to be able to copy DNA artificially when a sample contains very small quantities and larger amounts are needed for forensic analysis.

The polymerase chain reaction (PCR) is used for copying DNA artificially. DNA polymerase is used in this procedure to copy the original molecule again and again, doubling the quantity with each cycle of replication.

DNA is copied in small tubes called eppendorfs. By the end of PCR there could be more than a hundred million copies of a gene in a 0.2 ml eppendorf.

To speed up PCR it is carried out at high temperatures. A special type of heat-stable DNA polymerase has to be used: **Taq DNA polymerase**. This enzyme is obtained from *Thermus aquaticus*, a bacterium that is adapted to living in hot springs, so its enzymes are active at temperatures that would denature proteins from other organisms.

Millions of copies of the DNA can be produced by PCR in a few hours because of the high temperatures used. The details of the PCR procedure are described in Topic 3.

Cell respiration

ENERGY AND CELLS

All living cells need a continual supply of energy. This energy is used for a wide range of processes including active transport and protein synthesis. Most of these processes require energy in the form of ATP (adenosine triphosphate).

Every cell produces its own ATP, by a process called **cell respiration.** Carbon compounds (organic compounds) such as glucose or fat are carefully broken down and the energy released by doing this is used to make ATP. Cell respiration is defined as **controlled release of energy from organic compounds to produce ATP**.

The advantage of ATP is that it is immediately available as an energy source in the cell. It can diffuse to any part of the cell and release its energy within a fraction of a second.

AEROBIC AND ANAEROBIC CELL RESPIRATION

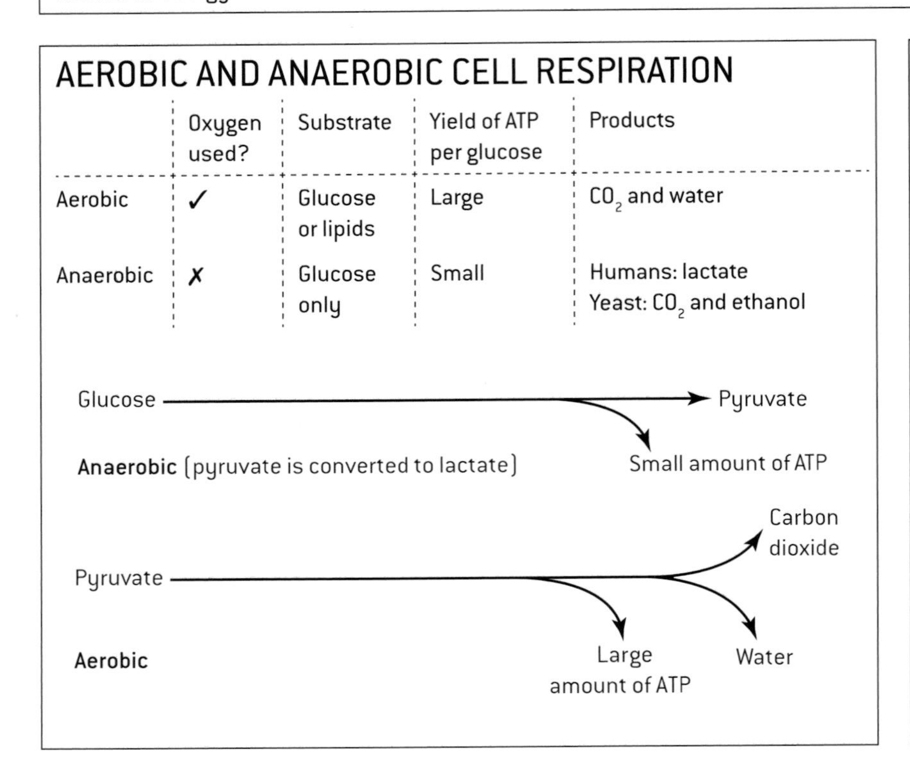

	Oxygen used?	Substrate	Yield of ATP per glucose	Products
Aerobic	✓	Glucose or lipids	Large	CO_2 and water
Anaerobic	✗	Glucose only	Small	Humans: lactate Yeast: CO_2 and ethanol

ENERGY FOR MUSCLES

Although anaerobic cell respiration produces fewer molecules of ATP per glucose, it can supply ATP at a more rapid overall rate for a short time, because it is not limited by how fast oxygen can be supplied. Anaerobic cell respiration is therefore used in muscles carrying out very vigorous exercise, for example muscles used for sprinting or weight lifting. Anaerobic cell respiration maximizes the power of muscle contractions. Lactate (lactic acid) and hydrogen ions are produced by this process. Anaerobic respiration can only be used to produce ATP for about two minutes. Beyond this duration, hydrogen ion concentrations would make the pH of the blood too low, so aerobic cell respiration must be used and high-intensity exercise cannot be continued.

USING YEAST IN BREWING AND BAKING

Both of the products of anaerobic respiration in yeast are used in industries.

1. Carbon dioxide and the baking industry

   Yeast is used in baking bread. It is mixed into the dough before baking. The yeast rapidly uses up all oxygen present in the dough and then produces ethanol and carbon dioxide by anaerobic cell respiration. The carbon dioxide forms bubbles making the dough rise – it increases in volume. This makes the dough less dense – it is leavened. When the dough is baked most of the ethanol evaporates and the carbon dioxide bubbles give the bread a light texture, which makes it more appetizing.

2. Ethanol and the brewing and biofuel industries

   Yeast can be used to produce ethanol by fermentation. The yeast is cultured in a liquid containing sugar and other nutrients, but not oxygen so it respires anaerobically. The ethanol concentration of the fluid around the yeast cells can rise to approximately 15% by volume, before it becomes toxic to the yeast and the fermentation ends. Most of the carbon dioxide bubbles out into the atmosphere. Beer, wine and other alcoholic drinks are brewed in this way. Ethanol is also produced by fermentation for use as a fuel.

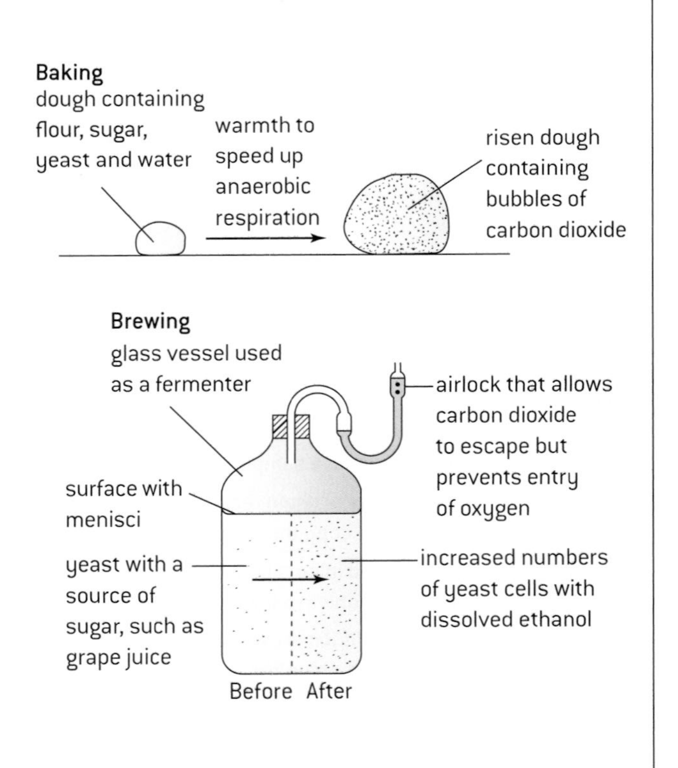

Respirometers

RESPIROMETERS AND RESPIRATION RATES

A respirometer is any device that is used to measure respiration rates. There are many possible designs. Most involve these parts:

- A sealed glass or plastic container in which the organism or tissue is placed.
- An **alkali**, such as potassium hydroxide, which absorbs carbon dioxide produced by cell respiration. The volume of air inside the respirometer should therefore reduce as a result of oxygen being used in cell respiration by the organisms in the respirometer.
- A capillary tube containing fluid, connected to the container, which allows the volume of air inside the respirometer to be monitored.

One possible design of respirometer is shown below.

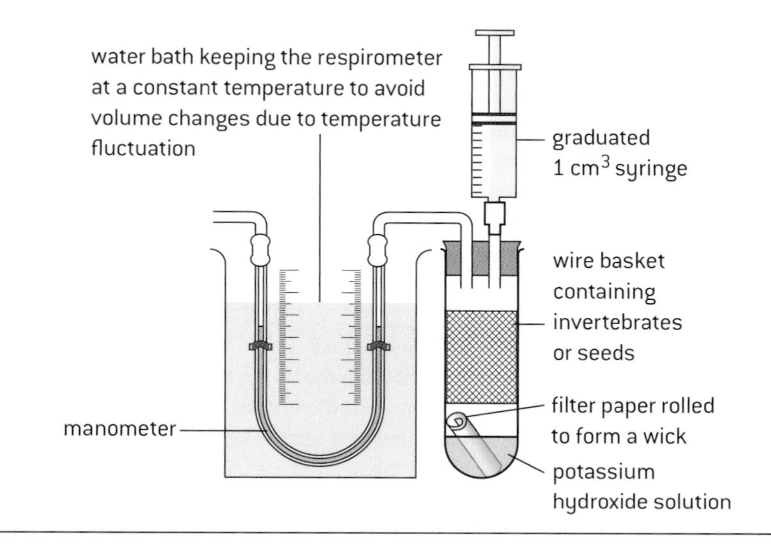

ETHICAL IMPLICATIONS

Fly larvae or other invertebrates are sometimes used in respirometer experiments. It is essential to assess the ethical implications of an experiment involving animals before doing it. In this case these questions should be asked:

1. Will the animals suffer pain or any other harm during the experiment? If the answer is 'yes' then the experiment should not be performed in an IB school.
2. Are there unacceptable risks to the animals, for example contact with the alkali? Again if the answer is 'yes', the experiment should not be done.
3. Will the animals have to be removed from their natural habitat and if so, can they be safely returned to it and continue to live natural lives?
4. Is it necessary to use animals in the experiment or could another organism be used, for example germinating seeds?

There are strict restrictions in most countries on the use of vertebrates in research, and fewer animals are now used than in the past.

ANALYSIS OF DATA FROM RESPIROMETER EXPERIMENTS

1. Calculating mean results

If the data includes repeats, mean results should be calculated. The mean is calculated by adding together all the results and dividing them by the number of results (n).

$$\text{Mean} = \frac{X_1 + X_2 + X_3 + \ldots X_n}{n}$$

2. Plotting a graph with range bars

A graph should be plotted of the mean results, with the **independent variable** on the horizontal ***x*-axis** (for example temperature) and the **dependent variable** (for example distance moved by fluid in capillary tube per minute) on the vertical ***y*-axis**. Vertical lines extending above and below the mean can be used to show the range from the lowest to the highest individual results. They are called **range bars**.

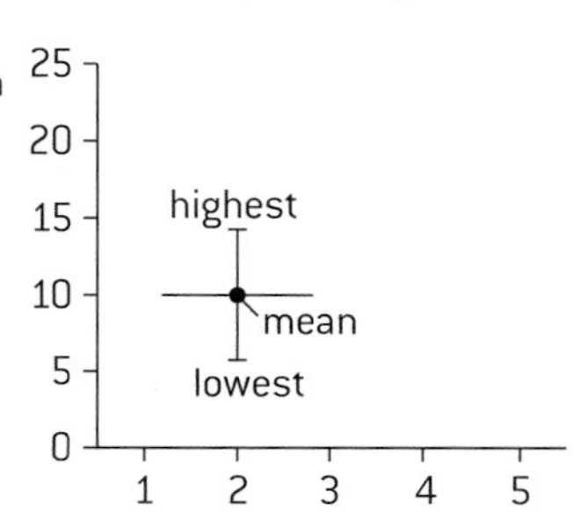

Graphs in scientific papers usually include **error bars**, which also give an indication of how widely spread the repeats are, but do this by showing a statistical measure of the variation, called the standard deviation or standard error. The error bar shows one standard deviation above and below the mean.

3. Describing the trend

The graph shows the results of a respirometer experiment using 100 g samples of pea seeds that were soaked in water for 24 hours to start germination.

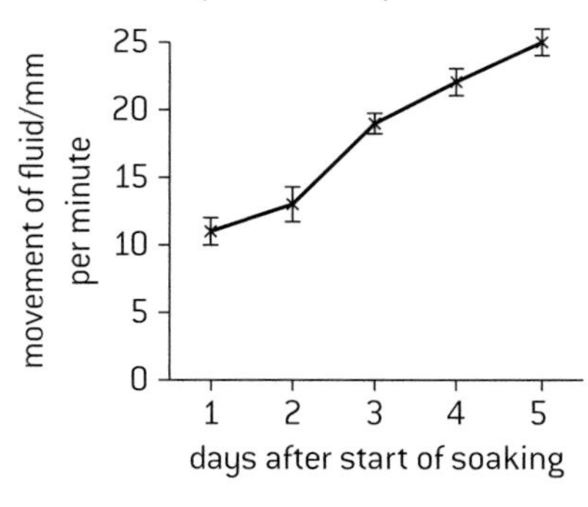

The movement of fluid in the manometer was due to oxygen consumption by aerobic cell respiration in the seeds. The rate of oxygen consumption in the dry ungerminated seeds was zero or too low to be measured. The rate of oxygen consumption increased during the first five days of germination.

4. Evaluating the data

The range bars show that there was some variation between the replicates at each stage of germination, but the variation within each treatment (day of germination) was mostly smaller than the variation between treatments, suggesting that there is a significant increase in respiration during germination.

Photosynthesis

INTRODUCING PHOTOSYNTHESIS

Photosynthesis is the production of carbon compounds in cells using light energy.

The substrates for photosynthesis are simple inorganic substances including carbon dioxide and water. For example, the synthesis of glucose can be summarized with this equation:

carbon dioxide + water + light energy ⟶ glucose + oxygen

Plants, algae and some bacteria produce all their carbon compounds by photosynthesis.

Photosynthesis includes these stages:

- Carbon dioxide is converted into carbohydrates and other carbon compounds. Energy is needed to do this.
- The energy is obtained in the form of light. The light is absorbed by photosynthetic pigments.
- Electrons are needed to convert carbon dioxide into carbohydrates. They are obtained by **photolysis**, which is the splitting of water molecules. Oxygen is a waste product from the photolysis of water.

PHOTOSYNTHESIS AND THE ATMOSPHERE

Oxygen is a waste product of photosynthesis. It is produced when water is split by photolysis to provide the electrons needed to convert carbon dioxide into carbohydrates and other carbon compounds.

The first organisms to release oxygen from photosynthesis into the atmosphere were bacteria, about 3.5 billion years ago. Before this there was little or no oxygen in the atmosphere. Between 2.4 and 2.2 billion years ago the oxygen content of the atmosphere rose from a very low level to 2%, due to photosynthesis. This caused dissolved iron in the oceans to precipitate as iron oxide. It sank to the ocean bed, forming deposits of rock called **banded iron formations**.

Oxygen levels in the atmosphere remained at about 2% until 750 million years ago, when they started to rise, reaching about 30% before dropping back down to today's level of 20%. The increases above 2% were probably due to the evolution of multicellular algae and land plants, which raised global photosynthesis rates.

ABSORPTION SPECTRA

A **spectrum** is a range of wavelengths of electromagnetic radiation. The spectrum of visible light is the range of wavelengths from 400 nm to 700 nm that are used in human vision. Violet light has the shortest wavelength and red the longest. The same range of wavelengths is used in photosynthesis, because the photosynthetic pigments do not absorb other wavelengths. A graph showing the range of wavelengths absorbed by a pigment is called an **absorption spectrum**.

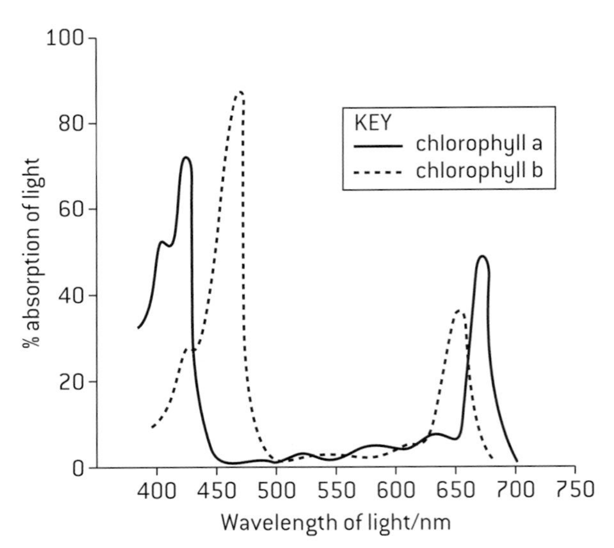

Chlorophyll is the main photosynthetic pigment. The graph above is the absorption spectrum for the two commonest forms of chlorophyll, a and b.

The absorption spectrum shows that chlorophyll absorbs red and blue light most effectively. Small amounts of green light are absorbed but most is reflected, making structures containing chlorophyll appear green to us.

ACTION SPECTRA

The efficiency of photosynthesis is not the same in all wavelengths of light. The efficiency is the percentage of light of a wavelength that is used in photosynthesis. The graph below shows the percentage use of the wavelengths of visible light in photosynthesis. This graph is called the **action spectrum of photosynthesis**. It shows that maximum photosynthesis rates are in blue light with another lower peak in red light. Green light is used less efficiently.

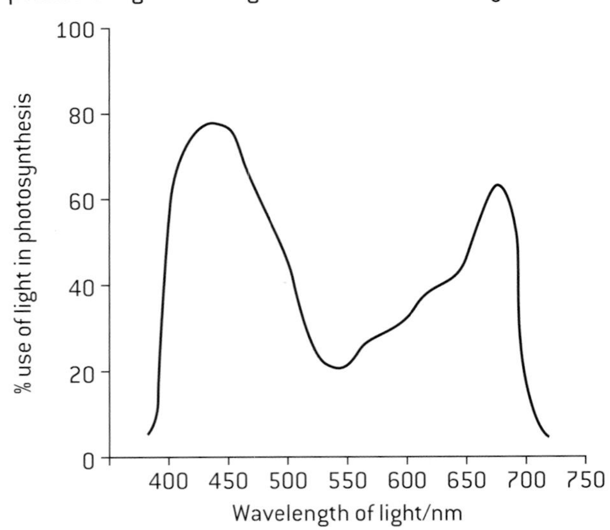

The absorption spectrum of chlorophyll (left) also has a maximum in blue light with a second lower peak in red, which explains these peaks in photosynthesis rates.

The action spectrum shows that there is some use of green light in photosynthesis, even though chlorophyll absorbs little of it. This is because accessory photosynthetic pigments are present, which absorb some green light that can be used in photosynthesis.

Investigating limiting factors

DESIGNING EXPERIMENTS TO INVESTIGATE LIMITING FACTORS

Processes such as photosynthesis are affected by various factors, but usually just one of these factors is actually limiting the rate at a particular time. This is the factor that is nearest to its minimum and is called the **limiting factor**. The three possible limiting factors for photosynthesis are **temperature, light intensity** and **carbon dioxide concentration**.

These principles should be remembered when designing an experiment to investigate the effect of a limiting factor on photosynthesis:

1. Only one limiting factor should be investigated at a time – this is the **independent variable**.
2. A suitable range for the independent variable should be chosen, from the lowest possible level, to a level at which the factor is no longer limiting.
3. An accurate method should be chosen for measuring the rate of photosynthesis. This is the **dependent variable** and is usually a measure of oxygen production per unit time.
4. Methods must be devised for keeping all factors constant, apart from the independent variable. These are the **control variables**. This part of experimental design is essential so it is certain that changes in the rate of photosynthesis are due only to the factor being investigated (the independent variable). Of the three factors temperature, light intensity and carbon dioxide concentration one will be the independent variable in the experiment and the other two will be control variables.

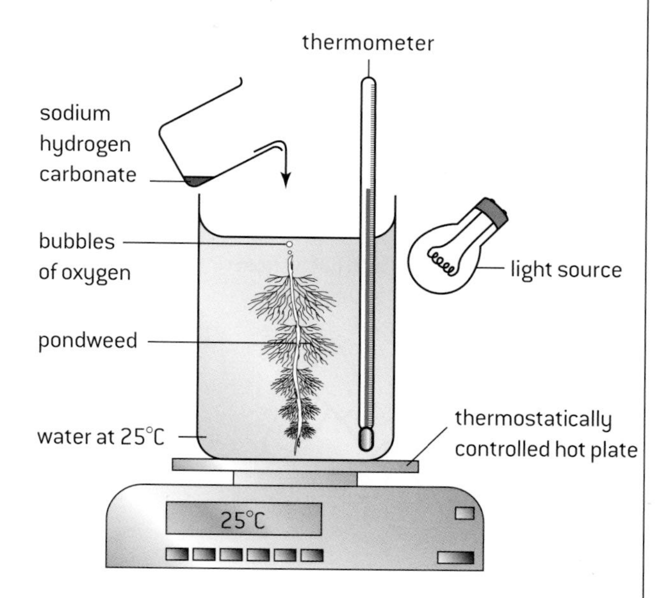

Limiting factor	Method of varying the factor	Suggested range	Controlling the factor
Temperature	Place pondweed in water in a thermostatically controlled water bath or on a hot plate to vary the temperature	5°C to 45°C in 5 or 10°C intervals	Set the thermostat at 25°C and keep it there throughout the experiment
Light intensity	Move light source to different distances and measure light intensity with a lux meter (light intensity = 1/(distance^2))	4, 5, 7, 10 and 14 cm and no light gives a good range of intensities	Keep the light source at a constant distance, such as 5 cm
Carbon dioxide concentration	Start with boiled, cooled water (no CO_2) then add measured quantities of $NaHCO_3$ to increase the CO_2 concentration	0 to 50 mmol dm^{-3} in 10 mmol dm^{-3} intervals	Add enough $NaHCO_3$ to give a high CO_2 concentration (50 mmol dm^{-3})

EFFECT OF LIGHT INTENSITY

At low light intensities, the rate of photolysis and therefore the production of oxygen is limited by the amount of light absorbed. As the light energy is used for the production of ATP and high energy electrons, which are needed for conversion of CO_2 into glucose, low light intensities limit the production of this sugar and other useful substances.

At high light intensities some other factor is limiting photosynthesis. Unless a plant is heavily shaded, or the sun is rising or setting, light intensity is not usually the limiting factor.

EFFECT OF CO_2 CONCENTRATION

Below 0.01% carbon dioxide the enzyme used to fix CO_2 (rubisco) is not effective and in many plants there is no net photosynthesis. Between about 0.01% and 0.04% the concentration of CO_2 is often the limiting factor, because the rate of successful collisions between CO_2 molecules and the active site of the enzyme that fixes it is still lower than any of other steps in photosynthesis. ATP and high energy electrons are not used as rapidly as they are produced, which restricts further photolysis and therefore oxygen production.

At very high CO_2 concentrations some other factor is limiting.

EFFECT OF TEMPERATURE

At low temperatures, all of the enzymes that catalyse the conversion of CO_2 into carbohydrate work slowly and below 5°C there is little or no photosynthesis in many plants. At temperatures above 30°C the enzyme used to fix carbon dioxide (rubisco) is decreasingly effective, even though it has not been denatured. Temperature is therefore the limiting factor at both low and high temperatures, with the low rate of use of ATP and high energy electrons restricting further photolysis and therefore oxygen production.

At intermediate temperatures, some other factor is limiting.

Chromatography

SEPARATING PHOTOSYNTHETIC PIGMENTS BY CHROMATOGRAHY

① Tear up a leaf into small fragments

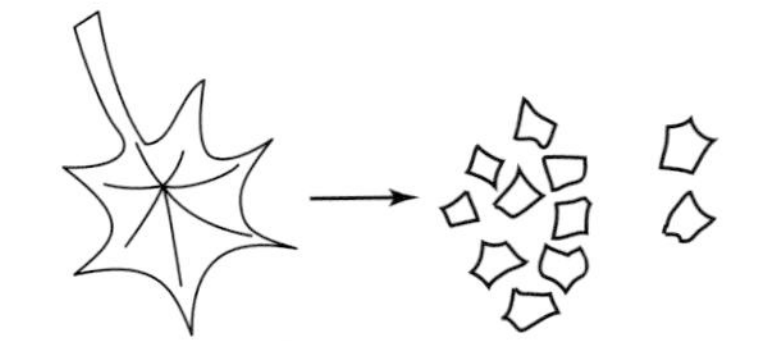

② Grind pieces of leaf with sharp sand and propanone to extract the leaf pigments

③ Transfer sample of extract to a watch glass

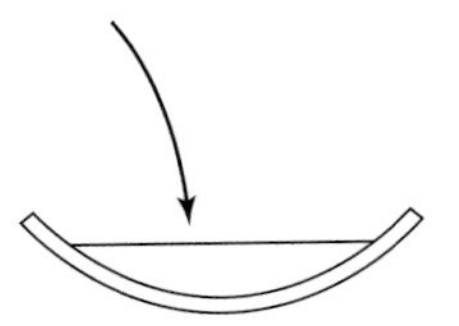

④ Evaporate to dryness with hot air from a hair-dryer

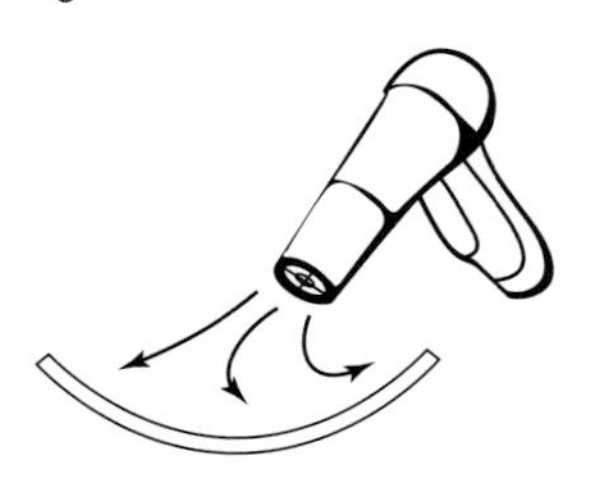

⑤ Add a few drops of propanone to dissolve the pigments

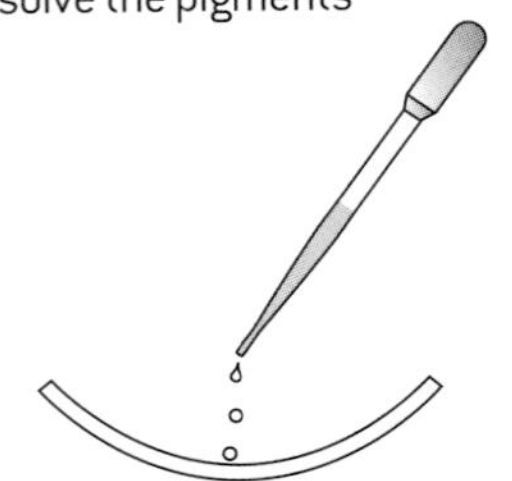

⑥ Build up a concentrated spot of pigment 10mm from the end of the strip of paper/TLC strip

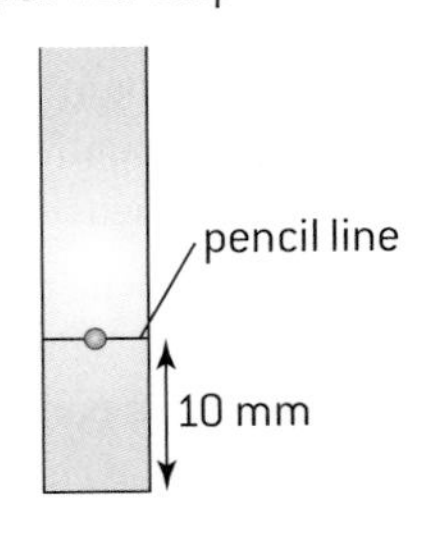

⑦ Suspend the strip in a tube with the base dipping into running solvent

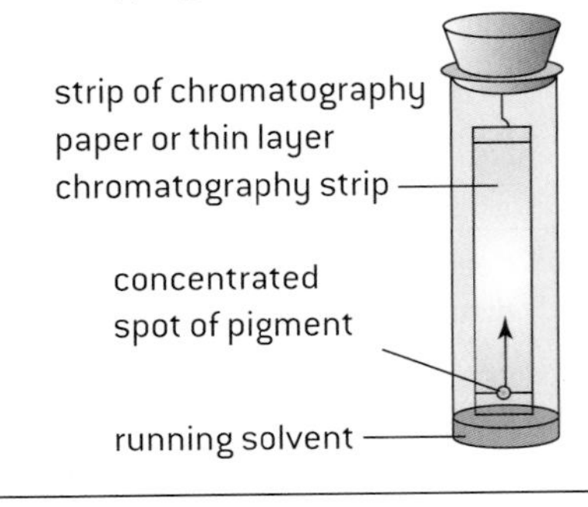

⑧ Remove the strip from the tube when the solvent has nearly reached the top

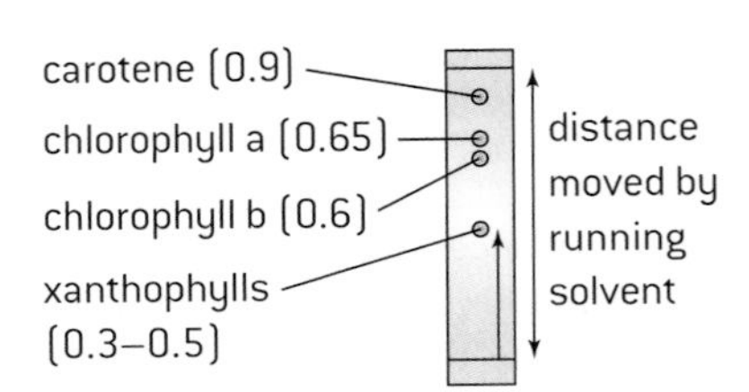

⑨ Calculate R_f values for each pigment spot

$$R_f = \frac{\text{distance moved by spot}}{\text{distance moved by solvent}}$$

approximate R_f values for the main pigments are shown left. The pigments separate because of their varying solubility in the running solvent

RESULTS OF AN INVESTIGATION INTO LIMITING FACTORS

The figure (right) shows the effects of light intensity on the rate of photosynthesis at two different temperatures and two carbon dioxide concentrations. It is possible to deduce which is the limiting factor at the point marked W–Z on each curve.

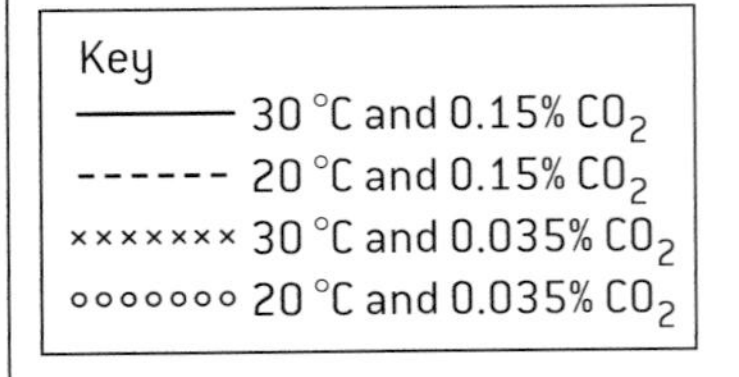

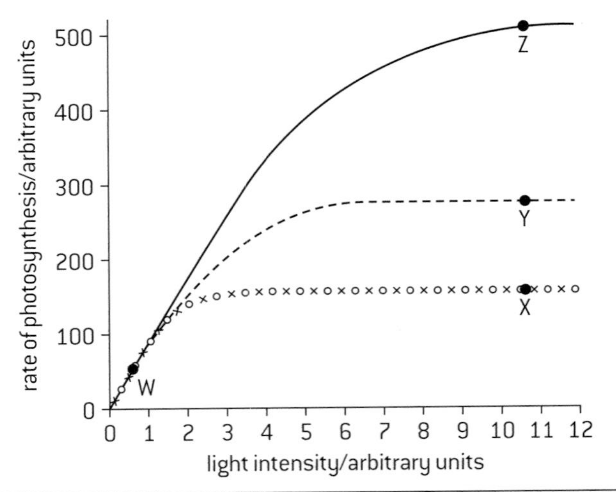

Questions – molecular biology

1. In an amino acid, the amine and carboxyl groups are bonded to the same carbon atom. What else is bonded to this atom?

   A. G and R   C. H and R

   B. G and P   D. H and P

2. In Meselson and Stahl's experiment, bands were detected at only three heights in the density gradient. What was an explanation for this?

   A. Each DNA strand had all ^{14}N or all ^{15}N bases.

   B. Transcription is semi-conservative.

   C. Samples were only taken after replication ended.

   D. *E. coli* has single-stranded DNA.

3. If ACUCGAGGUCUC was the base sequence of mRNA, what base sequence of DNA was transcribed?

   A. ACUCGAGGUCUC   C. UGAGCUCCAGAC

   B. ACTCGAGGTCTC   D. TGAGCTCCAGAG

4. If a person has a height of 200 cm and a mass of 80 kg, what is their BMI?

   A. 0.002   C. 20

   B. 0.005   D. 500

5. What is the range of wavelength of light in absorption spectra for photosynthetic pigments?

   A. 400–700 μm   C. 0–273 μm

   B. 400–700 nm   D. 0–273 nm

6. The table below shows the base composition of genetic material from ten sources.

Source of genetic material	Base composition (%)				
	A	C	G	T	U
Cattle thymus gland	28.2	22.5	21.5	27.8	0.0
Cattle spleen	27.9	22.1	22.7	27.3	0.0
Cattle sperm	28.7	22.0	22.2	27.2	0.0
Pig thymus gland	30.0	20.7	20.4	28.9	0.0
Salmon	29.7	20.4	20.8	29.1	0.0
Wheat	27.3	22.8	22.7	27.1	0.0
Yeast	31.3	17.1	18.7	32.9	0.0
E. coli (bacteria)	26.0	25.2	24.9	23.9	0.0
Human sperm	31.0	18.4	19.1	31.5	0.0
Influenza virus	23.0	24.5	20.0	0.0	32.5

   a) Deduce the type of genetic material used by

      (i) cattle [1]

      (ii) *E. coli* [1]

      (iii) influenza viruses. [1]

   b) Suggest a reason for the difference between thymus gland, spleen and sperm in the measurements of base composition in cattle. [1]

   c) (i) Explain the reasons for the total amount of adenine plus guanine being close to 50% in the genetic material of many of the species in the table. [3]

      (ii) Identify two other trends in the base composition of the species that have 50% adenine plus guanine. [2]

   d) (i) Identify a species shown in the table that does not follow trends in base composition described in (c). [1]

      (ii) Explain the reasons for the base composition of this species being different. [2]

7. The graph (below) shows the results of a data-logging experiment. *Chlorella*, a type of alga that is often used in photosynthesis experiments, was cultured in water in a large glass vessel. Light intensity, temperature and the pH of the water were monitored over a three-day period. Changes in pH were due to carbon dioxide concentration rising or falling. An increase in CO_2 concentration causes a pH decrease.

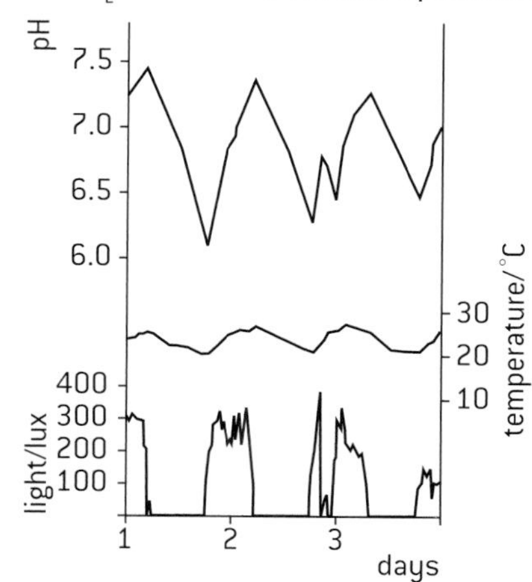

   a) State the relationship shown in the graph between

      (i) light intensity and CO_2 concentration [1]

      (ii) temperature and CO_2 concentration. [1]

   b) Deduce, from the data in the graph, whether the effect of light intensity or temperature on carbon dioxide concentration is greater. [2]

   c) The graph shows both rises and falls in CO_2 concentration. Explain the causes of

      (i) rises in CO_2 concentration [2]

      (ii) falls in CO_2 concentration. [2]

8. In an experiment into the activity of salivary amylase, 5 cm^3 samples of 0.1 mol dm^{-3} of starch solution and 0.1 cm^3 of samples of undiluted saliva are placed in block heaters at different temperatures and are mixed when they reach the target temperature. The time taken for each sample of starch to be fully digested was found by repeatedly testing drops of the starch–saliva mixture.

   a) (i) State the independent variable. [1]

      (ii) State the dependent variable. [1]

   b) The volume of the starch solution and saliva were control variables. State two other control variables in this experiment. [2]

   c) If 15 samples of starch and saliva were available, discuss whether it would be better to use five of them at each of 20 °C, 30 °C and 40 °C, or to use three of them at each of 20 °C, 30 °C, 40 °C, 50 °C and 60 °C or to use one of them at 5 °C intervals from 20 °C to 90 °C. [3]

   d) State the substrate in this experiment. [1]

   e) State the word equation for starch digestion. [2]

   f) Sketch a graph to show the expected relationship:

      (i) between temperature and time taken for all starch to be digested in a sample [3]

      (ii) between temperature and the rate of starch digestion. [2]

Chromosomes

PROKARYOTE AND EUKARYOTE CHROMOSOMES

In a prokaryote there is one chromosome consisting of a circular DNA molecule. The DNA is naked, meaning that it is not associated with proteins. Some prokaryotes have **plasmids**, which are much smaller extra loops of DNA.

There are four differences between the chromosomes of eukaryotes and prokaryotes:

Eukaryote chromosomes	Prokaryote chromosomes
contain a **linear** DNA molecule	consist of a **circular** DNA molecule
associated with **histone** proteins	**naked** – no associated proteins
no plasmids	plasmids often present
two or more different chromosomes	one chromosome only

AUTORADIOGRAPHY AND CHROMOSOMES

The technique of autoradiography combined with electron microscopy has been used by biologists from the 1940s onwards to find where radioactively labelled substances are located in cells. Thin sections of cells are coated with a photographic film. After having been left in darkness for days or weeks the film that is coating the section is developed. When viewed with a microscope both the structure of cells in the section and black dots in the photographic film are visible. Each black dot shows where a radioactive atom decayed and gave out radiation, which acts like light on the film.

John Cairns adapted this technique to research the chromosomes of *E. coli*, a prokaryote. He grew *E. coli* in a medium containing radioactively labelled thymine, so its DNA became labelled but not RNA. He placed cells on a membrane and digested their cell walls, allowing the DNA to spill out over the membrane. He coated the membrane with a photographic film and left it in the dark for two months. When the film was developed, lines of black dots showed the position of the DNA molecules from *E. coli*. A typical image is shown below, with a drawing to interpret it lower right. The DNA molecule is in the process of replicating.

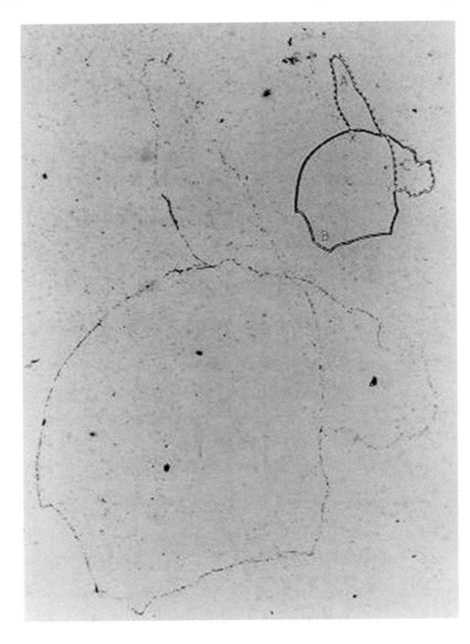

Cairns discovered that the DNA molecules were circular and 1,100 μm long, despite the *E. coli* cells only being 2 μm long. Other researchers then used similar techniques to investigate eukaryote chromosomes. They were found to contain linear rather than circular DNA and were much longer. For example, a chromosome from the fruit fly *D. melanogaster* was 12,000 μm long (12 millimetres). The lengths showed that eukaryote chromosomes contain one very long DNA molecule rather than a number of shorter molecules.

CHROMATIDS

Eukaryote chromosomes are only easily visible during mitosis. In prophase they condense and in metaphase reach their minimum length. The electron micrograph below shows chromosomes in metaphase.

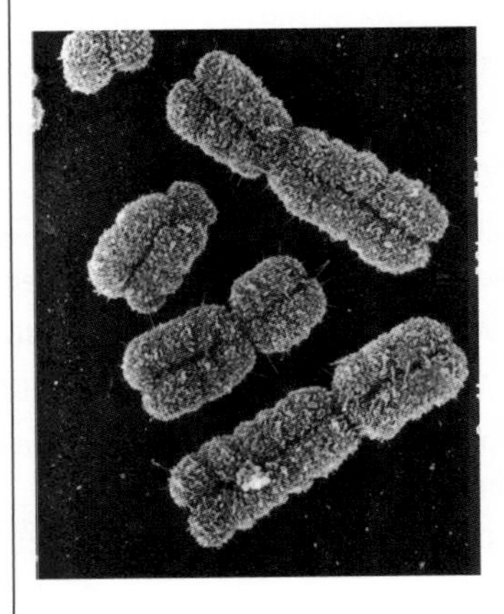

Each chromosome in prophase and metaphase of mitosis consists of two structures, known as **sister chromatids**. They each contain a DNA molecule that was produced by replication during interphase, so their base sequences are identical. Sister chromatids are held together by a centromere. At the start of anaphase the centromere divides allowing the chromatids to become separate chromosomes.

GENOMES

The **genome** is the whole of the genetic information of an organism. The size of a genome is therefore the total amount of DNA in one set of chromosomes in that species. It can be measured in millions of base pairs (bp) of DNA.

Genome sizes vary considerably. Five examples are shown below. *E. coli* is a gut bacterium and T2 phage is a virus that attacks *E. coli*. The fruit fly *Drosophila melanogaster* has been widely used in genetics research. *P. japonica* is a woodland plant with a remarkably large genome size.

Organism	Genome size (millions of bp)
T2 phage	0.18
Escherichia coli	5
D. melanogaster	140
Homo sapiens	3,000
Paris japonica	150,000

Karyograms

HOMOLOGOUS CHROMOSOMES

Prokaryotes only have one chromosome but eukaryotes have different chromosomes that carry different genes.

In humans, for example, there are 23 different chromosome types each of which carries a different group of genes. All the chromosomes of one particular type are **homologous**, which means that although they have the same genes in the same sequence they may not have the same alleles of those genes. Alleles are the different forms of a gene.

HAPLOID AND DIPLOID

Most plant and animal cells have a **diploid** nucleus. This means that the nucleus contain pairs of homologous chromosomes.

Some cells have a **haploid** nucleus, which has only one chromosome of each type. Gametes such as the sperm and egg cells of humans are haploid. Two haploid gametes fuse together during fertilization to produce one diploid cell – the zygote. This divides by mitosis to produce more diploid body cells with the same number of chromosomes.

CHROMOSOME NUMBERS

The number of chromosomes is a characteristic feature of members of a species. Usually the number quoted is the diploid number, as that is how many chromosomes are present in normal body cells.

The diploid number varies considerably – some species have fewer large chromosomes and others have a greater number of small chromosomes.

Five examples are given here:

Homo sapiens (humans)	46
Pan troglodytes (chimpanzee)	48
Canis familiaris (dog)	78
Oryza sativa (rice)	24
Parascaris equorum (horse threadworm)	4

SEX CHROMOSOMES

The twenty-third pair of chromosomes in humans determines whether an individual is male or female. There are two types of sex chromosome, a larger X and a smaller Y chromosome.

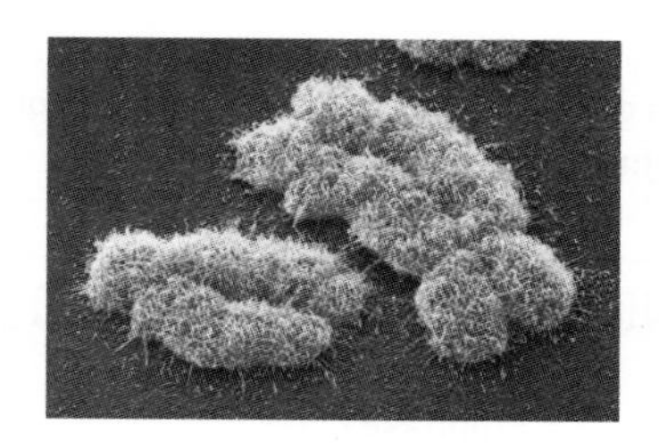

If two X chromosomes are present, a baby develops as a female and if one X and one Y are present, the baby develops as a male. The karyogram below (left) shows the karyotype of a male.

KARYOTYPES AND KARYOGRAMS

The **karyotype** is the number and type of chromosomes present in a cell or organism. A **karyogram** is a photograph or diagram in which the chromosomes of an organism are shown in homologous pairs of decreasing length. Karyograms are prepared so that the karyotype of an individual can be studied.

Example 1 – normal male (XY)

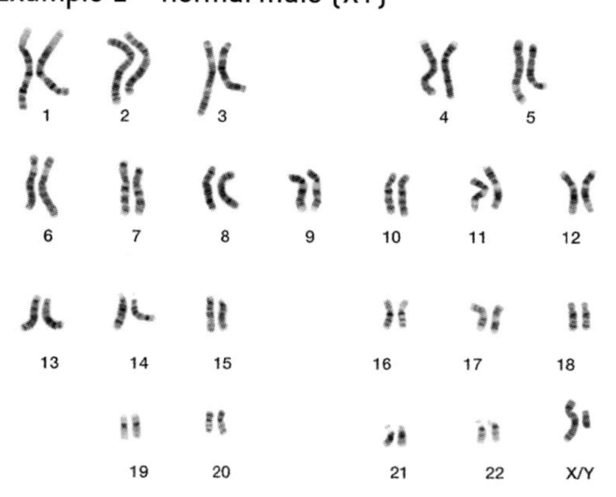

There are two common uses of karyograms in humans.

1. Deducing the sex of an individual: if there are two X chromosomes the person is female and if there is one X and one Y they are male.
2. Diagnosing conditions due to chromosome abnormalities: normal karyotypes have a pair of each chromosome type including a pair of sex chromosomes. If there are more or less than two of each pair, the person has a chromosome abnormality. The most common type is **Down syndrome** which is due to having 3 copies of chromosome 21.

Example 2 – female with Down syndrome

Example 3 – male with Klinefelter syndrome

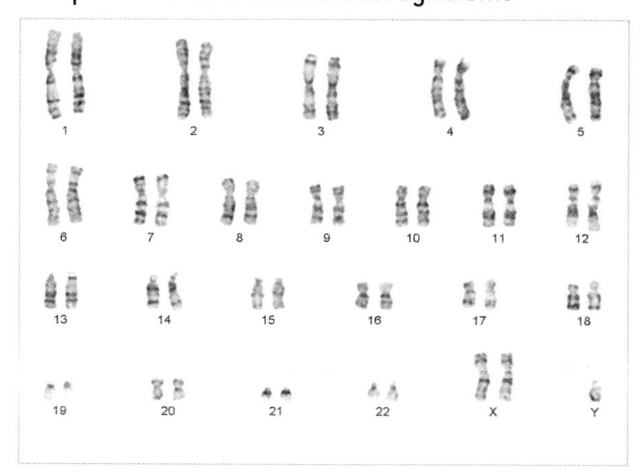

Meiosis

MEIOSIS AND SEXUAL LIFE CYCLES

All sexual life cycles include an event called **fertilization** in which a male and a female gamete fuse together to produce a **zygote**. The zygote has twice as many chromosomes as the gametes. At another stage in the life cycle the number of chromosomes per cell must be halved, or each generation would have twice as many chromosomes as the previous one.

Meiosis is the process that halves chromosome number and allows a sexual life cycle with fusion of gametes. The haploid number of chromosomes is represented by the letter *n* so the diploid number is $2n$.

In meiosis, a diploid nucleus divides twice to produce four haploid nuclei. The DNA of the chromosomes is replicated before the first division so each chromosome consists of two sister chromatids, but the DNA is not replicated between the first and second divisions. It is the separation of pairs of homologous chromosomes in the first division of meiosis that halves the chromosome number.

	Number of cells	Number of chromosomes	Chromatids per chromosome
Before the start of meiosis	1	$2n$	2
At the end of the first division	2	n	2
At the end of the second division	4	n	1

DRAWING THE STAGES OF MEIOSIS

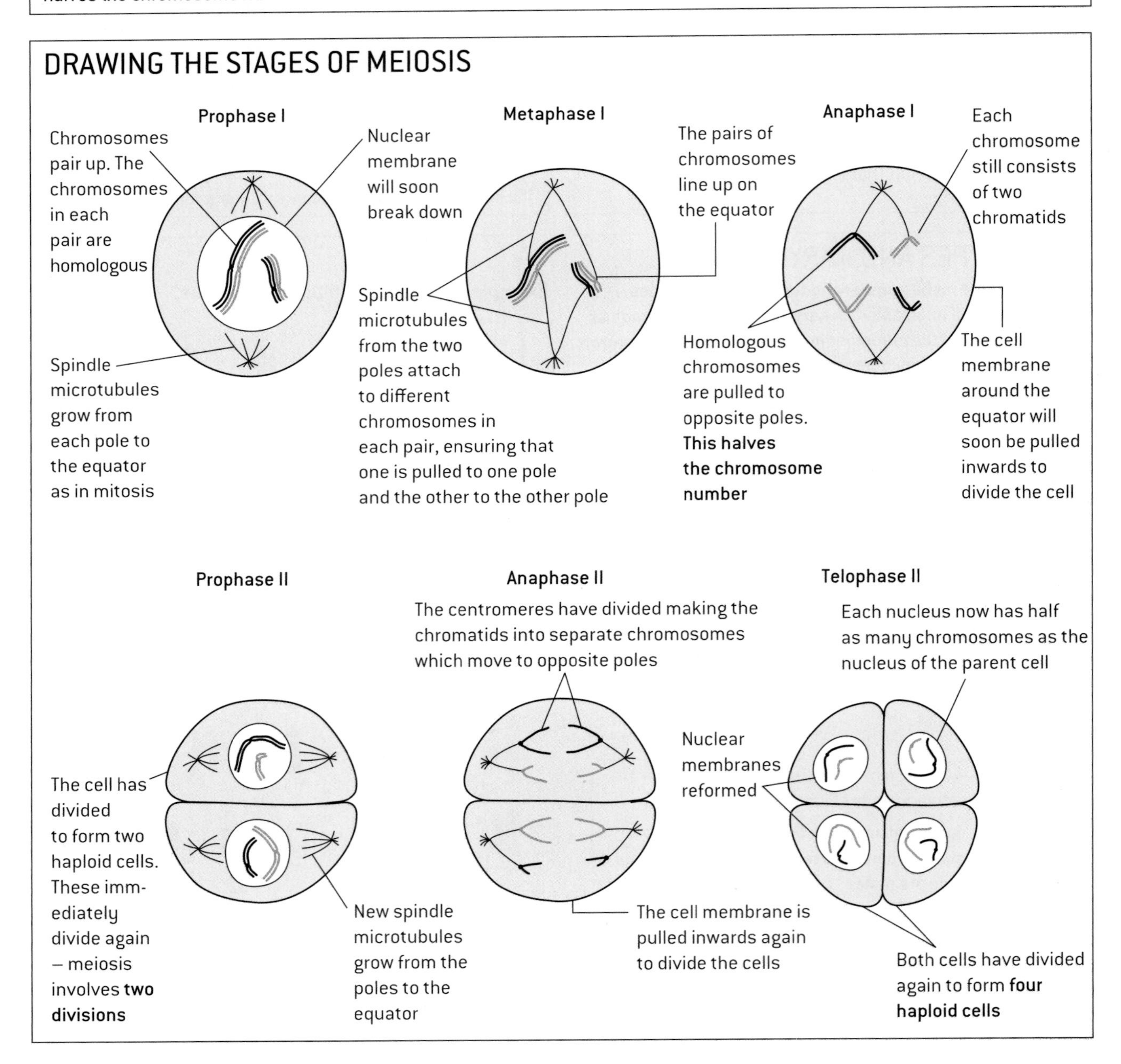

Meiosis and genetic variation

MEIOSIS AND GENETIC VARIATION

Two processes in meiosis promote genetic variation among the haploid cells produced by meiosis.

1. **Random orientation of pairs of homologous chromosomes in metaphase I**

For each pair of chromosomes there are two possible orientations that determine which chromosome moves to each of the two poles of the cell. Because the orientation of each pair is random and does not influence other pairs, different combinations of chromosomes can be produced and therefore different combinations of alleles. In the diagrams below one gene with two different alleles is shown on each chromosome.

The number of possible combinations of chromosomes produced by random orientation is 2^n in humans where n is 23. This is over 8 million combinations per parent.

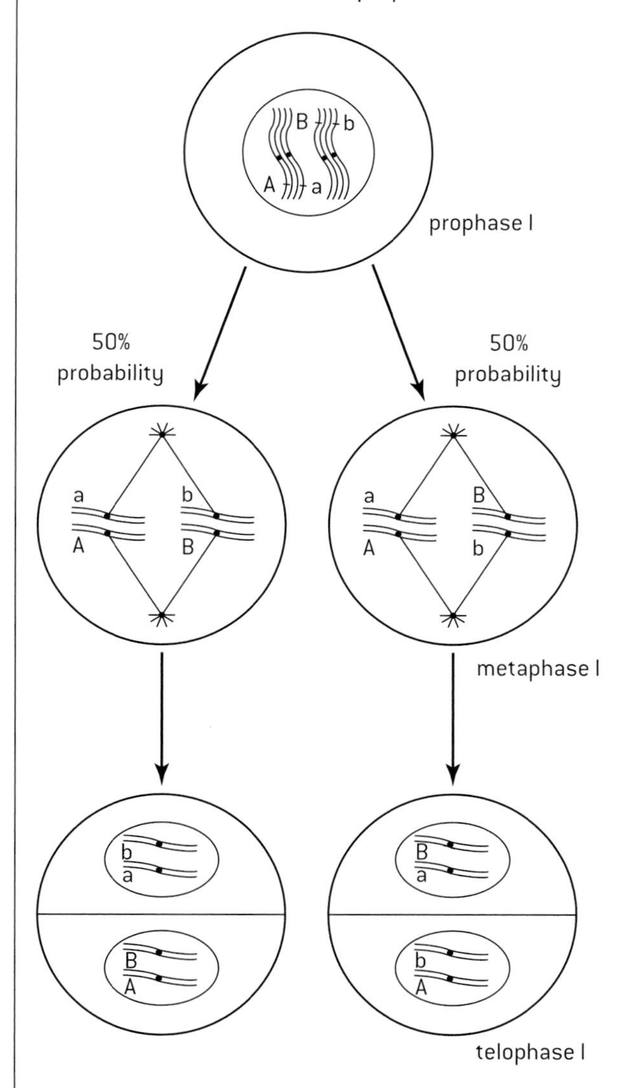

2. **Crossing over during prophase 1**

In the very early stages of meiosis homologous chromosomes pair up and parts of non-sister chromatids can be exchanged between them. This process is called **crossing over**. It produces chromatids with a new combination of alleles. It is a significant source of genetic variation because it is random where along the length of the chromosomes the exchange occurs.

FUSION OF GAMETES AND VARIATION

When gametes fuse together during fertilization, the alleles from two different parents are brought together in one new individual. This promotes **genetic variation**. Fertilization is a random process – any gamete produced by the father could fuse with any produced by the mother. Species that reproduce sexually thus generate genetic variation both by meiosis and by random fusion of gametes.

NON-DISJUNCTION AND DOWN SYNDROME

Sometimes chromosomes that should separate and move to opposite poles during meiosis do not and instead move to the same pole. This can happen in either the first (below left) or the second (below right) division of meiosis. Non-separation of chromosomes is called **non-disjunction**. The result is that gametes are produced with either one chromosome too many or too few.

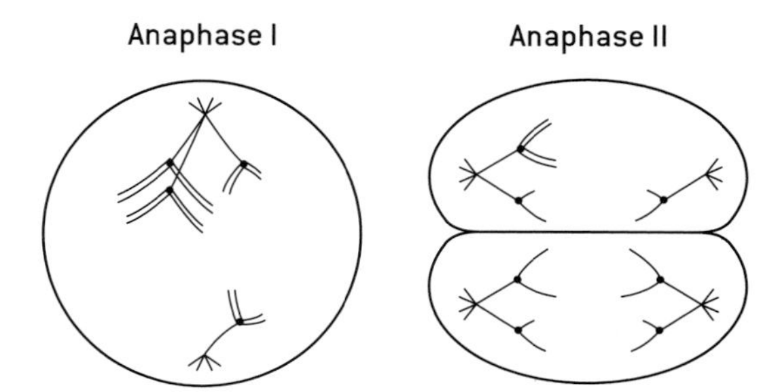

Gametes with one chromosome too few usually quickly die but gametes with one chromosome too many sometimes survive. When they are fertilized, a zygote is produced with three chromosomes of one type instead of two. This is called **trisomy**. For example, children are sometimes born with three chromosomes of type 21, rather than two. This causes **Down syndrome** or trisomy 21. It can be due to non-disjunction during the formation of the sperm or the egg. Many research studies have shown that the chance of Down syndrome increases with the age of the parents. These are typical figures for the mother:

Mother's age	25	30	35	40	45
Chance of baby with Down syndrome	1 in 1250	1 in 1000	1 in 400	1 in 100	1 in 30

There are two methods for obtaining cells of an unborn child for chromosome testing:

1. In **amniocentesis** a sample of amniotic fluid is removed from the amniotic sac around the fetus. To do this, a hypodermic needle is inserted through the wall of the mother's abdomen and wall of the uterus. Amniotic fluid is drawn out into a syringe. It contains cells from the fetus.
2. In **chorionic villus sampling** (CVS) cells are removed from fetal tissues in the placenta called chorionic villi. As with amniocentesis a hypodermic needle, inserted through the mother's abdomen and uterus wall, is used to obtain the cells. CVS has a slightly higher risk of miscarriage (2%) than amniocentesis (1%). Both methods have a very small risk of infections in maternal or fetal tissues.

Principles of inheritance

MENDEL AND QUANTITATIVE METHODS

Gregor Mendel is often regarded as the father of genetics. He crossed varieties of pea plants that had different characteristics and, from his results, he deduced the principles on which inheritance is based.

Mendel was not the first biologist to try to discover the principles of inheritance. His success depended on obtaining numerical results, rather than just descriptions of the outcomes, and on using large numbers of pea plants. It is important in scientific experiments to have enough replicates to ensure reliability and Mendel had very large numbers. For example in the cross shown below he counted a total of 7,324 seeds in the F_2 generation. There were 5,474 round seeds and 1,850 wrinkled. With such large numbers he could be very confident that the basic ratio was 3:1. He could be even more confident because he repeated his monohybrid cross with seven different traits and got the 3:1 ratio every time.

EXPLAINING THE 3:1 RATIO

Mendel crossed two varieties of pea together and found that all of the offspring (the **F_1 generation**) had the same characteristic as one of the parents. He allowed the F_1 generation to self-fertilize – each plant produced offspring by fertilizing its female gametes with its own male gametes. The offspring (the **F_2 generation**) contained both of the original parental types in a 3:1 ratio.

Using modern terms, Mendel's explanation is that each pea plant has two alleles of the gene that affects the character. The parents are **homozygous** because they have two of the same allele. The F_1 plants are **heterozygous** because they have two different alleles. The F_1 plants all have the character of one of the parents because that parent has the **dominant allele** and in a heterozygote it masks the effect of the other parent's **recessive allele**. One quarter of the F_2 generation have two recessive alleles and so show the character caused by this allele.

EXAMPLE OF A MONOHYBRID CROSS BETWEEN PEA PLANTS

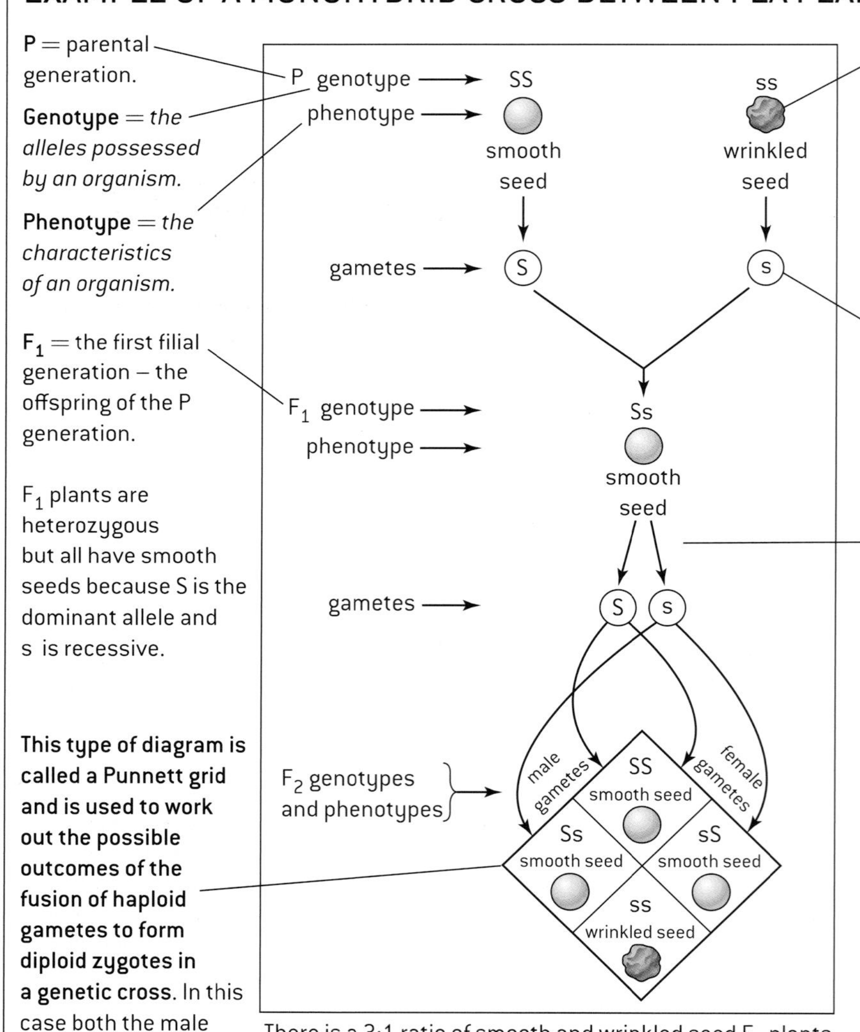

There is a 3:1 ratio of smooth and wrinkled seed F_2 plants. Crosses between two heterozygous individuals give a 3:1 ratio if one of the alleles is dominant and the other is recessive.

Seed shape is determined by a single gene. One allele of this gene (S) gives smooth seeds and the other (s) gives wrinkled seeds. The pea plants are diploid so they have two copies of each gene. The parental varieties are both homozygous.

Gametes are produced by meiosis so are haploid and only have one copy of each gene.

The two alleles of each gene separate into different haploid daughter nuclei during meiosis. This is called segregation. In this case each daughter nucleus and therefore each gamete will receive either S or s.

Segregation occurs during meiosis. The two alleles of a gene are located on homologous chromosomes which move to opposite poles, causing the segregation (see below).

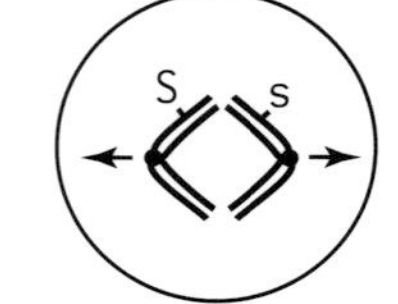

Autosomal genetic diseases

CYSTIC FIBROSIS AND HUNTINGTON'S DISEASE

The principles of inheritance discovered by Mendel in pea plants also operate in humans and help to predict the probability of inheritance of genetic diseases. Many genetic diseases have been identified, but most are very rare. A large proportion are due to recessive alleles of **autosomal** genes, e.g. **cystic fibrosis**. All chromosomes apart from sex chromosomes are autosomes, so any gene that has its locus on a non-sex chromosome is autosomal. A small proportion of genetic diseases are due to dominant alleles of autosomal genes, e.g. **Huntington's disease**.

Cystic fibrosis

This disease is caused by a recessive allele of a gene coding for a chloride channel. It is the commonest genetic disease in parts of western Europe. About 1 in 3,000 babies born in the US has cystic fibrosis. Usually neither parent has the disease, but they are both **carriers** of the recessive allele for the disease. A carrier has a recessive allele of a gene, but it does not affect their phenotype because a dominant allele is also present. The Punnett grid below shows that the probability of cystic fibrosis in a child of two carrier parents is 25%.

Key to alleles	Possible genotypes	Possible phenotypes
C normal allele	CC →	normal
c cystic fibrosis allele	Cc →	normal (carrier)
	cc →	cystic fibrosis

P phenotype: normal (carrier) × normal (carrier)

P genotype: Cc Cc

gametes: (C) (c) (C) (c)

Punnett grid to show possible outcomes:

	(C)	(c)
(C)	CC normal	Cc normal (carrier)
(c)	cC normal (carrier)	cc cystic fibrosis

Ratio: 3 normal : 1 cystic fibrosis

Huntington's disease

This neurodegenerative disease is caused by dominant alleles of the gene coding for huntingtin, a protein with an unknown function. The disease usually only develops during adulthood, by which time an individual who develops the disease may already have had children. Almost always one parent only develops the disease, so it is very unlikely for a child to be born with two copies of the dominant allele. The diagram below shows that the probability of a parent with Huntington's disease passing it on to a child is 50%.

Key to alleles	Possible genotypes	Possible phenotypes
HD Huntington's disease allele	HD HD →	Huntington's disease
hd normal allele	HD hd →	Huntington's disease
	hd hd →	normal

normal × Huntington's disease

hd hd × HD hd

gametes: (hd) (hd) (HD) (hd)

	(HD)	(hd)
(hd)	HD hd Huntington's disease	hd hd normal
(hd)	HD hd Huntington's disease	hd hd normal

Ratio: 1 normal : 1 Huntington's disease

PEDIGREE CHARTS FOR AUTOSOMAL GENES

Pedigree charts indicate whether a disease is caused by a dominant or recessive allele, and allow the genotypes of some individuals to be deduced. Parents are joined by a horizontal line with a vertical line leading to their children.

Key:
- ▨ affected male
- □ normal male
- ◍ affected female
- ○ unaffected female

Cystic fibrosis

I: 1 2
II: 1 2 3 4
III: 1 2 3 4 5

The parents I1 and I2 must be Cc. If we assume that II1 and II4 are CC, III1 and III2 have a 50% chance of being carriers and III4 and III5 a 100% chance.

Huntington's disease

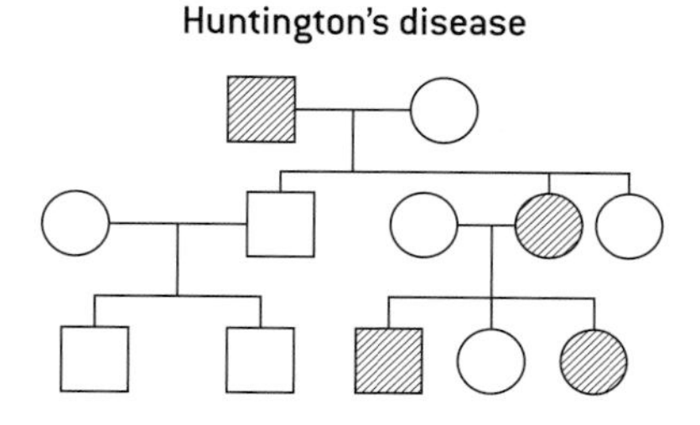

All the individuals with Huntington's disease must be HDhd and all the unaffected individuals are hdhd.

Sex-linkage

SEX DETERMINATION AND SEX-LINKED CONDITIONS

The sex chromosomes of a fetus determine whether it develops as a male or female. All normal egg cells carry an X chromosome so the sex of a child depends on whether the sperm is carrying an X or a Y chromosome. The diagram (right) shows the inheritance of sex.

♀ = Female
♂ = Male

♀ XX   ♂ XY
X X X Y
XX ♀   XY ♂   XX ♀   XY ♂

Sex-linkage is the association of a characteristic with the sex of the individual, because the gene controlling the characteristic is located only on a sex chromosome. There are very few genes on the Y chromosome, but the X chromosome is relatively large and has important genes on it. Sex-linkage is therefore almost always due to genes on the X chromosome. The pattern of inheritance of these genes differs in males and females because females have two X chromosomes and therefore two copies of each gene and males have only one. Only females can therefore be carriers of recessive alleles of sex-linked genes and conditions due to these alleles are much more frequent in males than in females. In humans, **hemophilia** and **red–green colour-blindness** are examples of conditions due to recessive alleles of sex-linked genes. The diagram below shows how two parents, neither of whom have hemophilia, could have a hemophiliac son.

A carrier has a recessive allele of a gene but it does not affect the phenotype because a dominant allele is also present.

The mother is heterozygous but is not hemophiliac because H is dominant and h is recessive. She is a **carrier** of the allele for hemophilia.

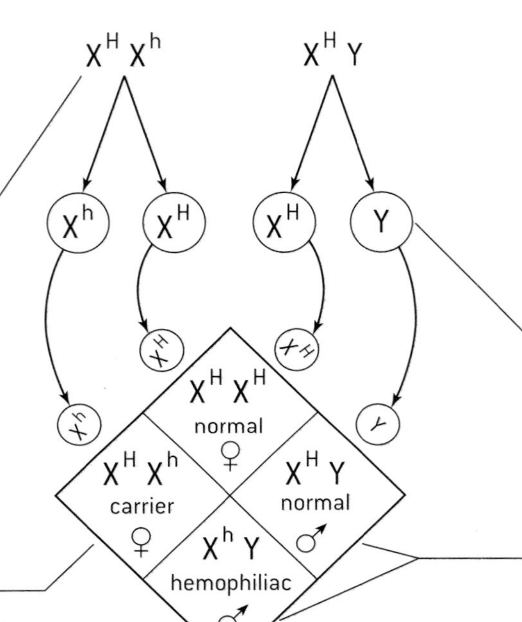

None of the female offspring are hemophiliac because they all inherited the father's X chromosome which carries the allele for normal blood clotting (H), but there is a 50% chance of a daughter being a carrier.

KEY

X^H X chromosome carrying the allele for normal blood clotting

X^h X chromosome carrying the allele for hemophilia

The Y chromosome does not carry either allele of the gene.

There is a 50% chance of a son being hemophiliac as half of the eggs produced by the mother carry X^h.
The chance of a daughter being hemophiliac is 0%, so the overall chance of offspring being hemophiliac is 25%.

PEDIGREE CHARTS FOR SEX-LINKED GENES

Below is part of a real pedigree for hemophilia. Many males but no females are affected, indicating sex-linkage. The genotypes of all males and any female with a hemophiliac son can be deduced. Right is a theoretical pedigree for red–green colour-blindness, in which the genotype of every individual can be deduced with certainty.

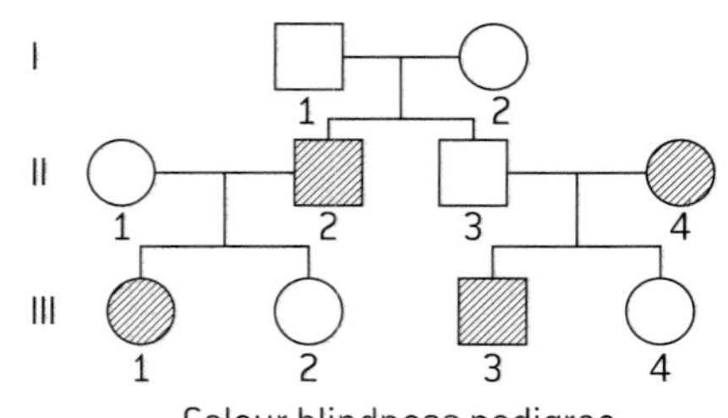

Colour blindness pedigree

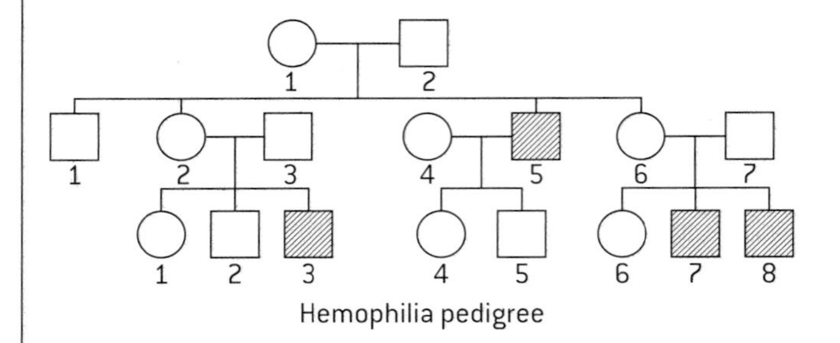

Hemophilia pedigree

Key:
affected male
affected female
unaffected male
unaffected female

Co-dominance

INHERITANCE OF BLOOD GROUPS

There are four blood groups in the ABO system: Group A, Group B, Group AB and Group O. The inheritance of ABO blood groups involves both co-dominance and multiple alleles.

1. **Co-dominance**

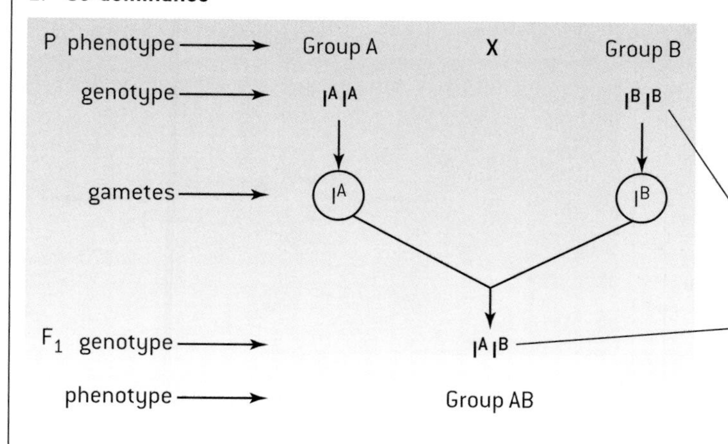

If dominant and recessive alleles are both present in a heterozygous individual, only the dominant allele has effects on the phenotype. If two alleles are **co-dominant**, they have joint effects on a heterozygous individual.

I^A is the allele for blood group A and I^B is the allele for blood group B. Neither allele is recessive, so both are given upper case letters as their symbol.

If I^A and I^B are present together, they both affect the phenotype because they are co-dominant. Co-dominant alleles are pairs of alleles that both affect the phenotype when present together in a heterozygote.

2. **Multiple alleles**

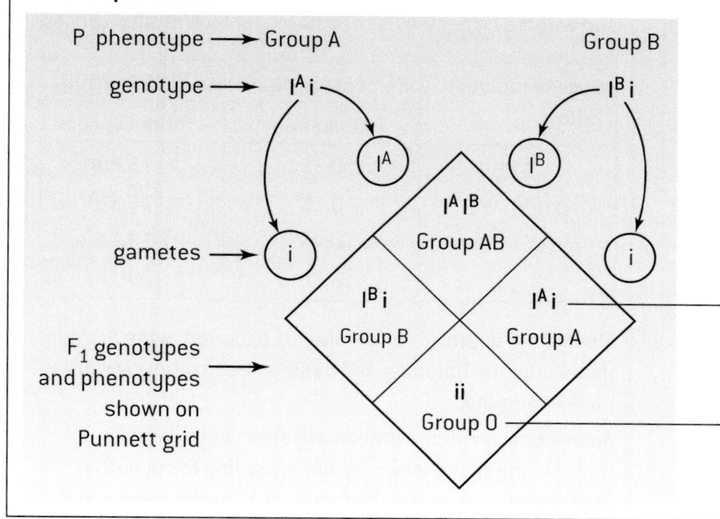

If there are more than two alleles of a gene they are called multiple alleles. The gene for ABO blood groups has three alleles, I^A, I^B and i.

i is recessive to both I^A and I^B so I^Ai gives blood group A and I^Bi gives blood group B.

Individuals who are homozygous for i are in blood group O.

PREDICTED AND ACTUAL OUTCOMES OF GENETIC CROSSES

Parents of blood groups O and AB, have genotypes ii and I^AI^B so offspring could be I^Ai or I^Bi with a 1:1 Group A to B ratio.

Predictions for AB × O cross

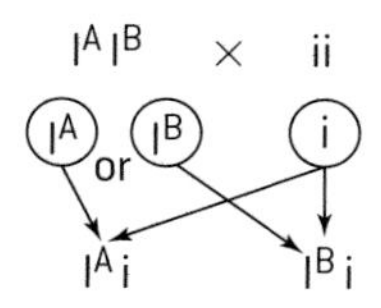

However, if the parents have two children, both may be Group A or both Group B. Predicted ratios are only expected when the number of offspring is large.

The allele for yellow coat colour in mice (Y) is dominant over the allele for grey coat (y). Yellow mice can either have the genotype YY or Yy. When crossed with grey mice (yy), any YY yellows were expected to produce all yellow offspring and any Yy mice a 1:1 ratio of yellow to grey. The actual outcome was a 1:1 ratio in every case so all the yellow mice must have been heterozygous.

In a cross between heterozygous yellow mice there are four equally likely outcomes: Yy, YY, yY, yy so a 3:1 ratio of yellow to grey is predicted. The table gives expected and actual results of 1598 offspring of this cross. The actual ratio was 2:1. The explanation is that the genotype YY is lethal, so among the surviving offspring $\frac{2}{3}$ are heterozygous yellow mice and $\frac{1}{3}$ are homozygous recessive greys.

	Predicted results	Actual results
Yellow	1198	1063
Grey	400	535

Mutation

CAUSES OF MUTATION

Mutations are random changes to the base sequence of a gene. A mutation that replaces one base in a gene with a different base is a **base substitution**. Mutations are important as they are a source of the genetic variation that is necessary for evolution to occur, but very few mutations prove to be beneficial and some cause genetic diseases or cancer. The mutation rate is increased by two types of mutagen:

- **high energy radiation** including X-rays, short or medium wave UV, gamma rays and alpha particles from radioactive isotopes
- **mutagenic chemicals** such as nitrosamines in tobacco, mustard gas that was used as a chemical weapon and the solvent benzene.

Because mutagens increase the mutation rate they are a cause of both genetic diseases and cancer.

The effects of radiation can be studied using two incidents, the nuclear accident at Chernobyl and the nuclear bombing of Hiroshima. The common feature of these incidents is that radioactive isotopes were released into the environment and as a result people were exposed to potentially dangerous levels of radiation. Chernobyl released far more radioactive material but will probably have caused fewer deaths than Hiroshima, because the isotopes released were spread over a wider area and have longer half-lives so the doses of radiation have been spread over a longer period.

NUCLEAR ACCIDENT AT CHERNOBYL

The accident at Chernobyl, Ukraine, in 1986 caused explosions and a fire in the core of a nuclear reactor. Radioactive iodine-131, caesium-134 and caesium-137 were released and spread over large parts of Europe. About six tonnes of uranium and other radioactive metals in fuel from the reactor were broken up into small particles by the explosions and escaped.

- 28 workers at the nuclear power plant died from the effects of radiation within three months. There have also been increased rates of leukemia in other workers exposed to high radiation doses.
- Concentrations of radioactive iodine in the environment rose and resulted in drinking water and milk with unacceptably high levels. Iodine is absorbed by the thyroid gland. More than 6,000 cases of thyroid cancer can be attributed to the radioactive iodine released. Horses and cattle near the plant died from damage to their thyroid glands.
- Bioaccumulation caused high levels of radioactive caesium in fish as far away as Scandinavia, Germany and Wales. Consumption of lamb contaminated with radioactive caesium was banned for many years in some areas due to the long half-life of caesium-137.
- There will almost certainly have been a small increase in the risk of cancer and genetic disease for large numbers of people in Europe due to radiation from Chernobyl, but it is hard to prove this.
- 4 km^2 of pine forest downwind of the reactor turned ginger brown and died due to high doses of radiation, but in the absence of humans some wildlife such as lynx and wild boar have thrived.

NUCLEAR BOMBING OF HIROSHIMA

The atomic bomb that was detonated over Hiroshima in 1945 killed 90,000–166,000; people either died directly or within a few months. The city was devastated with few buildings remaining.

The health of a large group of survivors of both the Hiroshima and Nagasaki nuclear bombs has been followed since then by the Radiation Effects Research Foundation in Japan. There have been long-term effects from the radiation with increased deaths due to cancer. The larger the dose of radiation received by a survivor, the higher the risk of both leukemia and other cancers.

Dose of radiation (GBq)	Percentage death rate (1950–2000)	
	Leukemia	Other cancers
<0.005	0.25	15.79
0.005–0.1	0.23	15.86
0.1–1.0	0.44	19.13
≥ 1.0	2.36	29.17

Most of the deaths due to leukemia occurred in the first ten years after the bombing, but deaths due to other cancers have continued.

Apart from cancer the other main effect of the radiation that was predicted was mutations, leading to stillbirths, malformation or death. The health of 10,000 children that were fetuses when the atomic bombs were detonated has been monitored. No evidence has been found of mutations caused by the radiation. There are likely to have been some mutations, but the number is too small for it to be statistically significant even with the large numbers of children in the study. Despite this lack of evidence of mutations due to the atomic bombs, survivors have sometimes felt that they were stigmatized. Some found that potential wives or husbands were reluctant to marry them for fear that their children might have genetic diseases.

Hiroshima Peace Memorial

Genes and alleles

GENES AND CHROMOSOMES

Genetics is the study of variation and inheritance. The basic unit of inheritance is the gene.

A gene is a heritable factor that consists of a length of DNA and influences a specific characteristic.

Every gene occupies a specific position on a chromosome. For example, in humans the gene for making the beta polypeptide of hemoglobin is located near the end of the short arm of chromosome 11.

locus of gene for
β polypeptide of hemoglobin
centromere
short arm of chromosome 11
long arm of chromosome 11

NUMBERS OF GENES

A typical animal or plant cell nucleus contains thousands of genes.

The total number of genes is not yet known precisely for humans or other species but these are current estimates:

Homo sapiens (humans)	23,000
E. coli (a gut bacterium)	3,200
Drosophila melanogaster (fruit fly)	14,000
Takifugu gambiae (puffer fish)	25,500
Oryza sativa (rice)	41,000

These estimates illustrate some trends:

- bacteria have fewer genes than eukaryotes
- some other animals have fewer genes than humans but some have more
- plants may seem less complex than humans but some have more genes.

ALLELES

A gene consists of a sequence of bases on a piece of DNA. There are different versions of some genes that have almost the same base sequence but differ in just one or a very small number of bases. These variant forms of a gene are called **alleles**.

Alleles are different forms of the same gene because they influence the same characteristic, occupy the same position on a type of chromosome and have base sequences that differ from each other by one or only a few bases.

For example, in the human gene for the beta polypeptide of hemoglobin, the second base in the sixth codon of the gene is adenine (A) in the commonest allele of the gene. There is a less common allele in which this base is thymine (T). This allele causes the genetic disease **sickle cell anemia.**

SICKLE-CELL ANEMIA

This is a genetic disease that demonstrates how a single base substitution mutation can have very significant consequences. The mutation occurred in HBB, the gene for the beta polypeptide of hemoglobin. This polypeptide consists of 146 amino acids.

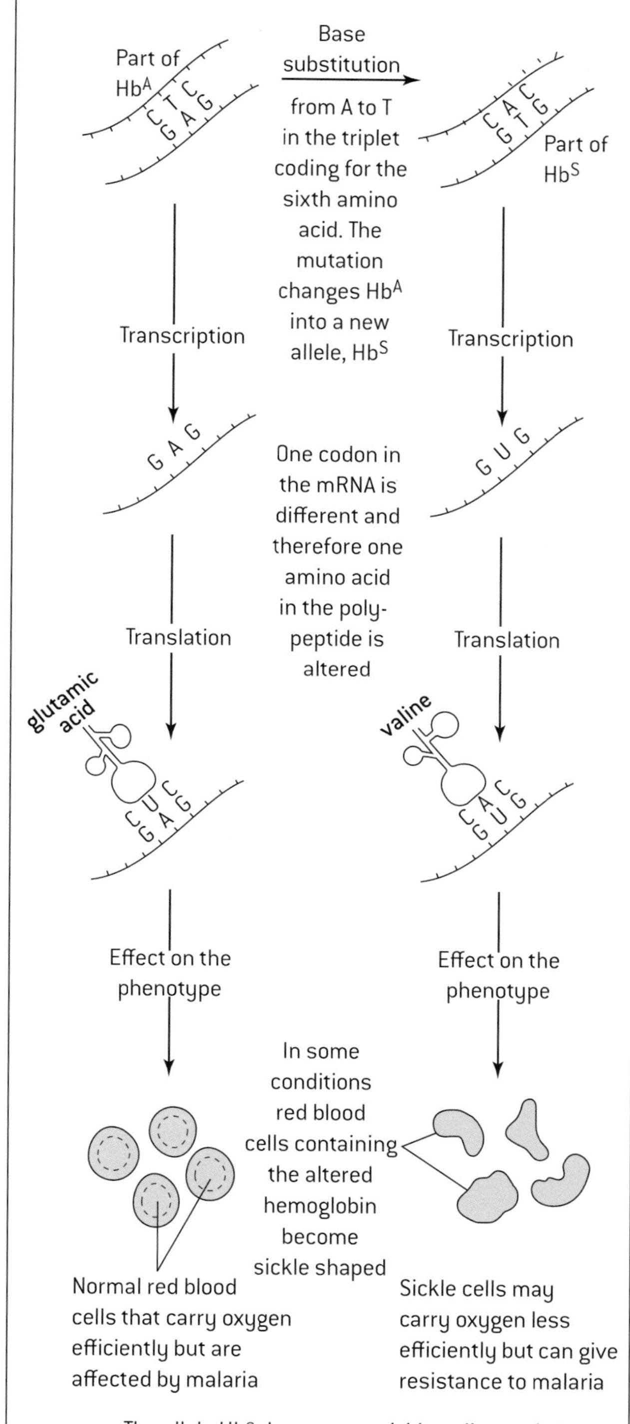

The allele Hb^S that causes sickle-cell anemia has become quite common in some parts of the world affected by malaria. In these regions the malaria resistance that it causes is an advantage

Gene sequencing

USING DATABASES

Databases have been developed since the 1960s to help store complex information and let researchers get access to it. They are ideally suited to storing the vast amount of base sequence data that is currently being generated by genome research. Exponential increases in the amount of data needing to be stored have been matched by increases in database capacity.

There have also been great improvements in the ease with which users can search databases and extract data. The development of the internet opened up access to databases from anywhere in the world, so data can be shared far more easily. Two uses of gene databases are explained below on this page.

GENE LOCI AND PROTEIN PRODUCTS

The locus of a gene is its particular position on homologous chromosomes. The loci of human genes can be found using the **OMIM** website (Online Mendelian Inheritance in Man).

- *Search for OMIM and open the home page.*
- *Choose OMIM Gene Map*
- *Enter the name of a gene into the Search gene map box.*

This should bring up a table with information about the gene, including its **locus** and the **protein product** of the gene.

Example: entering HBB brings up the information that this gene codes for the beta polypeptide of hemoglobin and the location is 11p15.4. The first number indicates that the gene is on chromosome 11, the letter p that it is on the short arm of this chromosome (q indicates the long arm) and 15.4 which region of the short arm the gene is in.

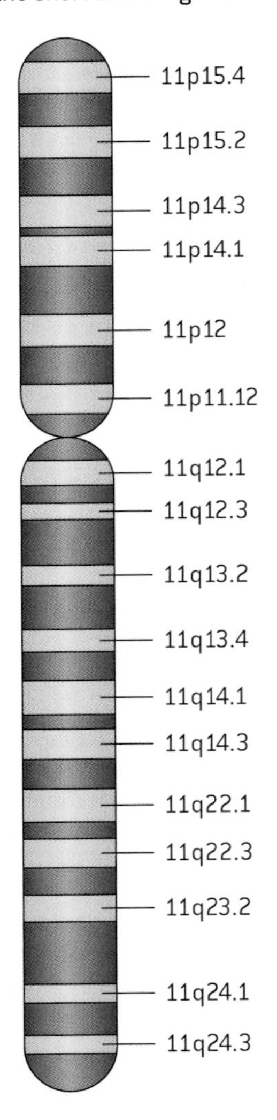

COMPARING BASE SEQUENCES OF GENES

Changes in the base sequence of genes occur over time. If a species splits to form two separate species, differences between the base sequences of the genes of those two species will gradually accumulate. The number of differences can give an indication of how long ago species diverged from the common ancestral species. It is therefore useful to be able to compare the base sequences of genes and find out how many differences there are. This can be done using base sequence data from the **GenBank** database and downloadable software.

- *Search for Genbank and open the home page.*
- *Choose 'Gene' from the search menu.*
- *Enter the name of a gene plus the organism, such as HBB human.*
- *Move your mouse over the section 'Genomic regions, transcripts and products' until Nucleotide links appears.*
- *Choose FASTA and the entire base sequence of the gene should appear.*

You can then search for similar sequences in the GenBank database using **BLAST** software. There may be other alleles in the same species or genes with similar sequences in other species.

Alternatively you can compare the sequence with other selected genes (for example HBB in chimpanzees) using **ClustalX** software which can either be downloaded or used online:

- Copy the whole sequence in the FASTA format (including the symbols starting with >) and paste it into a text file or notepad file.
- Repeat with a number of different species that you want to compare. Copy them to the same file, separating by pressing the return button on your keyboard and saving the file each time.
- Open ClustalX and enter the text file containing the sequences that you wish to compare. A base by base comparison of the sequences should appear with any differences highlighted.

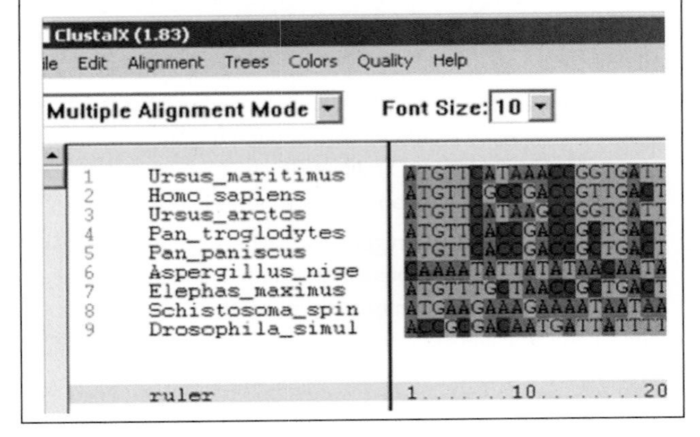

DNA technology

PCR – THE POLYMERASE CHAIN REACTION

PCR consists of a cycle of stages carried out again and again to produce many copies of a DNA molecule:

Millions of copies of the DNA can be produced in a few hours. This is very useful when very small quantities of DNA are found in a sample and larger amounts are needed for analysis. DNA from very small samples of semen, blood or other tissue or even from long-dead specimens can be amplified using PCR. (Reasons for the use of Taq DNA polymerase in PCR are described in Topic 2.)

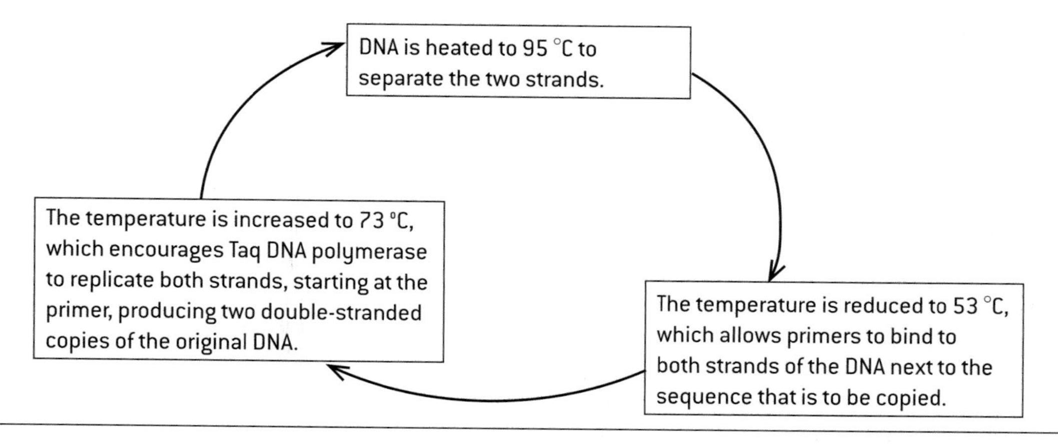

GEL ELECTROPHORESIS

Gel electrophoresis is a method of separating mixtures of proteins or fragments of DNA, which are charged. The mixture is placed on a thin sheet of gel, which acts like a molecular sieve. An electric field is applied to the gel by attaching electrodes to both ends. Depending on whether the particles are positively or negatively charged, they move towards one of the electrodes or the other. The rate of movement depends on the size of the molecules – small molecules move faster than larger ones.

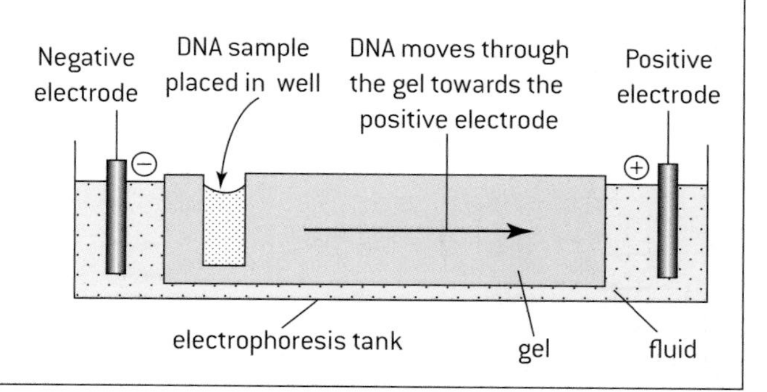

GENE SEQUENCING TECHNOLOGY

Many developments in scientific research follow improvements in technology. In some cases research projects stimulate improvements in technology.

Methods for finding the base sequence of genes were developed in the 1970s and the technology has been improved repeatedly since then.

The idea of sequencing the entire human genome seemed impossibly difficult at one time but improvements in sequencing technology towards the end of the 20th century made it possible, though still very ambitious. **The Human Genome Project** began in 1990 and was expected to take 15 years but improvements in technology continued once the project was underway and draft sequences were completed much sooner than expected in 2000.

Further advances are allowing the genomes of other species to be sequenced at an ever increasing rate and lower cost.

By 2008 the genomes of over a thousand different humans from all parts of the world had been sequenced, to study genetic variation, and by 2012 the cost of sequencing a human genome had dropped below $10,000. By 2014 the genomes of hundreds of prokaryotes had been sequenced and over a hundred eukaryotes. The 1,000 Plant Genomes Project was well on its way towards the planned sequencing of the genomes of a thousand different plant species.

DNA profiling

DNA PROFILING

In the DNA of humans and other organisms there are loci in the chromosomes where instead of a gene consisting of a long sequence of bases there are much shorter sequences of three, four or five bases that are repeated many times. The repeated sequences are called **short tandem repeats (STR)**. At these STR loci there are many different possible alleles that vary in the number of repeats. STR alleles are used in DNA profiling (also called DNA fingerprinting).

1. A sample of DNA is obtained from a person. It must not be contaminated with DNA from anyone else or another organism.
2. DNA from a selection of STR loci is copied by PCR. The DNA from between 11 and 13 loci is copied in commonly used DNA profiling methods. It is very unlikely for two individuals to have the same number of repeats at each of these loci.
3. The copies of STR alleles made by PCR from one person's DNA sample are separated by gel electrophoresis. The result is a pattern of bands. Two individuals are extremely unlikely to have the same pattern of bands unless they are identical twins.

DNA profiling is used in forensic investigations (obtaining evidence to use in court cases) and investigating paternity (who the father of a child is).

Forensic investigations

The first DNA profiles to be used in a forensic investigation (the Enderby double murder case) are shown below. Key: a = hair roots from the first victim, b = mixed semen and vaginal fluids from the first victim, c = blood of second victim, d = vaginal swab from second victim, e = semen stain on second victim, s = blood of prime suspect.

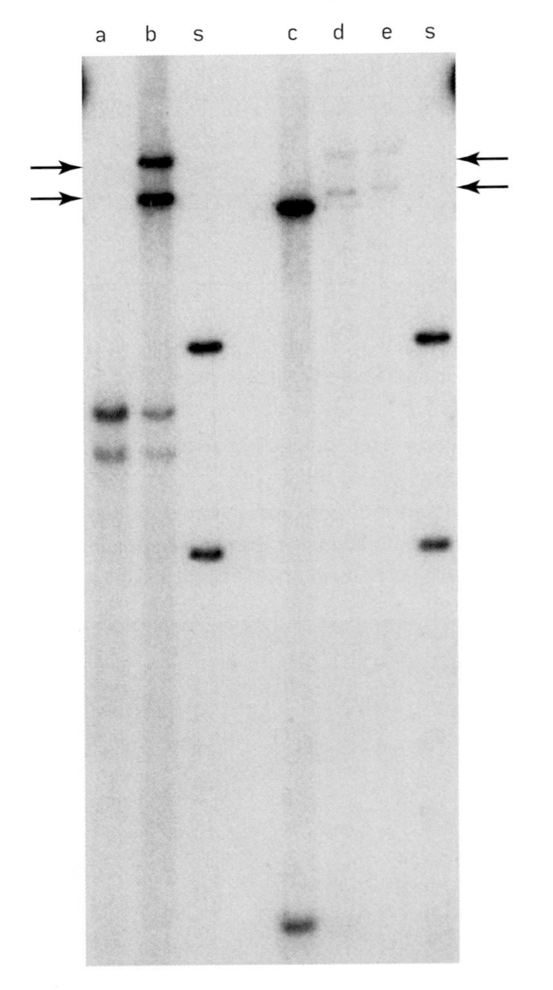

Two bands in track b indicated by arrows must be from DNA in the culprit's semen but are not present in DNA from the prime suspect, who was not guilty despite having confessed to the murders.

Paternity investigations

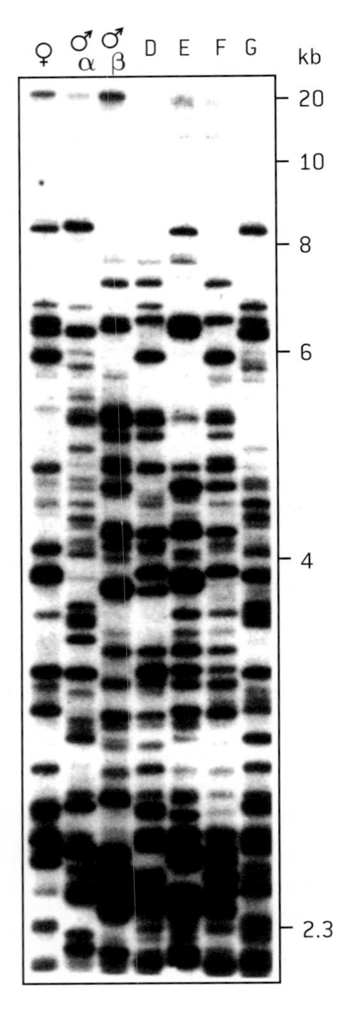

The DNA profiles of a family of dunnocks (*Prunella modularis*) are shown above. Dunnocks are small birds found in Europe. The tracks from left to right are: the mother, two resident males that might have been the father of the offspring and four offspring. There are bands in the profiles of three offspring (D, E and F) that are found in the profile of the β male, but not the α male or the mother, showing that the β male fathered them despite being less dominant than the α male.

Genetic modification

GENE TRANSFER USING PLASMIDS

Genetic modification is the transfer of genes from one species to another. Organisms that have had genes transferred to them are called **genetically modified organisms** (GMO) or transgenic organisms. The transfer of the gene for human insulin to bacteria was outlined in Topic 2. The methods used for gene transfer to bacteria are explained in the flow chart below. Genes are transferred between species using a **vector**. In this case the vector is a small loop of DNA called a **plasmid**. Two enzymes are used to insert genes into plasmids: **restriction endonucleases** cut DNA molecules at specific base sequences and **DNA ligase** makes sugar–phosphate bonds to link nucleotides together and form continuous strands of nucleotides. A plasmid with a gene from another species inserted is called a **recombinant plasmid**. The cell that receives the gene is a host cell. In the example below *E. coli* is the host cell.

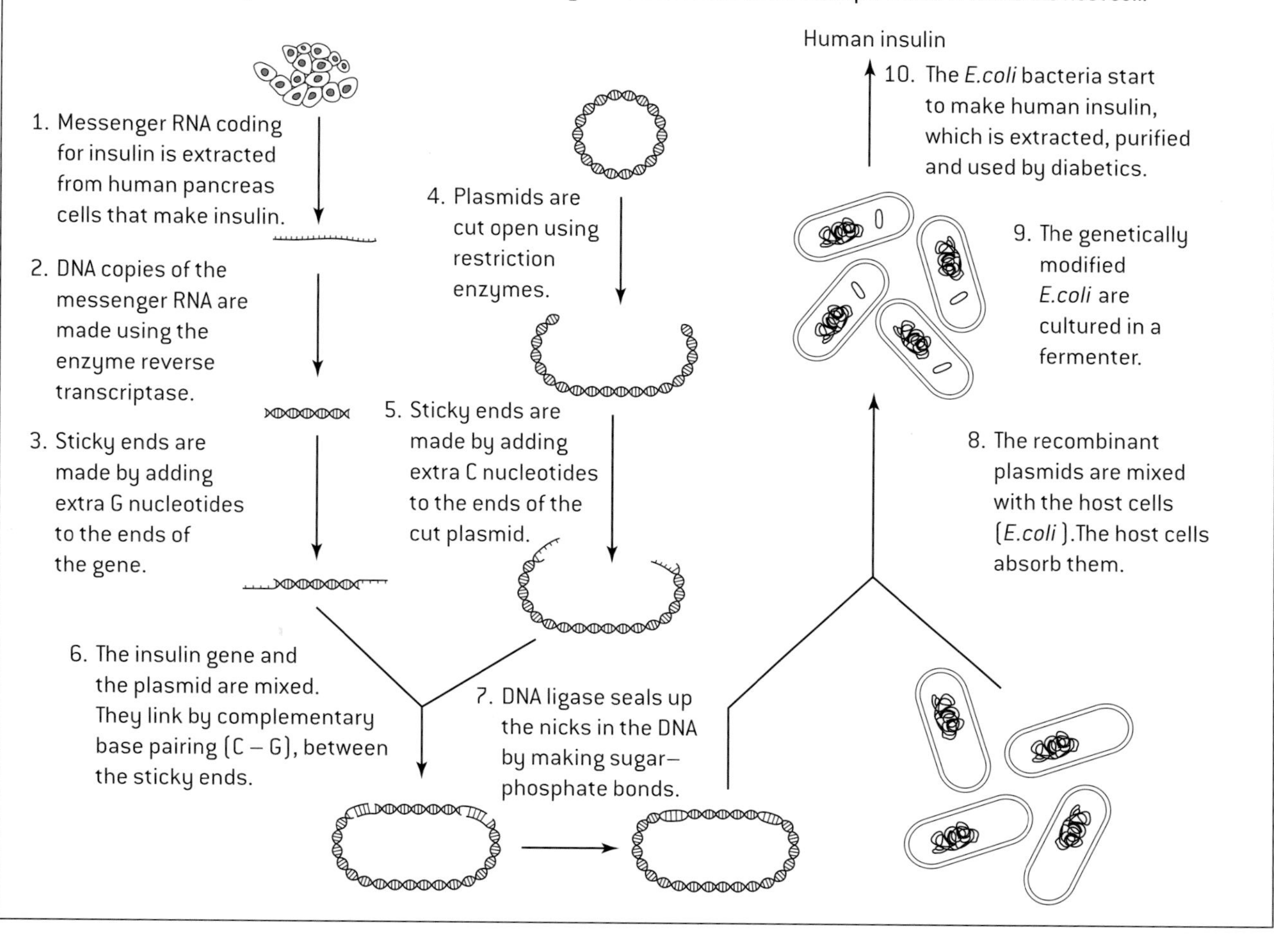

BENEFITS AND RISKS OF GENETIC MODIFICATION OF CROPS

The production of human insulin using bacteria has enormous benefits and no obvious harmful effects. Genetic modification of crop plants is more controversial. An example of this is corn or maize (*Zea mays*). A gene from a bacterium (*Bacillus thuringiensis*) has been transferred to some varieties. The gene codes for a bacterial protein called Bt toxin, which kills insect pests feeding on the crop, especially corn borers that can cause serious damage.

Potential benefits of Bt maize

1. Higher crop yields and thus more food for humans, due to less pest damage.
2. Less land needed for crop production, so some could become areas for wildlife conservation.
3. Less use of insecticide sprays, which are expensive and can be harmful to farm workers and to wildlife.

(Other GM crops are being produced with herbicide resistance, increased vitamin content, decreased allergen or toxin content, resistance to virus diseases and increased tolerance to drought, cold or saline soils.)

Possible harmful effects of Bt maize

1. Insects that are not pests could be killed. Maize pollen containing the toxin is blown onto wild plants growing near the maize. Insects feeding on the wild plants, including caterpillars of the Monarch butterfly (*Danaus plexippus*) are therefore affected even if they do not feed on the maize. Leaves and stems from the crop after harvest still contain the toxin which could harm insect detritivores in the soil and in streams.
2. The transferred gene might spread to populations of wild plants by cross-pollination, making them also toxic to insects feeding on them.
3. The insects pests of corn may develop resistance to the Bt toxin.

Cloning

CLONES AND CLONING

A group of genetically identical organisms derived from a single original parent cell is a **clone** and production of an organism that is genetically identical to another organism is **cloning**. Asexual reproduction is a natural form of cloning. Many plant species and some animal species can do this, using mitosis to produce the genetically identical cells required. For example, plants clone themselves by growing extra bulbs, tubers, runners or other structures. Female aphids (greenfly) can give birth to young formed asexually from their own cells.

There are also methods of artificial cloning for both plants and animals. Cloning is very useful if more organisms with a desirable combination of characteristics are wanted.

ARTIFICIAL CLONING OF ANIMALS

The simplest method of cloning an animal is to break up an embryo into more than one group of cells at an early stage when it consists entirely of embryonic stem cells. Each group of cells develops into a separate genetically identical individual. The drawback is that at the embryo stage the characteristics of an animal are mostly unknown.

It is much more difficult to clone an adult animal with known characteristics, but methods have been developed. One method is **somatic-cell nuclear transfer**, in which the nucleus is removed from an egg cell and replaced by a nucleus from a differentiated somatic (body) cell. This method (see below) was used to produce Dolly, the first mammal to be cloned from an adult somatic cell.

CLONING ADULT ANIMALS USING DIFFERENTIATED CELLS

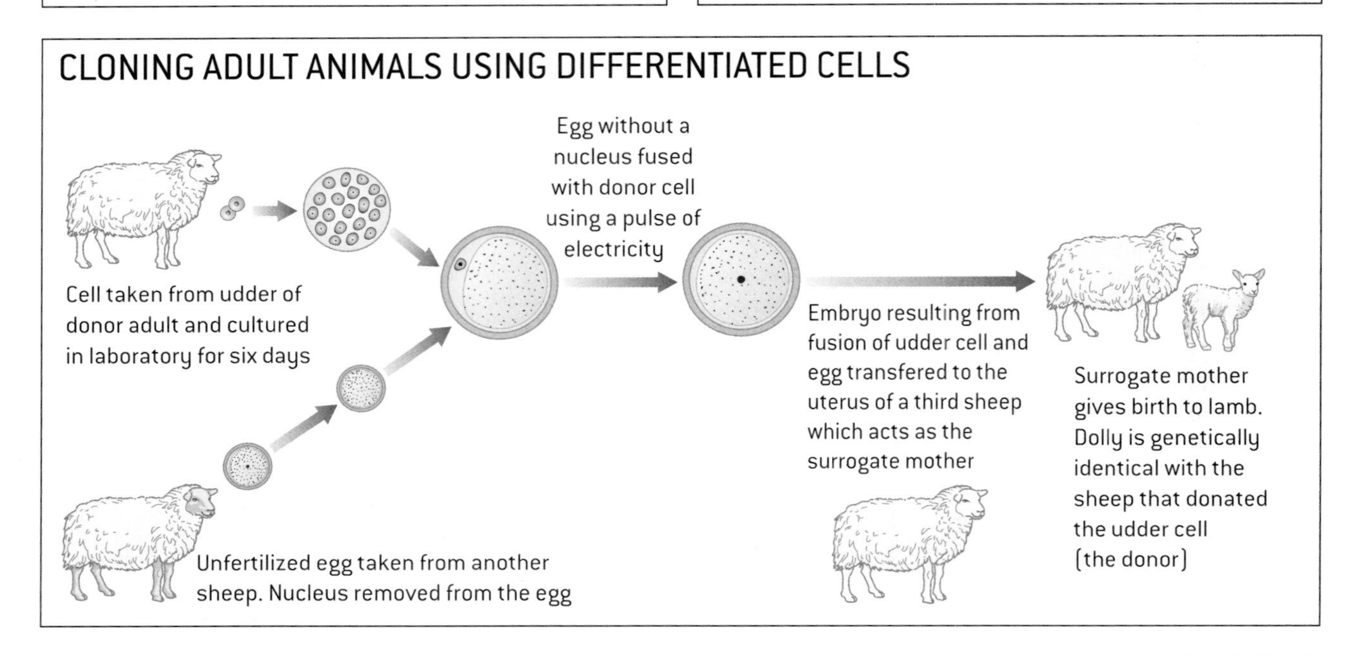

INVESTIGATING FACTORS AFFECTING ROOTING IN STEM CUTTINGS

Stem cuttings are short lengths of stem that are used to clone plants artificially. If roots develop from the stem, the cutting can become an independent new plant. Some plant species root when the base of the cutting is placed in water but others root better when it is inserted into a solid medium. The diagram (right) shows a basic method for rooting a cutting.

Many factors affect whether the cutting will form roots or not. One of these could be investigated – it is the **independent variable**. For example, the independent variable could be how many leaves are left on the cutting, whether a hormone rooting powder is used, how warm the cuttings are kept and whether a plastic bag is placed over the cuttings. The **dependent variable** could simply be whether any roots are formed or not, or to make the investigation quantitative the number of roots could be counted. All other factors that could affect rooting are **control variables** and must be kept the same. For example, cuttings from the same species of plant should be used for the whole investigation. There should be **repeats** to make the investigation reliable and avoid anomalous results leading to false conclusions.

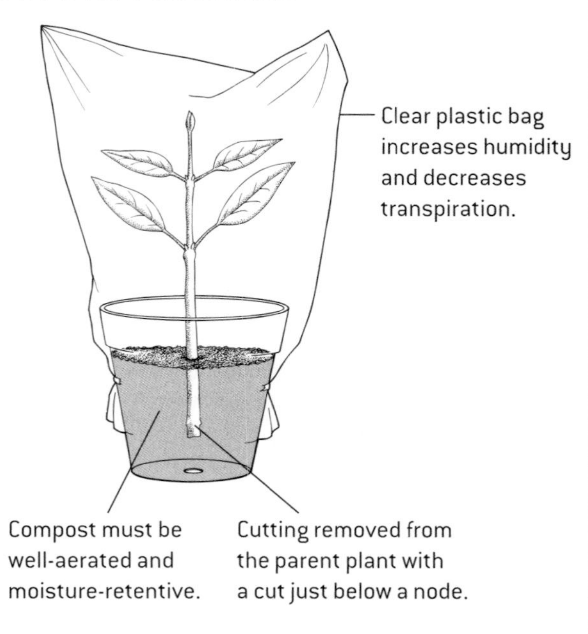

Questions – genetics

1. a) List the following in order of increasing genome size: *Drosophila melanogaster, Escherichia coli, Homo sapiens, Paris japonica* and T2 phage. [4]

   b) List the following in order of increasing numbers of chromosomes in the body cells: *Canis familiaris, Homo sapiens, Oryza sativa, Pan troglodytes* and *Parascaris equorum*. [4]

   c) (i) Explain the reasons for body cells in animals having even rather than odd chromosome numbers. [2]

   (ii) Suggest one cause of body cells having an odd chromosome number. [1]

   d) Analyse the closeness of relationship between complexity and genome size and between complexity and chromosome number in these organisms. [4]

2. The micrograph of bluebell anther below shows cells in meiosis.

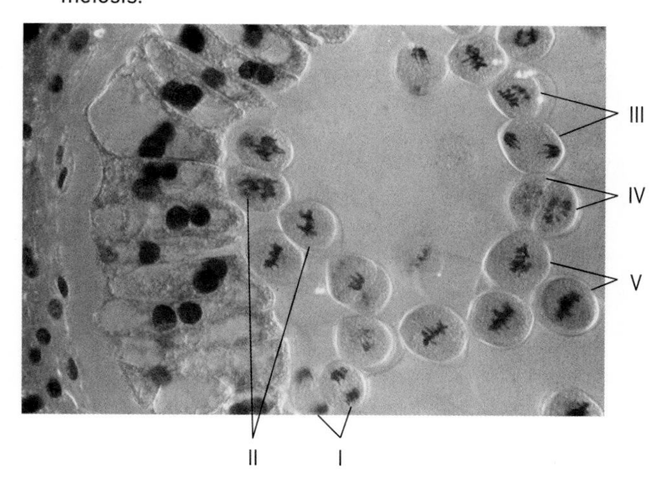

   a) Identify the stages of meiosis of cells I to V. [5]

   b) Draw diagrams to show each of these stages. [10]

3. The pedigree below shows the blood group of some individuals.

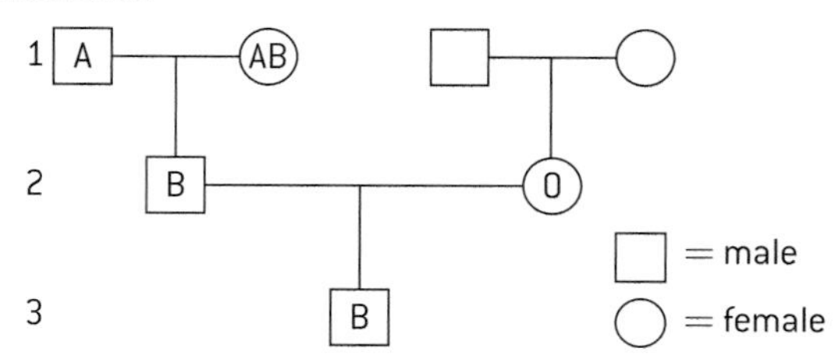

   a) Explain the conclusions that can be drawn about the genotypes of the individuals in the pedigree in generations 2 and 3. [3]

   b) Explain to which blood groups the parents of the blood group O female in the pedigree could have belonged. [3]

   c) Use a Punnett grid to determine the ratio between possible genotypes and blood groups for children whose parents both have blood group AB. [4]

4. a) State the name used by biologists for a group of genetically identical organisms derived from a single original parent cell. [1]

   b) Outline one technique for cloning adult animals, using differentiated cells. [2]

   The figure below shows DNA profiles of sheep that were involved the Dolly cloning experiment. U = differentiated cells taken from the udder of a sheep used in the experiments C = cells in a culture derived from the udder cells D = blood cells taken from Dolly the sheep 1–12 = results from other sheep.

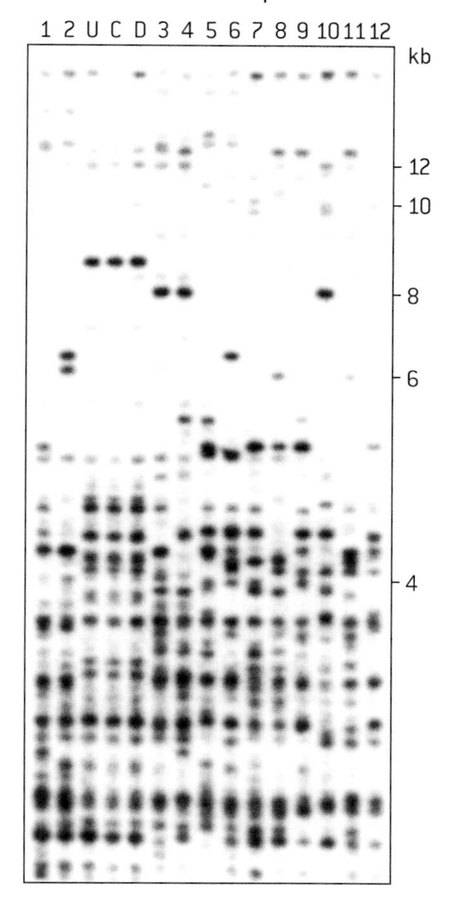

   c) (i) Explain whether DNA fragments in the profiles had moved upwards or downwards. [2]

   (ii) Explain the conclusions that can be drawn from the DNA profiles of the sheep. [3]

   d) State two uses of DNA profiling in humans. [2]

5. The table below shows the percentage of egg cells that failed to be fertilized in IVF, resulting from non-disjunction due to two possible causes.

Error in the first division of meiosis	Age of mother (years)		
	25–34	35–39	40–45
A bivalent fails to split	1.5%	7.4%	24.2%
A centromere divides	14.9%	20.6%	18.1%

   a) Explain how Down syndrome could be caused by

   (i) the chromosomes in a bivalent failing to separate [3]

   (ii) a centromere dividing during meiosis 1. [3]

   b) Evaluate the hypothesis that the chance of non-disjunction increases with maternal age using the data in the table. [4]

Modes of nutrition

POPULATIONS AND SPECIES

A **species** is a group of organisms with similar characteristics, which can potentially interbreed and produce fertile offspring.

A **population** is a group of organisms of the same species, who live in the same area at the same time.

Members of a species may be reproductively isolated in separate populations but, as long as they could still interbreed if the populations came together again, they are the same species. For example, the wood mouse (*Apodemus sylvaticus*) lives in Britain and on Iceland. These two populations do not actually interbreed but potentially could, so they are the same species.

AUTOTROPHS AND HETEROTROPHS

There are two main modes of nutrition: **autotrophic** and **heterotrophic**. Put simply, autotrophs make their own food and heterotrophs get food from other organisms. A fuller explanation is needed than this, however.

Autotrophs absorb carbon dioxide, water and inorganic nutrients such as nitrates from the abiotic (non-living) environment and use them to synthesize all the carbon compounds that they need. An external energy source such as light is needed to do this.

Example of an autotroph: a corn plant (*Zea mays*).

Heterotrophs cannot make all the carbon compounds that they need and instead obtain them from other organisms. Many carbon compounds including proteins or starch must be digested by heterotrophs before they can absorb and use them.

Example of a heterotroph: wood mouse (*A. sylvaticus*).

MODES OF HETEROTROPHIC NUTRITION

There are three main modes of heterotrophic nutrition:

Saprotrophs obtain organic nutrients from dead organisms by external digestion. They secrete digestive enzymes into material such as dead leaves or wood, dead animals and feces. Protein, cellulose and other carbon compounds are digested externally and the saprotrophs then absorb the substances that they need. Saprotrophs are also known as decomposers. Most saprotrophs are bacteria or fungi.

Consumers feed on living organisms by ingestion. This means that they take other organisms into their digestive system for digestion and absorption. The organism may be swallowed whole or in parts. It may still be alive or have recently been killed. For example, deer eat the leaves of plants which are still alive whereas a vulture eats parts of an animal that has been killed. The skull of the vulture *Gyps rueppellii* shows adaptation for picking meat off carcasses.

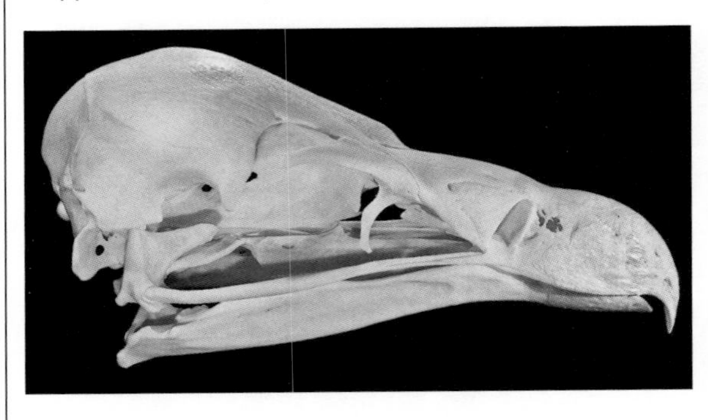

Detritivores obtain organic nutrients from detritus by internal digestion. Detritus is dead material from living organisms and includes dead leaves or roots, parts of decomposing animals and feces.

Honey bees secrete wax to make comb in their colonies. Larvae of the wax moth (*Achroia grisella*), shown left, are detritivores that feed on the wax comb. The moths prefer old comb as it has the protein-rich pupal cases of honey bees in it.

HETEROTROPHIC PLANTS AND ALGAE

Most plants and algae are autotrophs, but there are some exceptional species that no longer make food by photosynthesis and instead obtain carbon compounds from other organisms, They are therefore heterotrophs. Some obtain carbon compounds directly from plants, for example dodder (*Cuscuta europaea*), which feeds on the stems of other plants. Others obtain carbon compounds from fungi living on the roots of trees, for example the ghost orchid (*Epipogium aphyllum*) which lives entirely underground except when it flowers. Because of species such as this, we cannot assume that a plant or alga is autotrophic – not all are.

Communities and ecosystems

COMMUNITIES

Populations do not live in isolation – they live together with other populations in ecological communities. A **community** is a group of populations of different species living together and interacting with each other in an area.

There are many types of interaction between populations in a community. Trophic relationships are very important – where one population of organisms feeds on another population. The complex network of feeding relationships in a community is called a **food web**.

ECOSYSTEMS

Communities of living organisms interact in many ways with the soil, water and air that surround them. The non-living surroundings of a community are its **abiotic environment**.

A community forms an **ecosystem** by its interactions with the abiotic environment. There are many of these interactions, but particularly important are transfers of chemical elements between populations in the community and the abiotic environment because these are an essential part of nutrient recycling.

MESOCOSMS

Ecosystems have the potential to be sustainable over long periods of time. As long as nutrients are recycled, ecosystems only require a supply of energy, usually in the form of light, to continue indefinitely. This can be demonstrated by setting up mesocosms.

A **mesocosm** is a small experimental area set up in an ecological research programme. The apparatus below shows one design of mesocosm.

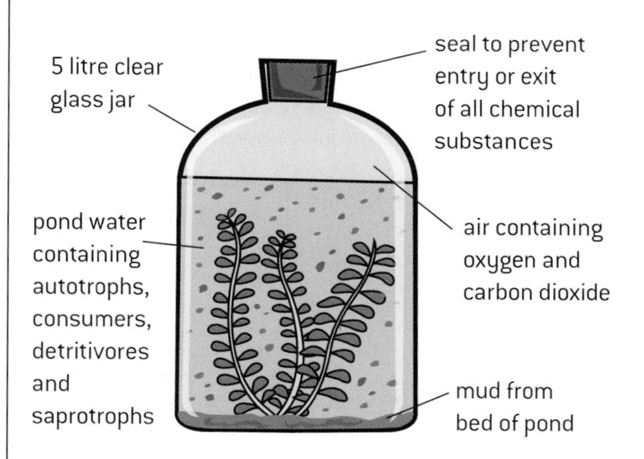

- Autotrophs are an essential component, to produce carbon compounds and regenerate oxygen used in cell respiration by organisms in the mesocosm.
- Saprotrophs are also essential, to decompose dead organic matter and recycle nutrients.
- Consumers and detritivores may not be essential, but are a normal part of ecosystems so are usually included. It is unethical to include large animals in mesocosms that cannot obtain enough food or oxygen.

QUADRAT SAMPLING OF COMMUNITIES

A **quadrat** is a square sample area used in ecological research. To carry out quadrat sampling of an area, first mark out gridlines along two edges of the area. Use a calculator or tables to generate two random numbers to use as coordinates, and place a quadrat on the ground with its corner at these coordinates.

Record the presence or absence of each species of interest inside the quadrat, or record the number of individuals. Repeat with as many quadrats as possible.

e.g. 14 and 7

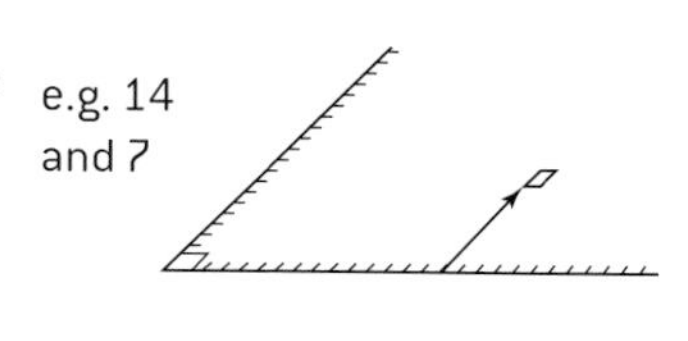

USING THE CHI-SQUARED TEST

If the presence or absence of two species is recorded in a large number of quadrats, a chi-squared test for association between the species can be performed.

Example: the presence or absence of two seaweeds was recorded in fifty 1 m^2 quadrats on a rocky sea shore at Musselwick on the Atlantic coast of Wales.

The contingency table below shows the results.

Expected results assuming no association are shown in brackets. They are calculated using this equation:

		Fucus vesiculosus	
		Present	Absent
Fucus serratus	Present	6 (10.9)	15 (10.1)
	Absent	20 (15.1)	9 (13.9)

$$\text{expected frequency} = \frac{\text{row total} \times \text{column total}}{\text{grand total}}$$

To calculate chi-squared (χ^2) this equation is used:

$$\chi^2 = \Sigma \frac{(f_o - f_e)^2}{f_e}$$

where f_o and f_e are the observed and expected frequencies

$$\chi^2 = 2.20 + 2.37 + 1.59 + 1.73 = 7.89$$

The calculated value of chi-squared (7.89) is compared with the critical region. This is found from a table of chi-squared values. The number of degrees of freedom must be known and also the significance level, which is usually 5%.

The number of degrees of freedom is calculated using this equation: degrees of freedom $= (m-1)(n-1)$, where m and n are the number of rows and number of columns in the contingency table. In this example, there is one degree of freedom.

The critical value for chi-squared with one degree of freedom and a significance level of 5% is 3.84, giving a critical region of $\chi^2 > 3.84$. The calculated value for χ^2 is 7.89, which is within the critical region. There is therefore evidence at the 5% level for a significant difference between the actual and expected results.

The results in the contingency table show that the two species of algae tend not to occur together in the same quadrats. This is because *Fucus serratus* mostly grows in a zone towards the bottom of the beach and *F. vesiculosus* in a zone further up the beach.

Energy flow

ENERGY SOURCES

The organisms in a community all need a supply of energy. Most organisms obtain their energy in one of two ways:

1. Plants, algae and some bacteria absorb light energy and convert it by photosynthesis into chemical energy in carbon compounds. Because these organisms make their own food they are called **producers**.
2. Consumers, detritivores and saprotrophs obtain energy from their food. There is chemical energy in carbon compounds in the food. Carbon compounds and the energy contained in them can pass from organism to organism along food chains, but all food chains start with a producer that originally made the carbon compounds by photosynthesis. Light is therefore the initial energy source for the whole community.

The flow chart (right) shows how light can provide energy for all the organisms in an ecosystem.

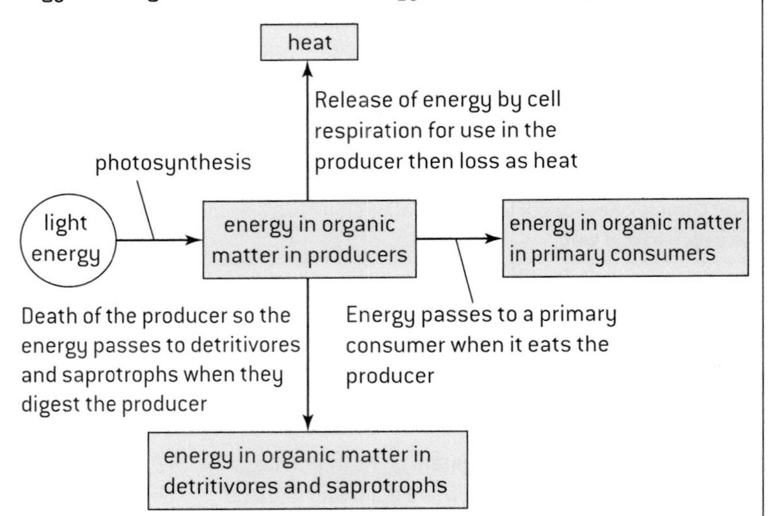

ENERGY LOSSES

Food containing energy is passed along food chains when the primary consumer feeds on the producer, the secondary consumer feeds on the primary consumer and so on. At each successive stage in the food chain less food is available and therefore less chemical energy. This is due to losses of food and energy between the stages in a food chain. There are three ways, shown in the flow chart (right) by which food and energy can be passed to detritivores or saprotrophs rather than to the next stage in the food chain:

1. Some organisms die before they are eaten by the next organism in the food chain. For example, foxes do not eat every rabbit in their community and some rabbits die from disease.
2. Some parts of organisms are not eaten, such as bones, hair and gall bladders.
3. Some parts of organisms are indigestible, such as cellulose in food eaten by humans. The undigested parts are egested in feces.

So, energy in the bodies of dead organisms, parts of organisms and in feces passes to detritivores or saprotrophs.

The other cause of energy losses is cell respiration. All organisms release energy from carbon compounds by cell respiration and use the energy for essential processes such as muscle contraction or active transport. Energy used in this way is converted into heat which is lost from the organism. No organisms can convert the heat energy back into chemical energy, and the heat is eventually lost from the ecosystem. For this reason, ecosystems need an energy source to replace energy lost. For most ecosystems the energy source is sunlight. This is summarized in the flow chart (rigtht).

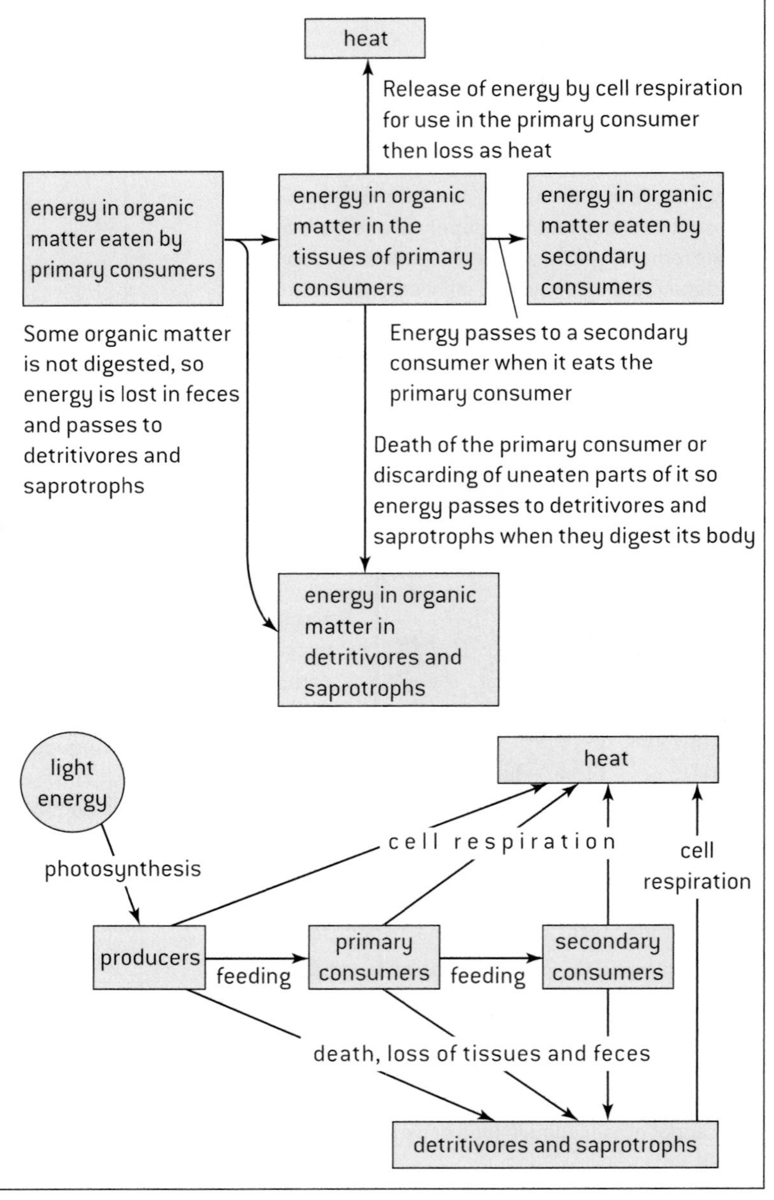

Food chains and energy pyramids

TROPHIC LEVELS AND FOOD CHAINS

Each species in a food chain feeds on the previous one, apart from the producer at the start that makes its own food by photosynthesis. The example of a food chain shown below is from rainforest at Iguazu in north-east Argentina.

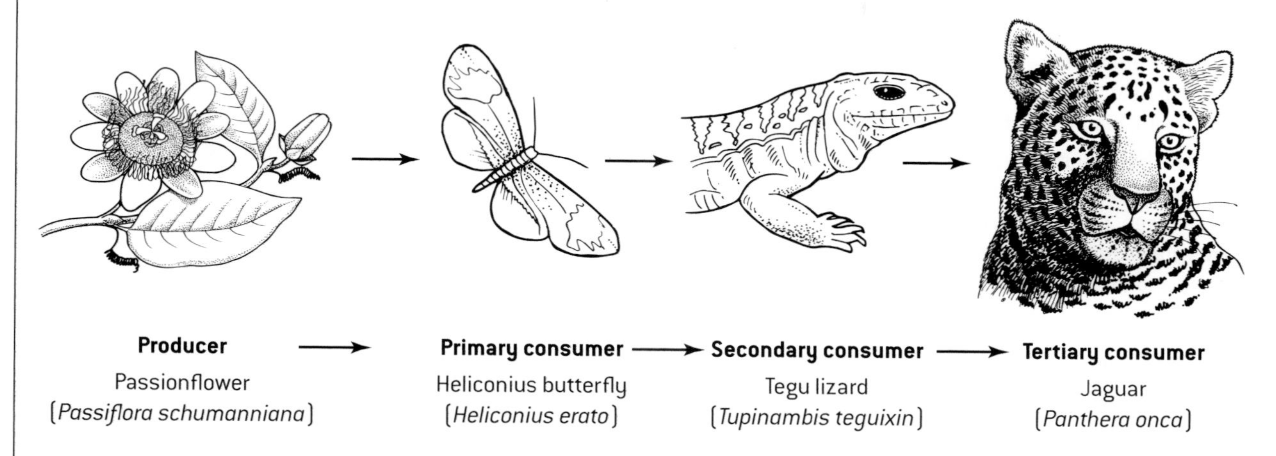

Producer, primary consumer, secondary consumer and tertiary consumer are **trophic levels**. The trophic level of an organism is its position in the food chain. Food chains commonly contain three or four trophic levels and rarely more than five.

The limited length of food chains can be explained by the theories of energy flow and energy losses. Only a small proportion of energy and biomass is passed on from one trophic level to the next. The percentage is very variable but is unlikely to be more than 10%.

Energy pyramids show clearly how the amount of energy drops along food chains and that once the fourth or fifth trophic level is reached, too little energy remains to sustain another level.

ENERGY PYRAMIDS

Energy pyramids are diagrams that show how much energy flows through each trophic level in a community. The amounts of energy are shown per square metre of area occupied by the community and per year (kJ m^{-2} $year^{-1}$).

The figure (right) is a pyramid of energy for Silver Springs, a stream in Florida.

The figure (below right) is a pyramid of energy for a salt marsh in Georgia.

Pyramids of energy are always pyramid shaped – each level is smaller than the one below it. This is because less energy flows through each successive trophic level. Energy is lost at each trophic level, so less remains for the next level. Biomass is also lost so the energy content per gram of the tissues of each successive trophic level is not lower. Biomass is lost when carbon compounds are broken down by cell respiration and the carbon dioxide produced is excreted. Removal of waste products of metabolism such as urea also causes loss of biomass.

A pyramid of energy is a type of bar chart with horizontal bars arranged symmetrically. The bars should all be drawn to scale. Triangular pyramids of biomass are not appropriate as they do not show the amounts of energy for the trophic levels accurately. Labels should indicate the trophic levels with producers at the base, then primary consumers, secondary consumers and so on. It is helpful to put the energy values alongside each bar on the pyramid.

Energy pyramid for a stream (kJ m^{-2} $year^{-1}$)

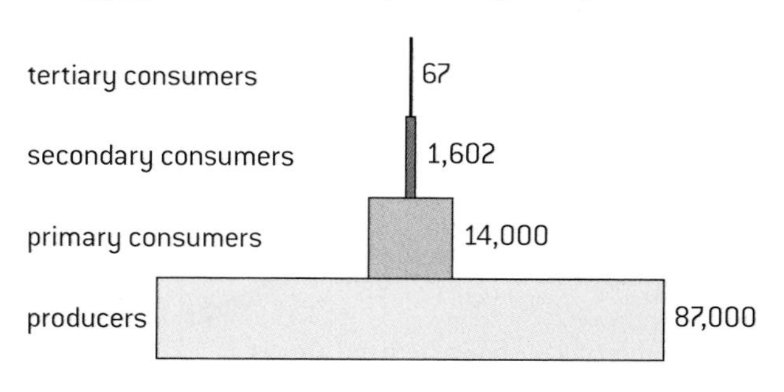

Energy pyramid for a salt marsh (kJ m^{-2} $year^{-1}$)

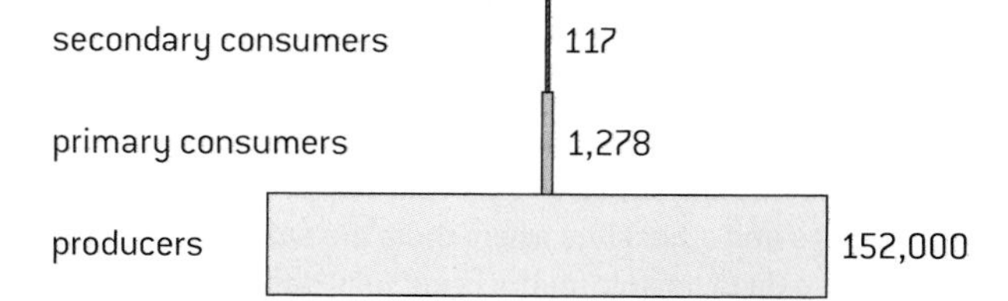

Nutrient cycles

NUTRIENT RECYCLING IN ECOSYSTEMS

There is an important difference between energy and inorganic nutrients in ecosystems:

Energy is supplied to ecosystems in the form of light and converted to chemical energy by producers (autotrophs). This chemical energy is eventually used in a living cell and converted to heat, which cannot be recycled and is lost from the ecosystem, but more light is received.

Ecosystems have limited supplies of nutrients, but these supplies do not run out because nutrients can be recycled. Carbon, nitrogen, phosphorus and all the other essential elements are absorbed from the environment, used by living organisms and then returned to the environment.

CARBON SOURCES IN AIR AND WATER

Autotrophs absorb carbon dioxide either from the atmosphere or from water. The carbon dioxide is absorbed into autotrophs by diffusion and converted into carbohydrates and other carbon compounds.

In aquatic ecosystems carbon is also present in the form of hydrogen carbonate ions (HCO_3^-), formed when water and carbon dioxide combine to form carbonic acid, which dissociates to produce hydrogen carbonate ions.

$$CO_2 + H_2O \rightarrow H_2CO_3 \rightarrow H^+ + HCO_3^-$$

Many aquatic autotrophs absorb and use both dissolved carbon dioxide and hydrogen carbonate ions in photosynthesis.

DRAWING THE CARBON CYCLE

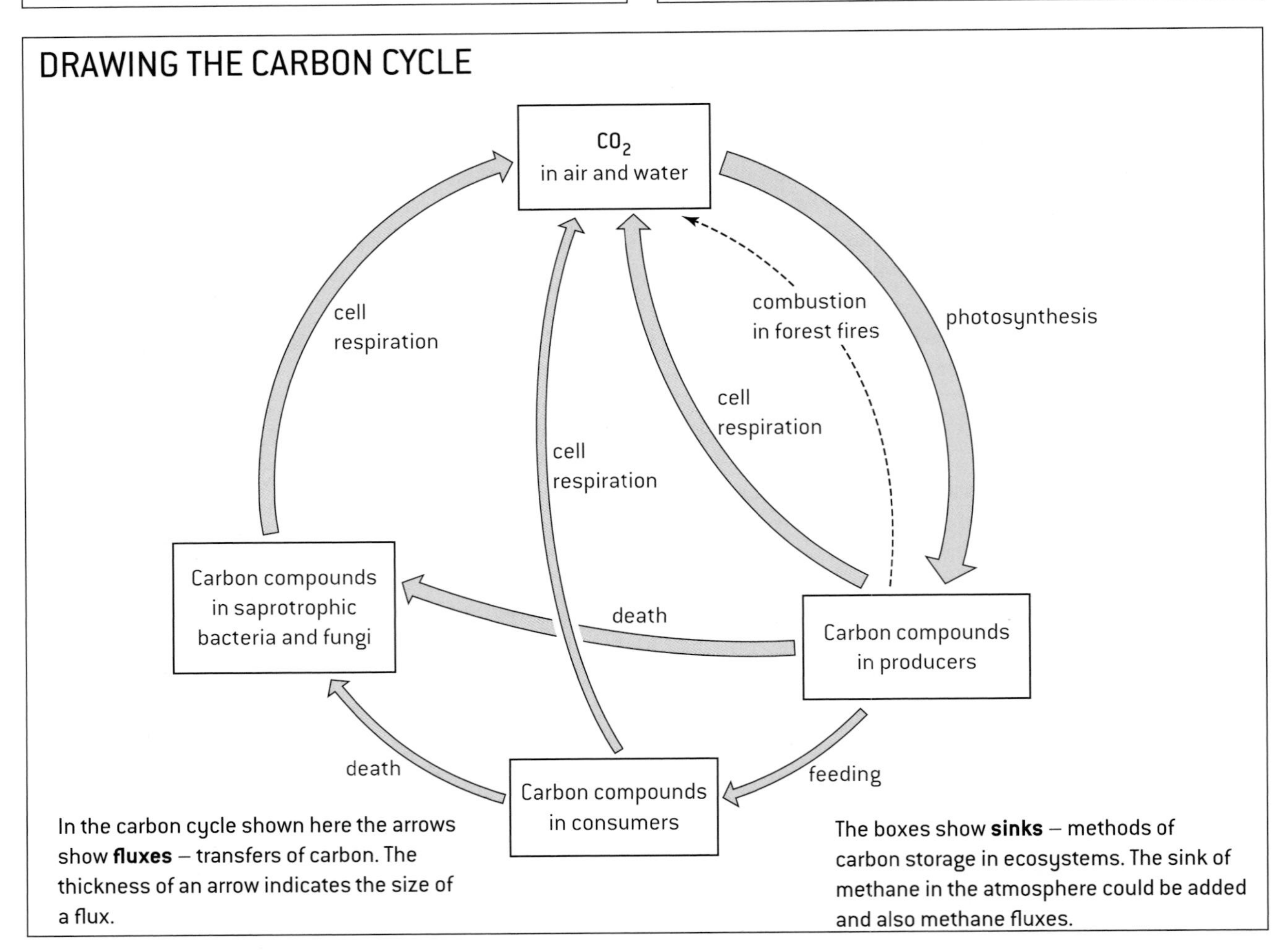

In the carbon cycle shown here the arrows show **fluxes** – transfers of carbon. The thickness of an arrow indicates the size of a flux.

The boxes show **sinks** – methods of carbon storage in ecosystems. The sink of methane in the atmosphere could be added and also methane fluxes.

METHANE IN THE CARBON CYCLE

The gas methane is produced naturally by a group of prokaryotes called **methanogenic archaeans**. They break down organic matter in anaerobic conditions and release methane as a waste product. This process happens in swamps, bogs and other sites where there are anaerobic conditions, so dead organic matter is not fully decomposed by saprotrophic bacteria and fungi. The methane may accumulate in the ground or diffuse into the atmosphere.

Methane is a relatively stable substance in the atmosphere, but is eventually oxidized to carbon dioxide, so concentrations of methane in the atmòsphere have remained low.

COMBUSTION IN THE CARBON CYCLE

Carbon dioxide is produced by the combustion of carbon compounds. Although it is a non-biological process because it is not carried out by living organisms, it occurs naturally in some ecosystems where lighting can set fire to forest or grassland. Biomass then burns, releasing carbon dioxide.

Over a million years ago, humans learned how to set fire to wood and control the process of combustion. During the industrial revolution, methods were developed for extracting coal, oil and gas and for generating energy from their combustion. This now releases large and increasing quantities of carbon dioxide into the atmosphere.

Carbon cycle

ATMOSPHERIC MONITORING

Air monitoring stations at various sites around the world measure concentrations of carbon dioxide, methane and other gases. The measurements are as accurate as possible so that reliable data is available to scientists.

CO_2 concentrations show an annual fluctuation. There is a drop from May to October and then a rise through to the next May. The drop is due to an excess of photosynthesis over cell respiration globally and vice versa for the rise. These changes follow northern hemisphere seasons, as the area of land is greater and CO_2 concentrations are greater on land than in the sea.

In addition to the annual fluctuation there is also a rising trend in atmospheric CO_2 concentrations due to human activities.

CARBON FLUXES IN THE CARBON CYCLE

It is not possible to measure global carbon fluxes precisely but scientists have produced estimates. These are based on many measurements in natural ecosystems and in mesocosms. Global fluxes are very large, so estimates are in gigatonnes. Ocean uptake is CO_2 from the atmosphere dissolving in sea water and ocean loss is the opposite.

Process	**Flux/gigatonnes year^{-1}**
Photosynthesis	−120
Cell respiration	+119.6
Ocean uptake	−92.2
Ocean loss	+90.6
Deforestation and land use changes	+1.6
Combustion of fossil fuels	+6.4

THE BIOGEOCHEMICAL CARBON CYCLE

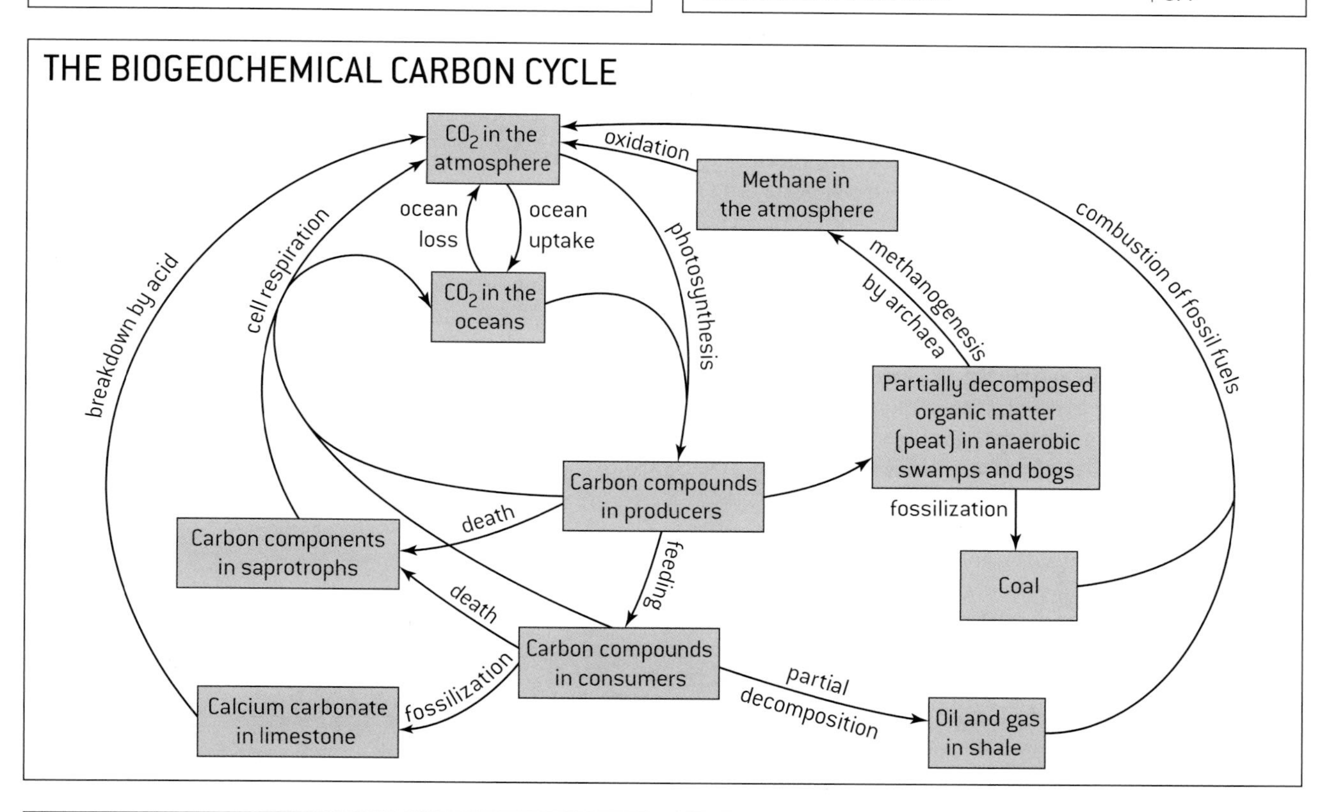

LIMESTONE IN THE CARBON CYCLE

Limestone is rock that consists mainly of calcium carbonate ($CaCO_3$). It often contains many fossils such as mollusc shells and skeletons of hard corals. These organisms absorb calcium and carbonate ions and secrete them as calcium carbonate. The shells of marine molluscs fall to the sea bed when they die and become part of limestone rock. Skeletons of hard corals accumulate over long periods of time gradually building coral reef, which consists of limestone. Huge amounts of carbon are locked up in limestone on Earth. This carbon can be released if the limestone reacts with acid. Rainwater contains carbonic acid (H_2CO_3) and can cause calcium carbonate to break down and release its carbon:

carbonic acid + calcium carbonate → Ca^{2+} (calcium ions) + HCO_3^- (hydrogen carbonate ions)

FORMATION OF FOSSIL FUELS

1. **Peat and coal**
   Saprotrophs cannot break down dead leaves and other organic matter in acidic and anaerobic conditions. These conditions are found in bogs and swamps, so partially decomposed plant matter accumulates to form thick deposits called **peat**. In past geological eras peat was crushed and converted into **coal**.
2. **Oil and gas**
   Silt is deposited on the bed of some shallow seas, together with remains of dead marine organisms. The organic matter is only partially decomposed because of anaerobic conditions. This process occurred in past geological areas. The silt on the sea bed was converted to shale, with compounds from the organic matter becoming oil or gas trapped in pores in the rock.

Global warming and the greenhouse effect

THE GREENHOUSE EFFECT

The Sun emits radiation and some of this reaches the Earth. The radiation is predominantly short wavelength. 25% of it is absorbed in the atmosphere, with ozone absorbing much of the ultraviolet. 75% of the solar radiation therefore reaches the Earth's surface, where most of it is absorbed and converted to heat. The surface of the Earth re-emits radiation, but at much longer wavelengths, mostly infrared (heat). A far higher percentage of this longer wavelength radiation is absorbed in the atmosphere before it has passed out to space. Between 70 and 85% is trapped by gases in the atmosphere. The gases re-emit the radiation and some of it passes back to the surface of the Earth, causing warming. It is called the greenhouse effect and the gases that trap the radiation are known as greenhouse gases.

Cause of the greenhouse effect

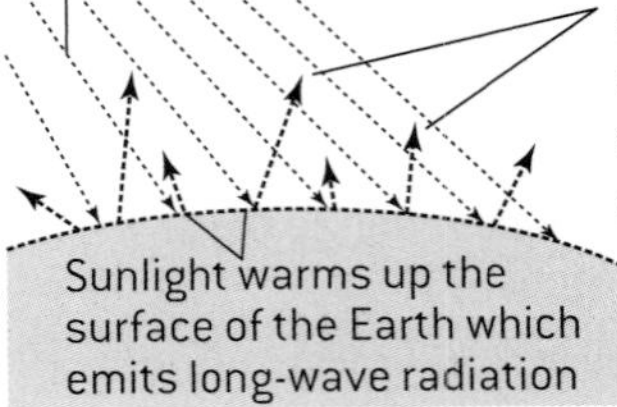

GREENHOUSE GASES

Only certain gases in the atmosphere have the ability to trap long-wave radiation and therefore act as a greenhouse gas. The impact of a gas depends both on its ability to absorb long-wave radiation and on its concentration in the atmosphere. For example, methane causes much more warming per molecule than carbon dioxide, but as it is at a much lower concentration in the atmosphere its impact on global warming is less. **Carbon dioxide** and **water vapour** are the most significant greenhouse gases. **Methane** and **nitrogen oxides** also have an effect, but it is smaller. It is important to note that stratospheric ozone is not a significant greenhouse gas. It intercepts much more incoming short-wave radiation than outgoing long-wave radiation, so ozone depletion does not therefore increase the greenhouse effect. The pie chart (right) shows the proportion of warming caused by carbon dioxide, methane, nitrous oxide and halocarbons produced by industry. Water vapour is not included, as its warming effects are difficult to assess.

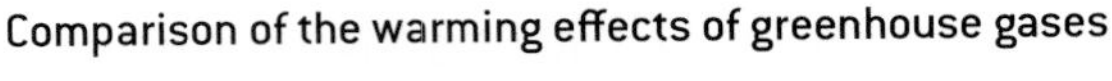
Comparison of the warming effects of greenhouse gases

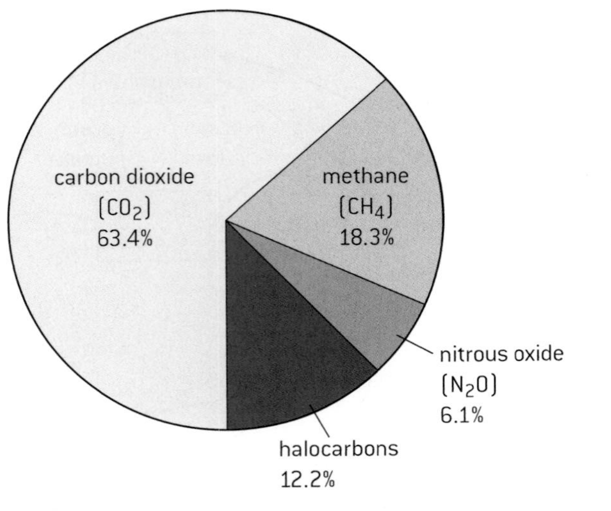

GLOBAL WARMING AND CLIMATE CHANGE

Physical scientists have calculated that without the greenhouse effect the mean temperature at the Earth's surface would be about −18 °C. The actual mean temperature is more than 30 °C higher, so it is beyond dispute that global temperatures and climate patterns are influenced by concentrations of greenhouse gases in the Earth's atmosphere. It is also clear that human activities are causing increases in the concentrations of carbon dioxide, methane and other greenhouse gases in the atmosphere. Furthermore, there is strong evidence that temperatures on Earth have increased over the last 200 years. The graph (right) shows mean global temperatures compared with the average from 1951–1980.

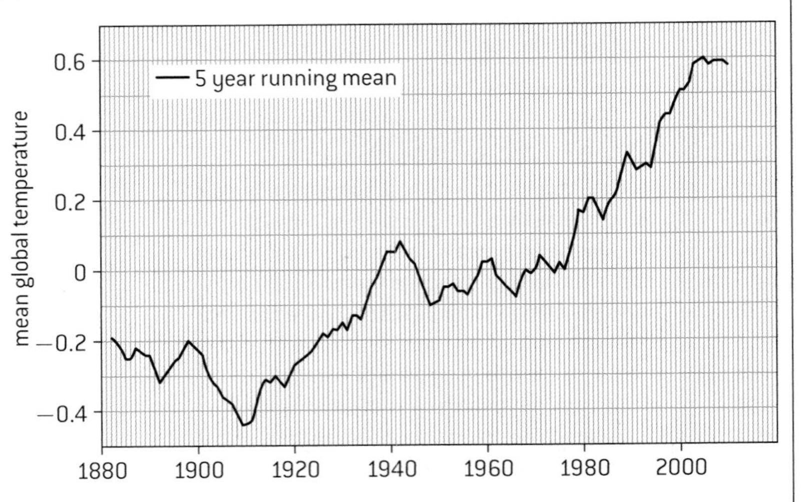

What is strongly disputed by some people is whether global warming and other climate changes are due to greenhouse gas emissions from human activity. This claim has been evaluated by many climate change scientists, who have almost all concluded that human activity is indeed influencing the global temperatures and climate patterns.

Rising carbon dioxide concentrations

GLOBAL TEMPERATURES AND CARBON DIOXIDE CONCENTRATIONS

To deduce CO_2 concentrations and temperatures in the past, columns of ice have been drilled in the Antarctic. The ice has built up over thousands of years, so ice from deeper down is older than ice near the surface. Bubbles of air trapped in the ice can be extracted and analysed to find the carbon dioxide concentration. Global temperatures can be deduced from ratios of hydrogen isotopes in the water molecules. The graph (below) shows results for an 800,000 year period before the present. During this period there has been a repeating pattern of rapid periods of warming followed by much longer periods of gradual cooling, which correlate very closely with changes in CO_2 concentrations.

It is important always to remember that correlation does not prove causation, but in this case we know from other research that carbon dioxide is a greenhouse gas. At least some of the temperature variation over the past 800,000 years must therefore have been due to rises and falls in atmospheric carbon dioxide concentrations.

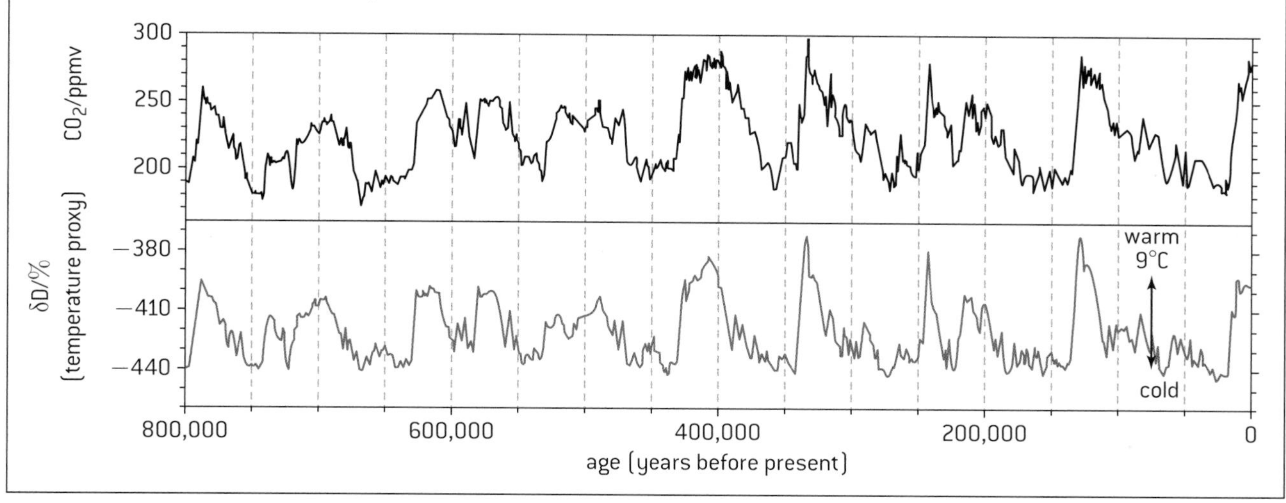

CARBON EMISSIONS AND GLOBAL WARMING

Over the last 150 years atmospheric carbon dioxide concentrations have risen above the range shown in the graph above. This is largely due to combustion of fossilized organic matter (coal, oil and gas). Since the start of the industrial revolution 200 years ago, both the burning of fossil fuels and the CO_2 concentration have increased faster and faster. Global temperatures have also increased over this period by about 0.8 °C, with most of the increase occurring since 1980 (see graph on previous page).

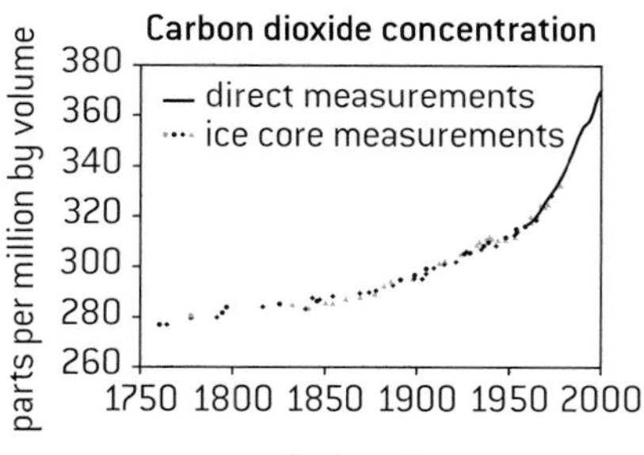

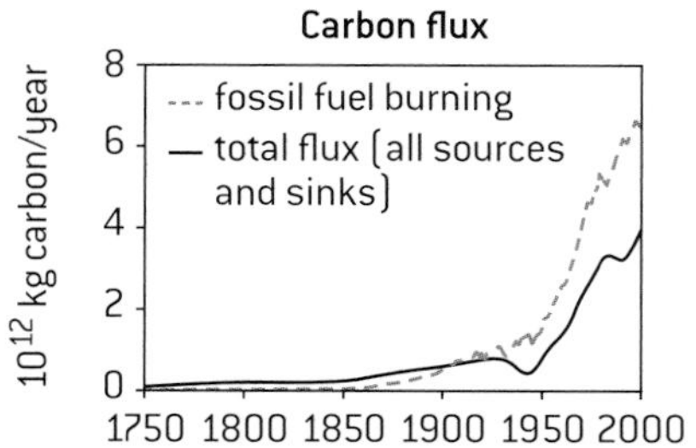

Carbon dioxide concentrations and global temperature are correlated but are not directly proportional as there are other variable factors that affect temperatures. As a result global warming is much more uneven year on year than rises in CO_2. There may be periods of slower temperature rise despite CO_2 increases but also periods of particularly rapid temperature increase.

CORAL REEFS AND CARBON DIOXIDE

In addition to its contribution to global warming, emissions of carbon dioxide are having effects on the oceans. Over 500 billion tonnes of CO_2 released by humans since the start of the industrial revolution have dissolved in the oceans. This has caused the pH to drop from about 8.25 to 8.14. This seemingly small change represents a 30% acidification. Ocean acidification will become more severe if the rise in the CO_2 concentration of the atmosphere continues.

Marine animals such as reef-building corals that deposit calcium carbonate in their skeletons (above) need to absorb carbonate ions from seawater. The concentration of carbonate ions in seawater is low, because they are not very soluble. Dissolved carbon dioxide makes the concentration even lower as a result of some interrelated chemical reactions. Carbon dioxide reacts with water to form carbonic acid, which dissociates into hydrogen ions and hydrogen carbonate ions. Hydrogen ions convert carbonate into hydrogen carbonate. With reduced carbonate concentrations in seawater not only can new calcium carbonate not be made, but it dissolves in existing corals, threatening the existence of all reef ecosystems.

Questions – ecology

In questions 1–5 choose answers from these nutritional classes.

A. autotrophs
B. consumers
C. detritivores
D. saprotrophs

1. *Penicillium* fungus growing on blue cheese secretes digestive enzymes into the cheese and then absorbs the products of digestion. What nutritional class is it in?

2. Cyanobacteria absorb carbon dioxide, mineral ions and light, and synthesize carbon compounds. What nutritional class are they in?

3. Marine iguanas have teeth and dive into the sea to eat algae. What nutritional class are they in?

4. Dung flies lay their eggs on feces and the larvae that hatch out from them ingest the feces. What nutritional class are they in?

5. Female *Anopheles* mosquitoes have piercing sucking mouthparts and can spread malaria when they feed on blood. What nutritional class are they in?

6. The diagram below shows in simplified form the transfers of energy in a generalized ecosystem.

   Each box represents a category of organisms, grouped together by their trophic position in the ecosystem.

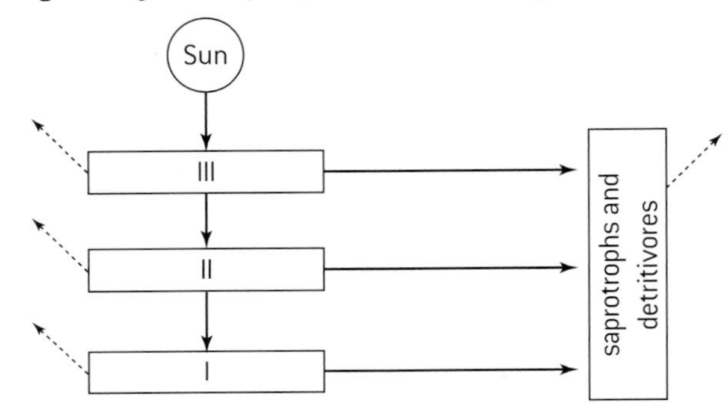

   a) Deduce trophic levels I, II and III. [3]
   b) State the form of energy entering box I. [1]
   c) Identify which arrow represents the greatest transfer of energy per unit of time. [1]
   d) Explain what the dotted arrows indicate. [3]

7. Methane acts as a greenhouse gas in the atmosphere. The main sources of methane are the digestive systems of cattle and sheep, activity of archaeans in swamps, marshes and rice paddies, burning of biomass (for example forest fires) and release of natural gas.

   a) Discuss whether methane emissions from these sources cause change in the Earth's temperature. [3]
   b) Discuss whether release of methane is a natural process or an example of a human impact on the environment. [3]
   c) Suggest measures that could be taken to reduce the emission of methane. [3]

8. The graph (top right) shows the Keeling Curve – the atmospheric carbon dioxide concentrations measured at Mauna Loa Observatory on Hawaii since 1958. Two curves are shown: the oscillating curve shows monthly average concentrations; the smoothed curve has been adjusted for seasonal variations.

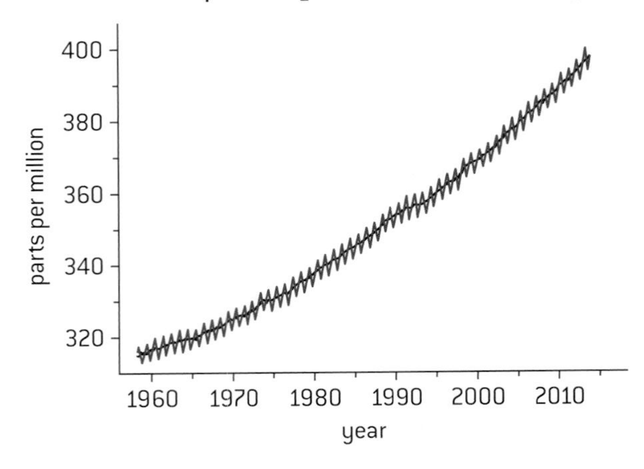

   a) Estimate the size of seasonal CO_2 fluctuations. [1]
   b) Explain the falls in CO_2 concentration from May to October and rises from November to April. [4]
   c) (i) Calculate the change in seasonally adjusted carbon dioxide concentration between the start of the data in March 1958 and the end in December 2013. [2]
      (ii) Calculate the mean increase in carbon dioxide concentration between 2000 and 2010. [2]
      (iii) Assuming this rate of increase continues, predict the carbon dioxide concentration in the year 2100. [3]
   d) Explain the increase in carbon dioxide concentration shown by the seasonally adjusted curve. [3]

9. Ten 17 m × 4 m plastic bags were used to create mesocosms in a fjord on the coast of Sweden. The mesocosms contained 55 m^3 of water and were open to the atmosphere at the top. Acidified seawater was added to five of them to simulate an atmospheric CO_2 concentration of 1000 ppm. The numbers of *Vibrio* bacteria in samples of water from the mesocosms were measured from March to May. Some *Vibrio* species are pathogens of fish and of humans. Maximum numbers reached during the trial period in each mesocosm are shown in the table below. The maximum in the open water of the fjord was 0.14 cm^{-3}. Mesocosms treated with acidified seawater are shaded grey.

Vibrio bacteria (cm^{-3} seawater)									
M1	M2	M3	M4	M5	M6	M7	M8	M9	M10
0.60	0.27	0.22	0.88	0.07	1.20	1.71	0.26	0.10	1.89

   a) Calculate the mean numbers of *Vibrio* bacteria in acidified and in control mesocosms. [4]
   b) Discuss evidence from the data for increased CO_2 concentration causing increased *Vibrio* numbers. [4]
   c) The standard error of the difference between the two means was 0.4406. When this value is used in a statistical test (t-test) the conclusion is that the difference is not significant at the 5% level. Explain the reasons for this and the implications. [3]
   d) Outline the potentially harmful consequences of ocean acidification, apart from increases in *Vibrio*. [4]

Introducing evolution

EVOLUTION IN BIOLOGY

The word 'evolution' has several meanings, all of which involve the gradual development of something. In biology, the word has come to mean the changes that occur in living organisms over many generations. Evolution happens in populations of living organisms. It only happens with characteristics that can be inherited. This is a useful summary:

Evolution occurs when heritable characteristics of a species change.

Although it is not possible to prove that organisms on Earth are the result of evolution, there is very strong evidence for this theory. One example of evolution is described on this page and four types of evidence for evolution are explained on this page and the next.

SPECIATION AND PATTERNS OF VARIATION

Populations of a species sometimes become separated and therefore unable to breed with each other. They are then able to evolve differently and diverge in their characteristics more and more. The change in the populations may be very gradual and take place over thousands of years or even longer, but eventually they are so different that they would not be able to interbreed even if they inhabited the same area again. The populations have therefore evolved into separate species. When taxonomists try to classify living organisms into species there is often much argument about whether populations in different geographical areas are part of the same species or are different species. This is because there is continuous variation in the amount of difference between populations from slight to very great. This is expected if populations gradually diverge by evolution to become separate species. It does not fit in with the idea of distinct species being created and not evolving.

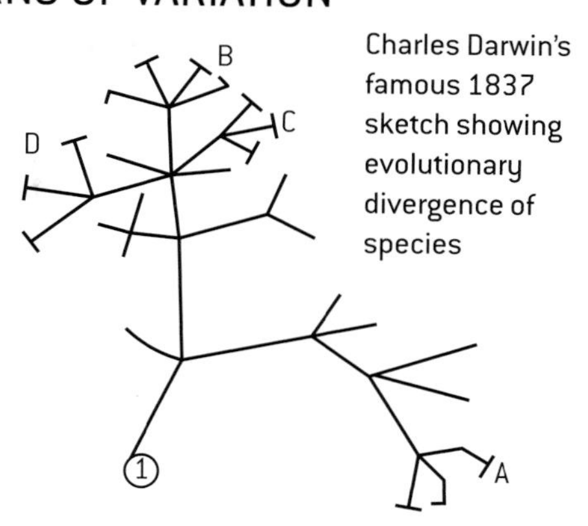

Charles Darwin's famous 1837 sketch showing evolutionary divergence of species

PENTADACTYL LIMBS AS EVIDENCE FOR EVOLUTION

Four groups of vertebrates have limbs: amphibians, reptiles, birds and mammals. These vertebrates use their limbs in a wide variety of ways, including walking, jumping, swimming, climbing and digging. Despite this, the basic bone structure is the same in all of them. The structure is known as the **pentadactyl limb**. The most plausible explanation is that all these vertebrates share an ancestor that had pentadactyl limbs. Many different groups have evolved from the common ancestor, but because they adopted different types of locomotion, the limbs developed in widely different ways, to suit the type of locomotion. This type of evolution is called **adaptive radiation**.

Structures like the pentadactyl limb that have evolved from the same part of a common ancestor are called **homologous structures**. They have similarities of structure despite the differences in their function, which would be difficult to explain in any way apart from evolution.

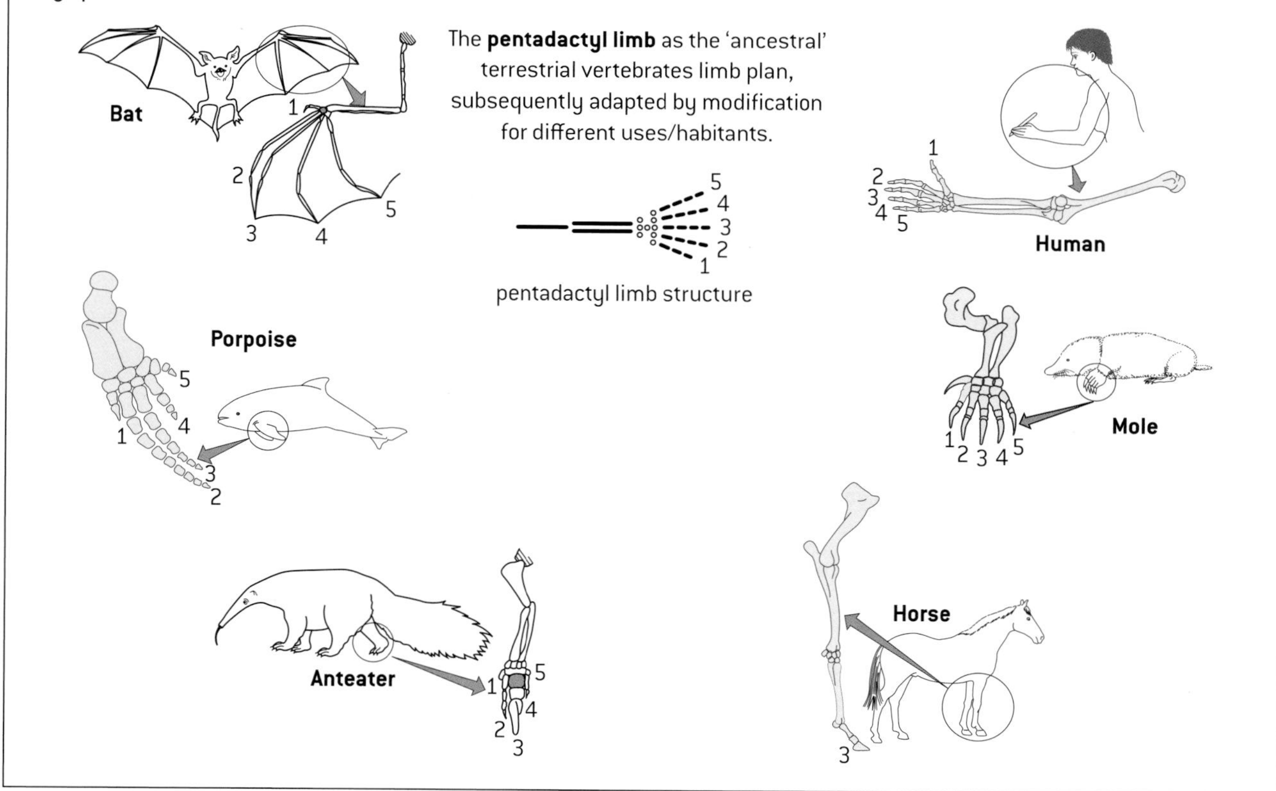

The **pentadactyl limb** as the 'ancestral' terrestrial vertebrates limb plan, subsequently adapted by modification for different uses/habitants.

Further evidence for evolution

SELECTIVE BREEDING OF DOMESTICATED ANIMALS

The breeds of animal that are reared for human use are clearly related to wild species and in many cases can still interbreed with them. These domesticated breeds have been developed from wild species, by selecting individuals with desirable traits, and breeding from them. This is known as **selective breeding**. The striking differences in the heritable characteristics of domesticated breeds give us evidence that species can evolve rapidly by artificial selection. The figure (right) shows a domesticated pig and its ancestor the wild boar from Darwin's *Animals and Plants Under Domestication*.

THE FOSSIL RECORD

Research into fossils has given us strong evidence for evolution. Fossils have been discovered of many types of organism that no longer exist, including trilobites and dinosaurs, and in most cases no fossils can be found of organisms that do exist today, suggesting that organisms change over time. Rocks can be dated, allowing the age of fossils within the rocks to be deduced and the times when those organisms lived on Earth. The sequence in which organisms appear in the fossil record matches their complexity, with bacteria and simple algae appearing first, fungi and worms later and land vertebrates later still. Many sequences of fossils are known, which link together existing organisms with their likely ancestors. For example, *Acanthostega* (above) is a 365-million-year-old fossil that has similarities to other vertebrates, but it has eight fingers and seven toes, so it is not identical to any existing organism. *Acanthostega* has four legs like most amphibians, reptiles and mammals but it also had gills and a fish-like tail and it lived in water. This shows that land vertebrates could have evolved from fish via an aquatic animal with legs.

Fossil of *Acanthostega*

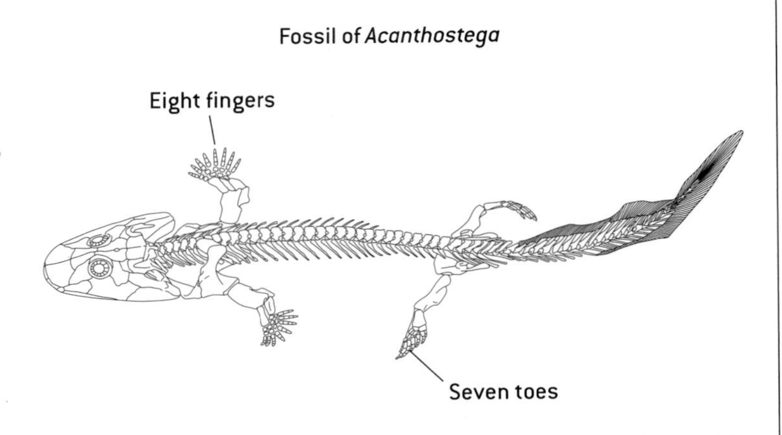

MELANISM – AN EXAMPLE OF EVOLUTION

Dark varieties of typically light coloured insects are called **melanistic**. The most famous example is *Biston betularia* (the peppered moth). The melanistic variety of this moth was originally very rare but in areas where industry developed in the 19th century in England and elsewhere it became much commoner, with the peppered form becoming much rarer. This is a change in the heritable characteristics of the species so it is an example of evolution. A simple explanation of industrial melanism is this:

- Adult *Biston betularia* moths fly at night to try to find a mate and reproduce.
- During the day they roost on the branches of trees.
- In unpolluted areas tree branches are covered in pale-coloured lichens, so peppered moths are well camouflaged against them.
- Sulphur dioxide pollution kills lichens on tree branches and soot from coal burning blackens them.
- Melanic moths are well camouflaged against dark tree branches.
- Birds and other animals that hunt in daylight predate moths if they find them.
- In areas that became industrial the peppered variety were mostly found and eaten and a higher proportion of the melanic variety survived to breed and pass on the dark wing colour, causing *Biston betularia* populations to evolve from being peppered to melanic.

Natural selection

THE DISCOVERY OF NATURAL SELECTION

There has been much discussion about which biologist was the first to realize that species evolve by natural selection. Charles Darwin is usually given credit, because he developed the theory and was the first to publish a detailed account of it.

Darwin did more than 20 years of research and, during this time, he amassed a wide range of evidence for natural selection. He delayed publication of his ideas for many years, fearing a hostile reaction, and might never have published them if another biologist, Alfred Wallace, had not written a letter to him in 1858 suggesting very similar ideas. Darwin then quickly wrote his pioneering work *The Origin of Species* and it was published in 1859. This book changed for ever the way that biologists think about the living world and the place of humans in it.

SOURCES OF VARIATION

Natural selection can only occur if there is variation among members of a species. There are three sources of variation:

1. **Mutation** is the original source of variation. New alleles are produced by gene mutation, which enlarges the gene pool of a population.
2. **Meiosis** produces new combinations of alleles by breaking up existing combinations in a diploid cell. Every cell produced by meiosis in an individual is likely to carry a different combination of alleles, because of crossing-over and the independent orientation of bivalents.
3. **Sexual reproduction** involves the fusion of male and female gametes. The gametes usually come from different parents, so the offspring has a combination of alleles from two individuals. This contributes to variation in a species.

EXPLAINING NATURAL SELECTION

The theory of evolution by natural selection can be explained in a series of observations and deductions:

1. Species tend to produce more offspring than the environment can support.
2. There is a struggle for existence in which some individuals survive and some die.
3. In natural populations there is variation between the individuals.
4. Some individuals are better adapted than others. An adaptation is a characteristic that makes an individual suited to its environment and way of life.
5. Individuals that are better adapted tend to survive and produce more offspring, while less well adapted individuals tend to die or produce fewer offspring so each generation contains more offspring of better adapted than less well adapted individuals.
6. Individuals that reproduce pass on characteristics to their offspring.
7. The frequency of characteristics that make individuals better adapted increases and the frequencies of other characteristics decrease, so species change and become better adapted.

Darwin aged 51, when he published *The Origin of Species*

HERITABILITY AND EVOLUTION

Living organisms acquire characteristics during their lifetimes, but such characteristics are not heritable so they are lost when the individual dies. Because acquired characteristics are not inherited by offspring they cannot increase in a species by natural selection. The arms of tennis players are an example of this. The muscles and bones of the dominant arms of tennis players increase in size as a result of being used intensively. Right-handed tennis players develop larger muscles and bones in their right arm than in their left arm, but because the genes that influence the size of the muscles and bones have not been altered, the left and right arms of the child of a right-handed tennis player are equal in size.

Natural selection in action

BEAKS OF FINCHES ON DAPHNE MAJOR

Daphne Major is a small island in the Galápagos archipelago. On this island there is a population of *Geospiza fortis* (medium ground finch), which feeds on seeds with a wide range of sizes, from small to large. The large seeds are harder and therefore more difficult to crack open. There is variation in the size of beaks, with some individuals having larger beaks than others. Beak size is a heritable characteristic. Every year from 1973 onwards *G. fortis* finches have been trapped on Daphne Major so their beak sizes can be measured. Changes in both the mean length and width of the beaks have been observed.

The climate of the Galápagos archipelago is very variable because of an oscillation between warm ocean temperatures (El Niño), which bring heavy rains, and cold temperatures (La Niña), which bring droughts. During droughts there are few small soft seeds available, but larger hard seeds are still produced.

From 1974 to 1977 La Niña conditions were experienced on Daphne Major, ending with a severe drought that resulted in the *G. fortis* population dropping from 1,300 to 300. The mean beak size of the finches that died during the drought was significantly smaller than the beak size of those that survived. This natural selection occurred because finches with large beaks are better adapted to feeding on large seeds. When the population started breeding again after the drought, the mean beak size remained raised because offspring produced inherited the larger beak size of their survivor parents. In 1983 there was a strong El Niño event with heavy rain and abundant supplies of small soft seeds. In the years that followed mean beak size dropped, as small beak sizes are better adapted to feeding on small seeds.

(a) *G. fortis (large beak)*

(b) *G. fortis (small beak)*

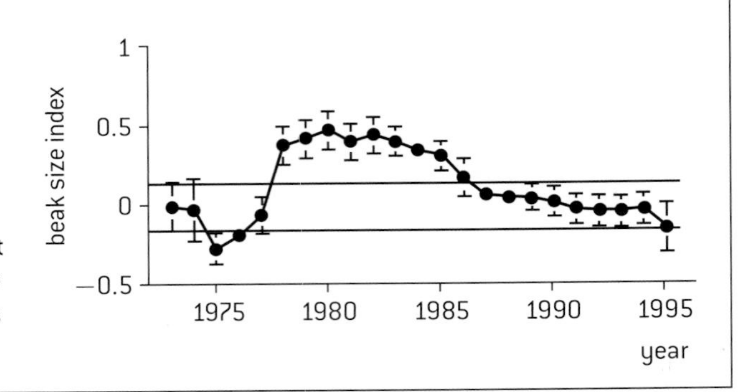

ANTIBIOTIC RESISTANCE IN BACTERIA

Antibiotics are used to control diseases caused by bacteria in humans. There have been increasing problems with disease-causing (pathogenic) bacteria being resistant to antibiotics. The graph (right) shows the percentage of pathogenic *E. coli* bacteria that were resistant to the antibiotic ciprofloxacin between 1990 and 2004. The trend with many other diseases has been similar.

The theory of evolution by natural selection can explain the development of antibiotic resistance in bacteria.

- Genes that give resistance to an antibiotic occur in the microorganisms that naturally make that antibiotic.
- These antibiotic resistance genes can be transferred to a bacterium by means of a plasmid or in some other way. There is then variation in this type of bacterium – some of them are resistant to the antibiotic and some are not.
- If doctors or vets use the antibiotic to control bacteria it will kill bacteria that are susceptible to the antibiotic, but not those that are resistant. This is an example of natural selection, even though it is caused by humans using antibiotics.
- The antibiotic-resistant bacteria reproduce and pass on the resistance gene to their offspring. These bacteria spread from person to person by cross-infection.
- The more an antibiotic is used, the more bacteria resistant to it there will be and the fewer that are non-resistant. As a result of excessive use of an antibiotic, most of the bacteria may eventually be resistant.

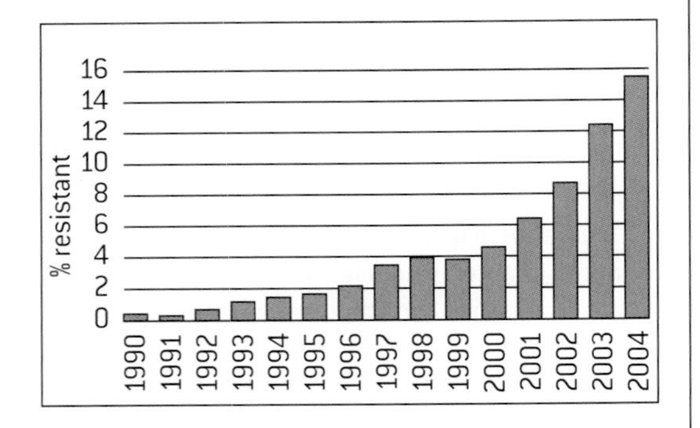

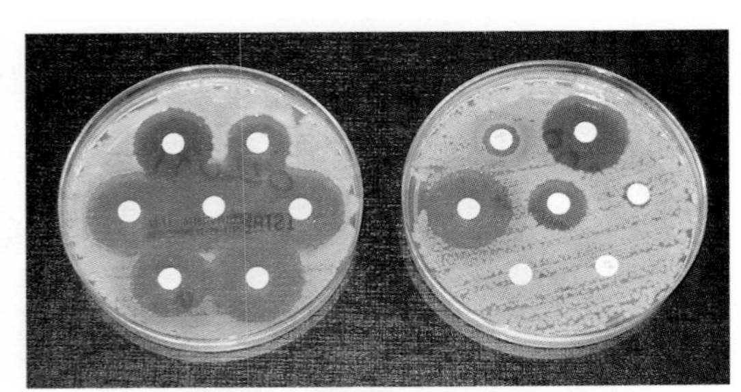

The two plates (above) were inoculated with different bacteria. Each disc contains a different antibiotic. The bacterium on the left plate was killed by all the antibiotics but the one on the right plate was resistant to four of the antibiotics.

Naming and identifying

NAMING SPECIES

When species are discovered they are given scientific names using the **binomial system**. This system is universal among biologists and has been agreed and developed at a series of international congresses. It avoids the confusion that would result from using the many different local names that can exist for a species. The binomial system is a very good example of cooperation and collaboration between groups of scientists. The binomial system has these features:

- The first name is the genus name. A genus is a group of closely related species.
- The genus name is given an upper case first letter.
- The second name is the species name.
- The species name is given a lower case first letter.
- Italics are used when a binomial appears in a printed or typed document.

Examples of binomials:

Homo sapiens – humans

Scrophularia landroveri – a plant discovered by an expedition that travelled in a Land Rover.

USING A DICHOTOMOUS KEY

Many aquatic plants in aquariums in biology laboratories belong to one of these four genera:

- *Cabomba*
- *Elodea*
- *Ceratophyllum*
- *Myriophyllum*

All of these plants have cylindrical stems with whorls of leaves. The shape of four leaves is shown in the figure (below). A key can be used to identify which of the four genera a plant belongs to, if it is known to be in one of them.

1. Simple undivided leaves .... *Elodea*
   Leaves forked or divided into segments 2
2. Leaves forked once or twice to form two or three segments .... *Ceratophyllum*
   Leaves divided into more than four segments 3
3. Leaves divided up into many flattened segments .... *Cabomba*
   Leaves divided into many filamentous segments .... *Myriophyllum*

Leaves of aquarium plants

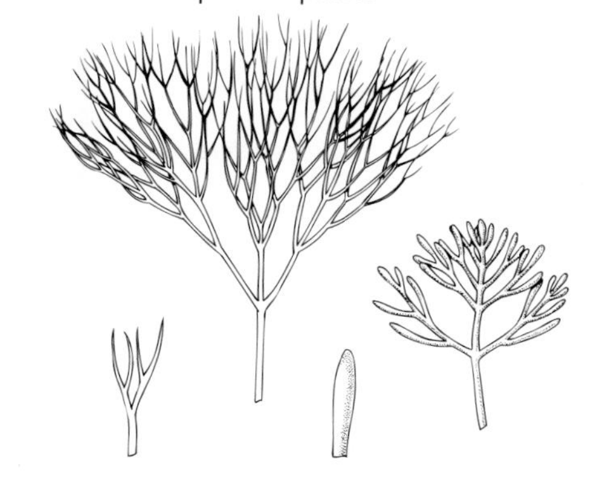

IDENTIFYING SPECIES

The first stage in many ecological investigations is to find out what species of organism there are in the area being studied. This is called **species identification**. This is done using **dichotomous keys**, which have these features:

- The key consists of a series of numbered stages.
- Each stage consists of a pair of alternative characteristics.
- Some alternatives give the next numbered stage of the key to go to.
- Eventually the identification of the species will be reached.

An example of a dichotomous key for identifying aquarium pondweeds is shown below left.

CONSTRUCTING A DICHOTOMOUS KEY

The five animals shown below are found in beehives. It would be useful to construct a dichotomous key to allow a beekeeper to identify them, as some of them are very harmful and others are harmless to honey bees.

The most useful keys use characteristics that are easy to observe and are reliable, because they are present in every member of the species.

Galleria mellonella is a species of moth and has three pairs of legs.

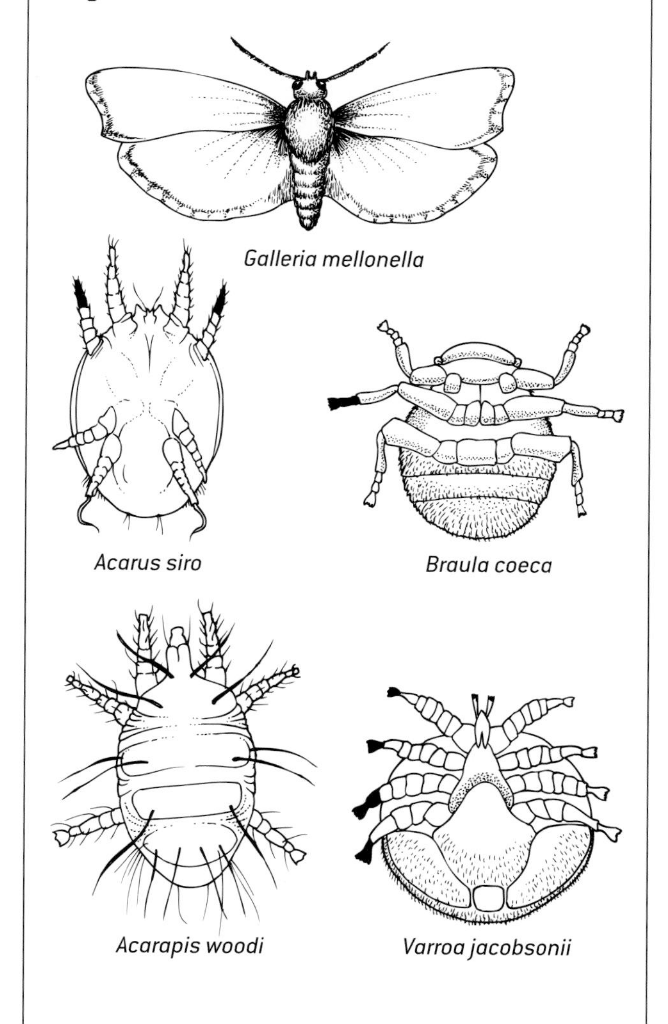

Classification of biodiversity

NATURAL CLASSIFICATION

Classification in biology is the process of putting species of living organisms into groups. It is an essential process because there are so many species and without a classification system it would be very hard to store and access information about them. The biologists that specialize in classification are **taxonomists**.

There are many possible ways of classifying organisms. For example, animals could be classified into two groups: those that have wings and those that do not. This would be an **artificial classification** because animals with wings (insects, birds and bats) are not similar in enough other ways to be grouped together. Even their wings are different in structure because they are not homologous and evolved separately.

Biologists try to devise a **natural classification**, in which the species in a group have evolved from one common ancestral species. The species in the group therefore share many characteristics that they have inherited from the ancestral species.

Natural classifications therefore allow the **prediction of characteristics** shared by species within a group. They also help in **identification of species**.

It is not always obvious what the pattern of evolution was in a group of species and therefore what the natural classification of them is. Taxonomists sometimes reclassify groups of species when new evidence shows that a previous taxon contains species that have evolved from different ancestral species. For example, a system of classification of living organisms into **five kingdoms** was developed in the second half of the 20th century. Biologists mostly accepted it. In this classification, all prokaryotes were placed in one kingdom and eukaryotes in four kingdoms. However, when the base sequence of nucleic acids was compared, two very different groups of prokaryotes were identified. These groups are as different from each other as from eukaryotes. A higher grade of taxonomic group was needed to reflect this, now called a **domain**.

THREE DOMAINS

Currently all organisms are classified into three domains.

- Archaea (referred to as archaeans)
- Eubacteria (referred to as bacteria)
- Eukaryota (referred to as eukaryotes)

The original evidence for this came from base sequences of ribosomal RNA, which is found in all organisms and evolves slowly, so it is suitable for studying the earliest evolutionary events.

The tree diagram below represents the likely evolution of a sample of species, based on ribosomal RNA sequences. It suggests that prokaryotes diverged into Eubacteria and Archaea early in the evolution of life, so it is not appropriate to classify them together in one kingdom.

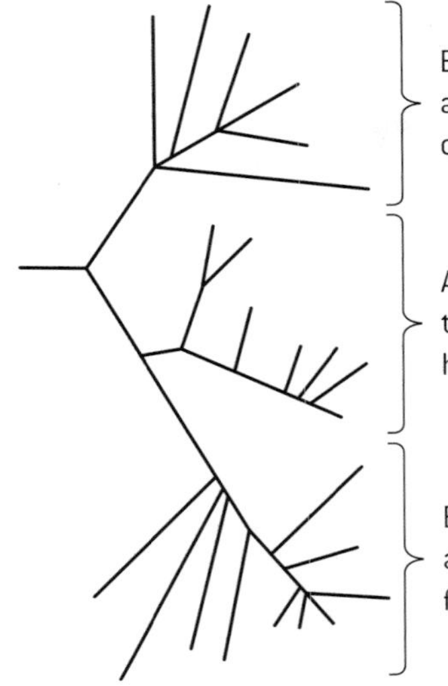

Note that viruses are not classified into any of the domains or into a domain of their own, as they are not considered by biologists to be living organisms.

CLASSIFICATION FROM SPECIES TO DOMAIN

A group of organisms, such as a species or a genus, is called a **taxon**. Species are classified into a series of taxa, each of which includes a wider range of species than the previous one. This is called the **hierarchy of taxa**.

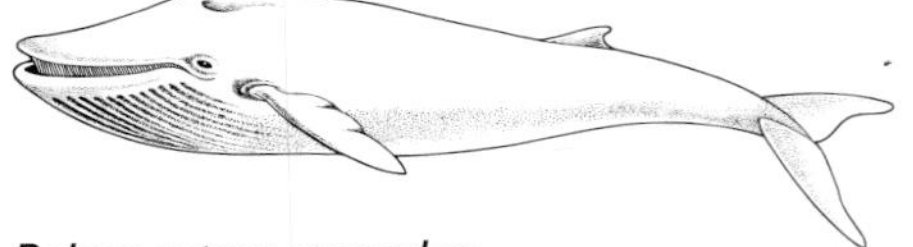

Balaenoptera musculus

	Animal example	Plant example
	Balaenoptera musculus – the blue whale (left)	*Sequoiadendron giganteum* – the giant redwood (right)
Species that are similar are grouped into a **genus**	Genus *Balaenoptera*	Genus *Sequoia*
Genera that are similar are grouped into a **family**	Family *Balaenopteridae*	Family *Cupressaceae*
Families that are similar are grouped into an **order**	Order *Cetacea*	Order *Pinales*
Orders that are similar are grouped into a **class**	Class *Mammalia*	Class *Pinopsida*
Classes that are similar are grouped into a **phylum**	Phylum *Chordata*	Phylum *Coniferophyta*
Phyla that are similar are grouped into a **kingdom**	Kingdom *Animalia*	Kingdom *Plantae*
Kingdoms that are similar are grouped into a **domain**	Domain *Eukaryota*	Domain *Eukaryota*

Sequoiadendron giganteum

Classification of eukaryotes

RECOGNITION FEATURES OF PLANTS

There are four main phyla of plants, which can be easily distinguished by studying their external structure.

	Roots, stems and leaves	Reproductive structures
Bryophytes (mosses)	Structures similar to root hairs called rhizoids but no roots. Mosses have simple leaves and stems; liverworts have a flattened thallus. No vascular tissue.	Spores are produced in a capsule. The capsule develops at the end of a stalk.
Filicinophytes (ferns)	Roots, leaves and short non-woody stems. Leaves curled up in bud and often divided into pairs of leaflets (pinnate). Vascular tissue is present.	Spores are produced in sporangia, usually on the underside of the leaves.
Coniferophytes (conifers)	Shrubs or trees with roots, leaves and woody stems. Leaves often narrow with a thick waxy cuticle. Vascular tissue is present.	Seeds, which develop from ovules on the surface of the scales of female cones. Male cones produce pollen.
Angiospermophytes (flowering plants)	Very variable, but usually have roots, leaves and stems. Stems of shrubs and trees are woody. Vascular tissue is present.	Seeds, which develop from ovules inside ovaries in flowers. Fruits develop from the ovaries, to disperse the seeds.

RECOGNITION FEATURES OF ANIMALS

There are over 30 phyla of animals. Recognition features of seven large phyla are shown here:

Porifera

- no clear symmetry
- attached to a surface
- pores through body
- no mouth or anus

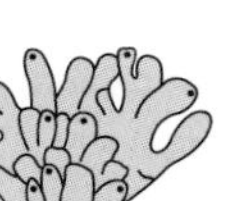

Cnidaria

- radially symmetric
- tentacles
- stinging cells
- mouth but no anus

Platyhelminths

- bilaterally symmetric
- flat bodies
- unsegmented
- mouth but no anus

Annelida

- bilaterally symmetric
- bristles often present
- segmented
- mouth and anus

Arthropoda

- bilaterally symmetric
- exoskeleton
- segmented
- jointed appendages

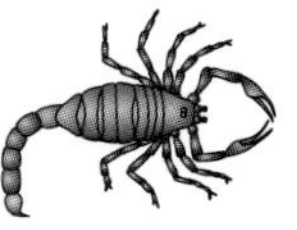

Mollusca

- muscular foot and mantle
- shell usually present
- segmentation not visible
- mouth and anus

Chordata

- notochord
- dorsal nerve cord
- pharyngeal gill slits
- post-anal tail

RECOGNITION FEATURES OF VERTEBRATES → kindom annimalia

Almost all chordates have a backbone consisting of vertebrae. Apart from fish, all these vertebrates are tetrapods with pentadactyl limbs, though in some species the limbs have become modified or lost through evolution. Recognition features of the five major classes of vertebrate are shown here:

Bony ray-finned fish

- Scales grow from the skin
- Gills with a single gill slit
- Fins supported by rays
- Swim bladder for buoyancy
- External fertilization

Amphibians

- Soft moist permeable skin
- Lungs with small internal folds
- External fertilization in water
- Protective gel around eggs
- Larval stage lives in water

Reptiles

- Dry scaly impermeable skin
- Lungs with extensive folding
- Internal fertilization
- Soft shells around eggs
- One type of teeth

Birds

- Feathers growing from skin
- Lungs with parabronchial tubes
- Wings instead of front legs
- Hard shells around the eggs
- Beak but no teeth

Mammals

- Hairs growing from the skin
- Lungs with alveoli
- Give birth to live young
- Mammary glands secrete milk
- Teeth of different types

Cladistics

CLADES

A **clade** is a group of organisms that evolved from a common ancestor. Clades can be large groups, with a common ancestor far back in evolution, or smaller groups with a more recent common ancestor. Evidence for which species are part of a clade can be obtained by looking at any characteristics, but anatomical features are now rarely used because it is sometimes hard to distinguish between homologous traits that were derived from a common ancestor and analogous characteristics that have developed by convergent evolution. Instead the base sequences of a gene are used, or the corresponding amino acid sequence of a protein. Sequence differences accumulate gradually, so there is a positive correlation between the number of differences between two species and the time since they diverged from a common ancestor.

CLADOGRAMS

Cladograms are tree diagrams that show the most probable sequence of divergence in clades. The term 'clade' is derived from the Greek word *klados* – a branch. On cladograms there are branching points (nodes) that show groups of organisms which are related, and therefore presumably had common ancestry.

Towards the end of the 20th century, both the amount of base and amino acid sequence data and the analytical power of computers grew exponentially. Cladograms could therefore be produced showing the probable evolutionary relationships of large groups of species. These cladograms have been used to re-evaluate the classification of many groups of organisms. The procedures used are very different from those previously used, so a new name has been given to this method of classification – **cladistics**.

CLADISTICS AND HUMAN CLASSIFICATION

Classifying humans is particularly difficult because the differences between humans and other species seem so huge to us. Cladistics can be used to produce an objective classification. For example, mitochondrial DNA from three humans and four related primates has been completely sequenced and used to construct a cladogram (below). Using the numbers of differences in base sequence as an evolutionary clock, the approximate dates for splits between groups are: 5 million years ago, human–chimpanzees split; 140,000 years ago, African–European/Japanese split; 70,000 years ago, Europeans–Japanese split.

Phylogenetic tree for humans and closely related apes

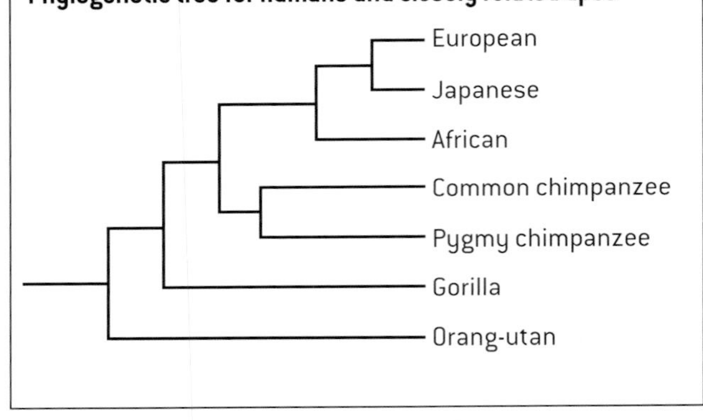

RECLASSIFICATION OF FIGWORTS

Evidence from cladistics has shown that classifications of some groups based on structure did not correspond with the evolutionary origins of a group or species. The figwort family of plants (*Scrophulariaceae*) is a good example of this. Cladograms showed that species in the family did not all share a recent common ancestor. Some genera have therefore been moved to the plantain and broomrape families and other genera have been transferred to two newly created families – the lindernia and calceolaria families. Two existing families, the buddleja and myoporum families were found to contain species that shared common ancestry with the figwort family so they were merged with it. This is not the end of research into the classification of these plants and just as the traditional classification was falsified and replaced as a result of cladistics, evidence may be discovered that shows further reclassification is needed.

Scrophularia neesii

ANALYSIS OF CLADOGRAMS TO DEDUCE EVOLUTIONARY RELATIONSHIPS

Cladograms can be analysed to find out how closely organisms are related to each other. They can also indicate the probable sequence in which groups split. The cladogram (below) was constructed using the number of differences in the amino acid sequence of hemoglobin. The scale above the cladogram shows the percentage difference in amino acid sequence and this has been used to add an estimated time scale below.

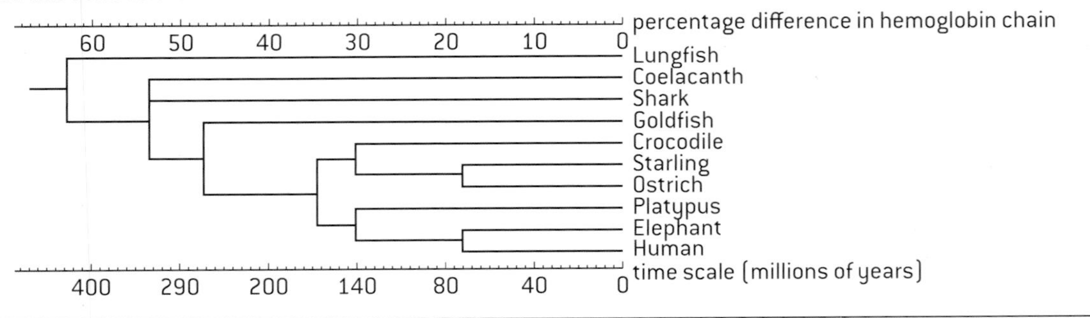

Questions – evolution and biodiversity

1. Are the pairs of traits analogous or homologous?

	Leg of ostrich and leg of orang-utan	Wing of hoverfly and wing of hummingbird
A	analogous	analogous
B	analogous	homologous
C	homologous	analogous
D	homologous	homologous

2. What is the cause of variation in animal species?
   A. mutation and meiosis only
   B. mutation and sexual reproduction only
   C. meiosis and sexual reproduction only
   D. mutation, meiosis and sexual reproduction

3. What is a feature that coniferophytes have but not filicinophytes?
   A. flowers
   B. leaves
   C. seeds
   D. vascular tissue

4. What is most help for identifying species?
   A. Cladograms
   B. Dichotomous keys
   C. Zoos and botanic gardens
   D. Natural selection

5. What causes increases in numbers of melanistic insects in polluted areas?
   A. Mutations caused by the pollution
   B. Predators eating more non-melanistic insects
   C. Pollution building up in the bodies of insects
   D. Melanistic insects using pollutants to camouflage themselves to prevent predator attacks

6. Identify the phylum in which each of the animals below is classified.

a) 

b) 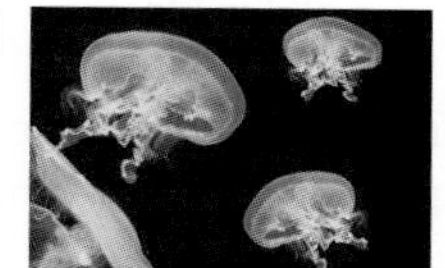

c) 

d) 

e) 

f) 

7. In the current system of classification, all living organisms are placed in one of three major groups.
   a) (i) State the name of this type of group in classification. [1]
      (ii) Eukaryotes are one of the major groups. State the names of the other two. [2]
   b) Explain the reasons for viruses not being placed in any of the three major groups. [2]
   c) State the seven levels in the hierarchy of taxa used to classify eukaryotes. [7]
   d) Some structures have evolved differently in related organisms and are now used for different functions. State the name used for this type of evolution. [1]
   e) Outline evidence for differences between organisms in the past and those alive today. [2]

8. The graph below shows the percentage of three infections that were resistant to fluoroquinolone in hospitals in Spain. During this period the percentage of patients treated with the antibiotic increased steadily from 5.8 to 10.2.

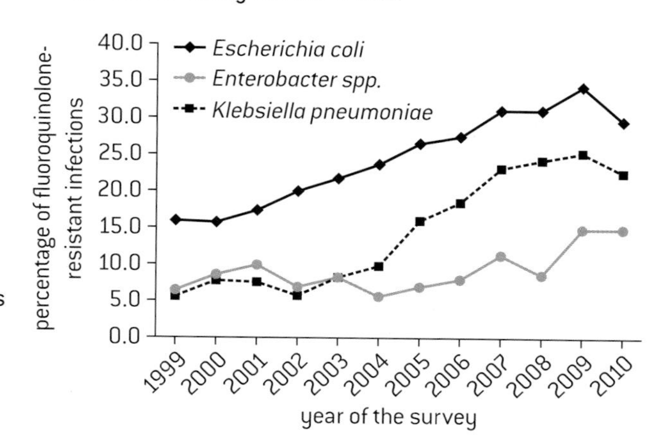

   a) Compare and contrast the trends in fluoroquinolone resistance in the three types of bacteria. [3]
   b) Suggest two ways in which resistance to an antibiotic can appear in a type of bacterium. [2]
   c) Explain the mechanism that causes the percentage of infections that are resistant to an antibiotic to rise. [3]
   d) Predict the consequences of a continued rise in the percentage of patients treated with fluoroquinolone. [2]

9. The figure below shows the base sequence of part of a hemoglobin gene in four species of mammal.

Human	TGACAAGAACA–GTTAGAG–TGTCCGA
Orang-utan	TCACGAGAACA–GTTAGAG–TGTCCGA
Lemur	TAACGATAACAGGATAGAG–TATCTGA
Rabbit	TGGTGATAACAAGACAGAGATATCCGA

   a) Determine the number of differences between the base sequences of:
      (i) humans and orang-utans
      (ii) humans and lemurs
      (iii) humans and rabbits
      (iv) orang-utans and lemurs
      (v) orang-utans and rabbits
      (vi) lemurs and rabbits. [6]
   b) Using the differences in base sequence between the four mammal species, construct a cladogram. [4]
   c) Deduce the evolutionary relationships of these four species from your cladogram. [2]

Digestion

THE HUMAN DIGESTIVE SYSTEM

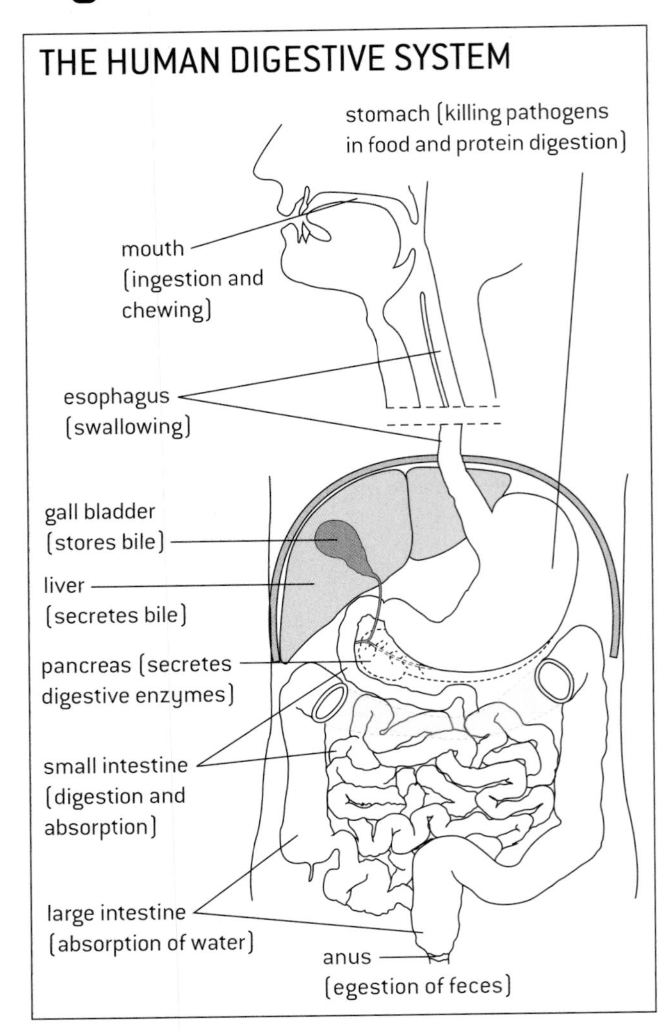

STRUCTURE OF THE SMALL INTESTINE

The micrograph below shows part of the wall of the small intestine. The diagram below it allows the tissue layers in the micrograph to be identified.

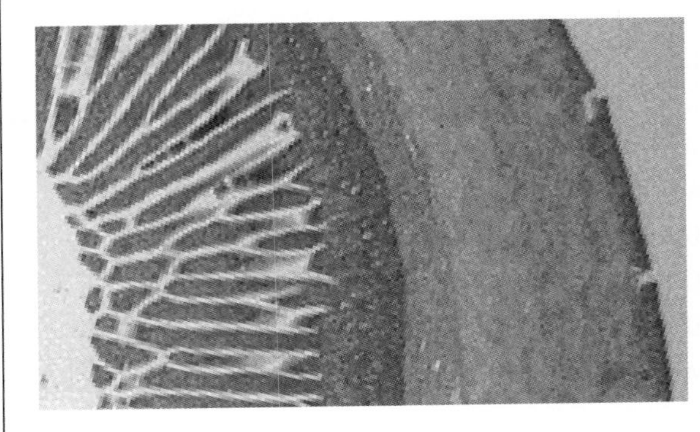

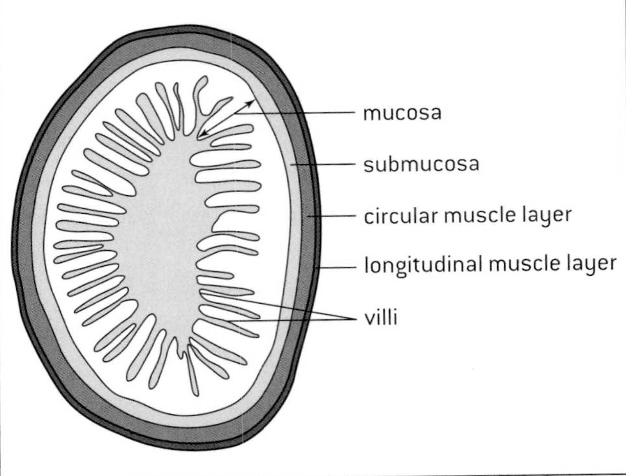

DIGESTION IN THE SMALL INTESTINE

Waves of muscle contraction, called **peristalsis**, pass along the intestine. Contraction of circular muscle behind the food constricts the gut to prevent food from being pushed back towards the mouth.

Contraction of longitudinal muscle where the food is located moves it on along the gut. Contraction of both layers of muscle mixes food with enzymes in the small intestine.

Enzymes digest most macromolecules in food into monomers in the small intestine. These macromolecules include proteins, starch, glycogen, lipids and nucleic acids. Cellulose remains undigested. The pancreas secretes three types of enzyme into the lumen of the small intestine:

lipids (fats and oils) $\xrightarrow{\textbf{lipase}}$ fatty acids + glycerol

polypeptides $\xrightarrow{\textbf{endopeptidase}}$ shorter peptides

starch $\xrightarrow{\textbf{amylase}}$ maltose

The details of starch digestion in the small intestine are explained (right).

DIGESTION OF STARCH

There are two types of molecule in starch: **amylose** and **amylopectin**. They are both polymers of α-glucose linked by 1,4 bonds but in amylose the chains are unbranched and in amylopectin there are some 1,6 bonds that make the molecule branched.

Amylase breaks 1,4 bonds in chains of four or more glucose monomers, so it can digest amylose into maltose but not glucose. Because of the specificity of its active site, amylase cannot break the 1,6 bonds in amylopectin. Fragments of the amylopectin molecule containing a 1,6 bond that amylase cannot digest are called **dextrins**.

Digestion of starch is completed by enzymes in the membranes of microvilli on villus epithelium cells: **maltase** and **dextrinase** digest maltose and dextrins into glucose. Also in the membranes of the microvilli are protein pumps that cause the absorption of the glucose produced by digesting starch.

Blood carrying glucose and other products of digestion flows though villus capillaries to venules in the submucosa of the wall of the small intestine. The blood in these venules is carried via the hepatic portal vein to the liver, where excess glucose can be absorbed by liver cells and converted to glycogen for storage.

Absorption

INTESTINAL VILLI

The process of taking substances into cells and the blood is called **absorption**. In the human digestive system nutrients are absorbed by the epithelium, which is the single layer of cells forming the inner lining of the mucosa. The rate of absorption depends on the surface area of this epithelium.

Absorption occurs principally in the small intestine. The small intestine in adults is about seven metres long and 25–30 millimetres wide and there are folds on its inner surface, giving a large surface area of epithelium.

The area of epithelium is further increased by the presence of **villi**, which are small finger-like projections of the mucosa on the inside of the intestine wall. A villus is between 0.5 and 1.5 mm long and there can be as many as 40 of them per square millimetre of small intestine wall. They increase the surface area by a factor of about ten.

The villi absorb mineral ions and vitamins and also monomers formed by digestion such as glucose.

Structure of a villus

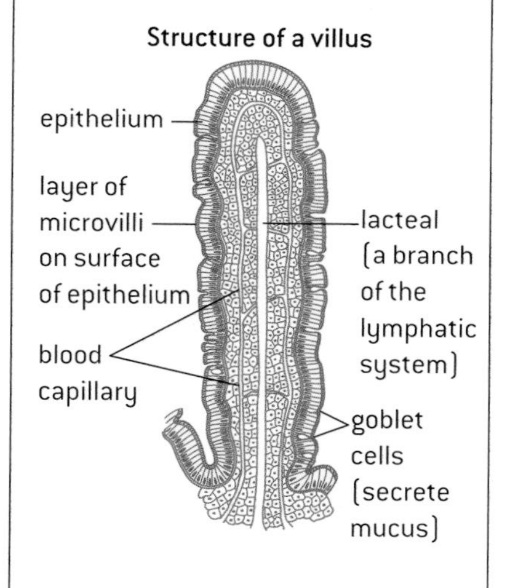

METHODS OF ABSORPTION

Different methods of membrane transport are used in epithelium cells to absorb different nutrients:

- **Simple diffusion**, in which nutrients pass down the concentration gradient between phospholipids in the membrane.

  Example – hydrophobic nutrients such as fatty acids and monoglycerides.
- **Facilitated diffusion**, in which nutrients pass down the concentration gradient through specific channel proteins in the membrane.

  Example – hydrophilic nutrients such as fructose.
- **Active transport**, in which nutrients are pumped through the membrane against the concentration gradient by specific pump proteins.

  Example – mineral ions such as sodium, calcium and iron.
- **Endocytosis** (pinocytosis), in which small droplets of the fluid are passed through the membrane by means of vesicles.

  Example – triglycerides and cholesterol in lipoprotein particles.

There are some more complex methods of transport. For example, glucose is absorbed by sodium co-transporter proteins which move a molecule of glucose together with a sodium ion across the membrane together into the epithelium cells. The glucose can be moved against its concentration gradient because the sodium ion is moving down its concentration gradient. The sodium gradient is generated by active transport of sodium out of the epithelium cell by a pump protein.

MODELLING ABSORPTION WITH DIALYSIS TUBING

Dialysis tubing can be used to model absorption by the epithelium of the intestine. The diagram shows one possible method. Cola drink contains a mixture of substances which can be used to model digested and undigested foods in the intestine. The water outside the bag is tested at intervals to see if substances in the cola have diffused through the dialysis tubing.

The expected result is that glucose and phosphoric acid, which have small-sized particles, diffuse through the tubing but caramel, which consists of larger polymers of sugar, does not.

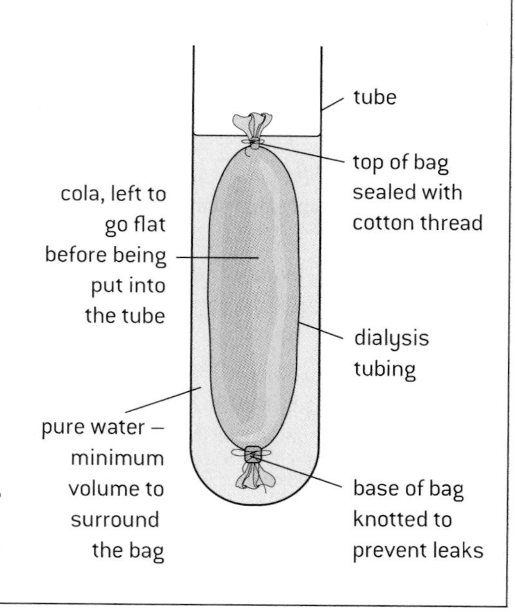

USING MODELS IN SCIENCE

A model in science is a theoretical representation of the real world. Models sometimes consist of mathematical equations but in biology they often represent a structure or process non-mathematically.

When a model has been proposed predictions are made using it, which are then tested. This is done with experiments or with observations of the real world. If predictions based on a model fit experimental data or observations, the model is trusted more. If the predictions are not as close as they could be, the model is modified.

Sometimes evidence shows that a model or theory is incorrect. This known as falsification. The model or theory must then be discarded and replaced.

Theoretical models used to explain the structure of biological membranes were described in Topic 1. An example of physical models is described above.

The cardiovascular system

HARVEY AND THE CIRCULATION OF BLOOD

Until the 17th century the doctrines of Galen, one of the ancient Greek philosophers, about blood were accepted with little questioning by doctors. Galen taught that blood is produced by the liver, pumped out by the heart and consumed in the other organs of the body.

William Harvey is usually credited with the discovery of the circulation of the blood. He had to overcome widespread opposition because Galen's theories were so well established. Harvey published his results and also toured Europe to demonstrate experiments that overturned previous theories and provided evidence for his theory. As a result, his theory that there is a circulation of blood became generally accepted.

Harvey demonstrated that blood flow through vessels is unidirectional with valves to prevent backflow and also that the rate of flow through major vessels is far too high for blood to be consumed in the body after being pumped out by the heart. He showed that the heart pumps blood out in the arteries and that it returns in veins.

William Harvey predicted the presence of numerous fine vessels, too small to be seen with contemporary equipment, that linked arteries to veins in the tissues of the body. Microscopes had not been invented by the time that he published his theory about the circulation of blood in 1628. It was not until 1660, after his death, that blood was seen flowing from arteries to veins though capillaries, as Harvey had predicted.

THE DOUBLE CIRCULATION

The circulation that Harvey discovered in humans is double: there are separate circulations for the lungs (**pulmonary circulation**) and for other organs of the body (**systemic circulation**).

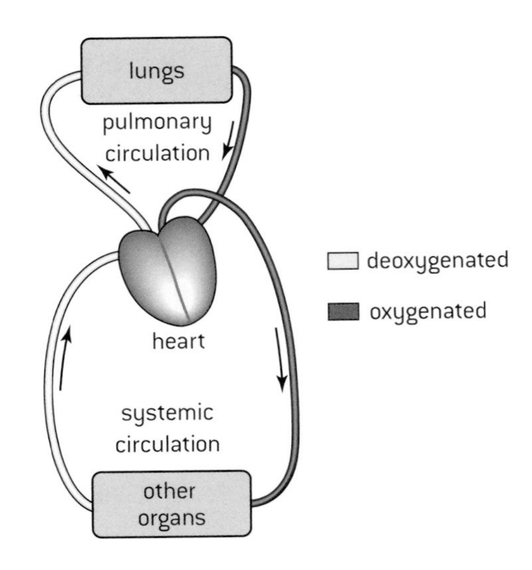

The heart is a double pump with left and right sides. The right side pumps deoxygenated blood to the lungs via the pulmonary artery. Oxygenated blood returns to the left side of the heart in the pulmonary vein. The left side pumps this blood via the aorta to all organs of the body apart from the lungs. Deoxygenated blood is carried back the right side of the heart in the vena cava.

STRUCTURE AND FUNCTION OF BLOOD VESSELS

Blood vessels are tubes that carry blood. There are three main types.

- **Arteries** convey blood pumped out at high pressure by the ventricles of the heart. They carry the blood to tissues of the body.
- **Capillaries** carry blood through tissues. They have permeable walls that allow exchange of materials between the cells of the tissue and the blood in the capillary.
- **Veins** collect blood at low pressure from the tissues of the body and return it to the atria of the heart.

Arteries

Tough outer coat

Thick layer containing elastic fibres that maintain high pressure between pumping cycles and muscle that contracts or relaxes to adjust the diameter of the lumen

Thick wall to withstand the high pressures

Narrow lumen to help maintain the high pressures

Capillaries

Wall consists of a single layer of thin cells so the distance for diffusion in or out is small

Pores between cells in the wall allow some of the plasma to leak out and form tissue fluid. Phagocytes can also squeeze out

Very narrow lumen – only about 10 µm across so that capillaries fit into small spaces. Many small capillaries have a larger surface area than fewer wider ones would

Veins

Thin layers of tissue with few or no elastic fibres or muscle as blood flow is not pulsatile

Thin wall allows the vein to be pressed flat by adjacent muscles, helping to move the blood

Wide lumen is needed to accommodate the low pressure, slow flowing blood.

Outer coat is thin as there is no danger of veins bursting

Valves are present at intervals in veins to prevent back-flow

The heart

CARDIAC MUSCLE

The walls of the heart are made of **cardiac muscle**, which has a special property – it can contract on its own without being stimulated by a nerve (myogenic contraction).

There are many capillaries in the muscular wall of the heart. The blood running through these capillaries is supplied by the coronary arteries, which branch off the aorta, close to the semilunar valve. The blood brought by the coronary arteries brings nutrients. It also brings oxygen for aerobic cell respiration, which provides energy for cardiac muscle contraction.

Valves in the heart ensure circulation of blood by preventing back-flow. The atria are collecting chambers and the ventricles are pumping chambers.

STRUCTURE OF THE HEART

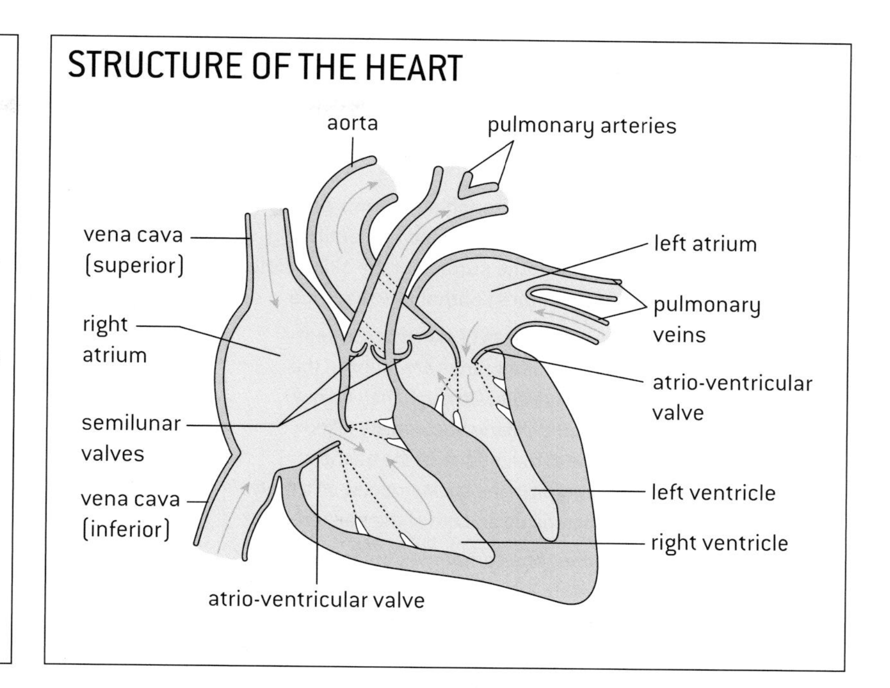

THE CARDIAC CYCLE

The beating of the heart consists of a cycle of actions:

1. The walls of the atria contract, pushing blood from the atria into the ventricles through the atrio-ventricular valves, which are open. The semilunar valves are closed, so the ventricles fill with blood.
2. The walls of the ventricles contract powerfully and the blood pressure rapidly rises inside them. This first causes the atrio-ventricular valves to close, preventing back-flow to the atria and then causes the semilunar valves to open, allowing blood to be pumped out into the arteries. At the same time the atria start to refill by collecting blood from the veins.
3. The ventricles stop contracting so pressure falls inside them. The semilunar valves close, preventing back-flow from the arteries to the ventricles. When the ventricular pressure drops below the atrial pressure, the atrio-ventricular valves open. Blood entering the atrium from the veins then flows on to start filling the ventricles.

   The next cardiac cycle begins when the walls of the atria contract again.

PRESSURES IN THE CARDIAC CYCLE

The graph below shows pressure changes in the left atrium, the left ventricle and the aorta during the cardiac cycle.

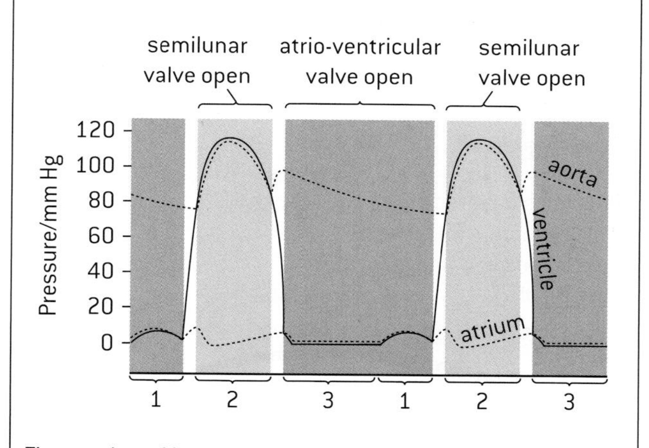

The numbered brackets indicate the three phases of the cardiac cycle described (left).

CONTROL OF HEART RATE

One region of specialized cardiac muscle cells in the wall of the right atrium acts as the pacemaker of the heart by initiating each contraction. This region is called the **sinoatrial (SA) node**. The SA node sends out an electrical signal that stimulates contraction as it is propagated first through the walls of the atria and then through the walls of the ventricles.

Messages can be carried to the SA node by nerves and hormones.

- Impulses brought from the medulla of the brain by two nerves can cause the SA node to change the heart rate. One nerve speeds up the rate and the other slows it down.
- The hormone epinephrine increases the heart rate to help to prepare the body for vigorous physical activity.

CORONARY ARTERY DISEASE

Coronary artery disease is caused by fatty plaque building up in the inner lining of coronary arteries, which become occluded (narrowed). As this becomes more severe blood flow to cardiac muscle is restricted, causing chest pain. Minerals often become deposited in the plaque making it hard and rough. Various factors have been shown by surveys to be associated with coronary artery disease and are likely causes of it:

- high blood cholesterol levels
- smoking
- high blood pressure (hypertension)
- high blood sugar levels, usually due to diabetes
- genetic factors (thus a family history of the disease).

Defence against infectious disease

BARRIERS TO INFECTION

A **pathogen** is an organism or virus that causes disease. The skin and mucous membranes are the primary defence against pathogens, by forming a barrier preventing entry.

- The outer layers of the skin are tough and form a physical barrier. Sebaceous glands in the skin secrete lactic acid and fatty acids, which make the surface of the skin acidic. This prevents the growth of most pathogenic bacteria.
- Mucous membranes are soft areas of skin that are kept moist with mucus. Mucous membranes are found in the nose, trachea, vagina and urethra. Although they do not form a strong physical barrier, many bacteria are killed by lysozyme, an enzyme in the mucus. In the trachea pathogens tend to get caught in the sticky mucus; cilia then push the mucus and bacteria up and out of the trachea.

Despite these barriers, pathogens sometimes enter the body so other defences are needed. Two types of white blood cell fight infections in the body: **phagocytes** and **lymphocytes**.

PHAGOCYTES

Phagocytes ingest pathogens by endocytosis. The pathogens are then killed and digested inside the cell by enzymes from lysosomes. Phagocytes can ingest pathogens in the blood. They can also squeeze out through the walls of blood capillaries and move through tissues to sites of infection. They then ingest the pathogens causing the infection. Large numbers of phagocytes at a site of infection form pus.

Phagocytes give us what is called **non-specific immunity** to diseases, because a phagocyte does not distinguish between pathogens – it ingests any pathogen if stimulated to do so.

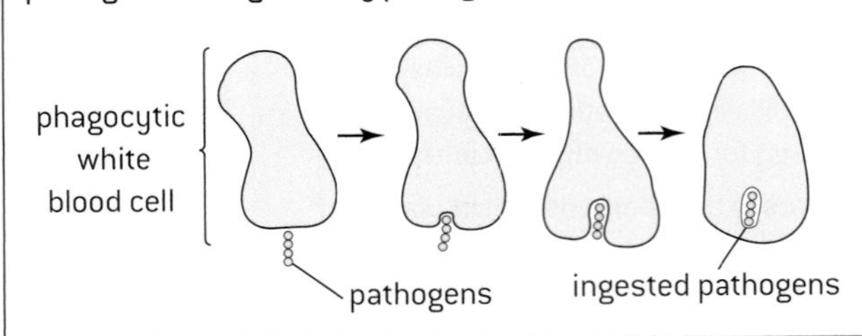

BLOOD CLOTTING

When the skin is cut and blood escapes from blood vessels, a semi-solid blood clot is formed from liquid blood to seal up the cut and prevent entry of pathogens. Platelets have an important role in clotting. Platelets are small cell fragments that circulate with red and white blood cells in blood plasma. The clotting process begins with the release of clotting factors either from damaged tissue cells or from platelets. These clotting factors set off a cascade of reactions in which the product of each reaction is the catalyst of the next reaction.

This system helps to ensure that clotting only happens when it is needed and also makes it a very rapid process. In the last reaction fibrinogen, a soluble plasma protein, is altered by the removal of sections of peptide that have many negative charges. This allows the remaining polypeptide to bind to others, forming long protein fibres called fibrin. Fibrin forms a mesh of fibres across wounds. Blood cells are caught in the mesh and soon form a semi-solid clot. If exposed to air the clot dries to form a protective scab, which remains until the wound has healed.

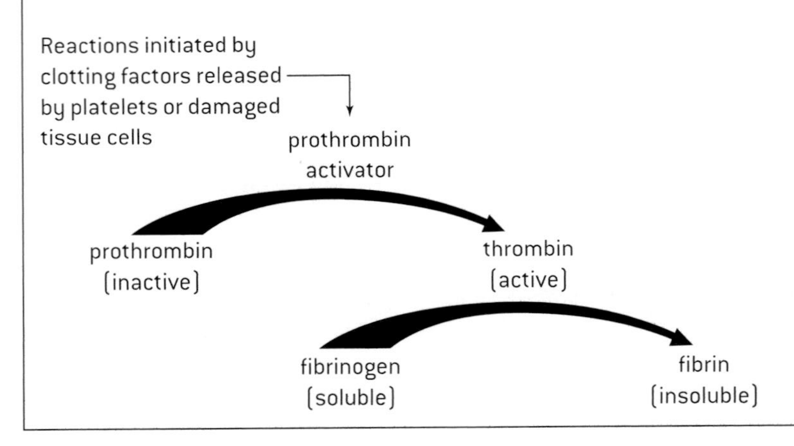

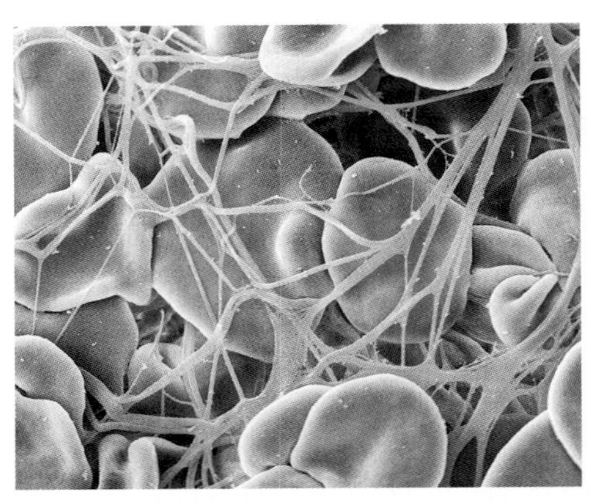

BLOOD CLOTS IN CORONARY ARTERIES

If the deposits of plaque in coronary arteries rupture, blood clots form (**coronary thrombosis**), which may completely block the artery. The consequence of this is that an area of cardiac muscle receives no oxygen and so stops beating in a coordinated way. This is often called a **heart attack**. Uncoordinated contraction of cardiac muscle is **fibrillation**. Sometimes the heart recovers and starts beating again, but severe heart attacks can be fatal as contractions of the heart stop completely.

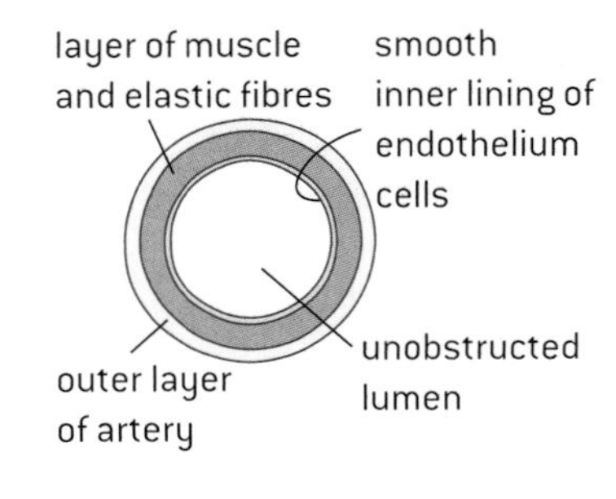

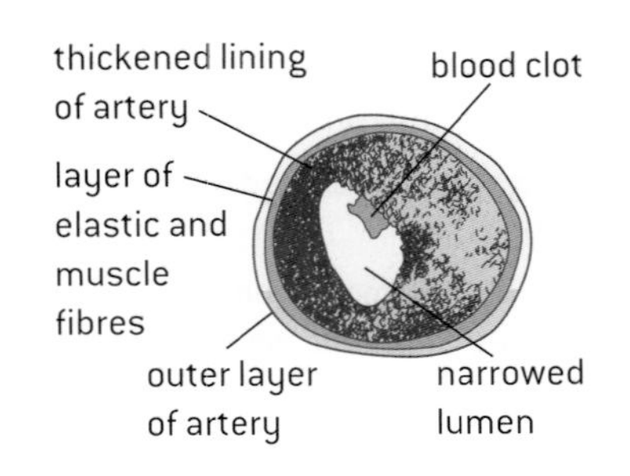

Antibodies and antibiotics

PRODUCTION OF ANTIBODIES

① Antibodies are made by lymphocytes, one of the two main types of white blood cell. Antigens are foreign substances that stimulate the production of antibodies.

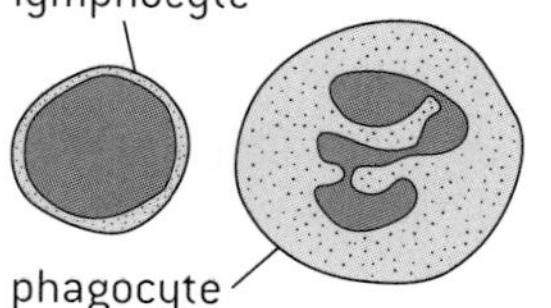

② A lymphocyte can only make one type of antibody so a huge number of different lymphocyte types is needed. Each lymphocyte puts some of the antibody that it can make into its cell surface membrane with the antigen-combining site projecting outwards.

③ When a pathogen enters the body, its antigens bind to the antibodies in the cell surface membrane of one type of lymphocyte.

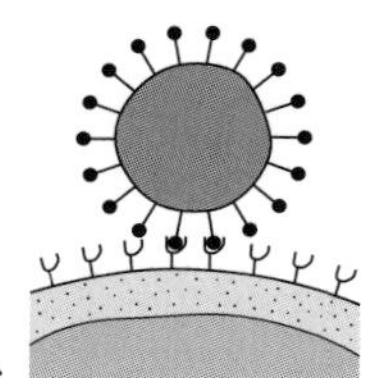

④ When antigens bind to the antibodies on the surface of a lymphocyte, this lymphocyte become active and divides by mitosis to produce a clone of many identical cells.

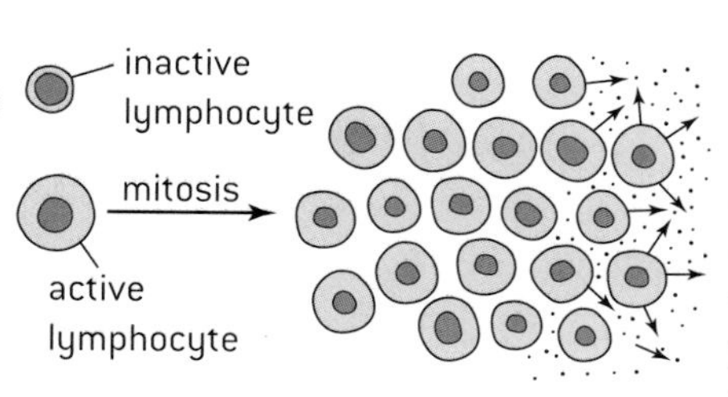

⑤ The cells produced by mitosis are plasma cells. They produce large quantities of the same antibody. The antibody binds to the antigens on the surface of the pathogen and stimulates its destruction. Production of antibodies by lymphocytes is known as specific immunity, because different antibodies are needed to defend against different pathogens. After an infection has been cleared from the body, most of the lymphocytes used to produce the antibodies disappear, but some persist as memory cells. These memory cells can quickly reproduce to form a clone of plasma cells if a pathogen carrying the same antigen is re-encountered.

HIV AND THE IMMUNE SYSTEM

HIV (human immunodeficiency virus) infects a type of lymphocyte that plays a vital role in antibody production. Over a period of years these lymphocytes are gradually destroyed. Without active lymphocytes, antibodies cannot be produced. This condition is called **AIDS (acquired immunodeficiency syndrome)** and, if untreated, leads to death from infections by a variety of pathogens that would normally be controlled easily.

HIV does not survive for long outside the body and cannot easily pass through the skin. Transmission involves the transfer of body fluids from an infected person to an uninfected one:

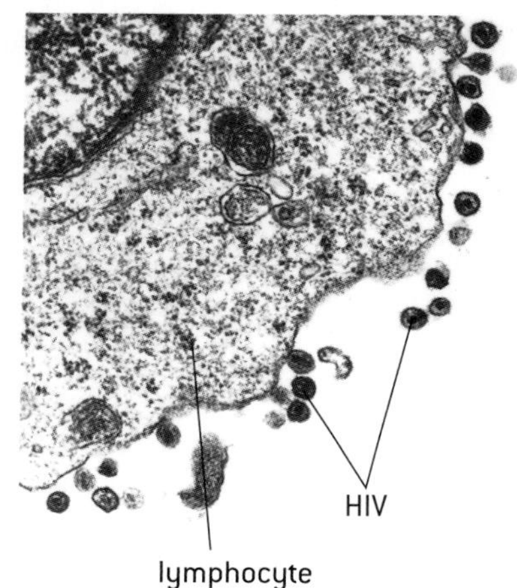

- Through small cuts or tears in the vagina, penis, mouth or intestine during vaginal, anal or oral sex.
- In traces of blood on hypodermic needles shared by intravenous drug abusers.
- Across the placenta from a mother to a baby, or through cuts during childbirth or in milk during breast-feeding.
- In transfused blood or with blood products such as Factor VIII used to treat hemophiliacs.

ANTIBIOTICS

Antibiotics are chemicals produced by microorganisms, to kill or control the growth of other organisms. For example, *Penicillium* fungus produces penicillin to kill bacteria. Antibiotics work by blocking processes that occur in prokaryotic cells but not eukaryotic cells. There are many differences between human and bacterial cells and each antibiotic blocks one of these processes in bacteria without causing any harm in humans.

Viruses lack a metabolism and instead rely on a host such as a human cell to carry out metabolic processes. It is not possible to block these processes using an antibiotic without also harming the human cells. For this reason viral diseases cannot be treated with antibiotics.

Most bacterial diseases in humans can be treated successfully with antibiotics, but some strains of bacteria have acquired genes that confer resistance to an antibiotic and some strains of bacteria now have multiple resistance.

TESTING PENICILLIN

Penicillin was developed as an antibiotic by Florey and Chain in the late 1930s. Their first test was on eight mice infected with a bacterium that causes a fatal pneumonia. All the four treated mice recovered but the untreated mice died. Initially they only had small quantities of relatively impure penicillin. They tested these on a man that was close to death from a bacterial infection. He started to recover but the antibiotic ran out. Five patients were then tested, all of whom were cured.

Florey and Chain's research would not be regarded as safe enough today. Extensive animal testing of new drugs is first done to check for harmful effects. After this small and then larger doses are tested on healthy, informed humans to see if the drug is tolerated. Only then is the drug tested on patients with the disease and if small scale trials suggest that it is effective, larger scale double-blind trials are carried out on patients to test the drug's effectiveness and look for rare side effects.

Ventilation

THE NEED FOR VENTILATION

Cell respiration happens in the cytoplasm and mitochondria and releases energy in the form of ATP for use inside the cell.

In humans, oxygen is used in cell respiration and carbon dioxide is produced. Humans therefore must take in oxygen from their surroundings and release carbon dioxide.

This process of swapping one gas for another is called **gas exchange**. It happens by diffusion in the alveoli of human lungs, so it depends on concentration gradients of oxygen and carbon dioxide between the air in the alveoli and blood flowing in the adjacent capillaries. To maintain these concentration gradients, the air in the alveoli must be refreshed frequently. The process of bringing fresh air to the alveoli and removing stale air is called **ventilation**.

The diagram of the ventilation system shows how air is carried to and from the alveoli in the trachea, bronchi and bronchioles.

MONITORING VENTILATION IN HUMANS

Ventilation rate is the number of inhalations or exhalations per minute.

Tidal volume is the volume of air taken in or out with each inhalation or exhalation. By monitoring ventilation rate and tidal volume at rest and then during mild and vigorous exercise the effect of ventilation can be investigated.

1. **Monitoring ventilation rate**

This can be done by simple observation or using data-logging:

- An inflatable chest belt is placed around the thorax and air is pumped in with a bladder.
- A differential pressure sensor is then used to measure pressure variations inside the chest belt due to chest expansions.
- The ventilation rate can be deduced and also the relative size of ventilations but not the absolute size.

2. **Monitoring tidal volumes**

Tidal volumes are measured using a **spirometer**. Simple spirometers can be made using a bell jar, with volumes marked on it, placed in a pneumatic trough. A tube is used to breathe out into the bell jar so the expired volume can be measured. There are many designs of electronic spirometer that doctors use.

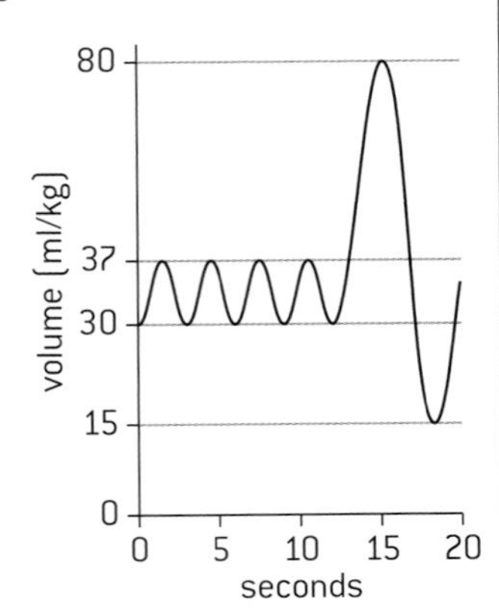

The graph (right) shows the type of data that is generated by monitoring ventilation with a spirometer. Tidal volume is deduced by how much the lung volume increases or decreases with each ventilation. Ventilation rate is deduced by counting the number of ventilations in a period on the graph and measuring the time period using the *x*-axis of the graph.

$$\text{Rate} = \frac{\text{number of ventilations}}{\text{time}}$$

THE VENTILATION SYSTEM

bronchus
intercostal muscles
trachea
left lung
ribs
bronchioles (ending in microscopic alveoli)
diaphragm

VENTILATION OF THE LUNGS

Muscle contractions cause the pressure changes inside the thorax that force air in and out of the lungs to ventilate them.

Different muscles are required for inspiration and expiration because muscles only do work when they contract. Muscles that cause the opposite movement from each other are **antagonistic muscles**.

Inhaling	Exhaling
The external intercostal muscles contract, moving the ribcage up and out	The internal intercostal muscles contract, moving the ribcage down and in
The diaphragm contracts, becoming flatter and moving down	The abdominal muscles contract, pushing the diaphragm up into a dome shape
These muscle movements increase the volume of the thorax	These muscle movements decrease the volume of the thorax
The pressure inside the thorax therefore drops below atmospheric pressure	The pressure inside the thorax therefore rises above atmospheric pressure
Air flows into the lungs from outside the body until the pressure inside the lungs rises to atmospheric pressure	Air flows out from the lungs to outside the body until the pressure inside the lungs falls to atmospheric pressure

Gas exchange

ADAPTATIONS OF AN ALVEOLUS FOR GAS EXCHANGE

Gas exchange surfaces have four properties (below). Although each alveolus is very small, the lungs contain hundreds of millions of alveoli, giving a huge overall surface area for gas exchange.

- permeable to oxygen and carbon dioxide
- a large surface area for diffusion
- thin, so the distance for diffusion is small
- moist, so oxygen can dissolve

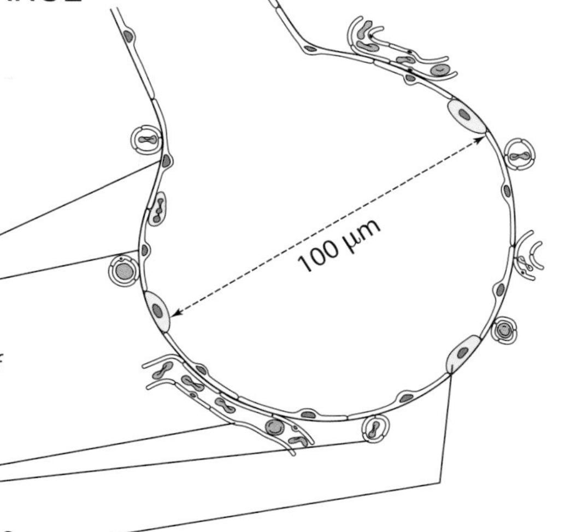

Type 1 pneumocytes
Extremely thin and permeable alveolar cells that are adapted to carry out gas exchange. Most of the wall of the alveolus consists of a single layer of these thin cells. Gases only have to diffuse a very short distance to pass through them.

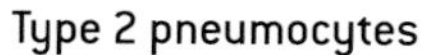

Blood capillaries
The alveolus is covered by a dense network of blood capillaries with low oxygen and high carbon dioxide concentrations. Oxygen therefore diffuses from the air in the alveolus to the blood and carbon dioxide diffuses in the opposite direction.

Type 2 pneumocytes
Cells in the alveolus wall that secrete a fluid to keep the inner surface of the alveolus moist and allow gases to dissolve. The fluid also contains a natural detergent (surfactant), to prevent the sides of the alveoli from sticking together by reducing surface tension.

LUNG CANCER

Epidemiology is the study of the incidence and causes of disease. Surveys are used to look for correlations between disease rates and factors that could be implicated. Correlation does not prove causation but careful analysis can show whether a factor actually causes a disease. The five main causes of lung cancer are these:

- **Smoking** – tobacco smoke contains many mutagens that cause tumours to develop. Smoking causes nearly 90% of lung cancer cases.
- **Passive smoking** – exhaled breath from smokers passes carcinogens on to others, both children and other adults. Smoking bans are reducing this.
- **Air pollution** – the many sources include diesel exhaust fumes, nitrogen oxides from vehicles and smoke from wood and coal fires.
- **Radon gas** – in some areas it leaks out of rocks, especially granite.
- **Asbestos and silica** – dust from these materials causes cancer if deposited in the lungs.

The consequences of lung cancer are:

- difficulties with breathing
- persistent coughing
- coughing up blood
- general fatigue
- chest pain
- loss of appetite
- weight loss

Lung cancer is usually fatal as it is only discovered at a late stage when the primary tumour is large and secondary tumours have already developed elsewhere in the body.

EMPHYSEMA

The main causes of emphysema are smoking and air pollution. Cilia that line the airways and expel mucus are damaged and cease to function, so mucus builds up in the lungs, causing infections. Toxins in cigarette smoke and polluted air cause inflammation and damage to the white blood cells that fight infections in the lungs. A protease (trypsin) is released from inflamed cells and damaged white blood cells. This enzyme digests elastic fibres in the lungs and eventually causes complete breakdown of alveolus walls. Microscopic alveoli (below left) are replaced by progressively larger air sacs with thicker, less permeable walls (below right).

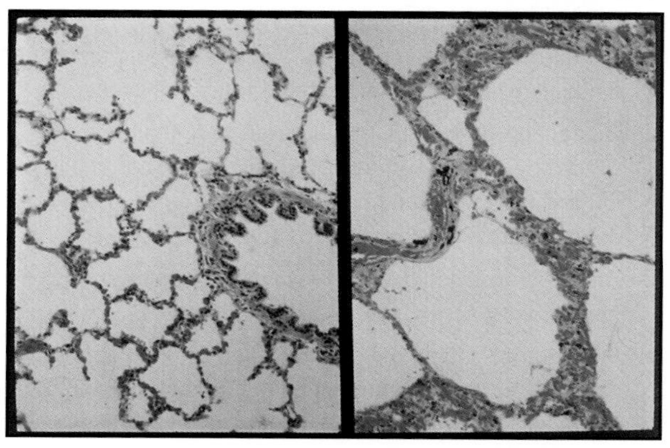

Emphysema is a chronic and progressive disease with serious consequences. The surface area for gas exchange reduces so the oxygen saturation of the blood falls and exercise is more and more difficult. The lungs lose their elasticity, making it increasingly difficult to exhale (shortness of breath). Mucus in the lungs causes coughing and wheezing.

Neurons and synapses

STRUCTURE AND FUNCTION OF NEURONS

The nervous system is composed of cells called **neurons**. These cells carry messages at high speed in the form of electrical impulses. Many neurons are very elongated and carry impulses long distances in a very short time. Myelinated nerve fibres have a myelin sheath with small gaps called **nodes of Ranvier**, allowing the nerve impulse to jump from node to node. This is known as **saltatory conduction** and speeds up the transmission.

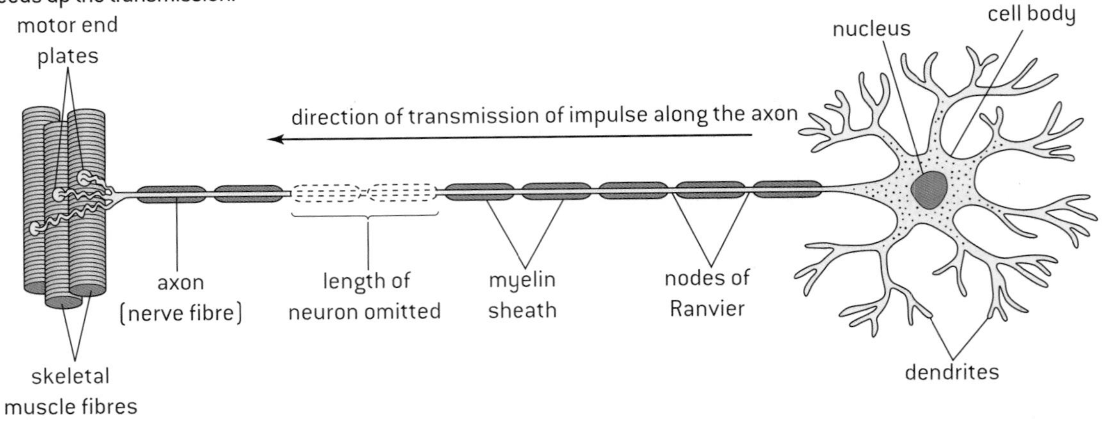

SYNAPSES

A **synapse** is a junction between two neurons or a junction between neurons and receptor or effector cells. The plasma membranes of the neurons are separated by a narrow fluid-filled gap called the synaptic cleft. Messages are passed across the synapse in the form of chemicals called neurotransmitters. The neurotransmitters always pass in the same direction from the pre-synaptic neuron to the post-synaptic neuron.

Many synapses function in the following way.

1. A nerve impulse reaches the end of the pre-synaptic neuron.
2. Depolarization of the pre-synaptic membrane causes vesicles of neurotransmitter to move to the pre-synaptic membrane and fuse with it, releasing the neurotransmitter into the synaptic cleft by **exocytosis**.
3. The neurotransmitter diffuses across the synaptic cleft and binds to receptors in the post-synaptic membrane.
4. The receptors are transmitter-gated sodium channels, which open when neurotransmitter binds. Sodium ions diffuse into the post-synaptic neuron. This causes depolarization of the post-synaptic membrane.
5. The depolarization passes on down the post-synaptic neuron as an action potential.
6. Neurotransmitter in the synaptic cleft is rapidly broken down, to prevent continuous synaptic transmission. The figure (right) shows the events that occur during synaptic transmission.

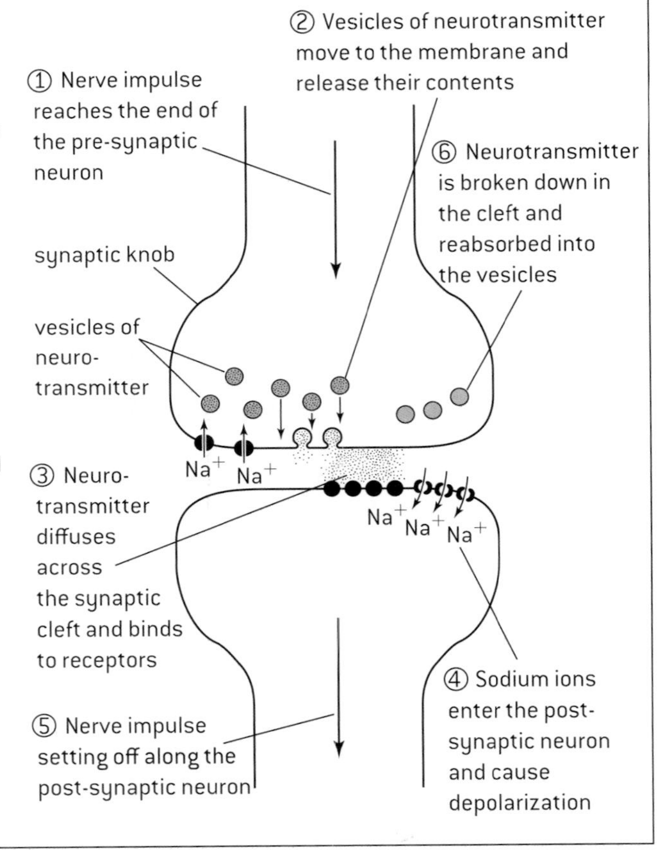

CHOLINERGIC SYNAPSES

Synapses do not all use the same neurotransmitter but many use **acetylcholine**. They are known as **cholinergic synapses**. The pre-synaptic neuron secretes acetylcholine into the synaptic cleft, which diffuses across the synapse and then binds to receptors in the post-synaptic membrane. The acetylcholine is broken down in the synaptic cleft by the enzyme **cholinesterase**, producing acetyl groups and choline. The choline is reabsorbed by the pre-synaptic neuron.

NEONICOTINOID PESTICIDES

Neonicotinoid pesticides bind to acetylcholine receptors in the post-synaptic membranes of cholinergic synapses in insects. Cholinesterase does not break down these pesticides so they remain bound to the receptors, preventing acetylcholine from binding. They therefore block synaptic transmission, which ultimately kills the insect. Unfortunately honeybees are killed along with insect pests that are the intended target of neonicotinoids.

Nerve impulses

RESTING POTENTIALS

A **resting potential** is the voltage (electrical potential) across the plasma membrane of a neuron when it is not conducting a nerve impulse. There are sodium–potassium pumps in the plasma membranes of axons. They pump sodium out and potassium in, by active transport. Concentration gradients of both sodium and potassium are established across the membrane. The inside of the neuron develops a net negative charge, compared with the outside, because of the presence of chloride and other negatively charged ions. There is therefore a potential (voltage) across the membrane. This is called the resting potential. A typical resting potential is −70mV.

ACTION POTENTIALS

An **action potential** is the depolarization and repolarization of a neuron, due to facilitated diffusion of ions across the membrane through voltage-gated ion channels. If the potential across the membrane rises from −70 to −50 mV, voltage-gated sodium channels open and sodium ions diffuse in down the concentration gradient. The entry of positively charged sodium ions causes the inside of the neuron to develop a net positive charge compared to the outside – the potential across the membrane is reversed. This is **depolarization**.

The reversal of membrane polarity causes potassium channels to open, allowing potassium ions to diffuse out down the concentration gradient. The exit of positively charged potassium ions causes the inside of the neuron to develop a net negative charge again compared with the outside – the potential across the membrane is restored. This is **repolarization**.

PROPAGATION OF NERVE IMPULSES

A nerve impulse is an action potential that travels along the axon of a neuron from one end to the other. There is an action potential whenever a part of the axon reaches the threshold potential of −50mV. An action potential in one part of the axon triggers an action potential in the next part. This is called the **propagation of the nerve impulse**. It is due to diffusion of sodium ions between a region with an action potential and the next region that is still at the resting potential. The diffusion of sodium ions along the axon, both inside and outside the membrane, is called **local currents**. It changes the voltage across the membrane from the resting potential of −70mV to the threshold potential of −50mV. This causes an action potential, because voltage-gated sodium channels open.

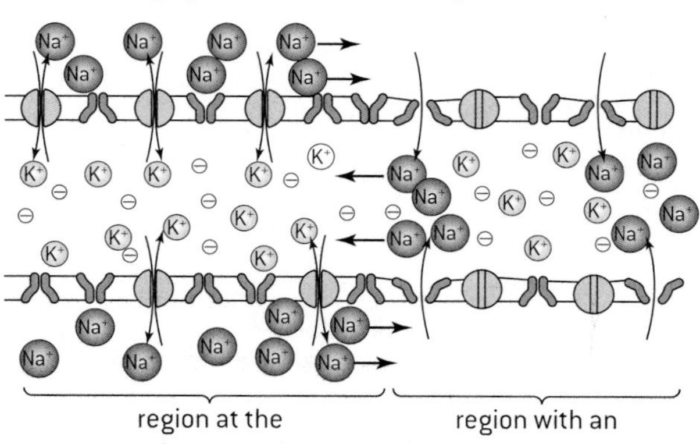

OSCILLOSCOPE TRACES

The changes in membrane potential in axons during an action potential can be measured using electrodes. The results are displayed on an oscilloscope. The figure below shows the type of trace that is obtained.

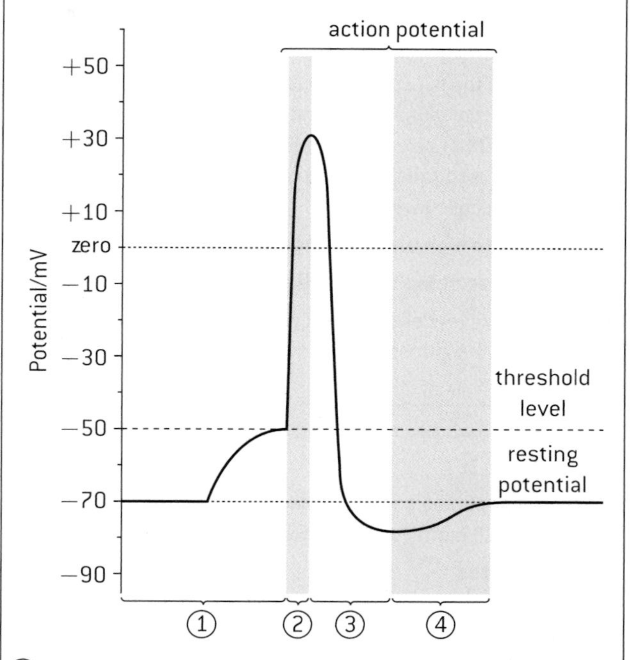

① The axon membrane is at a resting potential of −70 mV and then rises to the threshold potential of −50 mV, either due to local currents or to the binding of a neurotransmitter at a synapse.

② The membrane depolarizes due to voltage-gated Na^+ channels opening and Na^+ ions diffusing in.

③ The membrane repolarizes due to voltage-gated K^+ channels opening and K^+ ions diffusing out.

④ The membrane returns to the resting potential due to pumping of Na^+ ions out and K^+ ions in to the axon. This rebuilds concentration gradients of both types of ion, so another action potential could occur.

MEMORY AND LEARNING

Higher functions of the brain including memory and learning are only partly understood at present and are being researched very actively. They have traditionally been investigated by psychologists but increasingly the techniques of molecular biology and biochemistry are being used to unravel the mechanisms at work. Other branches of science are also making important contributions, including biophysics, medicine, pharmacology and computer science.

This is an excellent example of cooperation and collaboration between groups of scientists, which is an important aspect of the nature of science.

Research breakthroughs are often made in science when different techniques are combined to solve a problem. Scientists from different disciplines meet and exchange ideas both within universities and research institutes and also at international conferences and symposia.

Regulating blood glucose and body temperature

BLOOD GLUCOSE CONCENTRATION

Blood glucose concentration is usually kept between 4 and 8 millimoles per dm^3 of blood. Cells in the pancreas monitor the concentration and secrete the hormones **insulin** or **glucagon** when the level is high or low.

Responses to high blood glucose levels

Insulin is secreted by β (beta) cells.

It stimulates the liver and muscle cells to absorb glucose and convert it to glycogen. Granules of glycogen are stored in these cells. Other cells are stimulated to absorb glucose and use it in cell respiration instead of fat. These processes lower the blood glucose level.

Responses to high blood glucose levels

Glucagon is secreted by α (alpha) cells.

It stimulates liver cells to break glycogen down into glucose and release the glucose. This raises the blood glucose level.

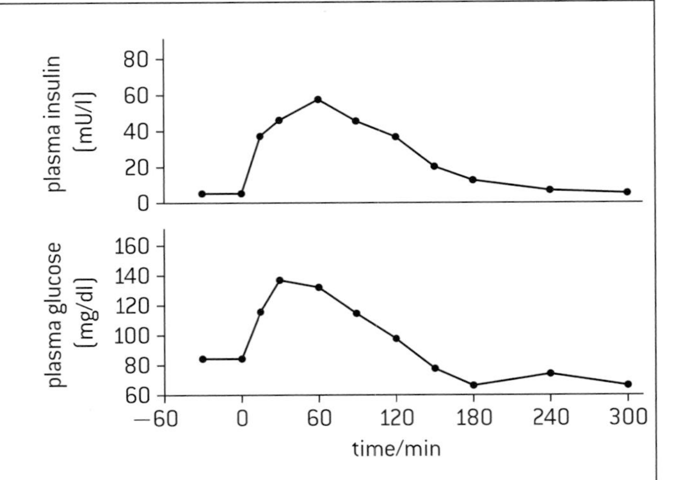

The graphs show the results of giving experimental subjects 75 g of glucose at time zero, after an overnight period of fasting.

DIABETES

In some people the control of blood glucose does not work effectively and the concentration can rise or fall beyond the normal limits. The full name for this condition is **diabetes mellitus**. There are two forms of this condition:

Type I diabetes

- The onset is usually during childhood.
- The immune system destroys β cells in the pancreas so the amount of insulin secreted becomes insufficient.
- Blood glucose levels have to be measured regularly and insulin injections, often before meals, are used to control glucose levels.
- Diet cannot by itself control this type of diabetes.

Type II diabetes

- The onset is usually after childhood.
- Target cells become insensitive to insulin, so insulin injections are not usually an effective treatment.
- Low carbohydrate diets can control the condition.
- Various risk factors increase the rate, particularly diets rich in fat and low in fibre, obesity due to over-eating and lack of exercise and genetic factors that affect fat metabolism.

THYROXIN

The hormone **thyroxin** is secreted by the thyroid gland in the neck. Its chemical structure is unusual as the thyroxin molecule contains four atoms of iodine. Prolonged deficiency of iodine in the diet therefore prevents the synthesis of thyroxin. This hormone is also unusual as almost all cells in the body are targets. Thyroxin regulates the body's metabolic rate, so all cells need to respond but the most metabolically active, such as liver, muscle and brain are the main targets. Higher metabolic rate supports more protein synthesis and growth and it increases the generation of body heat.

In addition, thyroxin is implicated in heat generation by shivering and by uncoupled cell respiration in brown adipose tissue (BAT).

In a person with normal physiology, cooling triggers increased thyroxin secretion by the thyroid gland, which stimulates heat production. Recent research has also suggested that thyroxin causes constriction of vessels that carry blood from the core to the skin, reducing heat loss. Thyroxin thus regulates the metabolic rate and also helps to control body temperature.

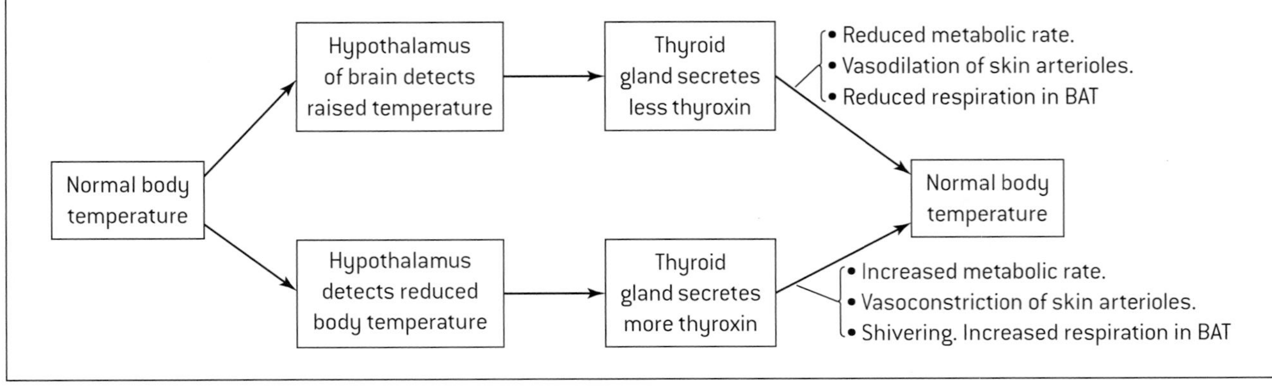

Leptin and melatonin

LEPTIN AND OBESITY

A strain of mice was discovered in the 1950s that feed ravenously, become inactive and gain mass, mainly through increased **adipose tissue**. They grow to a body mass of about 100 grams, compared with wild type mice of 20–25 grams.

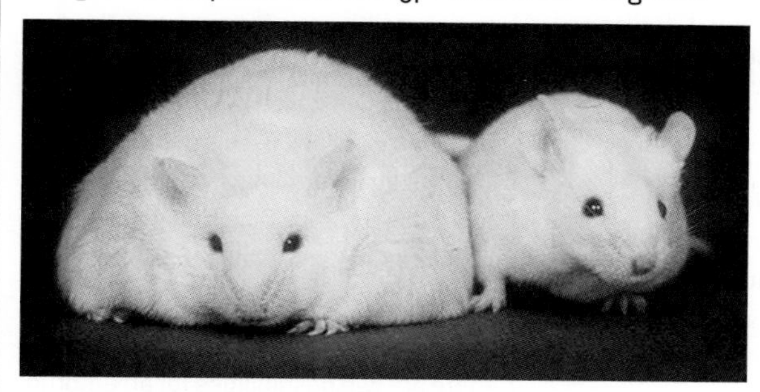

Breeding experiments showed that the obese mice had two copies of a recessive allele, *ob*.

In the early 1990s it was discovered that the wild-type allele of this gene supported the synthesis of a new hormone. It was named **leptin**, and was also found in humans. Leptin is a protein hormone secreted by **adipose cells** (fat storage cells). If the amount of adipose tissue in the body increases, the concentration of leptin in the blood rises. The target of this hormone is groups of cells in the **hypothalamus** of the brain that contribute to the control of appetite. Leptin binds to receptors in the membrane of these cells causing long-term **appetite inhibition** and reduced food intake.

When *ob/ob* mice were injected with leptin their appetite declined, energy expenditure increased and body mass dropped by 30% in a month. Trials were therefore done to see if leptin injections would control obesity in humans. A large clinical trial was carried out. 73 obese volunteers injected themselves either with one of several leptin doses or with a placebo. A double-blind procedure was used, so neither the researchers nor the volunteers knew who was injecting leptin until the results were analysed. The leptin injections induced skin irritation and swelling and only 47 patients completed the trial. The eight patients receiving the highest dose lost 7.1 kg of body mass on average compared with a loss of 1.3 kg in the 12 volunteers who were injecting the placebo. However the results of the group receiving the highest dose varied very widely from a loss of 15 kg to a gain of 5 kg. Also, any body mass lost during the trial was usually regained rapidly afterwards.

Such disappointing outcomes are frequent in drug research because human physiology differs from that of mice and other rodents. Further research has shown that most cases of obesity in humans are due not to insufficient leptin secretion but to target cells in the hypothalamus being resistant to leptin. They therefore fail to respond to it, even at high concentrations. Injections of extra leptin therefore fail to control obesity in these patients.

Obesity in humans is only due to mutations in the leptin gene in a very small proportion of cases. Trials in these obese people have shown significant weight loss while the leptin injections are being given. However, leptin is a short-lived protein and has to be injected several times a day, so most patients offered this treatment have refused it.

MELATONIN AND JET LAG

Humans are adapted to live in a 24-hour cycle and have circadian rhythms in behaviour that fit this cycle. Ganglion cells in the retina detect whether it is light or dark and send impulses to the supra-chiasmatic nuclei (SCN) in the hypothalamus. Neurones in the SCN control secretion of the hormone melatonin by the pineal gland. Melatonin secretion increases in the evening and drops to a low level at dawn. As the hormone is rapidly removed from the blood by the liver, concentrations rise and fall rapidly after a change in secretion.

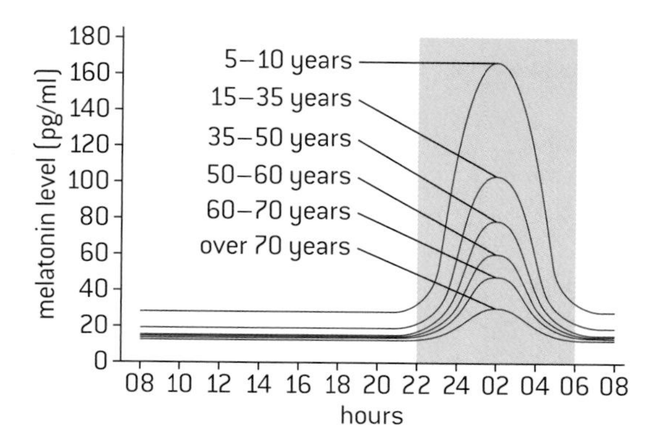

The graph shows that melatonin secretion declines with age, helping to explain how sleep patterns become more irregular as we grow older.

The body's circadian rhythms are disrupted by travelling rapidly between time zones. These symptoms are often experienced:

- sleep disturbance
- headaches
- fatigue
- irritability.

Together they are known as jet lag. They are caused by the SCN and pineal gland continuing to set a circadian rhythm to suit the timing of day and night at the point of departure rather than the destination. This only lasts for a few days, during which time impulses sent by ganglion cells to the SCN when they detect light help the body to adjust to the new regime.

Melatonin is sometimes used to try to prevent or reduce jet lag. It is taken orally at the time when sleep should ideally be commencing. Most trials of melatonin have shown that it is effective at promoting sleep and helping to reduce jet lag, especially if flying eastwards and crossing five or more time zones.

The graph below shows blood plasma concentrations of melatonin in the hours after ingesting different doses at time zero.

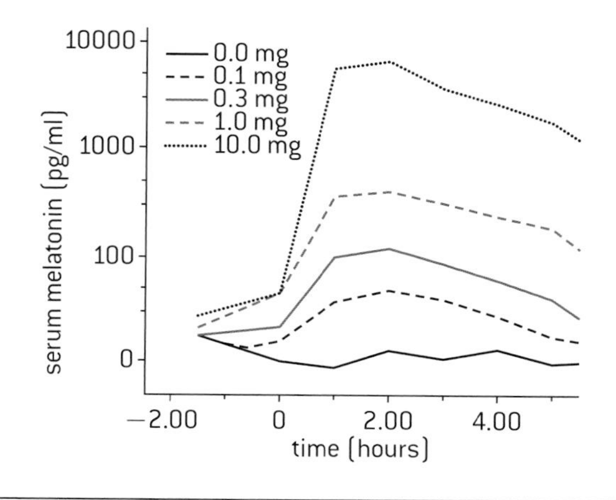

Reproductive systems

THE FEMALE REPRODUCTIVE SYSTEM

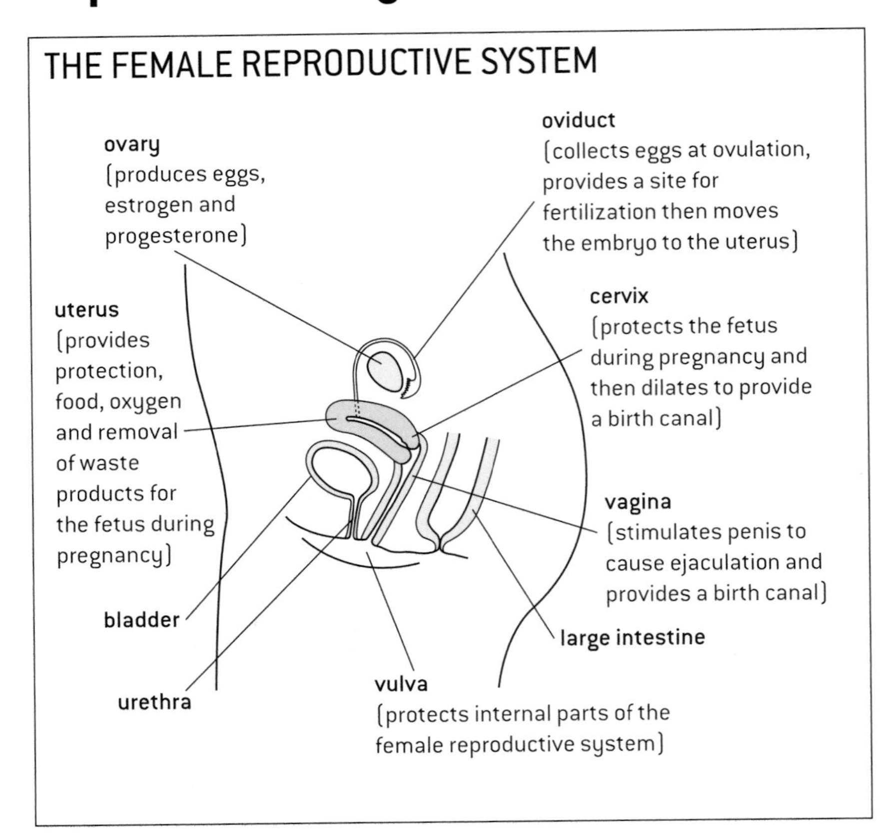

SEX DETERMINATION

Human reproduction involves the fusion of a sperm and an egg. Embryos all initially develop in a similar way. Embryonic gonads are formed that could become either ovaries or testes. The presence or absence of a single gene (SRY) decides which developmental pathway is followed. This gene codes for TDF (testis determining factor), a gene regulation protein. By binding to specific DNA sites TDF stimulates the expression of genes for testis development.

SRY is located on the Y chromosome, so there are two possibilities for an embryo:

♂ SRY is present in an embryo if the sex chromosomes are XY. The embryonic gonads therefore develop into testes and the fetus becomes male.

♀ SRY is absent in an embryo if the sex chromosomes are XX. TDF is therefore not produced, so the embryonic gonads develop as ovaries and the fetus becomes female.

STEROID HORMONES

Testosterone, estrogen and progesterone are all steroids.

Testosterone is produced by developing testes in the fetus. It causes pre-natal development of male genitalia, including the penis, sperm duct and prostate gland. During puberty testosterone production increases. It stimulates development of male secondary sexual characteristics during puberty, including growth of the testes, penis and pubic hair. Testosterone also stimulates sperm production from puberty onwards.

Estrogen causes pre-natal development of female reproductive organs if testosterone is not present. These organs include the oviduct, uterus and vagina. Raised levels of estrogen during puberty cause development of female secondary sexual characteristics, including growth of breasts and pubic hair.

Progesterone prepares the uterus during the menstrual cycle for the implantation of an embryo and has important roles in supporting a pregnancy.

THE MALE REPRODUCTIVE SYSTEM

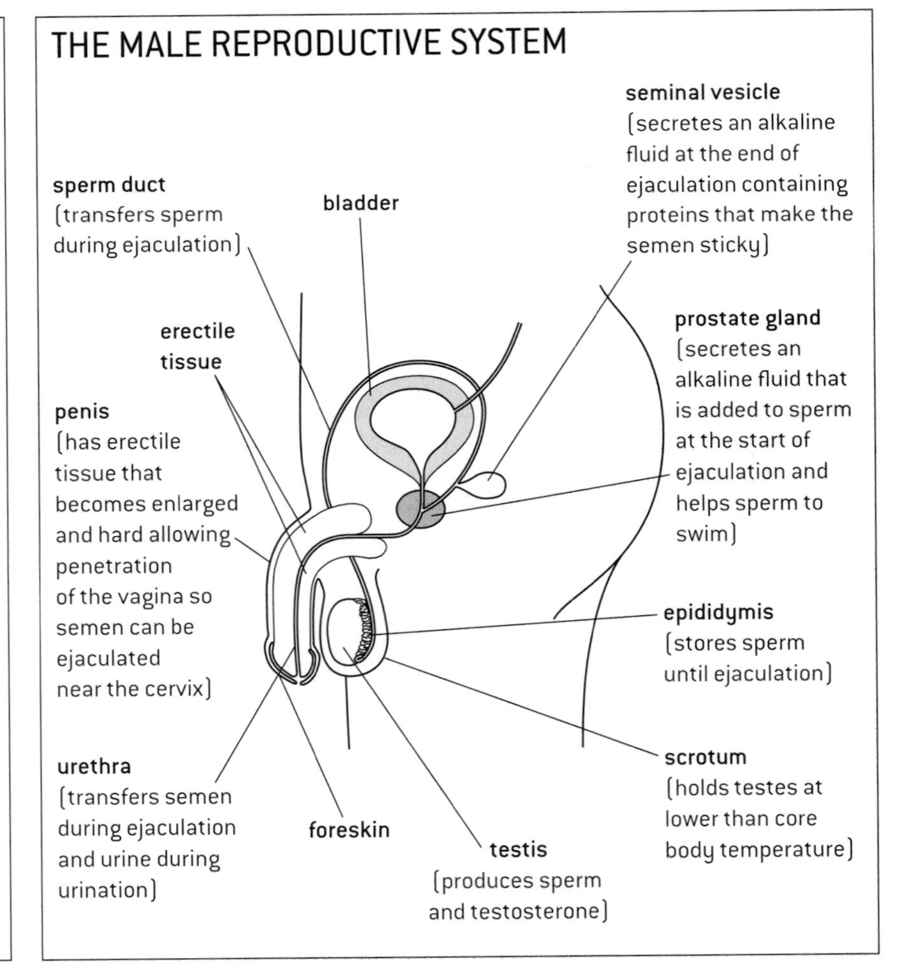

Conception and pregnancy

THE MENSTRUAL CYCLE

Between puberty and the menopause, women who are not pregnant follow a cycle called the menstrual cycle. This cycle is controlled by hormones FSH and LH, produced by the pituitary gland, and estrogen and progesterone, produced by the ovary. Both positive and negative feedback control is used in the menstrual cycle. During each menstrual cycle an **oocyte** (egg) matures inside a fluid-filled sac in the ovary called a **follicle**. The egg is released when the follicle bursts open during **ovulation**.

FEEDBACK CONTROL

In feedback systems, the level of a product feeds back to control the rate of its own production.

Negative feedback has a stabilizing effect because a change in levels always causes the opposite change. A rise in levels feeds back to decrease production and reduce the level. A decrease in levels feeds back to increase production and raise the level.

Positive feedback tends to lead to sudden rises or falls, because a rise causes further rises and a fall causes further falls.

STAGES OF THE MENSTRUAL CYCLE

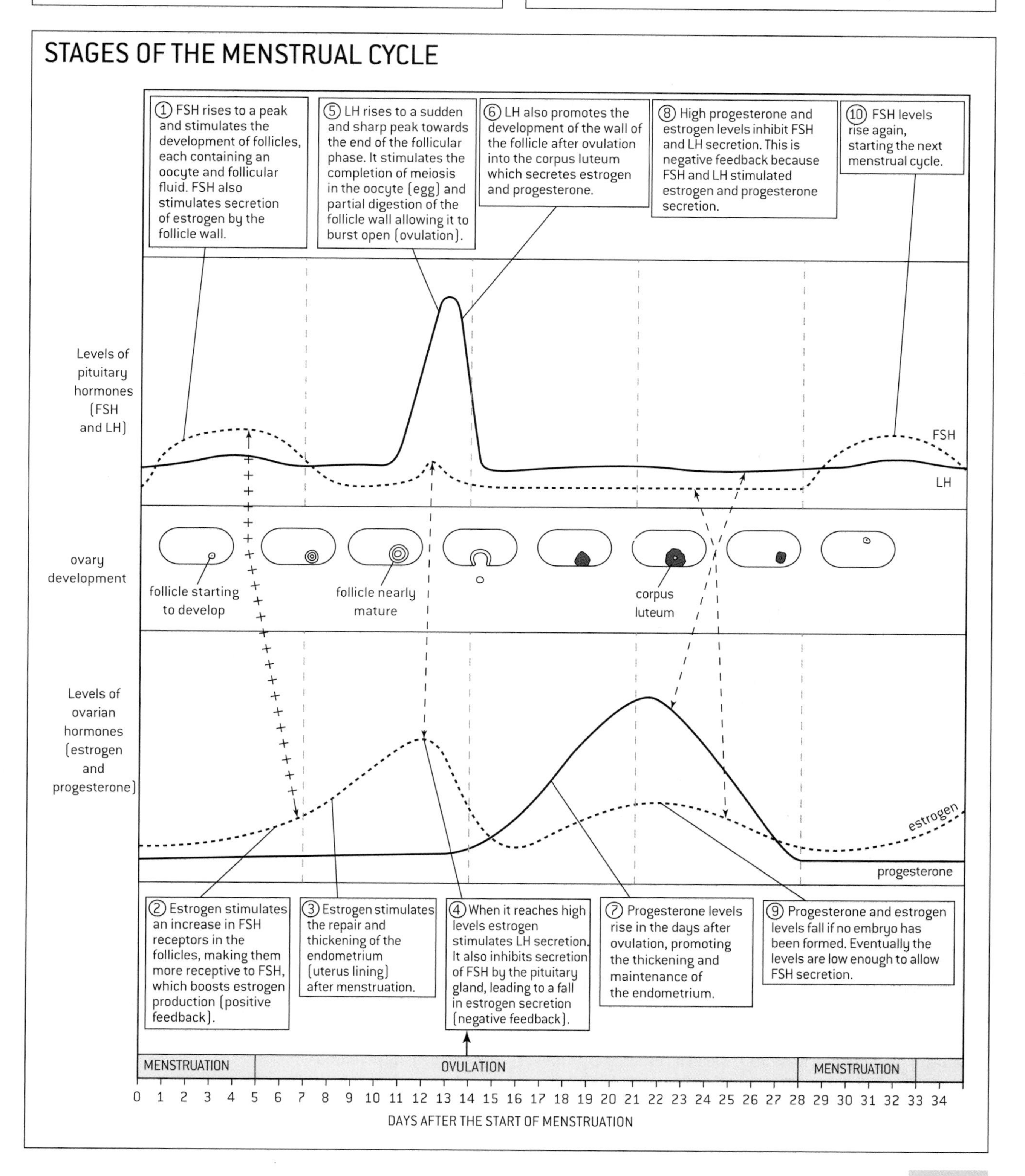

Research into reproduction

IN VITRO FERTILIZATION

Pioneering research in the second half of the 20th century led to the development of **in vitro fertilization**, often abbreviated to IVF. It has been used extensively to overcome fertility problems in either the male or female parent. The following procedures are usually used:

1. **Down-regulation**

The woman takes a drug each day, usually as a nasal spray, to stop her pituitary gland secreting FSH or LH. Secretion of estrogen and progesterone therefore also stops. This suspends the normal menstrual cycle and allows doctors to control the timing and amount of egg production in the woman's ovaries.

2. **Artificial doses of hormones**

Intramuscular injections of FSH and LH are then given daily for about ten days, to stimulate follicles to develop. The FSH injections give a much higher concentration than during a normal menstrual cycle, so far more follicles develop than usual. Twelve is not unusual and there can be as many as twenty follicles. This stage of IVF is therefore called superovulation.

3. **Egg retrieval and fertilization**

When the follicles are 18 mm in diameter they are stimulated to mature by an injection of hCG, another hormone that is normally secreted by the embryo. A micropipette mounted on an ultrasound scanner is passed through the uterus wall to wash eggs out of the follicles. Each egg is mixed with 50,000 to 100,000 sperm cells in sterile conditions in a shallow dish, which is then incubated at 37 °C until the next day.

4. **Establishing a pregnancy**

If fertilization is successful then one or more embryos are placed in the uterus when they are about 48 hours old. Because the woman has not gone through a normal menstrual cycle extra progesterone is usually given as a tablet placed in the vagina, to ensure that the uterus lining is maintained. If the embryos implant and continue to grow then the pregnancy that follows is no different from a pregnancy that began by natural conception.

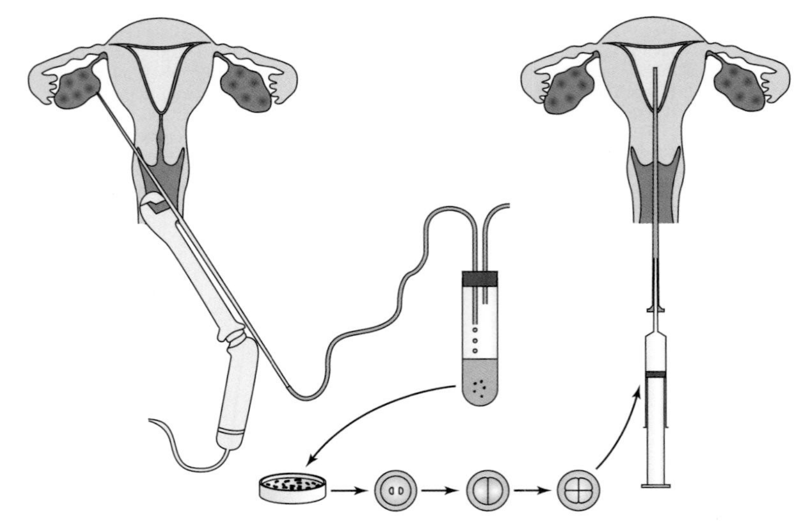

The diagrams above show egg retrieval from the ovaries, culture of eggs after in vitro fertilization and implantation of 4-cell embryos into the uterus.

HARVEY AND THE DISCOVERY OF SEXUAL REPRODUCTION

William Harvey's discovery of the circulation of blood in the 17th century shows that he was a brilliant research scientist and yet he made little progress in another area that interested him very much: reproduction in humans and other animals. He was taught the 'seed and soil' theory of Aristotle, according to which the male produces a seed, which forms an egg when it mixes with menstrual blood. The egg develops into a fetus inside the mother.

William Harvey tested Aristotle's theory using a natural experiment. Deer are seasonal breeders and only become sexually active during the autumn. Harvey examined the uterus of female deer during the mating season by slaughtering and dissecting them. He expected to find eggs developing in the uterus immediately after mating (copulation), but only found signs of anything developing in females two or more months after the start of the mating season.

He regarded his experiments with deer as proof that Aristotle's theory of reproduction was false, which it certainly is. However Harvey concluded that offspring cannot be the result of mating, which is also false. The problem for Harvey was that the gametes, the process of fertilization and early stages of embryo development are too small to see with the naked eye or a hand lens, and effective microscopes were not available when he was working. An effective microscope was not invented until 17 years after his death.

Harvey was understandably reluctant to publish his research into sexual reproduction, but he did eventually do so in 1651 when he was 73 years old in his work *Exercitationes de Generatione Animalium*. He knew that he had not solved the mystery of sexual reproduction. He was unlucky in his choice of experimental animal because embryos in the deer that he used remain microscopically small for an unusually long period.

Scientific research has often been hampered for a time by deficiencies in apparatus, with discoveries only being made following improvements. This will continue into the future and we can look forward to further transformations in our understanding of the natural world as new techniques and technology are invented.

Questions – human physiology

1. The graph shows oscilloscope traces for action potentials in a neuron and a cardiac muscle cell.

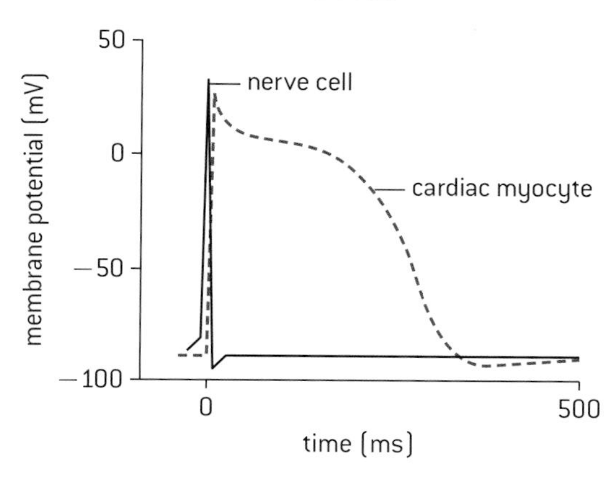

a) Estimate the resting potential of both cells. [2]

b) Compare and contrast the traces for the two cells. [3]

c) Annotate the trace for the cardiac muscle cell to show when depolarization and repolarization occur. [2]

d) Measure the time taken to repolarize each cell. [2]

e) Explain the very different repolarization times. [3]

2. The micrograph below shows a scan of a fetus at a level immediately below the diaphragm.

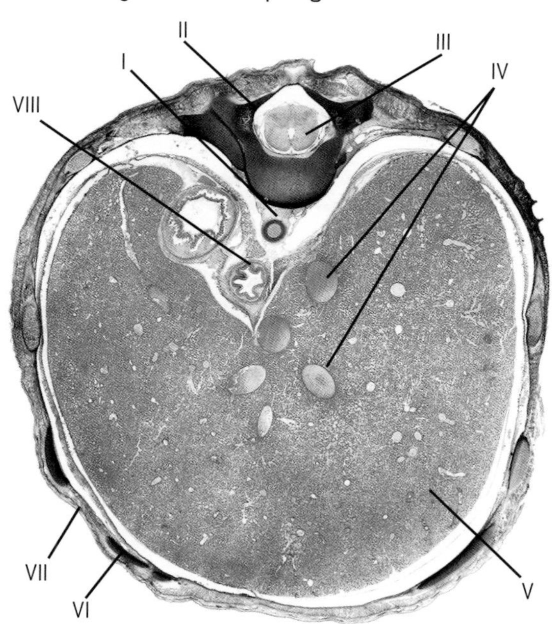

a) Deduce, with reasons, whether I and IV are arteries or veins. [4]

b) The structure labelled II is a vertebra. From your knowledge of chordates, identify structure III. [2]

c) From your knowledge of the digestive system, identify the organ labelled V. [1]

d) The structure labelled VI is a rib. Deduce, with reasons, what structure VII is. [2]

e) Structure VIII is the esophagus. It has the same layers in its wall as the small intestine. State the names of these layers from the outside inwards. [4]

f) Suggest, with a reason, which organ appears in the scan above and to the left of the esophagus. [2]

3. The diagram shows the gas exchange system.

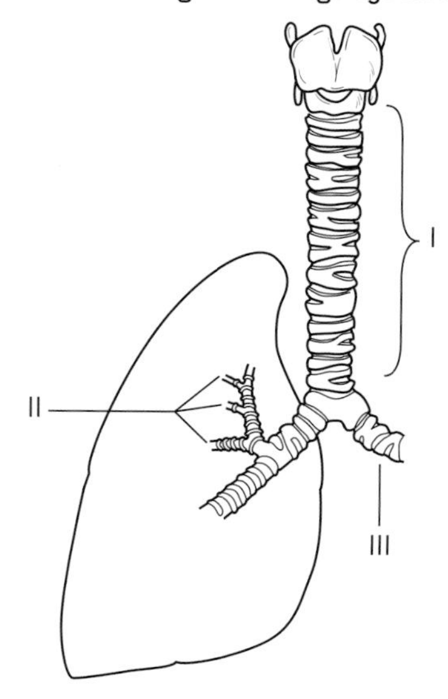

a) State the name of the parts labelled I, II and III. [3]

b) I, II and III allow the lungs to be ventilated. Explain briefly the need for ventilation. [2]

c) Draw and label a diagram of an alveolus and adjacent blood capillaries. [5]

4. Florey and Chain gave eight mice a lethal dose of *Streptococcus* bacteria and injected penicillin into four of them. Those four mice survived but the other four all died within hours. Use the instructions in Topic 4 to do a chi-squared test of association on these results:

a) Construct a contingency table to show the actual frequencies of survival and death for the treated and untreated mice and also the expected frequencies assuming no association. [4]

b) Calculate the statistic chi-squared. [4]

c) Identify the critical values for chi-squared at 5% and 1% significance levels. (see page 126 in Topic 10) [2]

d) Evaluate the hypothesis that there is no association between death or survival and whether mice were treated with penicillin, at both significance levels. [2]

e) The chi-squared test is in fact invalid with this data because the expected frequencies are not all 5 or more. Calculate the number of mice needed to give expected frequencies of 5 or more. [2]

f) Suggest two possible reasons for Florey and Chain not using larger numbers of mice in the experiment. [2]

5. a) Vitamin K and the pesticide DDT are both hydrophobic and dissolve in lipid droplets. Suggest how they are absorbed into the body in the ileum. [2]

b) A riboflavin (vitamin B_2) transport protein has recently been discovered in the membrane of small intestine epithelium cells. Outline two methods of riboflavin absorption that this protein might carry out. [4]

c) Experiments have shown that zinc absorption in the small intestine increases as the concentration of zinc in digested foods increases, until a plateau at which further increases in zinc concentration do not alter the rate of absorption. Explain these results. [4]

Landmarks in DNA research

DNA AS THE GENETIC MATERIAL

In the early 1950s it was still unclear whether genes were made of DNA or protein. **Hershey and Chase** used a virus that infects cells of the bacterium *E. coli* to investigate this. Viral proteins start being made in the cytoplasm of *E. coli* soon after the virus comes into contact with it, showing that the viral genes have entered the bacterium. The virus was T2.

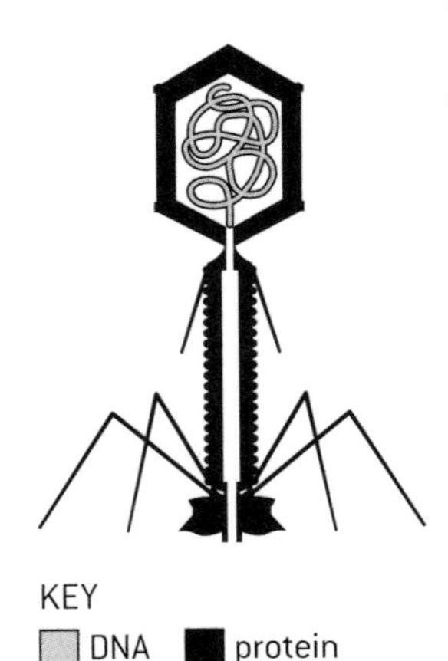

Viruses such as T2 consist only of DNA inside a protein coat. DNA contains phosphorus but not sulphur, and protein contains sulphur but not phosphorus. Hershey and Chase used this difference to prepare two strains of T2, one having its DNA radioactively labelled with ^{32}P and the other having its protein labelled with ^{35}S.

These two strains of labelled T2 were each mixed with *E. coli*. After leaving enough time for the bacteria to be infected, the mixture was agitated in a high-speed mixer and then centrifuged at 10,000 rpm to separate into a solid pellet containing the bacteria and a liquid supernatant. A Geiger counter was used to locate the radioactivity. The results are shown in the diagram.

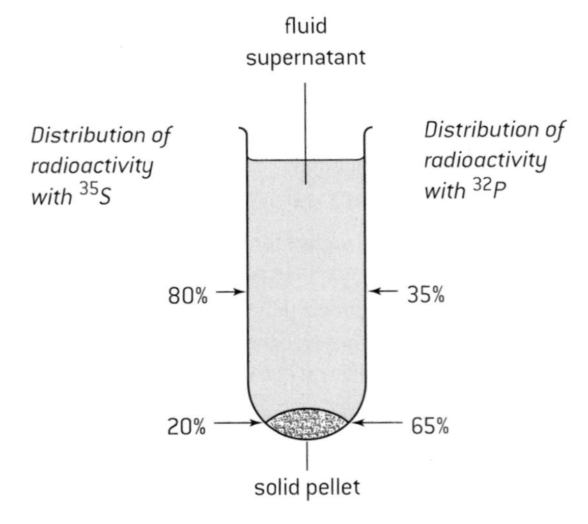

Analysis of results

T2 binds to the surface of *E. coli* and injects its DNA into the bacterium. This explains the high proportion of radioactivity with the bacteria in the pellet when ^{32}P was used. Agitation shakes many of the protein coats of the viruses off the outside of the bacteria and these coats remain in the supernatant. This explains the very high proportion of radioactivity in the supernatant when ^{35}S was used.

The small proportion of radioactivity in the pellet can be explained by the protein coats that remain attached to the bacteria and also the presence of some fluid containing protein coats in the pellet. This and other experiments carried out by Hershey and Chase give strong evidence for genes being composed of DNA rather than protein.

THE HELICAL STRUCTURE OF DNA

If a beam of X-rays is directed at a material, most of it passes through but some is scattered by the particles in the material. This scattering is called **diffraction**. The wavelength of X-rays makes them particularly sensitive to diffraction by the particles in biological molecules including DNA.

In a crystal the particles are arranged in a regular repeating pattern, so the diffraction occurs in a regular way. An X-ray detector is placed close to the sample to collect the scattered rays. The sample can be rotated in three different dimensions to investigate the pattern of scattering. Diffraction patterns can be recorded using X-ray film.

DNA cannot be crystallized, but in 1950 **Maurice Wilkins** developed a method of producing arrays of DNA molecules that were orderly enough for a diffraction pattern to be obtained, rather than random scattering.

Rosalind Franklin came to work in the same research department as Wilkins. She developed a high resolution detector that produced very clear images of diffraction patterns from DNA. The figure below shows the most famous of the diffraction patterns that she obtained.

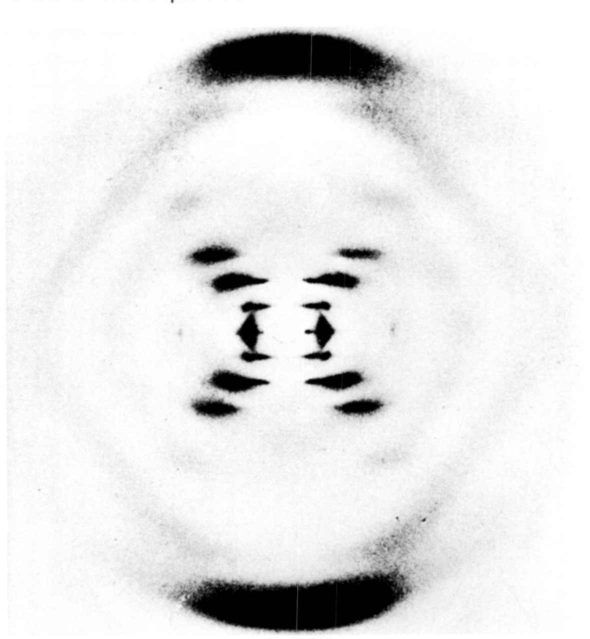

Analysis of results

From this diffraction pattern Franklin was able to make a series of deductions about the structure of DNA:

- The cross in the centre of the pattern indicated that the molecule was helical in shape.
- The angle of the cross shape showed the pitch (steepness of angle) of the helix.
- The distance between the horizontal bars showed turns of the helix to be 3.4 nm apart.

Rosalind Franklin's research is an excellent example of the importance of making careful observations in science. She was painstaking in her methods of obtaining X-ray diffraction images of DNA and in her analysis of the patterns in them. Here observations were critically important in the discovery of the double helix structure of DNA by Crick and Watson.

DNA replication

LEADING AND LAGGING STRANDS

The two ends of a strand of nucleotides in DNA or RNA are different. They are known as the 3' and 5' ends (3 prime and 5 prime). The **3' end** in DNA has a deoxyribose to which the phosphate of another nucleotide could be linked. The phosphate would bond with the –OH group on the C3 of the deoxyribose. The **5' end** in DNA has a phosphate that is attached to C5 of deoxyribose.

Nucleotides are linked to the end of a DNA strand during replication by one of a group of enzymes called **DNA polymerases.** These enzymes always add the phosphate of a free nucleotide to the deoxyribose at the 3' end of the strand. The direction of replication is therefore **5' to 3'**.

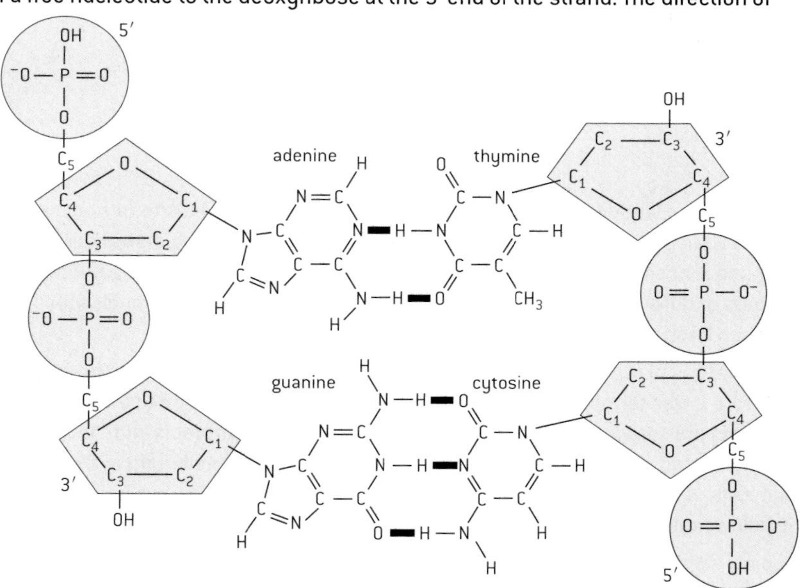

The two strands in a DNA molecule are **antiparallel** because they run in opposite directions. Each end of a DNA double helix therefore has one strand with a 3' end and one with a 5' end.

Because of the antiparallel structure of DNA, the two strands have to be replicated in different ways.

- On one strand DNA polymerases can move in the same direction as the replication fork so replication is continuous. This is the **leading strand.**
- On the other strand DNA polymerases have to move in the opposite direction to the replication fork, so replication is discontinuous. This is the **lagging strand.**

ROLES OF ENZYMES IN PROKARYOTIC DNA REPLICATION

Semi-conservative replication is carried out by a complex system of enzymes. There are differences between prokaryotes and eukaryotes in the mechanism of replication, though the basic principles are the same. The system used in prokaryotes is shown below.

① **DNA gyrase** moves in advance of helicase and relieves strains in the DNA molecule that are created when the double helix is uncoiled. Without this action the separated strands would form tight supercoils.

② **Helicase** uncoils the DNA double helix and splits it into two template strands. **Single-stranded binding proteins** keep the strands apart long enough to allow the template strand to be copied.

③ **DNA polymerase III** adds nucleotides in a 5′ to 3′ direction. On the **leading strand** it moves in the same direction as the replication fork, close to helicase.

3′
5′
primer

⑥ Short lengths of DNA are formed between RNA primers on the lagging strand, called **Okazaki fragments.**

5′
3′
2

⑧ **DNA ligase** seals up the nick by making another sugar-phosphate bond.

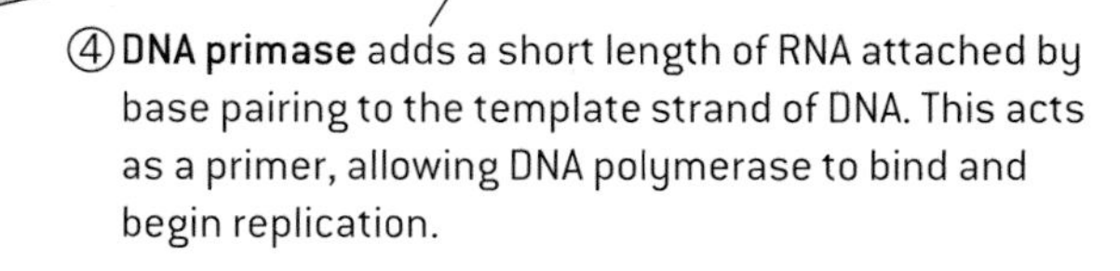

④ **DNA primase** adds a short length of RNA attached by base pairing to the template strand of DNA. This acts as a primer, allowing DNA polymerase to bind and begin replication.

3′
5′

⑦ **DNA polymerase I** removes the RNA primer and replaces it with DNA. A nick is left in the sugar-phosphate backbone of the molecule where two nucleotides are still unconnected.

⑤ DNA polymerase III starts replication next to the RNA primer and adds nucleotides in a 5′ to 3′ direction. It therefore moves away from the replication fork on the **lagging strand.**

Base sequences in DNA

SANGER SEQUENCING

Frederick Sanger developed a method of base sequencing that was used very widely for 25 years. It is based on nucleotides of dideoxyribonucleic acid (ddNA). These contain **dideoxyribose** instead of deoxyribose, so have no –OH group on carbon atom 3.

Deoxynucleotide

Dideoxynucleotide

If a dideoxynucleotide is at the end of a strand of DNA, there is no site to which another nucleotide can be added by a 5' to 3' linkage. In the sequencing machine single-stranded copies of the DNA being sequenced are mixed with DNA polymerase and normal DNA nucleotides, plus smaller numbers of ddNA nucleotides. The replication is repeated four times, once with dideoxynucleotides with each base, A, C, G and T.

The fragments of replicated DNA that are produced vary in length depending on how far replication got before it was terminated because a ddNA nucleotide was added to the end of the chain. The fragments are separated according to length by **gel electrophoresis** with four tracks, one for each base in the ddNA nucleotide that terminated replication. Each band in the gel represents one length of DNA fragment produced by replication. All the fragments of the same length end in the same base, so there is only a band in one of the four tracks for each length of fragment. This allows the base sequence of the DNA to be deduced quite easily from the gel.

A typical section of gel is shown (right). Part of the base sequence is indicated.

The whole base sequence can easily be deduced. This was initially done by hand but fluorescent markers were introduced that allowed the base sequence to be read by a machine.

FUNCTIONS OF DNA BASE SEQUENCES

There are thousands of sequences of bases that code for proteins in the DNA of a species. These **coding sequences** are transcribed and translated when a cell requires the protein that they code for.

There are also **non-coding sequences**. Some non-coding sequences have important functions.

- **Regulating gene expression** – some base sequences are sites where proteins can bind that either promote or repress the transcription of an adjacent gene.
- **Introns** – in many eukaryote genes the coding sequence is interrupted by one or more non-coding sequences. These introns are removed from mRNA before it is translated. Introns have numerous functions associated with mRNA processing.
- **Telomeres** – these are repetitive base sequences at the ends of chromosomes. When the DNA of a eukaryote chromosome is replicated, the end of the molecule cannot be replicated, so a small section of the base sequence is lost. The presence of the telomere prevents parts of important genes at the ends of the chromosomes from being lost each time DNA is replicated.
- **Genes for tRNA and rRNA** – transcription of these genes produces the transfer RNA used during translation and also the ribosomal RNA that forms much of the structure of the ribosome.

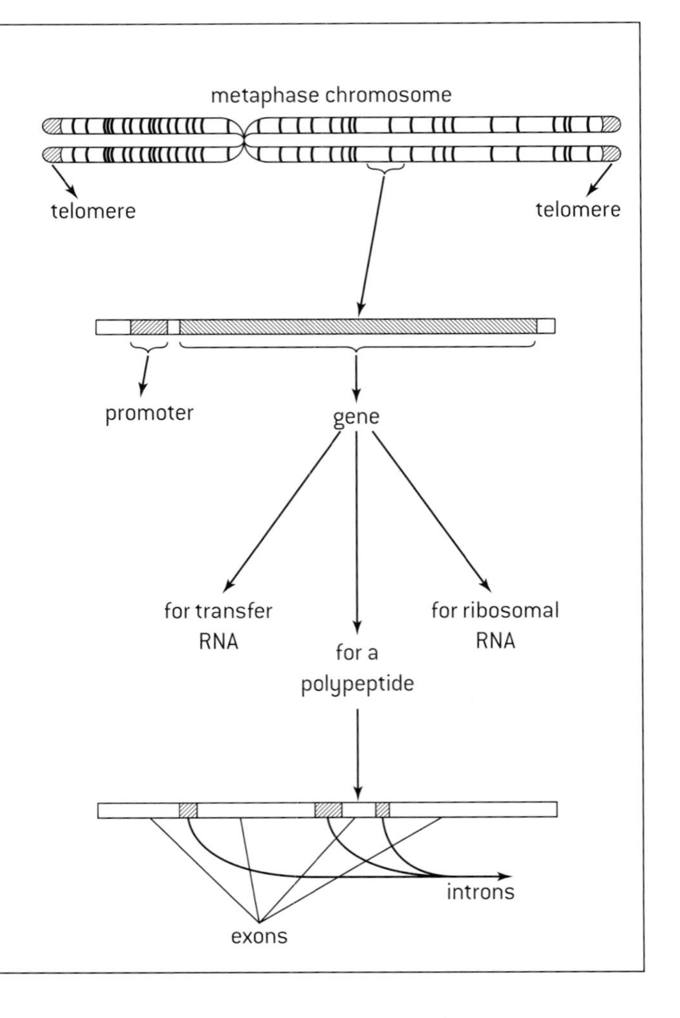

Bioinformatics and nucleosomes

BIOINFORMATICS

Computers now allow huge amounts of data to be stored and analysed, allowing the branch of biology called bioinformatics to develop. Base sequences are the main type of data stored and analysed in bioinformatics. Sequencing was at first only possible with short lengths of DNA such as individual genes, but now whole genomes can be sequenced and the amount of data generated is growing exponentially.

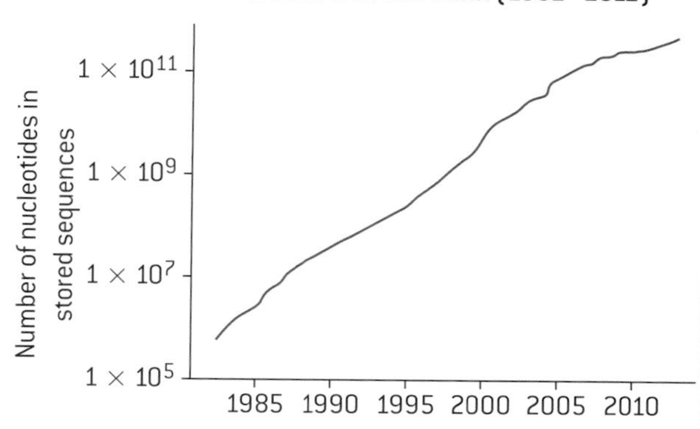

One of the main types of analysis in bioinformatics is locating genes that code for polypeptides within genomes. This is done using computers to search for ORFs (open reading frames). The details of this procedure are described in Option B.

Another type of analysis is to search for conserved sequences in the genomes of different organisms. These are base sequences similar enough for them to have been most likely inherited from a common ancestral gene. The conserved sequences are analysed to find differences in the base sequences (see Topic 3). Classification of living organisms has been revolutionized by these techniques (Topic 5).

TANDEM REPEATS

Within the genomes of humans and other species there are regions where adjacent sections of DNA have the same base sequence. These are called **tandem repeats**.

The length of the repeated sequence can be anything from two bases to 60 or more.

Examples:

ACACACAC

– two nucleotide repeat (dimeric)

GATAGATAGATAGATAGATA

– four base repeat (tetrameric)

The number of repeats varies between different individuals with some tandem repeats. These are therefore known as **variable number tandem repeats**.

DNA profiling (fingerprinting) is based on variable number tandem repeats. The methods used are described in Topic 3.

NUCLEOSOMES

DNA in eukaryotes is associated with proteins to form nucleosomes. These are globular structures that have a core of eight histone proteins with DNA wrapped around. Another histone protein called H1 binds the DNA to the core. A short section of linker DNA connects one nucleosome to the next.

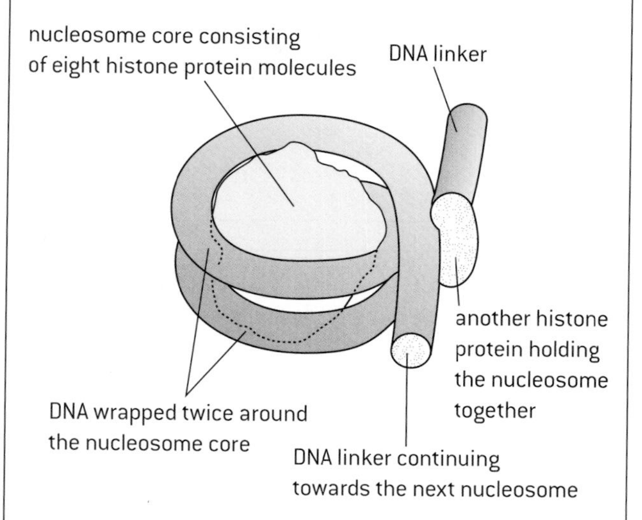

The eight histones in the core have **N-terminal tails** that extend outwards from the nucleosome. During the condensation of chromosomes in the early stages of mitosis and meiosis the tails of histones in adjacent nucleosomes link up and pull the nucleosomes together. This is part of the process of **supercoiling**.

During interphase, changes to the nucleosomes allow chromosomes to decondense (uncoil). The N-terminal tails are reversibly modified by adding acetyl or methyl groups. This prevents adjacent nucleosomes from packing together. The H1 histone protein is removed so the binding of DNA to the nucleosome core is loosened. The DNA then resembles a string of beads. Where these changes occur they allow access to the DNA by polymerase enzymes that carry out replication and transcription. Some sections of chromosomes remain condensed during interphase and genes in these sections are therefore not transcribed. Nucleosomes thus help to regulate transcription in eukaryotes, by controlling which sections of the chromosomes are condensed or decondensed during interphase.

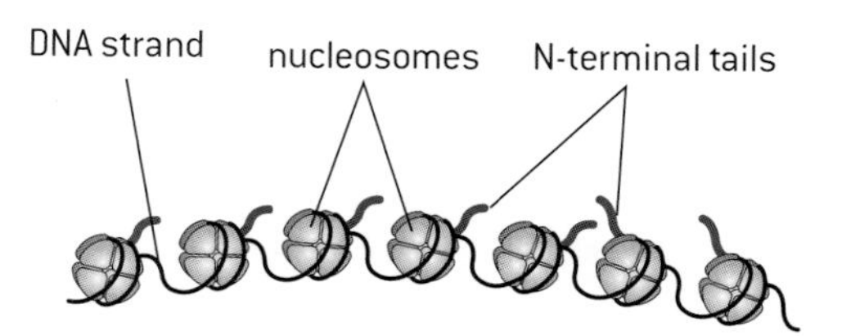

JMol molecular visualization software can be used to analyse the association between protein and DNA within nucleosomes. Go to 'Molecule of the Month' on the Protein Data Base (**PDB**). Select 'Nucleosome' and then 'DNA in a nucleosome' in the list of discussed structures. The JMol image of a nucleosome can be rotated and coloured in different ways to show components.

Gene expression

STAGES IN GENE EXPRESSION

Gene expression is the production of mRNA by transcription of a gene and then the production of polypeptides by translation of the mRNA. In prokaryotes translation can occur immediately after transcription, because there is no nuclear membrane. Translation can even begin before an mRNA molecule has been fully transcribed. In eukaryotes the mRNA is produced by transcription in the nucleus. It is modified while still in the nucleus, then passes out to the cytoplasm via nuclear pores and is translated in the cytoplasm.

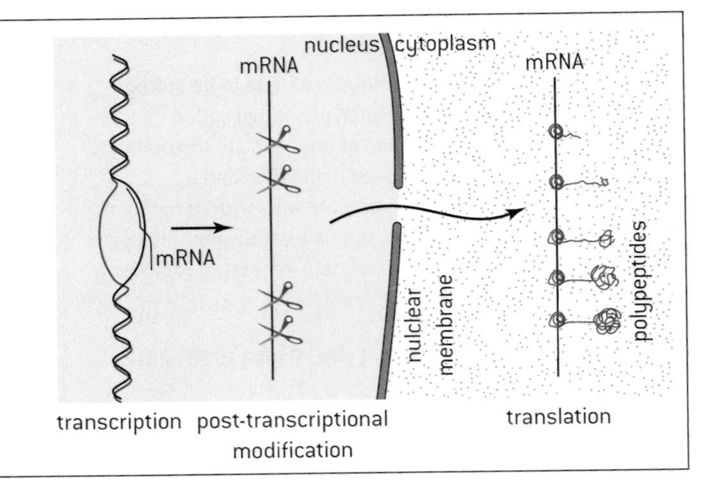

PROMOTERS AND TRANSCRIPTION

Gene expression can be controlled at the transcription stage – at any time in the life of a cell some genes in the nucleus are being transcribed and others are not. Control of gene expression involves a **promoter**. This is a base sequence close to the start of a gene. Every gene has a promoter, but the base sequences vary, allowing particular genes to be transcribed and not others. The promoter is not itself transcribed and does not code for an amino acid sequence, so it is an example of **non-coding DNA** with a function.

RNA polymerase (RNAP) binds directly to the promoter in prokaryotes and then starts transcribing. Repressor proteins can bind to the promoter and prevent transcription. The control of gene expression is more complicated in eukaryotes. Proteins called **transcription factors** bind to the promoter, which allows RNAP to bind and then initiate transcription. Several transcription factors are required, some of which may need to be activated by the binding of a hormone or other chemical signal. Repressor proteins can prevent transcription.

After transcription has been initiated RNAP moves along the gene, assembling an RNA molecule one nucleotide at a time. RNAP adds the 5' end of the free RNA nucleotide to the 3' end of the growing mRNA molecule, so transcription occurs in a 5' to 3' direction. The elongation of RNA by transcription was described in Topic 2. Transcription is terminated at the end of the gene and the DNA, RNA and RNAP separate.

IDENTIFYING POLYSOMES

The figure below is an electron micrograph showing groups of ribosomes called polysomes (or polyribosomes). A polysome is a group of ribosomes moving along the same mRNA, as they simultaneously translate it.

In the micrograph below the arrow indicates where transcription of a prokaryote gene is being initiated. Along the DNA of the gene are nine polysomes. Only in prokaryotes can translation begin before transcription is finished.

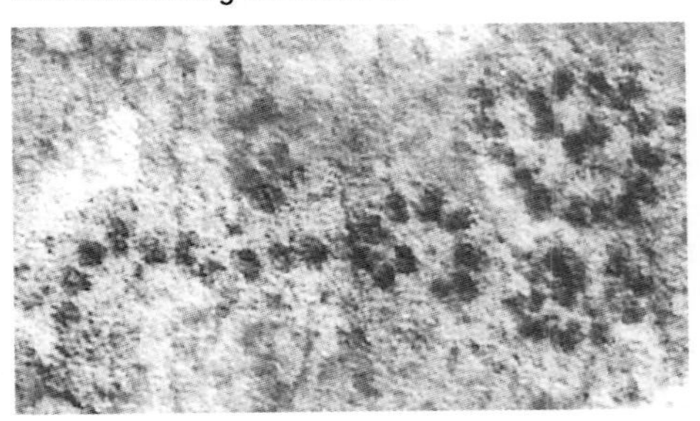

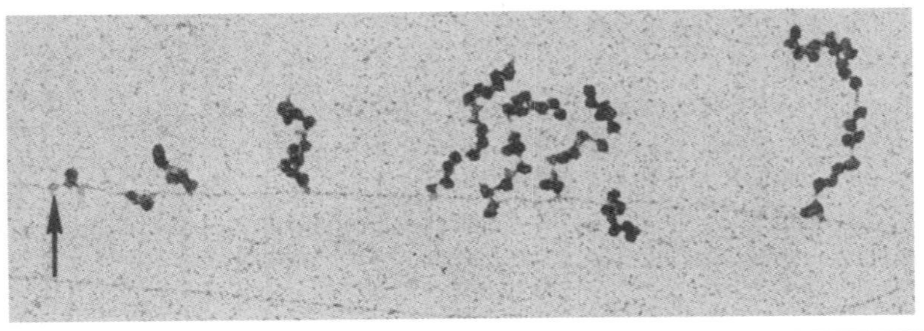

Epigenetics

INHERITANCE OF ACQUIRED CHARACTERISTICS AND EPIGENETICS

A fundamental theory of modern biology is that characteristics acquired during an individual's lifetime cannot be inherited by their offspring. The alternative theory that acquired characteristics can be inherited was propounded by the French biologist Lamarck, so is referred to as **Lamarckism**.

Evidence has sometimes been presented for inheritance of acquired characteristics, but has been falsified and Lamarckism was dismissed as tantamount to heresy. The discovery that DNA is the genetic material added to the evidence against Lamarckism – the environment of an individual during its lifetime cannot cause specific changes to the base sequences of their genes.

Nevertheless there is mounting evidence that the environment can indeed trigger heritable changes. One explanation involves small chemical markers that are attached to DNA in the nucleus of a cell to fix the pattern of gene expression. These markers are usually passed to daughter cells formed by mitosis, and help to establish tissues with common patterns of differentiation, but they are mostly erased during the gamete formation. However a small percentage of markers persists and is inherited by offspring.

The pattern of chemical markers established in the DNA of a cell is the **epigenome** and research into it is **epigenetics**.

Example of epigenetic inheritance: Methylation is one type of chemical marker. Variations in the pattern of methylation that affect height and flowering time in the model organism *Arabidopsis thaliana* (left) have been shown to be inherited over at least eight generations.

METHYLATION AND EPIGENETICS

Cytosine in DNA can be converted to methylcytosine by the addition of a methyl group ($-CH_3$). This change is catalysed by an enzyme and only happens where there is guanine on the 3' side of the cytosine in the base sequence. In some eukaryotes there is widespread **methylation** in parts of the genome.

Methylation inhibits transcription, so is a means of switching off expression of certain genes. The cells in a tissue can be expected to have the same pattern of methylation and this pattern can be inherited in daughter cells produced by mitosis. Environmental factors can influence the pattern of methylation and gene expression.

Fluorescent markers can be used to detect patterns of methylation in the chromosomes. Analysis of the patterns has revealed some trends:

1. Patterns of methylation are established during embryo development and the percentage of C-G sites that are methylated reaches a maximum at birth in humans but then decreases during the rest of an individual's life.

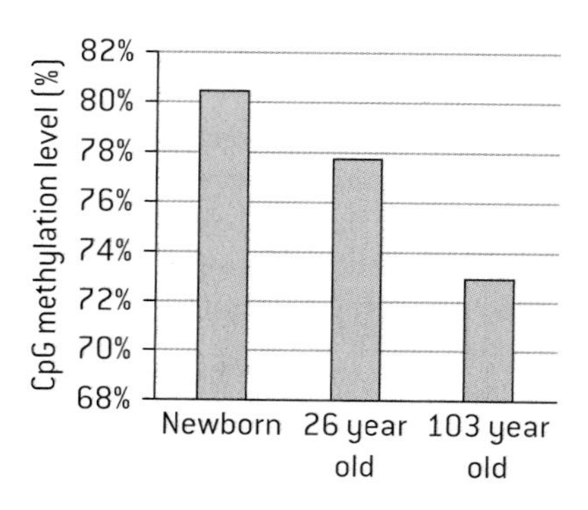

2. At birth identical twins have a very similar pattern of methylation, but differences accumulate during their lifetimes, presumably due to environmental differences. This is reflected in the decreasing similarity between identical twins as they grow older.

POST-TRANSCRIPTIONAL MODIFICATION

Eukaryotic cells modify mRNA after transcription. This happens before the mRNA exits the nucleus. In many eukaryote genes the coding sequence is interrupted by one or more non-coding sequences. These **introns** are removed from mRNA before it is translated. The remaining parts of the mRNA are **exons**. They are spliced together to form mature mRNA.

Some genes have many exons and different combinations of them can be spliced together to produce different proteins. This increases the total number of proteins an organism can produce from its genes.

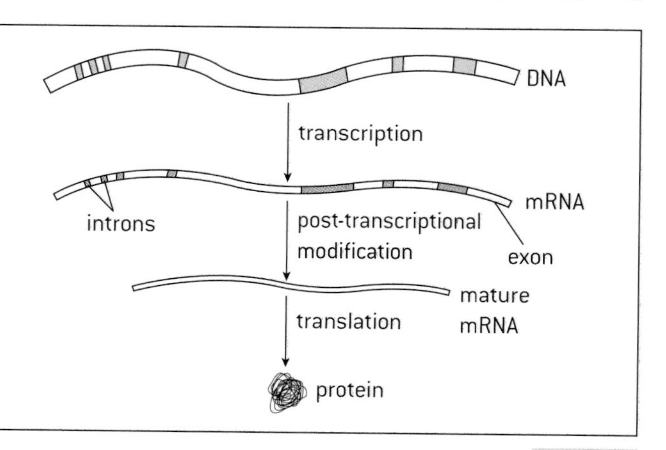

Ribosomes and transfer RNA

TRANSFER RNA

All transfer RNA molecules have:

- double-stranded sections with base pairing
- a triplet of bases called the **anticodon**, in a loop of seven bases, plus two other loops
- the base sequence CCA at the 3' terminal, which forms a site for attaching an amino acid.

These features allow all tRNA molecules to bind to three sites on the ribosome – the A, P and E sites.

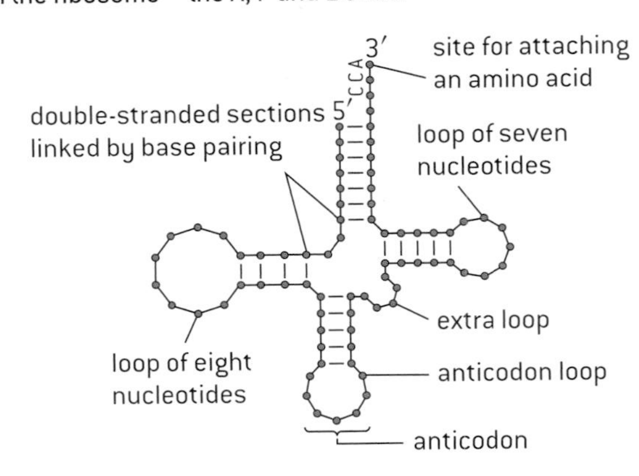

The base sequence of tRNA molecules varies and this causes some variable features in its structure. These give each type of tRNA a distinctive three-dimensional shape and distinctive chemical properties. This allows the correct amino acid to be attached to the 3' terminal by an enzyme called a tRNA activating enzyme. There are twenty different tRNA activating enzymes – one for each of the twenty different amino acids. Each of these enzymes attaches one particular amino acid to all of the tRNA molecules that have an anticodon corresponding to that amino acid. The tRNA activating enzymes recognize these tRNA molecules by their shape and chemical properties. This is an excellent example of **enzyme–substrate specificity**.

Energy from ATP is needed for the attachment of amino acids to tRNA. ATP and the appropriate amino acid and tRNA bind to the active site of the activating enzyme. A pair of phosphates is released from ATP and the remaining AMP bonds to the amino acid, raising its energy level. This energy allows the amino acid to bond to the tRNA. The energy from ATP later allows the amino acid to be linked to the growing polypeptide chain during translation.

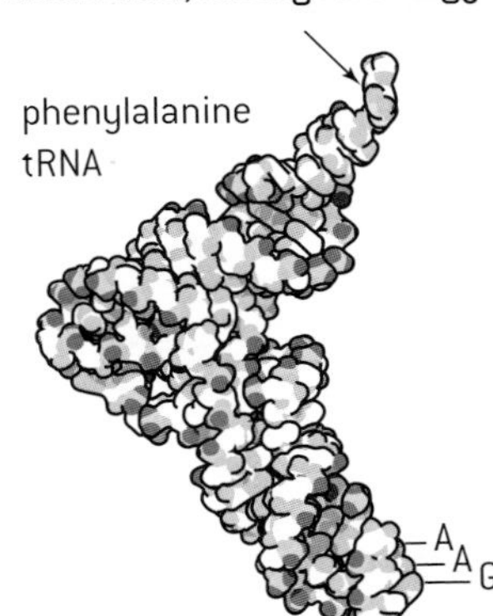

Images of tRNA molecules made using molecular visualization software can be obtained from the Protein Data Bank and viewed with JMol.

The image (above) shows a space-filling model of a tRNA for the amino acid phenylalanine. The position of the amino acid is indicated by the arrow and the anticodon by the three letters near the base.

RIBOSOMES

Ribosomes have a complex structure, with these features.

- Proteins and ribosomal RNA molecules (rRNA) both form part of the structure.
- There are two sub-units, one large and one small.
- There is a binding site for mRNA on the small sub-unit.
- There are three binding sites for tRNA on the large sub-unit:
  - A site for tRNA bringing in an **a**mino acid
  - P site for the tRNA carrying the growing **p**olypeptide
  - E site for the tRNA about to **e**xit the ribosome.

The structure of a ribosome is shown in outline (below).

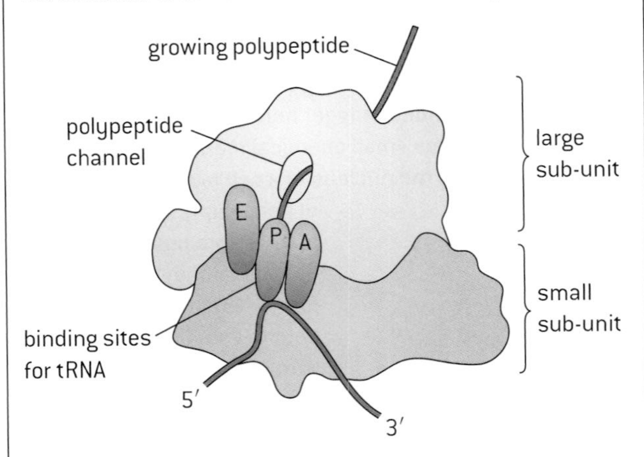

Much more accurate images of ribosome structure can be made using molecular visualization software. The image below is from the Protein Data Bank. This site allows a variety of three-dimensional coloured images of ribosomes to be produced and viewed from any angle.

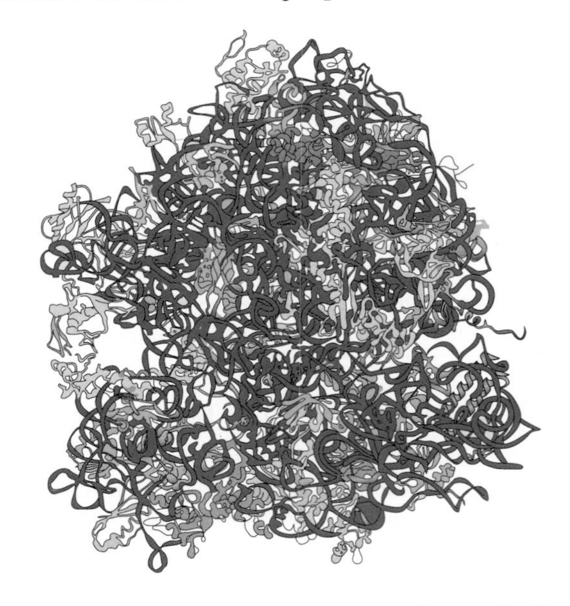

In the cytoplasm there are **free ribosomes** that synthesize proteins primarily for use within the cell. There are also ribosomes attached to membranes of the endoplasmic reticulum. They are called **bound ribosomes** and synthesize proteins for secretion from the cell or for use in lysosomes.

Translation

INITIATION OF TRANSLATION

A sequence of events occurs once, to start the process of translation:

1. The small sub-unit of the ribosome binds to mRNA with the start codon in a specific position on the mRNA binding site of the small sub-unit.
2. A tRNA with an anticodon complementary to the start codon binds. The start codon is usually AUG, so a tRNA with the anticodon UAC binds. This tRNA carries the amino acid methionine.

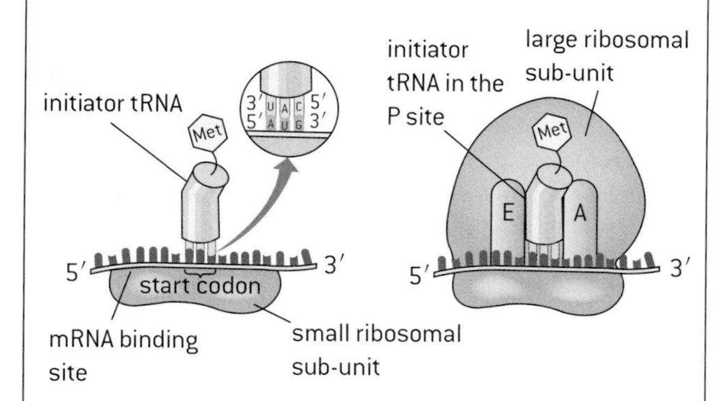

3. The large sub-unit of the ribosome binds to the small unit. The mRNA is positioned so that the initiator tRNA carrying methionine is in the P site. The E and A sites are vacant.
4. A tRNA with an anticodon complementary to the codon adjacent to the start codon binds to the A site.
5. A peptide bond forms between the amino acids held by the tRNAs in the P and A sites.

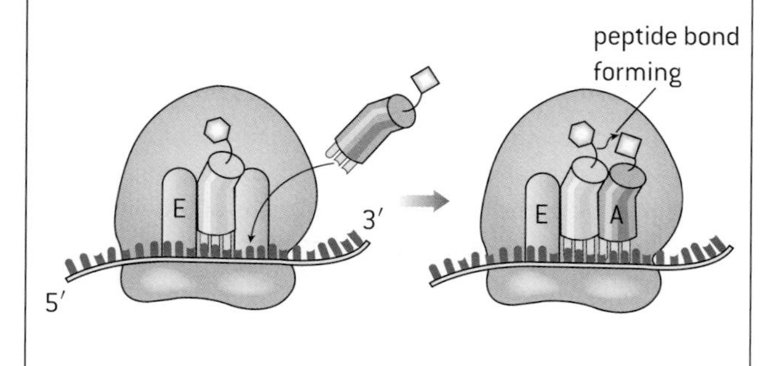

ELONGATION

The elongation of polypeptides involves a repeated cycle of events.

1. The ribosome moves three bases on along the mRNA towards the 3' end. This moves the tRNA in the P site to the E site and the tRNA carrying the growing polypeptide from the A to the P site, so the A site becomes vacant.
2. The tRNA in the E site detaches and moves away so this site is also vacant.
3. A tRNA with an anticodon complementary to the next codon on the mRNA binds to the A site.
4. The growing polypeptide that is attached to the tRNA in the P site is linked to the amino acid on the tRNA in the A site by the formation of a peptide bond.

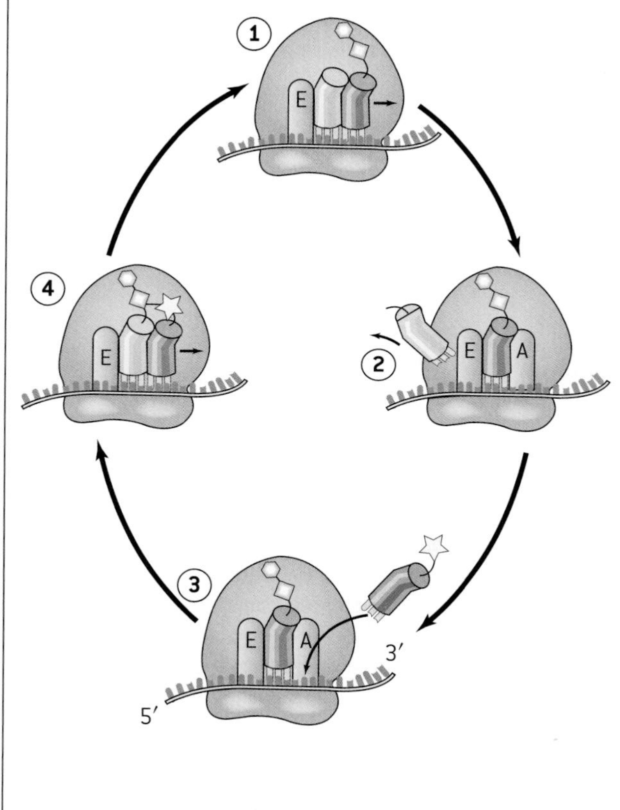

TERMINATION OF TRANSLATION

1. The ribosome moves along the mRNA in a 5' to 3' direction, translating each codon into an amino acid on the elongating polypeptide, until it reaches a stop codon.
2. No tRNA molecule has the complementary anticodon and instead release factors bind to the A site, causing the release of the polypeptide from the tRNA in the P site.
3. The tRNA detaches from the P site, the mRNA detaches from the small sub-unit, and the large and small sub-units of the ribosome separate.

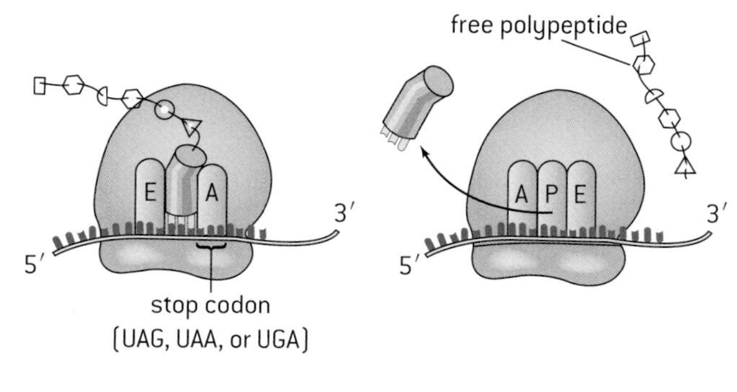

Primary and secondary structure of proteins

PRIMARY STRUCTURE

Protein structure is complex and is most easily understood by thinking about it in terms of four levels of structure, from primary to quaternary.

A molecule of **protein** contains one or more polypeptides.

A **polypeptide** is an unbranched chain of amino acids, linked by peptide bonds.

Primary structure is the **number** and **sequence of amino acids** in a polypeptide.

Most polypeptides consist of between 50 and 1,000 amino acids. The primary structure is determined by the base sequence of the gene that codes for the polypeptide.

An example of primary structure is given below: beta-endorphin, a protein consisting of a single polypeptide of 31 amino acids that acts as a neurotransmitter in the brain.

N-terminal
Tyrosine
Glycine
Glycine
Phenylalanine
Methionine
Threonine
Serine
Glutamic acid
Lysine
Serine
Glutamine
Threonine
Proline
Leucine
Valine
Threonine
Leucine
Phenylalanine
Lysine
Asparagine
Alanine
Isoleucine
Isoleucine
Lysine
Asparagine
Alanine
Tyrosine
Lysine
Lysine
Glycine
Glutamic acid
C-terminal

SECONDARY STRUCTURE

Polypeptides have a main chain consisting of a repeating sequence of covalently bonded carbon and nitrogen atoms: N – C – C – N – C – C and so on. Each nitrogen atom has a hydrogen atom bonded to it (N–H). Every second carbon atom has an oxygen atom bonded to it (C=O). This can be seen in the molecular diagram of beta-endorphin (below). Individual carbon atoms are not shown but occur at each point where lines indicating bonds meet. (The chain is shown folded so it fits across the page.)

Hydrogen bonds can form between the N – H and C = O groups in a polypeptide if they are brought close together. For example, if sections of polypeptide run parallel, hydrogen bonds can form between them. The structure that develops is called a **beta-pleated sheet**. If the polypeptide is wound into a right-handed helix, hydrogen bonds can form between adjacent turns of the helix. The structure that develops is called an **alpha helix**. Because the groups forming hydrogen bonds are regularly spaced, alpha helices and beta-pleated sheets always have the same dimensions. The formation of alpha helices and beta-pleated sheets stabilized by hydrogen bonding is the **secondary structure** of a polypeptide.

The diagrams left and below show the structure of an α-helix, a β-pleated sheet and also the position of secondary structures in lysozyme, using the ribbon model. Sections of α-helix are represented by helical ribbons and sections of β-pleated sheet by arrows.

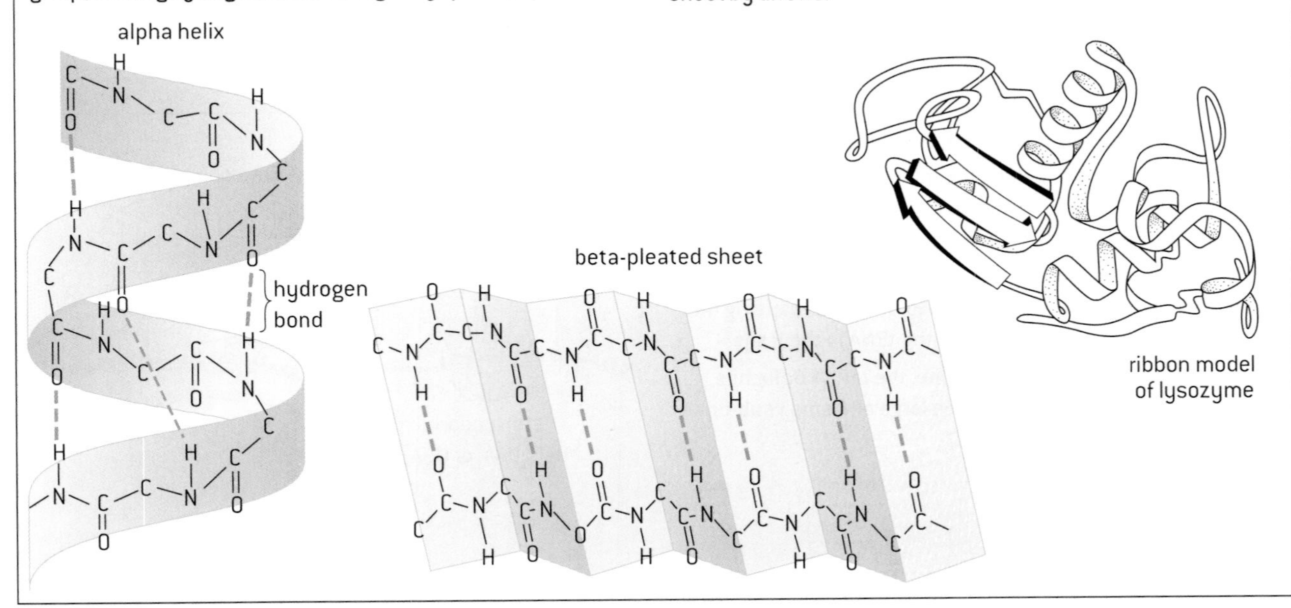

Tertiary and quaternary structure of proteins

TERTIARY STRUCTURE

Tertiary structure is the three-dimensional conformation of a polypeptide. It is formed when a polypeptide folds up after being produced by translation. The conformation is stabilized by intramolecular bonds and interactions that form between amino acids in the polypeptide, especially between their R groups. Intramolecular bonds are often formed between amino acids that are widely separated in the primary structure but which are brought together during the folding process. In water-soluble proteins non-polar amino acids are often in the centre, with hydrophobic interactions between them. Polar amino acids are on the surface where they bond to each other and come into contact with water. The figure below shows the tertiary structure of lysozyme using the sausage model.

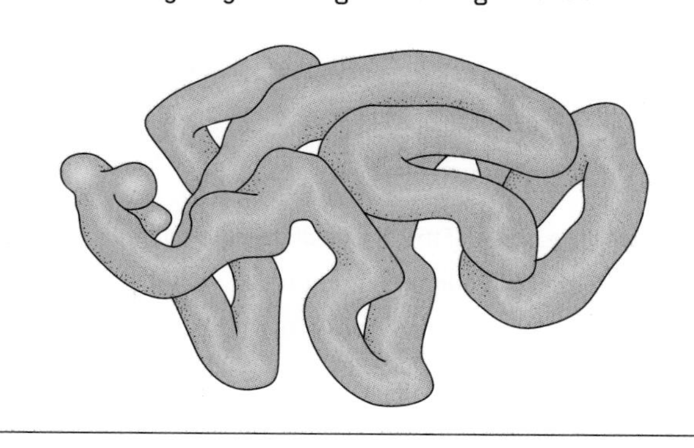

QUATERNARY STRUCTURE

Quaternary structure is the linking of two or more polypeptides to form a single protein. For example, insulin consists of two polypeptides linked together, collagen consists of three and hemoglobin consists of four. The same types of intramolecular bonding are used as in tertiary structure, including ionic bonds, hydrogen bonds, hydrophobic interactions and disulphide bridges. In some cases proteins also contain a non-polypeptide structure called a prosthetic group. For example, each polypeptide in hemoglobin is linked to a heme group, which is not made of amino acids. Proteins with a prosthetic group are called conjugated proteins. The figure below shows the quaternary structure of hemoglobin.

INTRAMOLECULAR BONDING IN TERTIARY AND QUATERNARY STRUCTURE

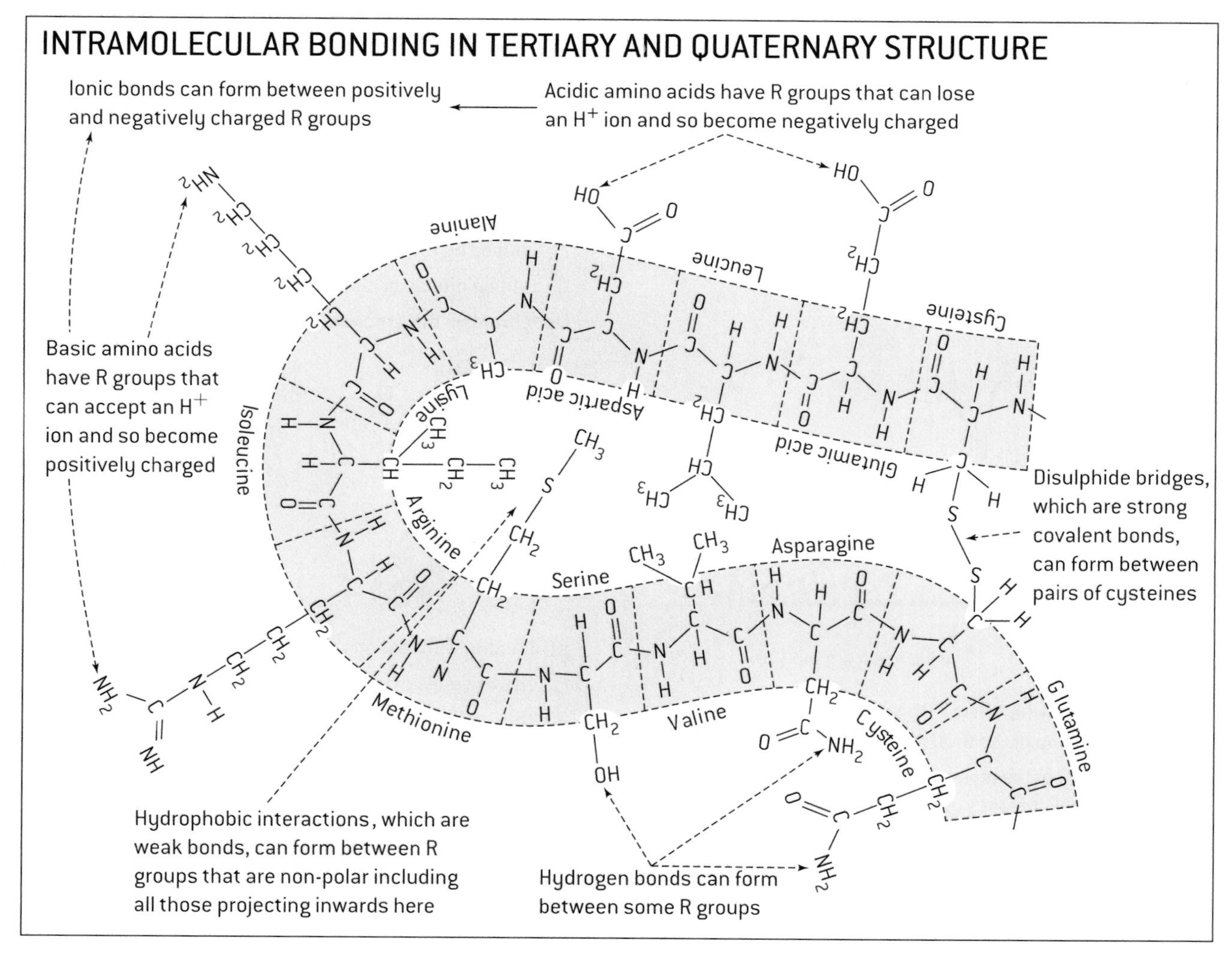

Questions – nucleic acids

1. To where do DNA polymerases start adding nucleotides during replication?

   A. 3′ end of an RNA primer   C. 3′ end of a DNA primer

   B. 5′ end of an RNA primer   D. 5′ end of a DNA primer

2. Which enzymes uncoil DNA during replication?

   A. helicase and gyrase   C. primase and ligase

   B. gyrase and primase   D. ligase and helicase

3. What are the functions of nucleosomes?

   I regulating transcription

   II making ribosomes

   III supercoiling DNA

   A. I and II only   C. I and III only

   B. II and III only   D. I, II and III

4. What protein structure includes prosthetic groups?

   A. primary   B. secondary

   C. tertiary   D. quaternary

5. *E. coli* were infected with T4 viruses and then started replicating T4 DNA. Radioactively labelled DNA nucleotides were given for between 2 and 120 seconds. DNA was extracted, split into single strands and separated by centrifugation according to the length of the strands. The shorter the strand of DNA, the closer it was to the top of the centrifuge tube. The graph shows amounts of DNA at each level in the tube measured by the radioactivity.

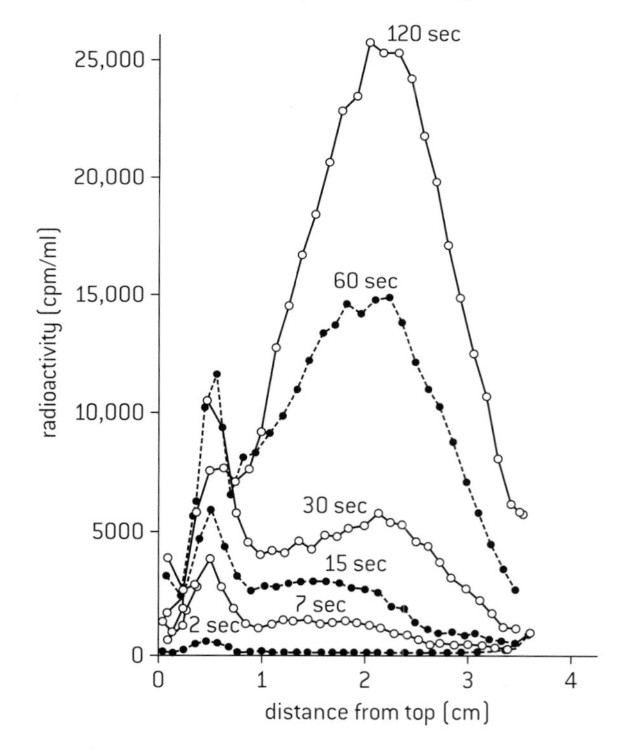

   a) (i) State whether the DNA strands at 0.5 cm from the top of the tube were short or long. [1]

   (ii) Suggest reasons for short strands of DNA even with increases in replication time. [2]

   (iii) Deduce the name for these strands of DNA. [1]

   b) (i) Distinguish between the 60 s and 120 s results. [3]

   (ii) Explain the differences between these results by the activity of DNA polymerases and ligase. [3]

6. The diagram below represents the structure of a methyltransferase enzyme in a ribbon/surface model. The enzyme's two substrates, DNA and the amino acid cysteine, are shown bound to the active site.

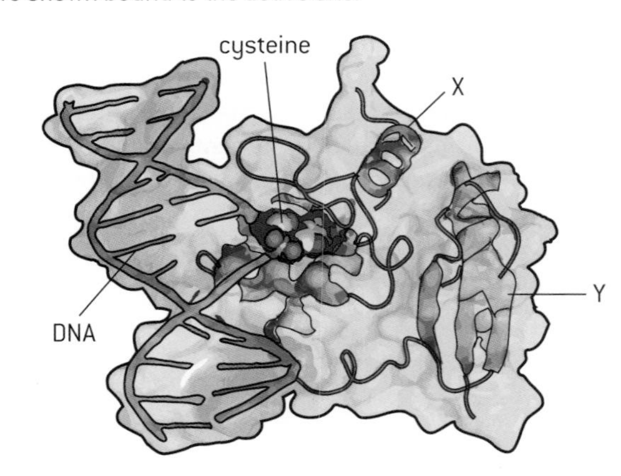

   a) State the name of the shape of this type of protein. [1]

   b) State what the primary structure of a protein is. [1]

   c) In the regions labelled X and Y two different types of secondary structure are found.

   (i) Identify each type of secondary structure. [2]

   (ii) State the type of bonding that is used to stabilize these structures. [1]

   d) Explain the importance of the tertiary structure of this protein to its function. [2]

   e) This enzyme removes methyl groups that have become attached to guanine in DNA and transfers them to cysteine. This type of DNA repair prevents mismatches of bases during DNA replication. Explain the harm caused by incorrect pairing of bases in replication. [2]

   f) Outline effects of cytosine methylation in DNA. [3]

7. Models of two tRNA molecules are shown below.

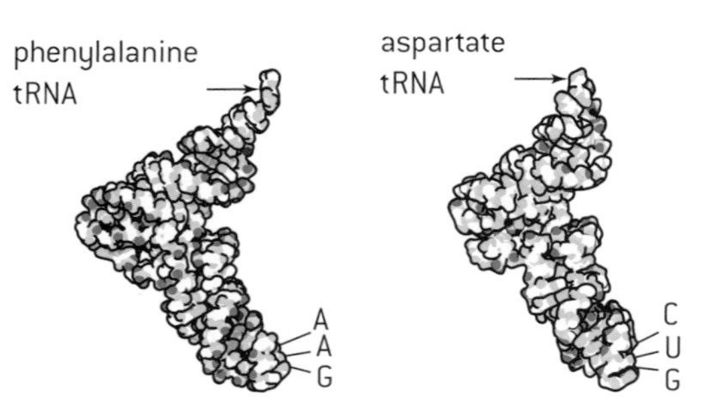

   a) Identify with reasons the parts of tRNA labelled with

   (i) a series of three letters. [3]

   (ii) an arrow. [2]

   b) Outline the relationship between tRNA, translation and transcription. [2]

   c) Explain the importance of the both differences and similarities between structure of the two tRNA molecules. [3]

Enzymes and activation energy

ENERGY CHANGES IN CHEMICAL REACTIONS

During chemical reactions, reactants are converted into products. Before a molecule of the reactant can take part in the reaction, it has to gain some energy. This is called the **activation energy** of the reaction. The energy is needed to break bonds within the reactant.

Later during the progress of the reaction, energy is given out as new bonds are made. Most biological reactions are exothermic – the energy released is greater than the activation energy.

Enzymes reduce the activation energy of the reactions that they catalyse and therefore make it easier for these reactions to occur. The graph (right) shows energy changes during uncatalysed and catalysed exothermic reactions.

The chemical environment provided by the active site for the substrate causes changes within the substrate molecule, which weakens its bonds. The substrate is changed into a transition state, which is different from the transition state during the reaction when an enzyme is not involved. The transition state achieved during binding to the active site has less energy, and this is how enzymes are able to reduce the activation energy of reactions.

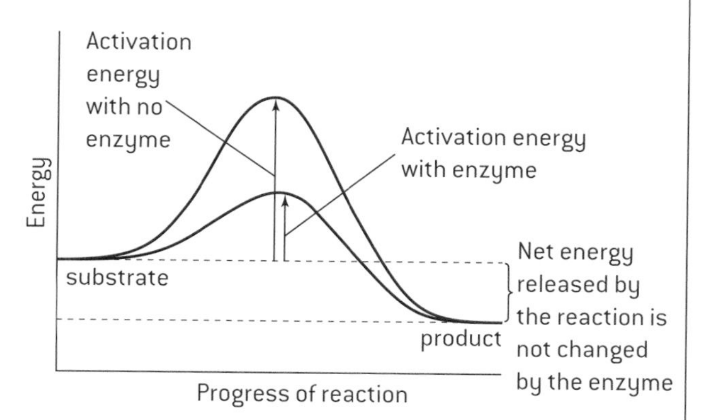

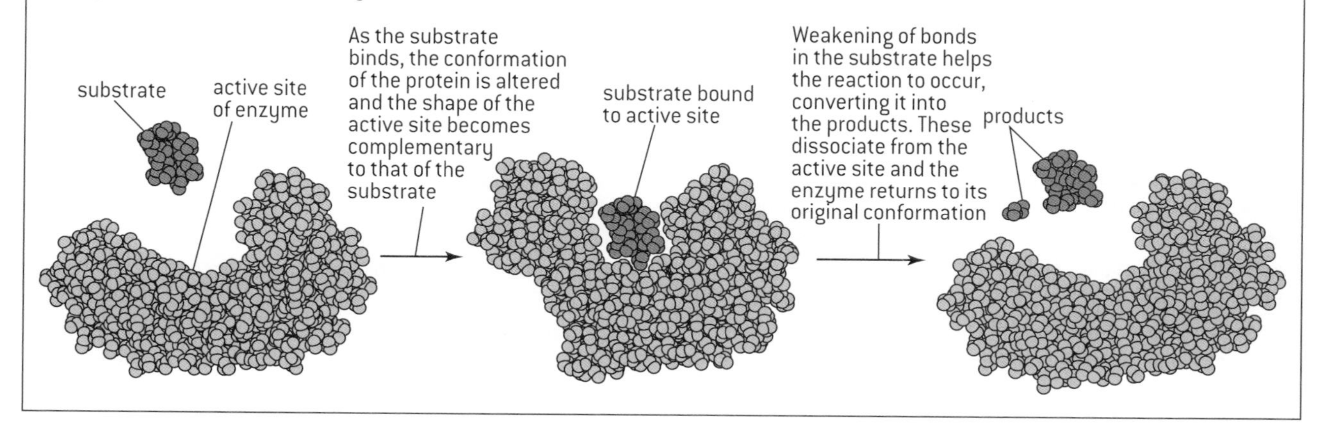

CALCULATING RATES OF REACTION

The rate of a reaction catalysed by an enzyme can be assessed by measuring the quantity of substrate used per unit time or the quantity of a product formed per unit time. These quantities can be measured as a mass or volume. The SI unit of time for rates is per second (s^{-1}).

Example

Slices of potato were added to 50cm^3 of hydrogen peroxide. The mass of the mixture was measured every two minutes. The catalase in the potato tissue catalysed the conversion of hydrogen peroxide to water plus oxygen. The oxygen was given off from the mixture, so the mass of the mixture decreased. The tables below show the raw results. The mass decreases were calculated by subtracting each mass from the previous one and the rate of mass decrease per second was calculated by dividing the decreases by the time periods in seconds (120 seconds).

Time after potato added (min)	0	2	4	6	8	10	12
Mass of mixture (g)	54.49	54.31	54.16	54.03	53.92	53.83	53.75

Time interval (min)	0–2	2–4	4–6	6–8	8–10	10–12
Mass decrease (g)	0.18	0.15	0.13	0,11	0.09	0.08
Mass decrease (mg)	180	150	130	110	90	80
Rate of mass decrease (mg s^{-1})	1.50	1.25	1.08	0.92	0.075	0.067

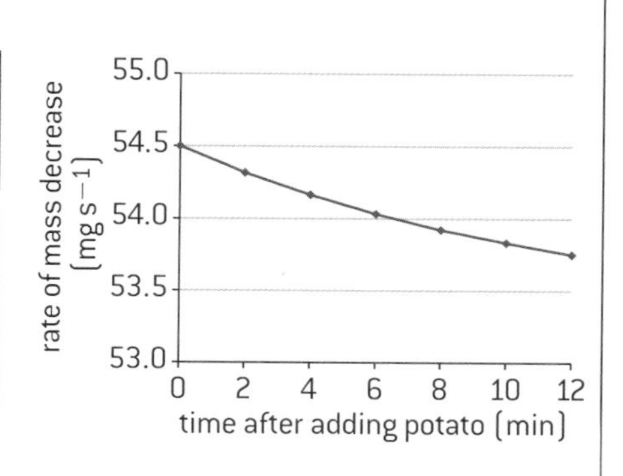

The graph (right) shows the rate of mass decrease over time.

Enzyme inhibition

COMPETITIVE AND NON-COMPETITIVE INHIBITORS

Enzyme inhibitors are chemical substances that reduce the activity of enzymes or even prevent it completely. There are two main types, competitive and non-competitive.

Competitive inhibition

- The substrate and inhibitor are chemically very similar.
- The inhibitor binds to the active site of the enzyme.
- While the inhibitor occupies the active site, it prevents the substrate from binding and so the activity of the enzyme is prevented until the inhibitor dissociates.

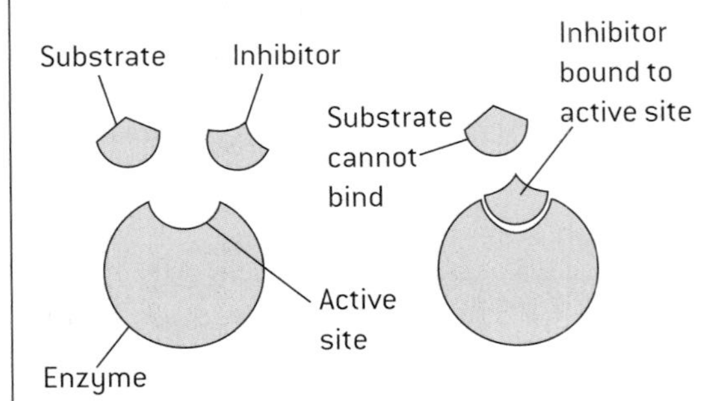

The activity of an enzyme is reduced if a fixed low concentration of a competitive inhibitor is added, but as the substrate concentration rises, the effect of the inhibitor becomes less and less until eventually it is negligible.

Explanation

The inhibitor and substrate compete for the active site. When the substrate binds to the active site, the inhibitor cannot bind and vice versa. As the substrate concentration rises, a substrate rather than an inhibitor molecule is increasingly likely to bind to a vacant active site. At very high substrate concentrations and low inhibitor concentrations, the substrate almost always wins the competition and binds to the active site so enzyme activity rate is nearly as high as when there is no inhibitor.

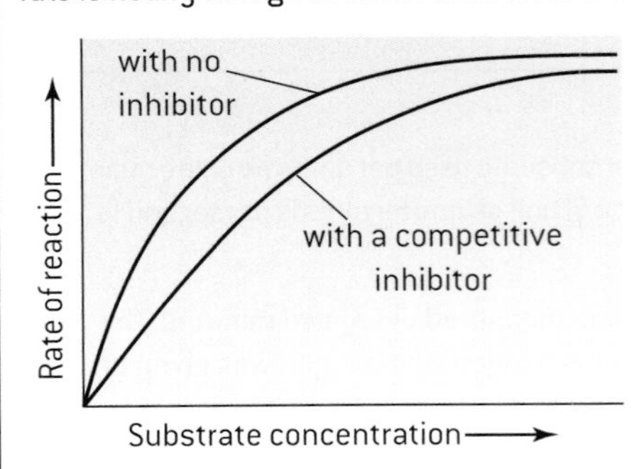

Example of competitive inhibition

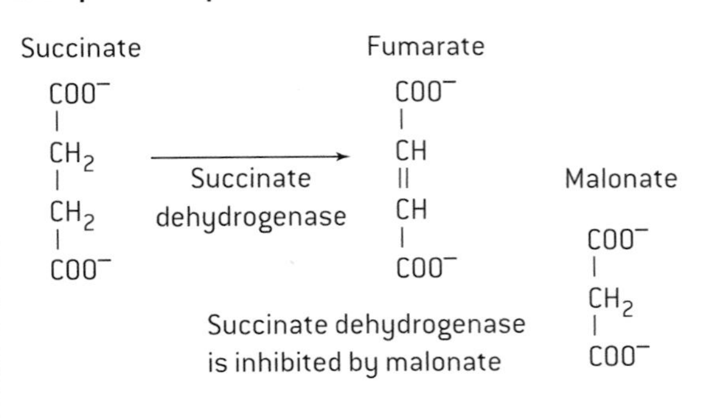

Non-competitive inhibition

- The substrate and are not similar.
- The inhibitor binds to the enzyme at a different site from the active site.
- The inhibitor changes the conformation of the enzyme. The substrate may still be able to bind, but the active site does not catalyse the reaction, or catalyses it at a slower rate.

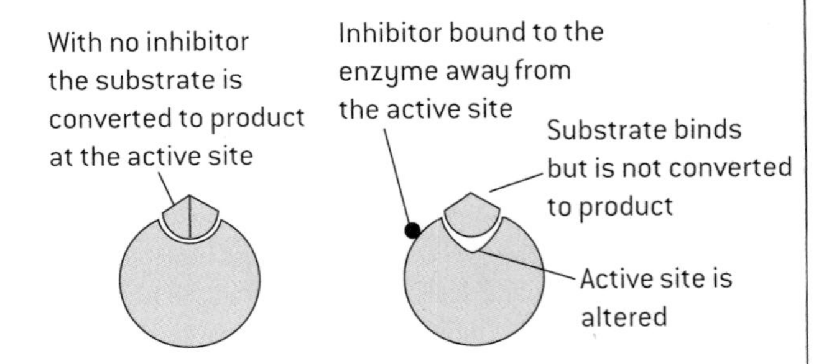

The activity of the enzyme is reduced at all substrate concentrations if a fixed low concentration of non-competitive inhibitor is added and the percentage reduction is the same at all substrate concentrations.

Explanation

The substrate and inhibitor are not competing for the same site, because the inhibitor binds somewhere on the enzyme other than the active site. The substrate cannot prevent the binding of the inhibitor, even at very high substrate concentrations. The same proportion of enzyme molecules is inhibited at all substrate concentrations. Even at very high substrate concentrations the enzyme activity rate is lower than when there is no inhibitor.

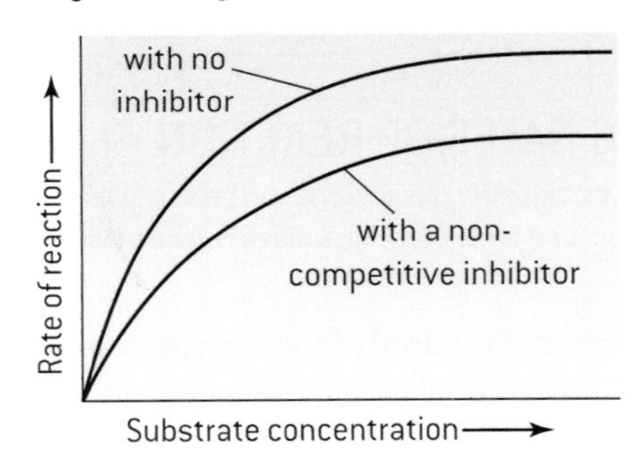

Example of non-competitive inhibition

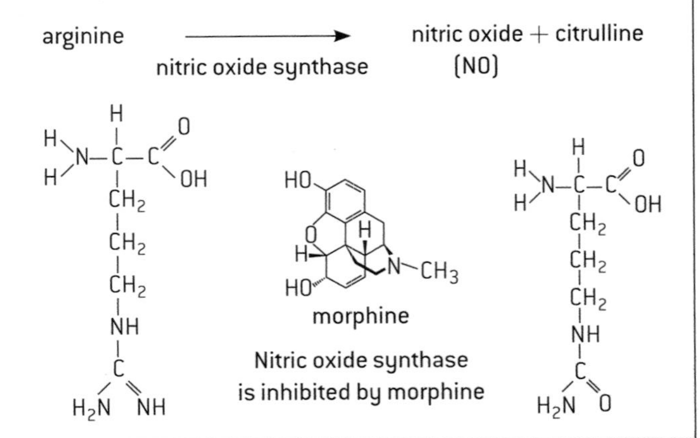

Controlling metabolic pathways

METABOLIC PATHWAYS

Metabolic pathways have these features:

- An enzyme catalyses each reaction in the pathway.
- All the reactions occur inside cells.
- Some pathways build up organic compounds (**anabolic** pathways) and some break them down (**catabolic** pathways).
- Some metabolic pathways consist of **chains** of reactions. Glycolysis is an example of a chain of reactions – a chain of ten enzyme-controlled reactions converts glucose into pyruvate.
- Some metabolic pathways consist of **cycles** of reactions, where a substrate of the cycle is continually regenerated by the cycle. The Krebs cycle is an example.

The figure (right) shows the general pattern of reactions in a chain and a cycle.

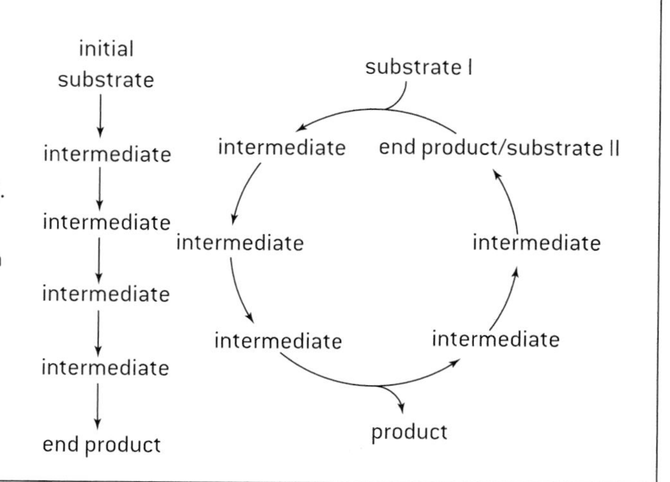

END-PRODUCT INHIBITION

In many metabolic pathways, the product of the last reaction in the pathway inhibits the enzyme that catalyses the first reaction. This is called **end-product inhibition**. The enzyme that is inhibited by the end products is an example of an allosteric enzyme. Allosteric enzymes have two non-overlapping binding sites. One of these is the active site. The other is the allosteric site. With end-product inhibition the allosteric site is a binding site for the end product. When it binds, the structure of the enzyme is altered so that the substrate is less likely to bind to the active site. This is how the end product acts as an inhibitor. Binding of the inhibitor is reversible and if it detaches, the enzyme returns to its original conformation, so the active site can bind the substrate easily again (right). The advantage of this method of controlling metabolic pathways is that if there is an excess of the end product the whole pathway is switched off and intermediates do not build up. Conversely, as the level of the end product falls, more and more of the enzymes that catalyse the first reaction will start to work and the whole pathway will become activated. End-product inhibition is an example of negative feedback. The inhibition of **threonine dehydratase** by **isoleucine** is an example of end-product inhibition.

Substrate binds to the active site and is converted to the product.

Substrate could bind to the active site as the allosteric site is empty.

Substrate is not likely to bind to the active site as the inhibitor has bound to the allosteric site.

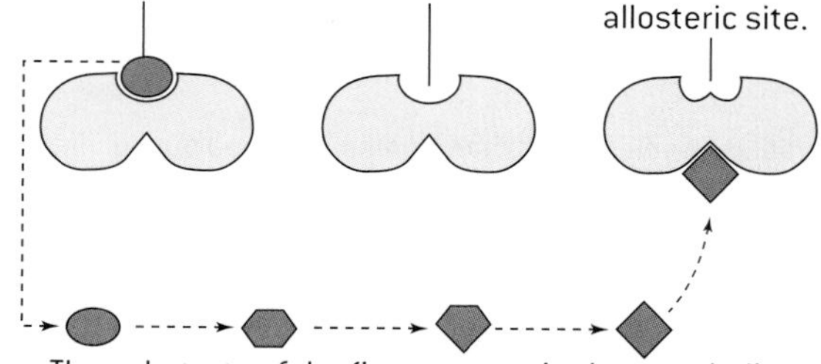

The substrate of the first enzyme in the metabolic pathway is converted by the pathway into an inhibitor of the enzyme.

threonine

threonine dehydratase

isoleucine

isoleucine is the end product of the pathway and inhibits threonine dehydratase which catalyses the first step

FINDING NEW ANTI-MALARIAL DRUGS

The malarial parasite (*Plasmodium*) has evolved resistance to most anti-malarial drugs so there is an urgent need for new drugs. The search is made easier by the huge bioinformatics databases that are held on computers. In a recent study 5,655 chemicals that might act as an enzyme inhibitor in *Plasmodium* were identified from a database of low molecular weight compounds. These were tested with nine *Plasmodium* enzymes identified from a database of metabolic pathways in the parasite. Inhibitors were found for six of the nine enzymes and these are now being researched as potential anti-malarial drugs.

Glycolysis

INTRODUCING GLYCOLYSIS

- Cell respiration involves the production of ATP using energy released by the oxidation of glucose, fat or other substrates.
- If glucose is the substrate, the first stage of cell respiration is a metabolic pathway called **glycolysis**.
- The pathway is catalysed by enzymes in the **cytoplasm**.
- Glucose is partially oxidized in the pathway and a small amount of ATP is produced.
- This partial oxidation is achieved without the use of oxygen, so glycolysis can form part of both aerobic and anaerobic respiration.

OXIDATION AND REDUCTION

Cell respiration involves many oxidation and reduction reactions. These reactions are the reverse of each other and can occur in different ways:

Oxidation reactions	**Reduction reactions**
Addition of oxygen atoms to a substance.	Removal of oxygen atoms from a substance.
Removal of hydrogen atoms from a substance.	Addition of hydrogen atoms to a substance.
Loss of electrons from a substance.	Addition of electrons to a substance.

In respiration, the oxidation of substrates is carried out by removing pairs of hydrogen atoms. Each hydrogen atom has one electron, so this method of oxidation is the removal of both hydrogen atoms and at the same time electrons. The hydrogen is accepted by a hydrogen carrier which is therefore reduced.

The most commonly used hydrogen carrier is NAD (nicotinamide adenine dinucleotide).

$$\text{NAD} + 2\text{H} \longrightarrow \text{reduced NAD}$$

An alternative form of notation is sometimes used for NAD and the equation is then different:

$$\text{NAD}^+ + 2\text{H} \longrightarrow \text{NADH} + \text{H}^+$$

PHOSPHORYLATION

In some metabolic reactions a phosphate group (PO_4^{3-}) is added to an organic molecule. This is called **phosphorylation**.

The effect of **phosphorylation** is to make the organic molecule **less stable** and therefore more likely to react in the next stage in a metabolic pathway. Phosphorylation can turn an endothermic reaction that will only occur at a very slow rate into an exothermic reaction that can proceed rapidly.

The phosphate group is usually transferred from **ATP**.

Example:

glucose + ATP ⟶ glucose 6-phosphate + ADP

STAGES IN GLYCOLYSIS

There are four main stages in glycolysis.

1. Two phosphate groups are added to a molecule of glucose to form hexose biphosphate. Adding a phosphate group is called **phosphorylation**. Two molecules of ATP provide the phosphate groups. The energy level of the hexose is raised by phosphorylation, so it is less stable and the subsequent reactions are possible.
2. The hexose biphosphate is split to form two molecules of triose phosphate. Splitting molecules is called **lysis**.
3. Two atoms of hydrogen are removed from each triose phosphate molecule. This is an oxidation. The energy released by the oxidation of each triose phosphate molecule is used to convert two ADP molecules to ATP. The end product of glycolysis is pyruvate.

The figure (below) shows the main stages of glycolysis.

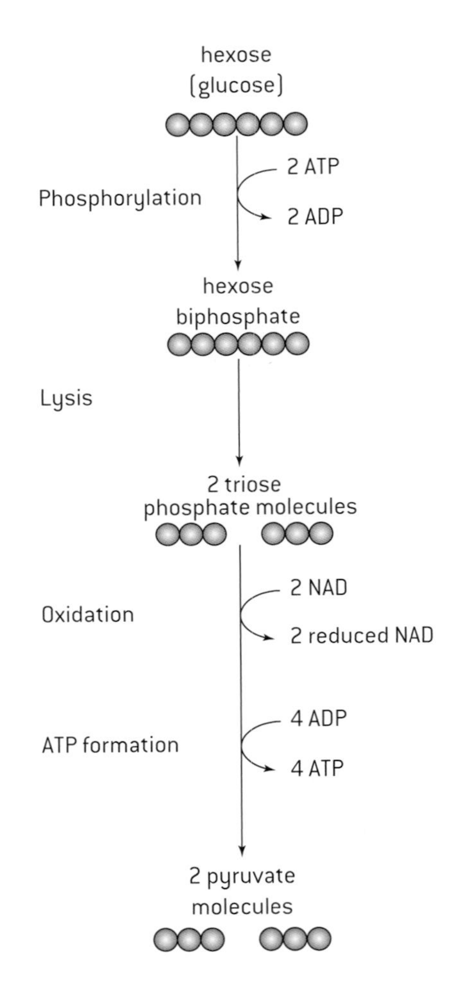

Summary of glycolysis:

- One glucose is converted into two pyruvates.
- Two ATP molecules are used per glucose but four are produced so there is a net yield of two ATP molecules. **This is a small yield of ATP per glucose, but it can be achieved without the use of any oxygen.**
- Two NADs are converted into two reduced NADs.

Krebs cycle

ANAEROBIC AND AEROBIC RESPIRATION

Glycolysis can occur without oxygen, so it forms the basis of anaerobic cell respiration. Pyruvate produced in glycolysis can only be oxidized further with the release of more energy from it if oxygen is available (right). This occurs in the mitochondrion. The first stage is called the **link reaction**. Enzymes in the matrix of the mitochondrion then catalyse a cycle of reactions called the **Krebs cycle**.

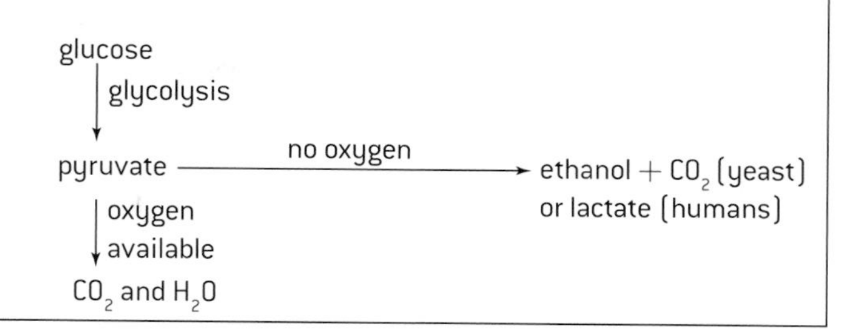

THE LINK REACTION

Pyruvate from glycolysis is absorbed by the mitochondrion. Enzymes in the matrix of the mitochondrion remove hydrogen and carbon dioxide from the pyruvate. The hydrogen is accepted by NAD. Removal of hydrogen is oxidation. Removal of carbon dioxide is decarboxylation. The whole conversion is therefore **oxidative decarboxylation**. The product of oxidative decarboxylation of pyruvate is an acetyl group, which is attached to coenzyme A to form acetyl coenzyme A (right).

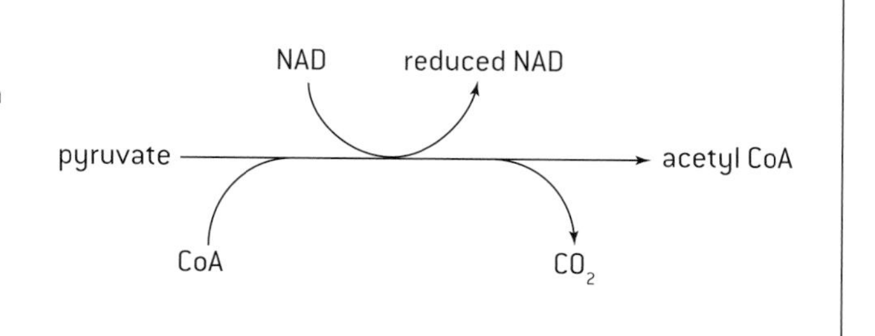

THE KREBS CYCLE

Acetyl groups from the link reaction are fed into the Krebs cycle. In the first reaction of the cycle an acetyl group is transferred from acetyl CoA to a four-carbon compound (oxaloacetate) to form a six-carbon compound (citrate). Citrate is converted back into oxaloacetate in the other reactions of the cycle.

Three types of reaction are involved.

- Carbon dioxide is removed in two of the reactions. These reactions are **decarboxylations**. The carbon dioxide is a waste product and is excreted together with the carbon dioxide from the link reaction.
- Hydrogen is removed in four of the reactions. These reactions are **oxidations**. The hydrogen is accepted by hydrogen carriers, which become reduced. In three of the oxidations the hydrogen is accepted by NAD. In the other oxidation FAD accepts it. These oxidation reactions release energy, much of which is stored by the carriers when they accept hydrogen. This energy is later released by the electron transport chain and used to make ATP.
- ATP is produced directly in one of the reactions. This reaction is **substrate-level phosphorylation**.

The figure (right) includes a summary of the Krebs cycle.

OXIDATIONS AND DECARBOXYLATIONS

In the summary of respiration below 3 decarboxylation and 6 oxidation reactions can be identified:

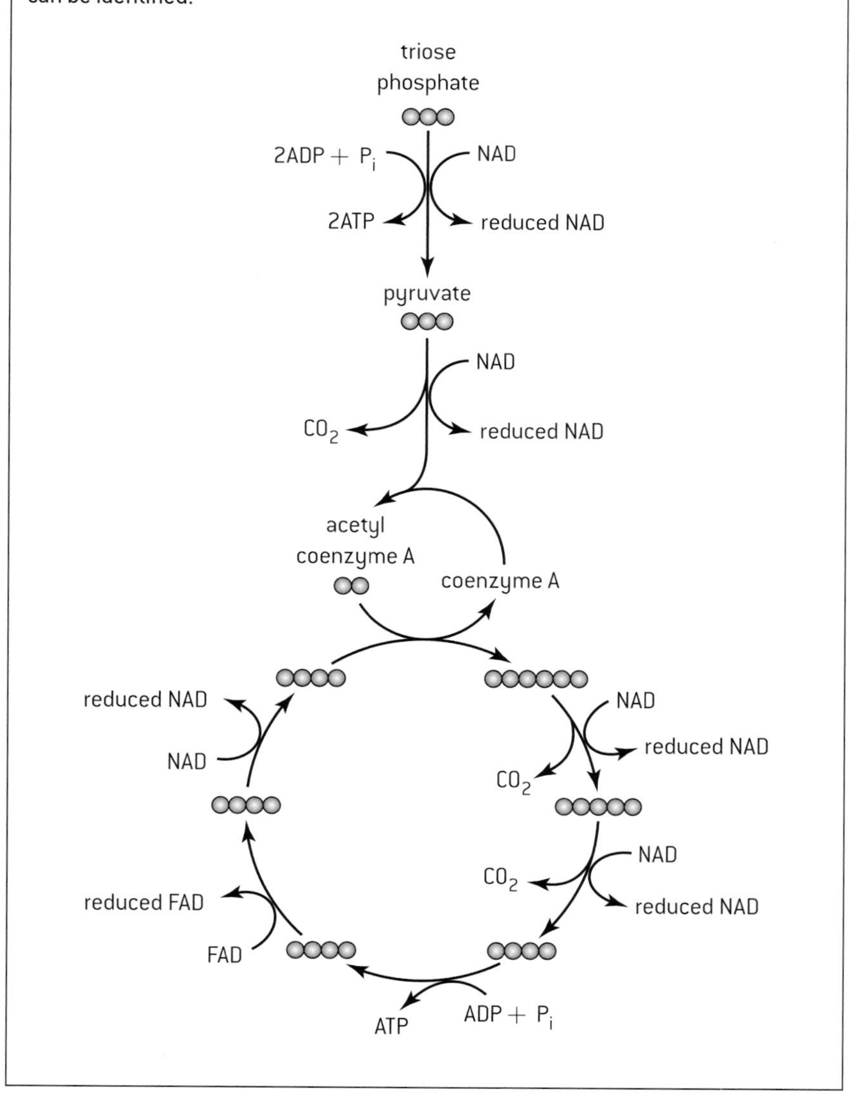

ATP production by oxidative phosphorylation

THE ELECTRON TRANSPORT CHAIN

The electron transport chain is a series of electron carriers, located in the inner membrane of the mitochondrion including the cristae.

Reduced NAD supplies two electrons to the first carrier in the chain. The electrons come from oxidation reactions in earlier stages of cell respiration and bring energy released by these oxidations.

As the electrons pass along the chain from one carrier to the next they give up energy. Some of the electron carriers act as proton pumps and use this energy to pump protons (H^+) against the concentration gradient from the matrix of the mitochondrion to the intermembrane space.

Reduced FAD also feeds electrons in to the electron transport chain, but at a slightly later stage than reduced NAD. Whereas the electrons from reduced NAD cause proton pumping at three stages in the electron transport chain, the electrons from reduced FAD cause proton pumping at only two stages.

THE ROLE OF OXYGEN

At the end of the electron transport chain the electrons are given to oxygen. This happens in the matrix, on the surface of the inner membrane. At the same time oxygen accepts free protons to form water. The use of protons in this reaction contributes to the proton gradient across the inner mitochondrial membrane.

The use of oxygen as the **terminal electron acceptor** at the end of the electron transport chain is the only stage where oxygen is used in cell respiration.

If oxygen is not available, electron flow along the electron transport chain stops and reduced NAD cannot be converted back to NAD. Supplies of NAD in the mitochondrion run out and the link reaction and Krebs cycle cannot continue. The only part of cell respiration that can continue is glycolysis, with a relatively small yield of ATP. Oxygen thus greatly increases the ATP yield, per glucose, of cell respiration.

CHEMIOSMOSIS IN THE MITOCHONDRION

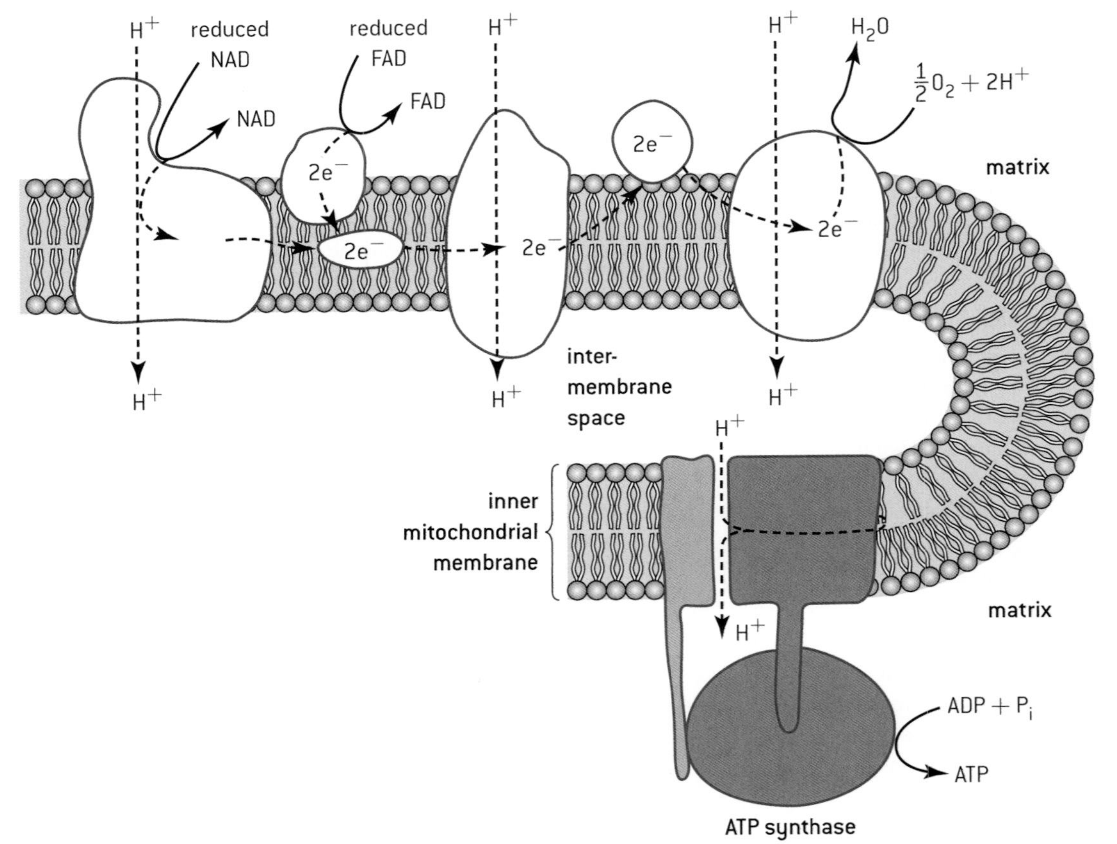

Energy released as electrons pass along the electron transport chain is used to pump protons (H^+) across the inner mitochondrial membrane into the space between the inner and outer membranes, including the space inside the cristae. A concentration gradient is formed, which is a store of potential energy.

ATP synthase, also located in the inner mitochondrial membrane, allows the protons to diffuse back across the membrane to the matrix. ATP synthase uses the energy that the protons release as they diffuse down the concentration gradient to produce ATP.

The generation of ATP using energy released by the movement of hydrogen ions across a membrane is called **chemiosmosis**.

Although this theory was proposed by Peter Mitchell in the 1960s it was not widely accepted until much later. The theory represented a **paradigm shift** in the field of bioenergetics and, as so often in science, it takes time for other scientists working in a field to accept paradigm shifts, even when there is strong evidence.

Mitochondria

STRUCTURE AND FUNCTION OF THE MITOCHONDRION

One of the recurring themes in biology is that **structure** and **function** are closely related in living organisms. This is known as **adaptation** and is the result of evolution by **natural selection**.

The figure (right) is an electron micrograph of a whole mitochondrion. The figure (below) is a drawing of the same mitochondrion, labelled to show how it is adapted to carry out its function.

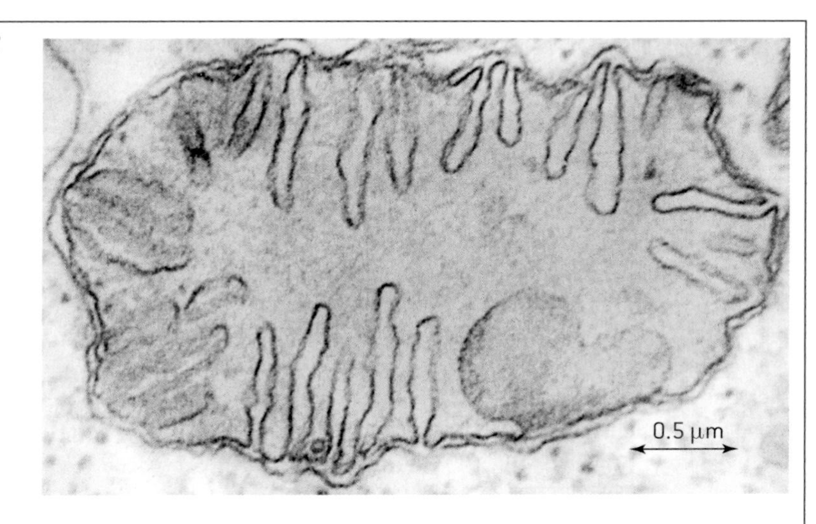

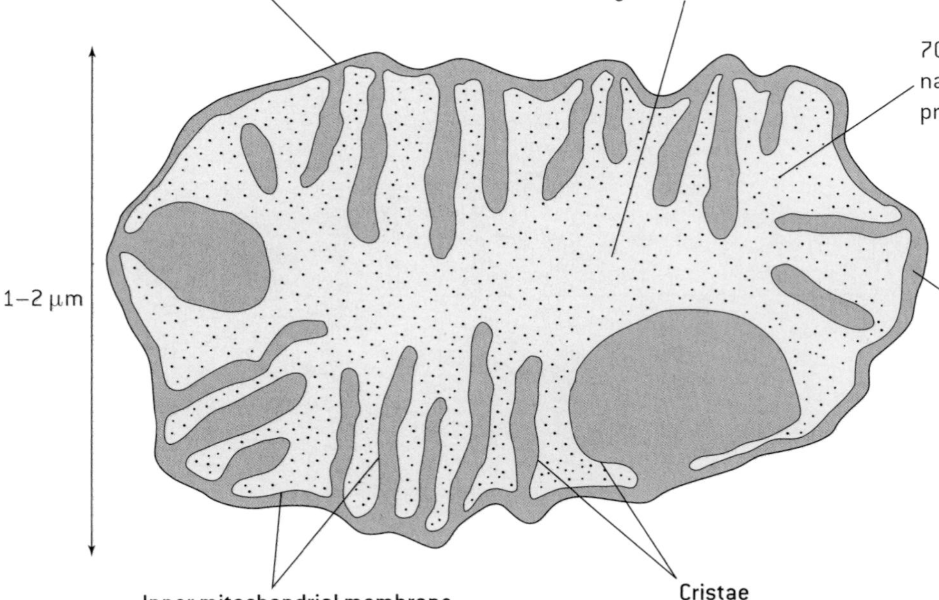

ELECTRON TOMOGRAPHY OF MITOCHONDRIA

The technique of electron tomography was developed relatively recently. It can be used to obtain three-dimensional images of active mitochondria. The image (right) shows a conventional thin section electron micrograph with an image of cristae superimposed that was produced by electron tomography. Electron tomography has revealed that cristae are connected with the intermembrane space between the inner and outer membranes via narrow openings (shown with arrows). The shape and volume of the cristae change when a mitochondrion is active in ways that are still being investigated.

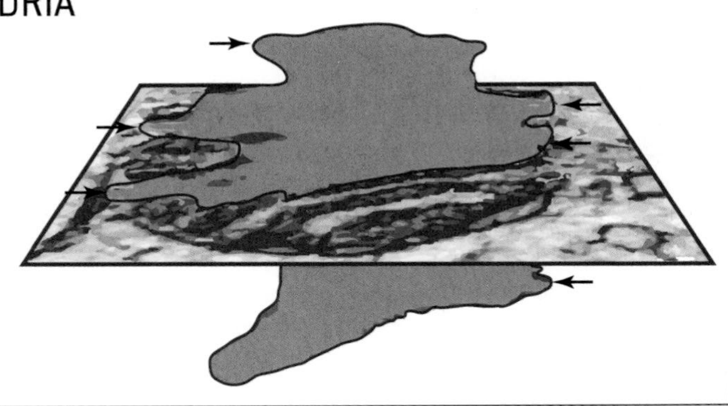

Light-dependent reactions of photosynthesis

LIGHT ABSORPTION

Pigments such as chlorophyll absorb certain wavelengths of light because they cause an electron in the pigment molecule to be raised to a higher energy level. The light energy is converted to chemical energy held by the excited electron. The main photosynthetic pigment is chlorophyll. Chlorophyll molecules in the chloroplast are part of large groups of pigment molecules, called **photosystems**, which work together to harvest light energy. Any of the pigments in a photosystem can absorb photons of light by an electron becoming excited. The excited electrons are then passed from pigment to pigment until they reach a special chlorophyll molecule at the reaction centre of the photosystem. This chlorophyll can pass pairs of excited electrons away to electron acceptors in the thylakoid membrane. The two types of photosystem, Photosystems I and II, are located in different parts of the thylakoid membranes.

PHOTOSYSTEM II AND ATP PRODUCTION

A pair of excited electrons from the reaction centre of Photosystem II is passed to a chain of carriers. The electrons give up energy as they pass from one carrier to the next. At one stage, enough energy is released to pump protons across the thylakoid membrane from the stroma into the space inside the thylakoid. This contributes to a proton gradient.

ATP synthase, also located in the thylakoid membranes, allows the protons to diffuse back across the membrane to the stroma and uses the energy that the protons release as they diffuse down the concentration gradient to produce ATP. The generation of ATP using energy released by the movement of hydrogen ions across a membrane is called **chemiosmosis**. Production of ATP in chloroplasts is called **photophosphorylation** because the energy needed for it is obtained by absorption of light. At the end of the chain of carriers the electrons are passed to Photosystem I.

LOCATION OF THE LIGHT-DEPENDENT REACTIONS IN THE THYLAKOID MEMBRANES

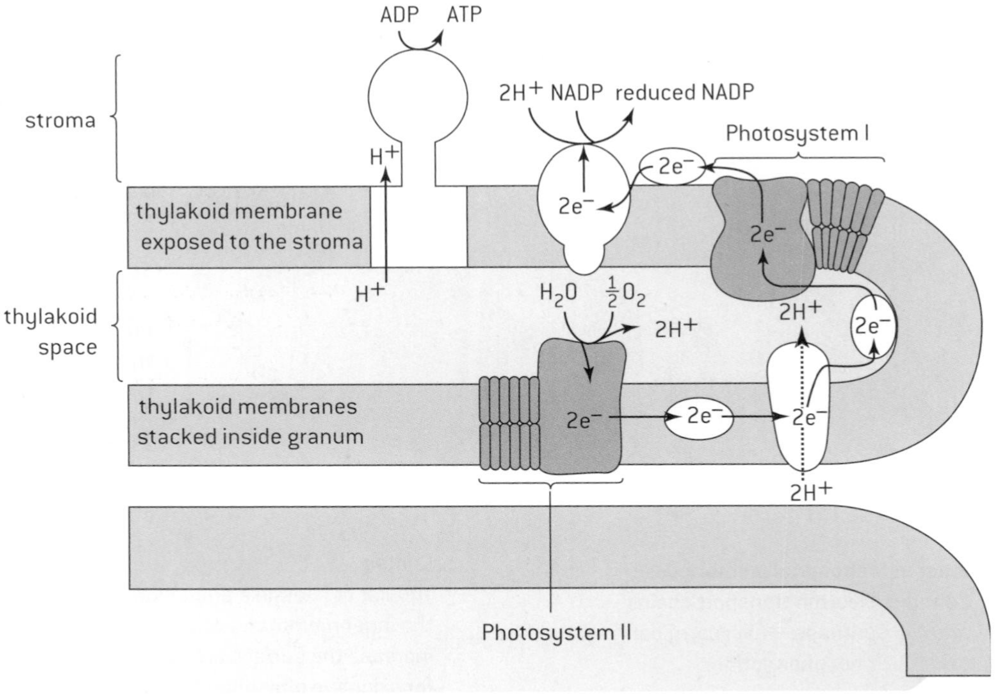

PHOTOSYSTEM I AND REDUCTION OF NADP

A pair of excited electrons is emitted from the reaction centre of Photosystem I and passes along a short chain of electron acceptors. At the end of this chain the electrons are passed to NADP in the stroma. NADP is converted to reduced NADP by accepting two electrons emitted by Photosystem I plus two protons from the stroma.

The electrons given away by Photosystem I are replaced by electrons that were emitted by Photosystem II and passed along the chain of electron carriers. Photosystem I can then absorb more photons of light to produce more excited electrons.

PHOTOLYSIS

Photosystem II must replace excited electrons given away by the chlorophyll at its reaction centre, before any more photons of light can be absorbed. With the help of an enzyme at the reaction centre, water molecules in the thylakoid space are split and electrons from them are given to the chlorophyll at the reaction centre. Oxygen and H^+ ions are formed as by-products. The splitting of water molecules only happens in the light so it is called **photolysis**.

All of the oxygen produced in photosynthesis is from photolysis of water. Oxygen is a waste product and is excreted. H^+ contributes to the proton gradient.

Chloroplast structure

ELECTRON MICROSCOPY AND CHLOROPLAST STRUCTURE

Chloroplasts were discovered using light microscopes. They are visible as small green blobs. In the clearest images darker green blobs can be seen inside chloroplasts. These were named **grana**. The invention of electron microscopes revealed that grana consist of stacks of membrane-bound structures, called **thylakoids**. Other structures were revealed: the stroma, starch grains, oil droplets and an envelope of two membranes. The electron micrograph below shows these chloroplast structures and also parts of the cell wall, plasma membrane, rER, ribosomes, polysomes and nucleus.

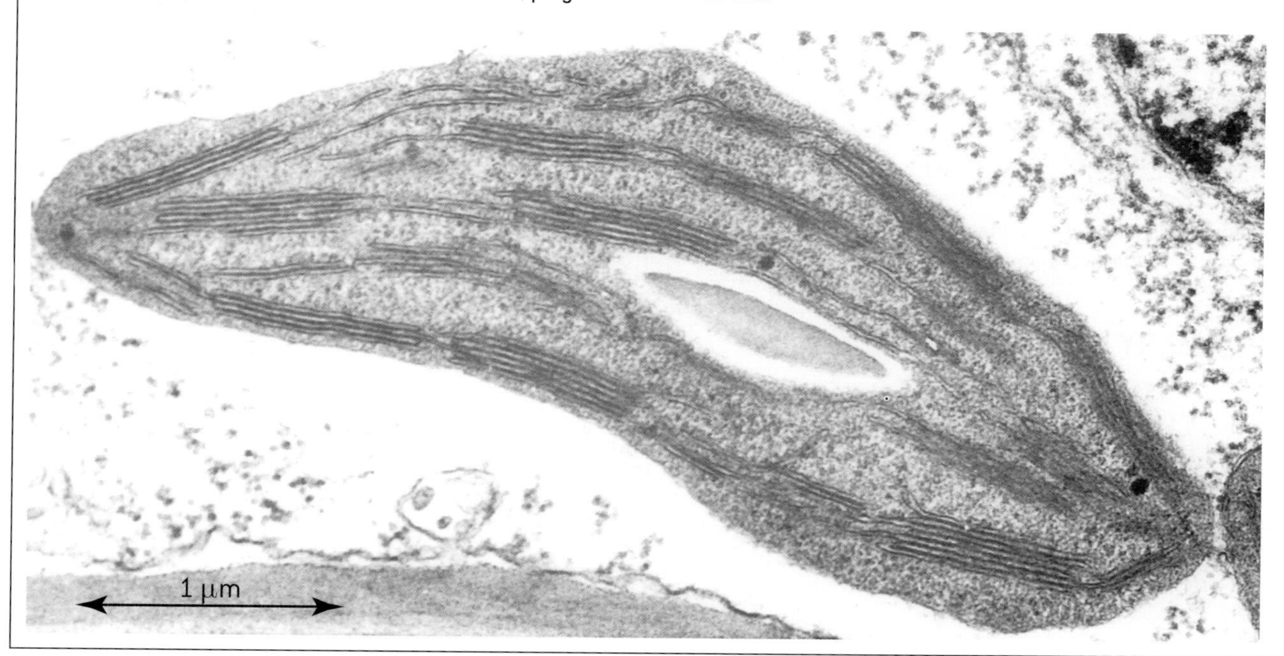

STRUCTURE AND FUNCTION OF THE CHLOROPLAST

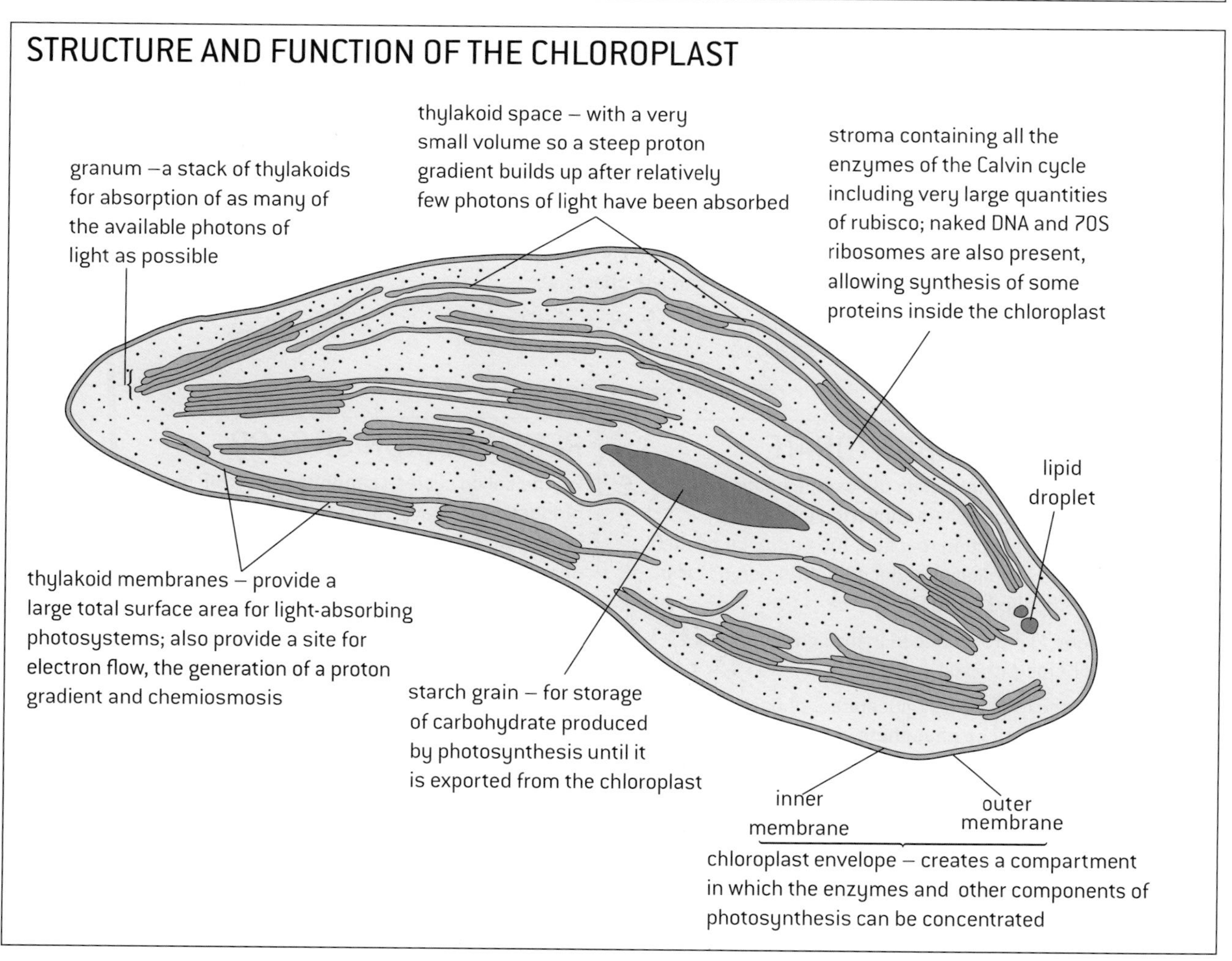

Light-independent reactions of photosynthesis

THE CALVIN CYCLE

The light-independent reactions take place in the stroma of the chloroplast.

The first reaction involves a five-carbon sugar, ribulose bisphosphate (RuBP), which is regenerated by the light-independent reactions. They therefore form a cycle, called the Calvin cycle. The reactions of the cycle are summarized in the diagram below.

The Calvin cycle was discovered by a team of biochemists led by the eponymous Melvin Calvin. The research methods used are described on the next page.

There are many alternative names for the intermediate compounds in the Calvin cycle, several of which have the initials GP, so this abbreviation GP should be avoided.

CARBOXYLATION OF RuBP

Carbon dioxide is an essential substrate in the light-independent reactions. It enters the chloroplast by diffusion. In the stroma of the chloroplast carbon dioxide combines with ribulose bisphosphate (RuBP), a five-carbon sugar, in a carboxylation reaction. The reaction is catalysed by the enzyme rubisco. (The full name of this enzyme is ribulose-1,5-bisphosphate carboxylase oxygenase, but it is much more convenient to use the abbreviation!)

The product of the carboxylation of RuBP is an unstable six-carbon compound, which immediately splits to form two molecules of **glycerate 3-phosphate**. This is therefore the first product of carbon fixation – the conversion of carbon dioxide into organic compounds.

SUMMARY OF THE CALVIN CYCLE

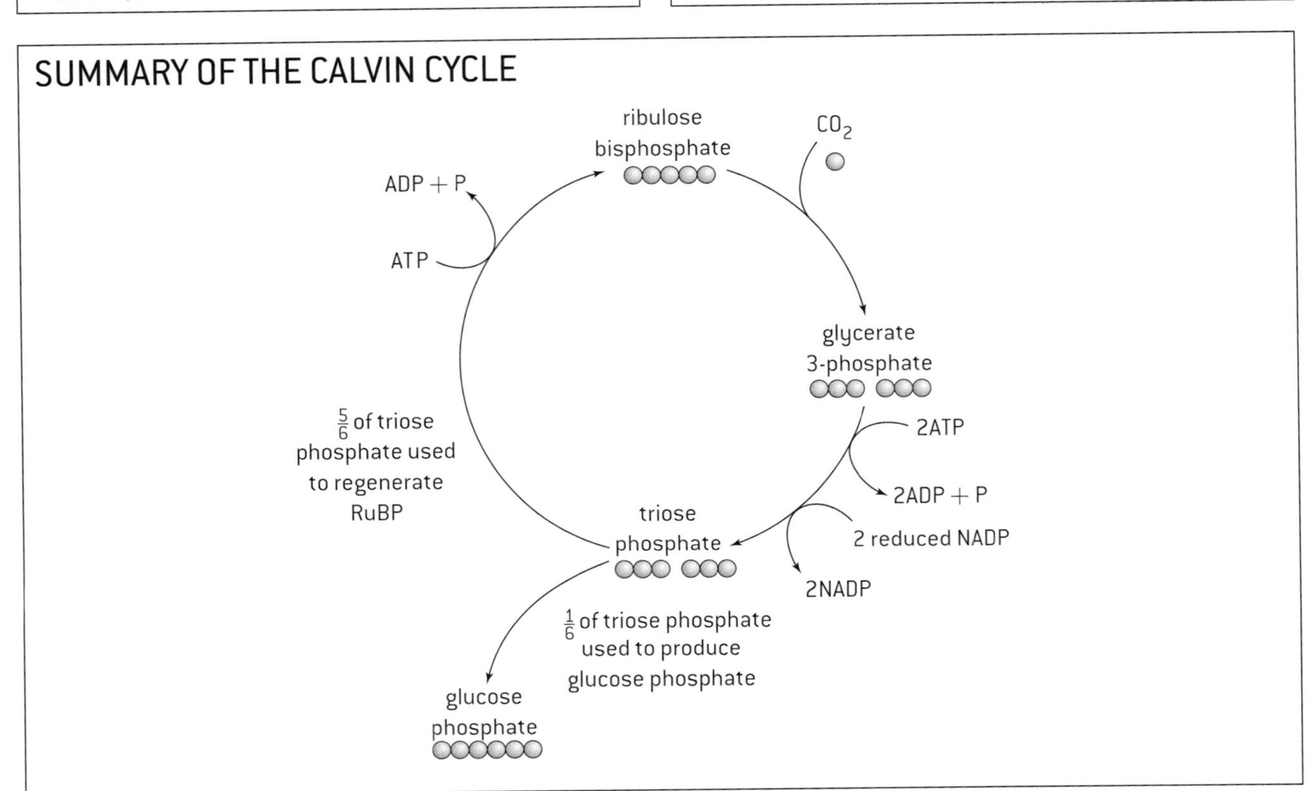

SYNTHESIS OF CARBOHYDRATE

Glycerate 3-phosphate, formed in the carbon fixation reaction, is an organic acid. It is converted into a carbohydrate by a **reduction** reaction. The hydrogen needed to carry this out is supplied by reduced NADP. Energy is also needed and is supplied by ATP.

Both NADPH and ATP are produced in the light-dependent reactions of photosynthesis.

The product of the reduction of glycerate 3-phosphate is a three-carbon sugar, **triose phosphate**.

Triose phosphate can be converted into a variety of other carbohydrates. Glucose phosphate is produced by linking together two triose phosphates. Starch, the storage form of carbohydrate in plants, is formed in the stroma by linking together many molecules of glucose phosphate by condensation reactions.

REGENERATION OF RuBP

For the Calvin cycle to continue, one RuBP molecule must be produced to replace each one that is used. Triose phosphate is used to regenerate RuBP. Five molecules of triose phosphate are converted by a series of reactions into three molecules of RuBP. This process requires the use of energy in the form of ATP. The reactions can be summarized using equations where only the number of carbon atoms in each sugar molecule is shown.

$$C_3 + C_3 \longrightarrow C_6$$
$$C_6 + C_3 \longrightarrow C_4 + C_5$$
$$C_4 + C_3 \longrightarrow C_7$$
$$C_7 + C_3 \longrightarrow C_5 + C_5$$

For every six molecules of triose phosphate formed in the light-independent reactions, five must be converted to RuBP.

Calvin's experiments

IMPROVEMENTS IN APPARATUS

Calvin's discovery of the mechanism used to fix CO_2 depended on three new experimental techniques:

1. **Radioactive labelling**

   Radioisotopes of elements have the same chemical properties as other isotopes of an element but can be distinguished by being radioactive. They can therefore be used to label organic compounds in biochemistry experiments. The radioactive isotope ^{14}C, discovered in 1940, is particularly suitable. Sources of ^{14}C (also known as carbon-14) were developed, so carbon dioxide and hydrogen carbonate labelled with ^{14}C could be produced and made available to researchers such as Calvin.

2. **Double-way paper chromatography**

   The technique of separating and identifying compounds by paper chromatography was discovered in 1943 and double-way chromatography for separating small organic compounds was developed after this. A spot of the mixture is placed in one corner of a large sheet of chromatography paper. A first solvent is run up through the paper to separate the mixture partially in one direction. The paper is dried and then a second solvent is run up at 90° to the first, spreading the mixture in a second dimension. This procedure was ideal for separating and identifying the initial products of carbon fixation.

3. **Autoradiography**

   Biologists used X-ray film from the 1940s onwards to find the location of radioisotopes. When atoms of ^{14}C decay they give off radiation, which makes a small spot in an adjacent X-ray film. To find radioisotopes in a sheet of chromatography paper it is placed next to a sheet of film that is the same size. The two sheets are kept together in darkness for several weeks and the X-ray film is then developed. Black patches appear in areas where the adjacent chromatography paper contained radioisotopes.

CALVIN'S EXPERIMENT

The figure below shows the apparatus used in the 1950s by Melvin Calvin and Andrew Benson to discover the Calvin cycle.

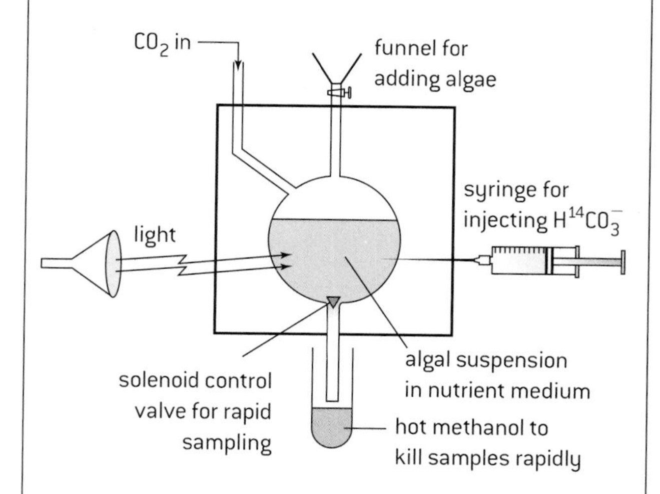

A suspension of *Chlorella* was placed in a thin glass vessel (called the lollipop vessel) and was brightly illuminated. *Chlorella* is a unicellular alga. The *Chlorella* was supplied with both carbon dioxide (CO_2) and hydrogen carbonate (HCO_3^-). Before the start of the experiment the carbon in both of these carbon sources was ^{12}C, but at the start of the experiment this was replaced with ^{14}C.

Calvin and his team took samples of the algae at very short time intervals and immediately killed and fixed them with hot methanol. They extracted the carbon compounds, separated them by double-way paper chromatography and then found which carbon compounds in the algae contained radioactive ^{14}C by autoradiography. The results are shown below. The amount of radioactivity of each carbon compound is shown in the graph as a percentage of the total amount of radioactivity.

CALVIN'S RESULTS

The autoradiogram for samples of *Chlorella* exposed to radioactive carbon dioxide and hydrogen for 5 seconds (above right) shows that there was more labelled glycerate 3-phosphate than any other compound, indicating that it is the first product of carbon fixation. The autoradiogram for 30 seconds (below right) shows that by then many carbon compounds were labelled. The amount of radioactivity in the different compounds was measured. Changes in the amounts are shown in the graph below. Again there is evidence for glycerate 3-phosphate as the first product with triose phosphate formed next.

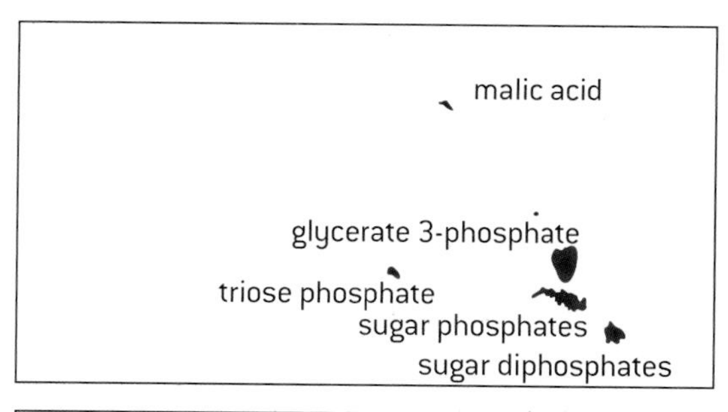

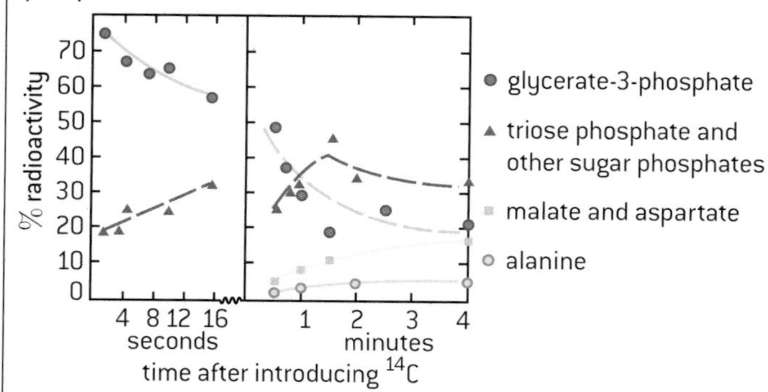

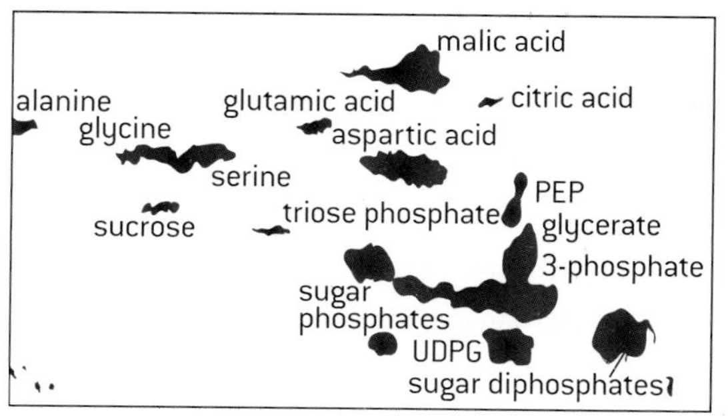

Questions – metabolism, cell respiration and photosynthesis

1. An enzyme experiment was conducted at three different temperatures. The graph shows the amount of substrate remaining each minute after the enzyme was added to the substrate. W shows the results obtained at a temperature of 40 °C.

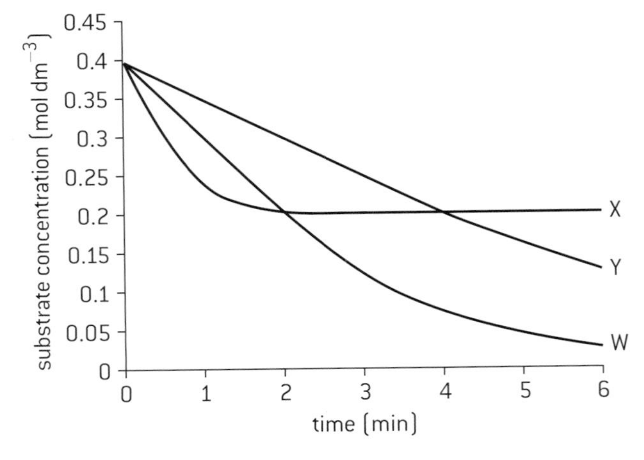

   a) (i) Explain whether the temperature used for X was higher or lower than 40 °C. [3]

   (ii) Estimate the temperature that was used for Y. [2]

   b) Draw a curve on the graph to show the expected results of repeating the experiment at 40 °C with

   (i) a higher concentration of enzyme; [2]

   (ii) a pH further from the optimum. [2]

2. Discs of tissue were cut from horse chestnut seeds and were placed in solutions of hydrogen peroxide. The enzyme catalase released from cut cells caused this reaction:

$$2H_2O_2 \xrightarrow{\text{catalase}} 2H_2O + O_2$$

The oxygen released by the reaction formed foam on the surface of the hydrogen peroxide. The volume of the foam was measured after five minutes using various hydrogen peroxide concentrations, both with and without a fixed low concentration of copper ions. The results are shown below.

Concentration of H_2O_2 (%)	Volume of oxygen (ml)	
	No Cu	With Cu
0	0.0	0.0
10	7.3	3.8
20	10.3	5.4
30	11.4	6.3
40	11.8	6.5
50	11.9	6.6

   a) Calculate the rate of reaction for each of the twelve results. [5]

   b) Plot a graph to show the effect of hydrogen peroxide concentration on the rate of reaction both with and without copper ions. [6]

   c) Deduce, with reasons, the effect of copper ions on catalase. [4]

3. a) Identify two metabolic processes that involve chemiosmosis. [2]

   b) Explain the need for membranes in chemiosmosis. [3]

   c) Suggest a location for chemiosmosis to occur in prokaryotes. [1]

4. The electron micrograph below shows part of a plant root cell, including mitochondria.

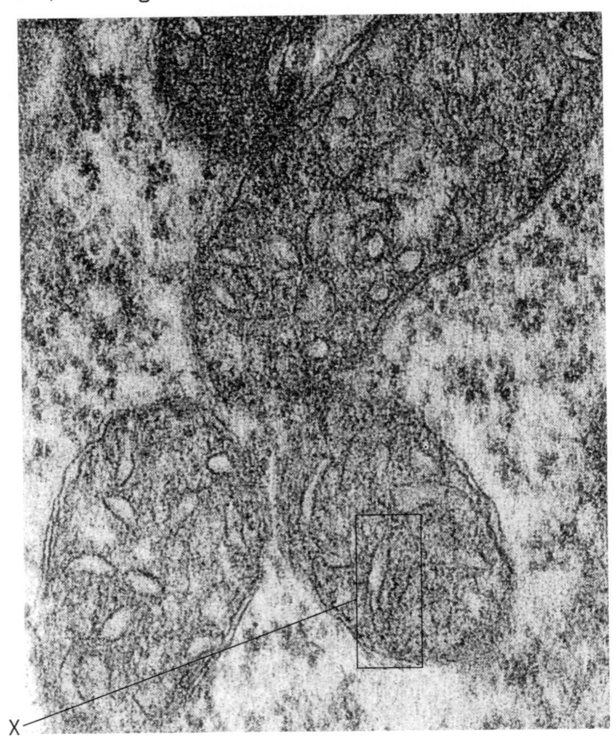

   a) Explain briefly two features that allow the mitochondria in the micrograph to be identified. [2]

   b) Draw the structure of region marked X. [2]

   c) Annotate the micrograph to show one example of

   (i) a region where the Krebs cycle takes place

   (ii) a location of ATP synthase

   (iii) a region where glycolysis takes place. [3]

5. a) Draw a curve of the action spectrum for photosynthesis on the axes below. [2]

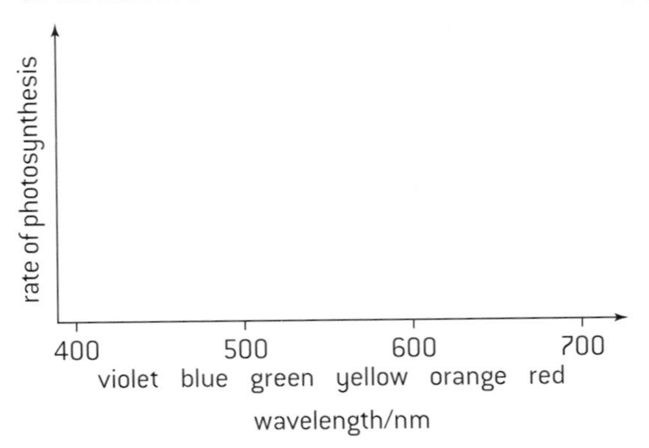

   b) Explain the relationship between the action spectrum for photosynthesis and the absorption spectra of photosynthetic pigments. [3]

Transpiration

WATER LOSS BY TRANSPIRATION

Transpiration is the loss of water vapour from the stems and leaves of plants. It is the inevitable consequence of gas exchange in the leaf.

Leaves must absorb carbon dioxide for use in photosynthesis and excrete oxygen (a waste product). Gas exchange requires a large area of moist surface. This is provided by the mesophyll. In many leaves there is spongy mesophyll in the lower part of the leaf with a network of air spaces that increases the surface area of moist cell walls exposed to air.

Unless the air spaces are fully saturated, water evaporates from the moist cell walls. This ensures that the air spaces have a high relative humidity so water vapour tends to diffuse from them to the air outside the leaf.

The epidermis of most plant leaves secretes wax to form a waterproof coating to the leaf (waxy cuticle). This prevents excessive transpiration, but also blocks gas exchange. Pores are therefore needed in the epidermis for CO_2 to enter the leaf and O_2 to leave. If the pores (stomata) are opened to allow gas exchange, they also usually allow water vapour to escape, which is transpiration.

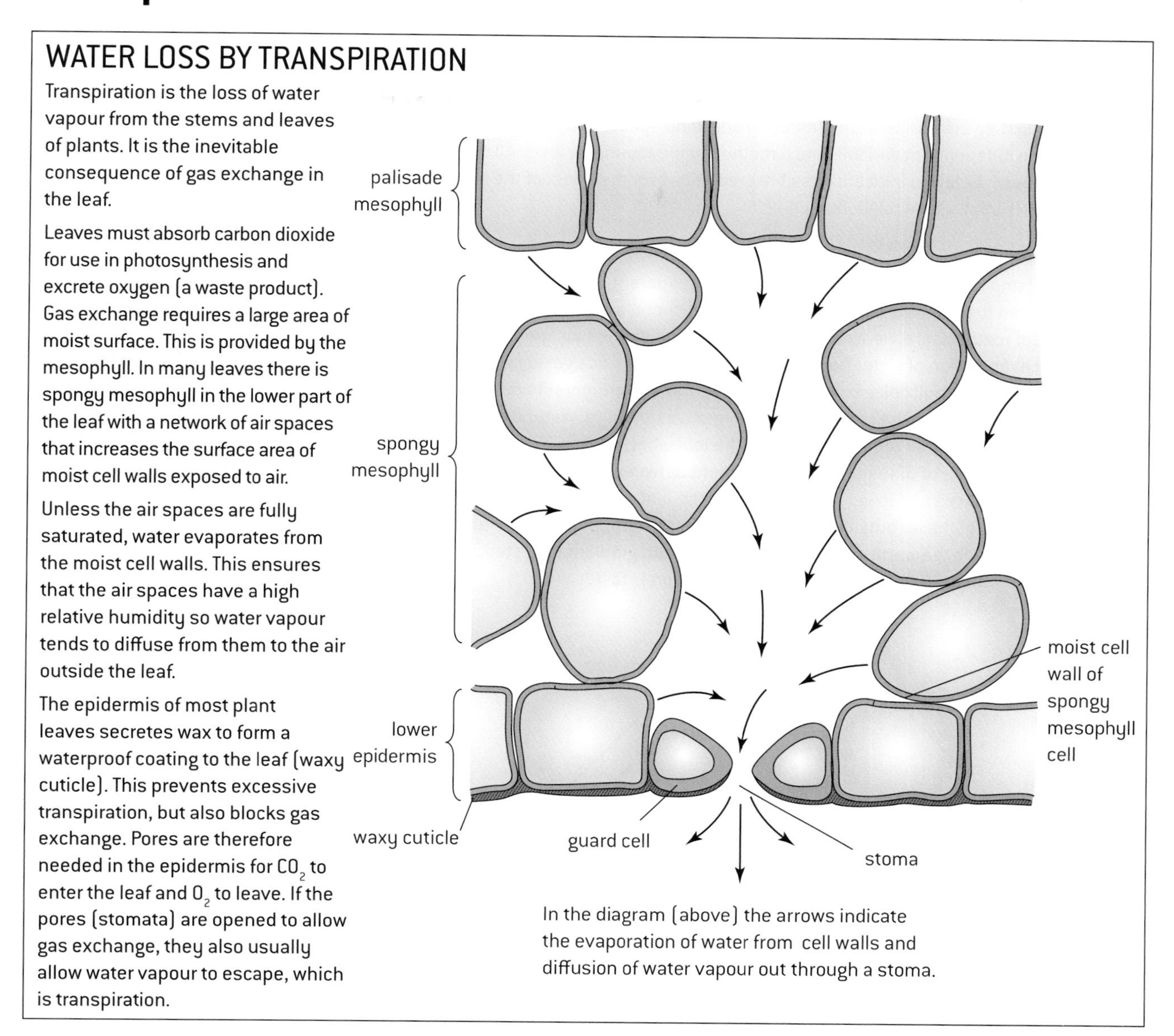

In the diagram (above) the arrows indicate the evaporation of water from cell walls and diffusion of water vapour out through a stoma.

MEASURING TRANSPIRATION RATES

The rate of transpiration is difficult to measure directly and instead the rate of water uptake is usually measured using a **potometer**.

The figure (right) shows one design of potometer.

As the plant transpires it draws water out of the capillary tube to replace the losses.

Because the capillary tube is narrow, small losses of water from the plant give measurable movements of the air bubble.

Repeat measurements of the distance moved in one minute are needed to ensure that the results are reliable.

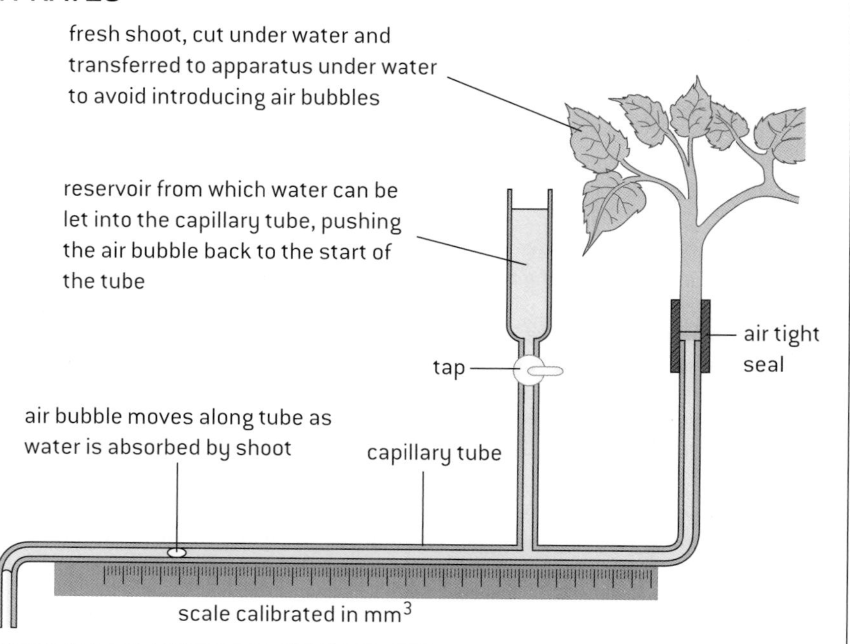

Investigating transpiration

INVESTIGATING FACTORS AFFECTING TRANSPIRATION RATES

Experiments can be designed using potometers to test the effect of external variables on the rate of transpiration. Three factors worth investigating are temperature, humidity and wind speed. In each case a method is needed to vary the external factor chosen as the independent variable and a method of measuring its level. All other factors should be kept constant. Sketch graphs are shown below to indicate possible predictions for the effect of the three factors. Your results either may or may not support these hypothetical relationships.

Temperature
Use a heat lamp to vary the temperature and an infrared thermometer to measure leaf temperature. Heat is needed for evaporation of water from the surface of mesophyll cells, so as temperature is increased the rate of transpiration rises. Higher temperatures also increase the rate of diffusion through the air spaces and lower the relative humidity of the air outside the leaf. In very high temperatures the stomata may close.

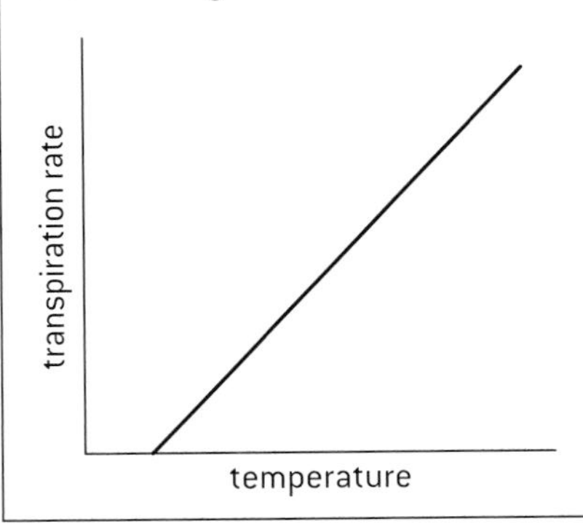

Humidity
Use a transparent plastic bag to enclose the leafy shoot, a mist sprayer to raise humidity inside the bag and desiccant bags containing silica gel to lower it. Use an electronic hygrometer to measure the relative humidity. Water diffuses out of the leaf when there is a concentration gradient between the humid air spaces inside the leaf and the air outside. As atmospheric humidity is reduced, the concentration gradient gets steeper and transpiration is faster.

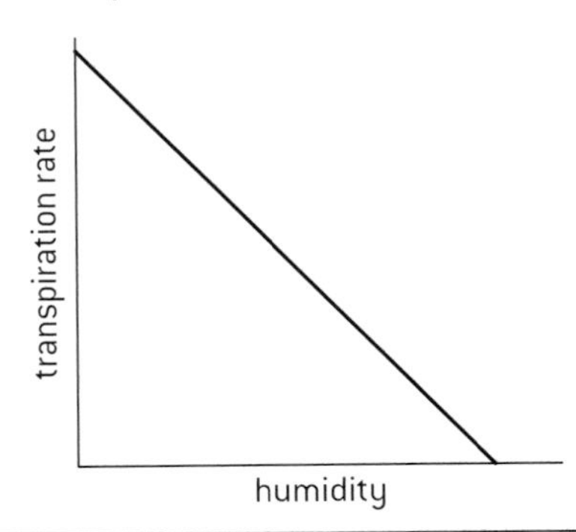

Wind speed
Use an electric fan to generate air movement, varying velocity by changing the distance of the fan or the rate of rotation. Use an anemometer to measure the speed of the air moving across the plant leaves. In still air, humidity builds up around the leaf, reducing the concentration gradient of water vapour and therefore reducing transpiration. Moderate wind velocities reduce or prevent this but high velocities can cause stomata to close.

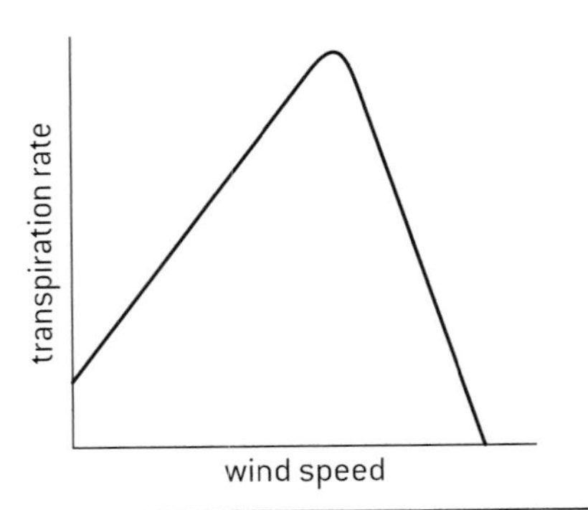

SUMMARY OF COMPULSORY PRACTICALS

The measurement of transpiration rates using potometers is a compulsory practical for HL biology students. In the core there are six practicals that you should have done before you take your IB Biology exams, whether you are an SL or HL biology student. Your teacher may have called them compulsory labs or compulsory practicals. You should be able to answer questions about details of these practicals or labs that you have learned by doing them.

Practical 1 in Topic 1 – see page 4
Use of a light microscope to investigate the structure of cells and tissues, with drawing of cells. Calculation of the magnification of drawings and the actual size of structures and ultrastructures shown in drawings or micrographs.

Practical 2 in Topic 1 – see page 11
Estimation of osmolarity in tissues by bathing samples in hypotonic and hypertonic solutions.

Practical 3 in Topic 2 – see page 27
Experimental investigation of a factor affecting enzyme activity.

Practical 4 in Topic 2 – see page 36
Separation of photosynthetic pigments by chromatography.

Practical 5 in Topic 4 – see page 55
Setting up sealed mesocosms to try to establish sustainability.

Practical 6 in Topic 6 – see page 78
Monitoring of ventilation in humans at rest and after mild and vigorous exercise.

Practical 7 in Topic 9 (HL only) – pages 111–112
Measurement of transpiration rates using potometers

Water uptake and water conservation

WATER UPTAKE IN ROOTS

Plants absorb water and also mineral ions from the soil using their roots. The surface area for this is increased by branching of roots and the growth of root hairs from epidermis cells. Plants absorb potassium, phosphate, nitrate and other mineral ions. The concentration of these ions in the soil is usually much lower than inside root cells, so they are absorbed by active transport. Root hair cells have mitochondria and protein pumps in their plasma membranes. Most roots only absorb mineral ions if they have a supply of oxygen, because they produce ATP for active transport, by aerobic cell respiration. As a result of active transport the cytoplasm of root cells has a higher overall solute concentration than the water in the soil. Root cells therefore absorb water from the soil by osmosis.

ADAPTATIONS OF PLANTS TO SALINE SOILS

Saline soils are found in coastal habitats and in arid areas where water moves up in soil and evaporates leaving dissolved ions at the surface. In saline soils the concentration of ions such as Na^+ and Cl^- is so high that most plants are unable to grow, but some specially adapted plants thrive (halophytes). The graph shows the growth of halophytes and non-halophytes in a range of NaCl concentrations.

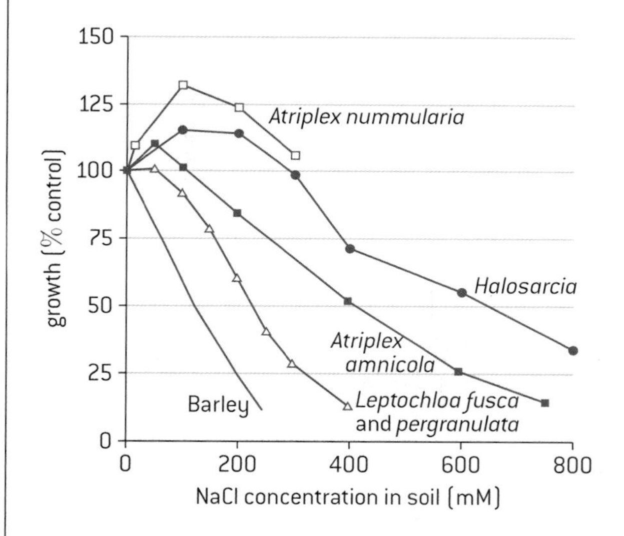

To prevent water moving by osmosis from halophytes to saline soils the solute concentration inside the plant must be higher than in the saline soil. This cannot be done simply by raising the Na^+ concentration because high concentration of this ion can have adverse effects on cell activities such as protein synthesis. High concentrations of other solutes such as sugars or K^+ are maintained in the cytoplasm instead. However, concentrations of Na^+ and Cl^- above those of the saline soil can be maintained in the vacuoles of cells as metabolic activities do not occur there.

Halophytes use a range of methods to get rid of excess Na^+ such as active transport back into the soil, excretion from special glands in the leaf, and accumulating the ion in certain leaves and then shedding them. Many halophytes also have adaptations for water conservation similar to those of xerophytes. Some have water storage tissue so are succulents.

ADAPTATIONS OF PLANTS IN DESERTS

Plants that are adapted to grow in very dry habitats such as deserts are called xerophytes. *Cereus giganteus*, the saguaro or giant cactus, is an example of a xerophyte. It grows in deserts in Mexico and Arizona and shows many xerophytic adaptations, which help to conserve water by reducing transpiration.

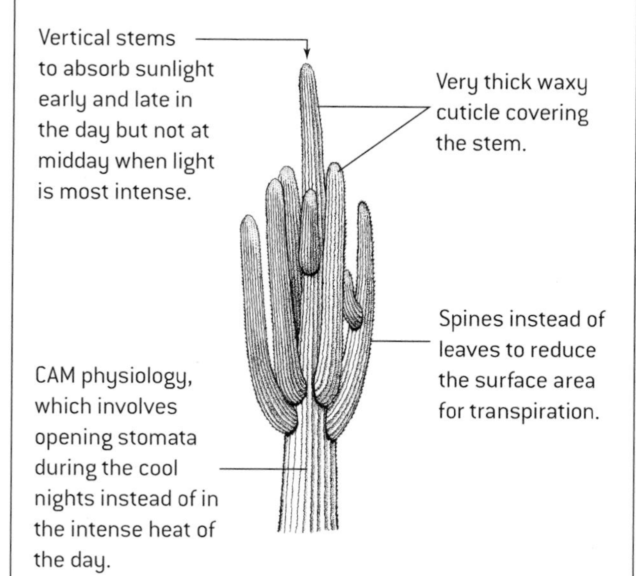

Plants that grow in sand dunes have xerophytic adaptations for water conservation. An example is marram grass (*Ammophila arenaria*). A horizontal section of a leaf of this plant is shown below.

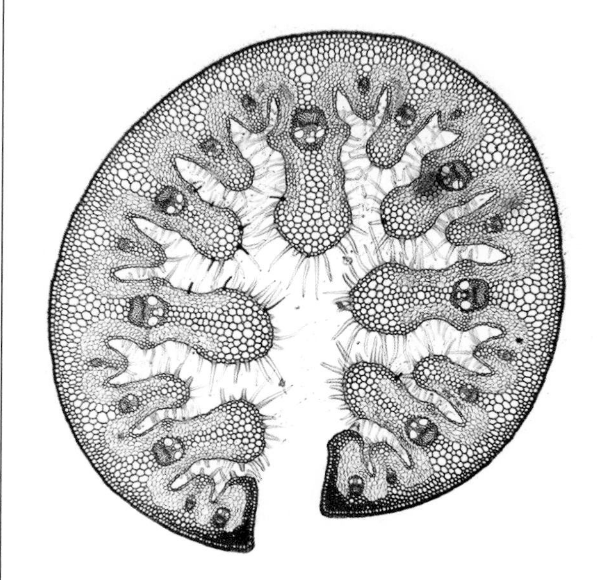

The leaf of *Ammophila arenaria* has:

- a thick waxy cuticle covering the leaf
- hairs on the underside of the leaf
- smaller air spaces in the mesophyll than other plants
- few stomata, that are sunk in pits
- cells that can change shape to make the leaf roll up, with the lower epidermis and stomata on the inside.

Vascular tissue in plants

XYLEM AND PHLOEM IN STEMS

Vascular tissue contains vessels used for transporting materials. The two types of vascular tissue in plants are xylem and phloem. These two tissues occur in stems.

The figure (below) is a plan diagram to show the position of the tissues in the stem of a young dicotyledonous plant.

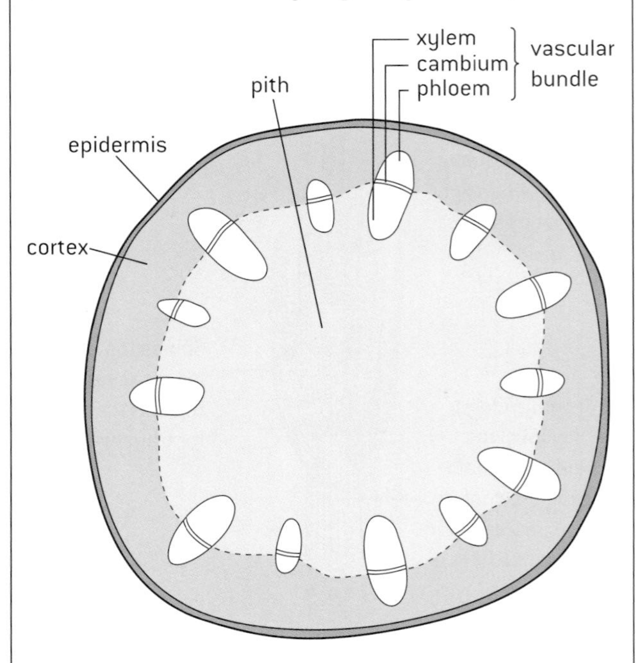

As stems grow thicker they develop more xylem and phloem tissue. The figure below shows a light micrograph of the stem of *Clematis flammula*.

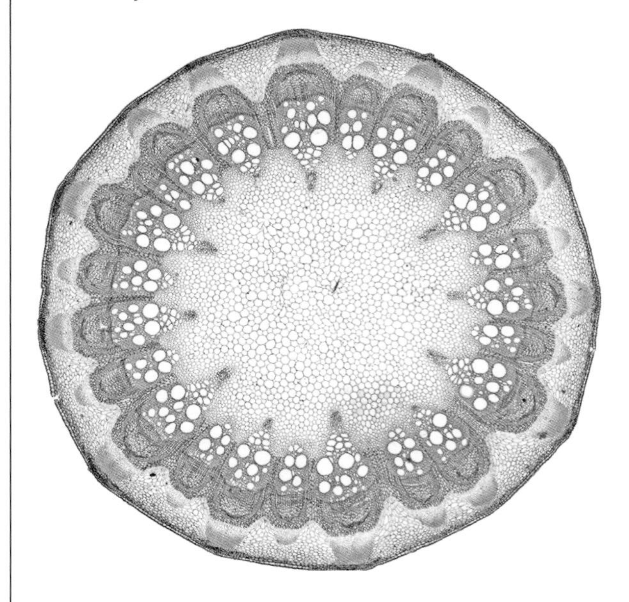

Xylem tissue is easily identified by the presence of large open xylem vessels.

Phloem tissue consists of areas of much smaller cells close to the xylem on the side nearer the epidermis.

Between the phloem and the epidermis in this stem are C-shaped regions of tough lignified cells (fibres) that provide support but are not used for transport.

XYLEM AND PHLOEM IN ROOTS

In roots the xylem and phloem tissue are in different positions from those in the stem. The figure (below) is a plan diagram to show the position of the tissues in a root.

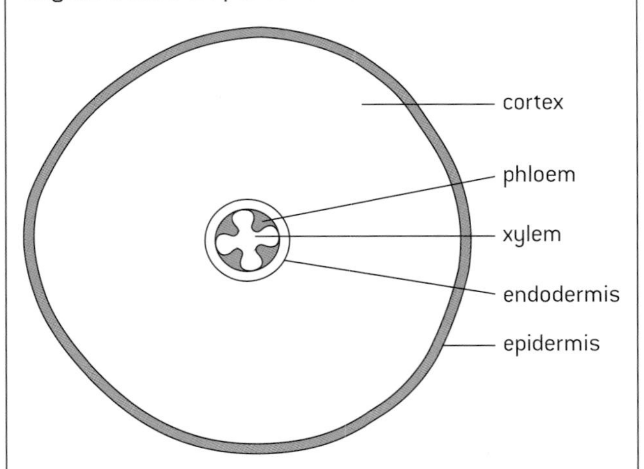

The light micrograph below shows the centre of a *Ranunculus* root in transverse section. A star-shaped area of xylem is clearly visible with phloem tissue between the points of the star and a single layer of endodermis on the outside of the vascular tissue.

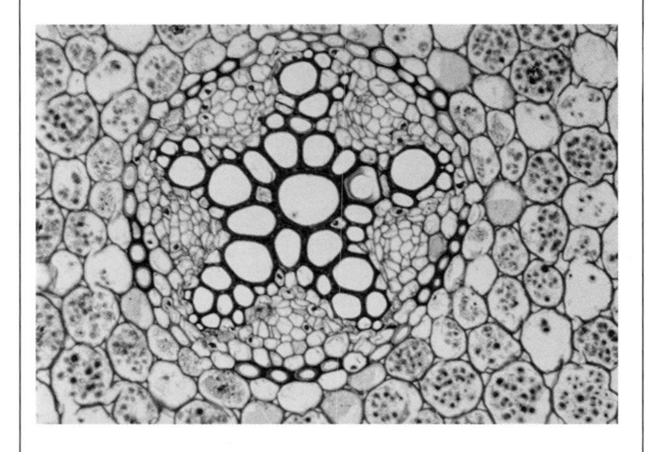

The light micrograph below shows xylem and phloem tissue in a stem of *Ricinus communis* in vertical section.

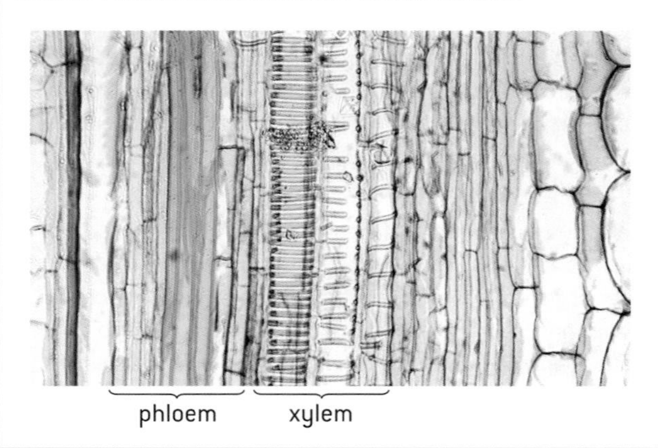

Water transport in xylem

STRUCTURE AND FUNCTION OF XYLEM

- Xylem is a tissue in plants that provides support and transports water. In flowering plants **xylem vessels** are the main transport route for water. These are long tubular structures, with strong side walls and very few cross walls.
- The main movement in xylem is from the roots to the leaves, to replace water losses from transpiration. This flow of water is called the **transpiration stream**.
- Pulling forces (tension) cause the water to move up to the leaves. These tension forces are generated in the leaves by transpiration and are due to the **adhesive** property of water. Water adheres strongly to cellulose in plant cell walls. When water evaporates from mesophyll cell walls in the leaf, more water is drawn through narrow cellulose-lined pores in leaf cell walls from the nearest xylem vessels to replace it, generating the tension.
- Tension can be transmitted from one water molecule to the next because of the **cohesive** property of water molecules that results from **hydrogen bonding**. The tension generated in the leaves is transmitted all the way down the columns of water in xylem vessels to the roots.
- At times of maximum transpiration the pressures in xylem vessels can be extremely low and the side walls have to be very strong to prevent inward collapse. This is achieved by strengthening the walls on the inside by depositing more cellulose and by impregnating this thickening of the wall with **lignin**. Thickened cell walls that have been impregnated with lignin are much harder and are **woody**.
- The first xylem formed by a shoot or root tip is primary xylem. The walls of primary xylem vessels are thickened in a helical or annular (ring-shaped) pattern. This allows the vessel to elongate as the root or shoot grows in length.

MODELS OF WATER TRANSPORT IN XYLEM

Simple models can be used to test theories about water transport in plants.

1. Water has adhesive properties

water adheres to glass so rises up the capillary tube

glass capillary tube

mercury does not adhere to glass so does not rise

water

2. Water is drawn through capillaries in cell walls

strip of paper (blotting, filter or chromatography)

water

paper is made of cellulose cell walls so water rises up through it against gravity in pores in the paper

3. Evaporation of water can cause tension

porous pot – is similar to leaf cell walls as water adheres to it and there are many narrow pores (capillaries) running through

water evaporates from the surface of the pot

more water is drawn into the pot to replace losses

water rises up the tube

DRAWING PRIMARY XYLEM VESSELS

The drawing (right) is based on microscope images such as the scanning electron micrograph (below left) and the light micrograph of xylem in pumpkin tissue (below right).

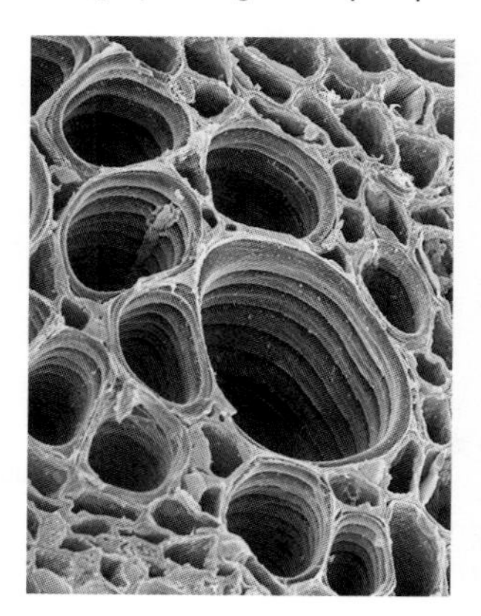

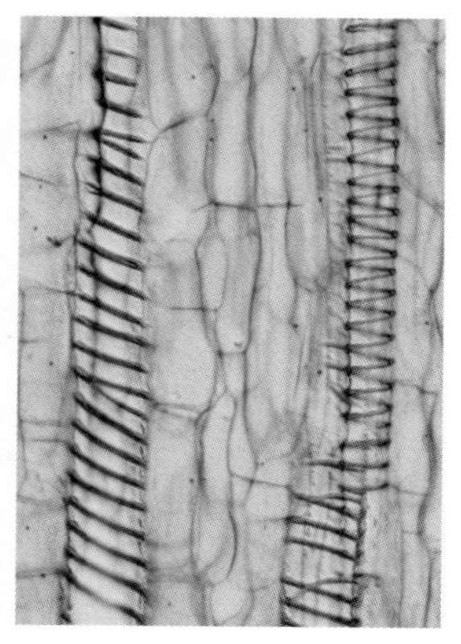

No plasma membranes are present in mature xylem vessels, so water can move in and out freely.

Lumen of the xylem vessel is filled with sap, as the cytoplasm and the nuclei of the original cells break down. End walls also break down to form a continuous tube.

Helical or ring-shaped thickenings of the cellulose cell wall are impregnated with lignin. This makes them hard, so that they can resist inward pressures.

Pores in the outer cellulose cell wall conduct water out of the xylem vessel and into cell walls of adjacent leaf cells.

Phloem transport

THE FUNCTION OF PHLOEM

Plants need to transport organic compounds, such as sugars and amino acids, from one part of the plant to another. This is the function of phloem. There are several cell types in **phloem** tissue. The movement of organic compounds takes place in **phloem sieve tubes**.

Sugars and amino acids are loaded into phloem sieve tubes by **active transport** in parts of the plant called **sources**. Examples of sources are parts of the plant where photosynthesis is occurring (stems and leaves) and storage organs where the stores are being mobilized.

Sugars and other organic compounds are unloaded from phloem sieve tubes in parts of the plant called **sinks**. Examples of sinks are roots, storage organs such as potato tubers and growing fruits including the seeds developing inside them. These are all parts of the plant where organic compounds cannot be produced but they are needed for immediate use or for storage.

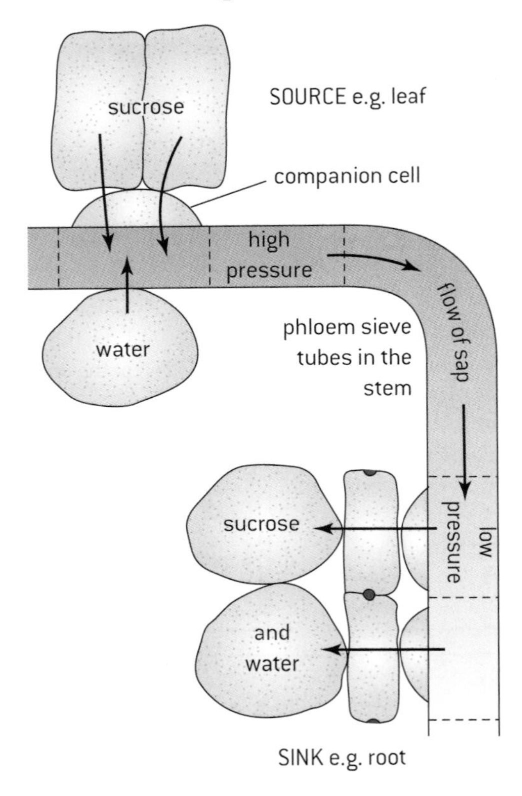

The incompressibility of water allows transport along **hydrostatic pressure gradients**.

- Hydrostatic pressure is pressure in a liquid.
- The high concentrations of solutes such as sugars in the phloem sieve tubes at the source lead to water uptake by osmosis and high hydrostatic pressure.
- The low solute concentrations of phloem sieve tubes at the sink lead to exit of water by osmosis and low hydrostatic pressure.
- There is therefore a pressure gradient that makes sap inside phloem sieve tubes flow from sources to sinks.

LOADING PHLOEM SIEVE TUBES

The main sugar carried by phloem sieve tubes is sucrose. Active transport is used to load it into the phloem but not by pumping sucrose molecules directly. Instead protons are pumped out of phloem cells by active transport to create a **proton gradient**. Co-transporter proteins in the membrane of phloem cells then use this gradient to move a sucrose molecule into the cell by simultaneously allowing protons out down the concentration gradient.

Some sucrose is loaded directly into phloem sieve tubes by this process. To speed up the process adjacent phloem cells also absorb sucrose by co-transport and then pass it to sieve tubes via narrow cytoplasmic connections (plasmodesmata).

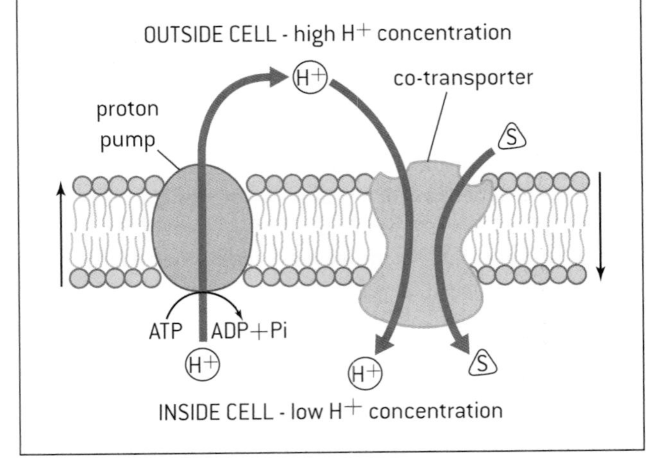

THE STRUCTURE OF PHLOEM SIEVE TUBES

Phloem sieve tubes develop from columns of cells that break down their nuclei and almost all of their cytoplasmic organelles, but remain alive. Large pores develop in the cross walls between the cells, creating the sieve plates that allow sap to flow. The diagram below shows the structure–function relationships of sieve tubes formed in this way.

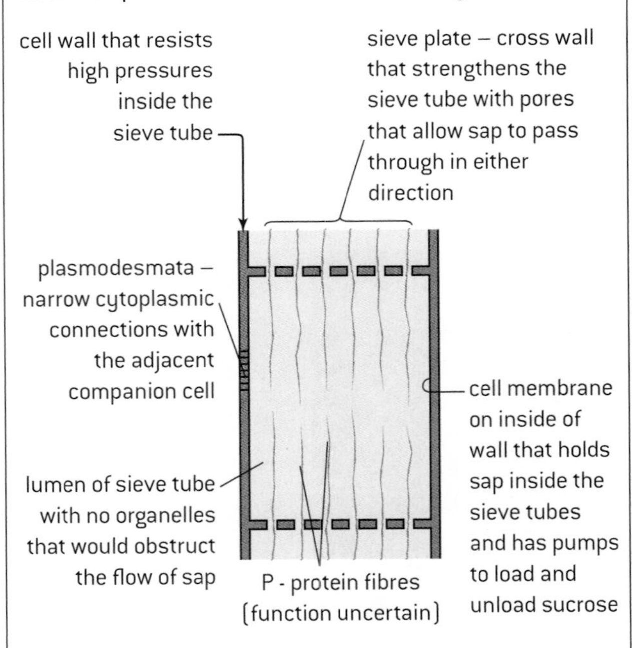

Research in plant physiology

MEASURING PHLOEM TRANSPORT RATES

Plant physiologists have developed a method using aphids of obtaining samples of phloem sap from single sieve tubes. Aphids have long piercing mouthparts called **stylets**, which they insert into stems or leaves and push inwards through the plant tissues until the stylet pierces a sieve tube. The drawing (right) shows an aphid feeding on phloem sap through its stylet. The high pressure inside the sieve tube pushes phloem sap out through the stylet into the gut of the aphid. To sample phloem sap, the aphid is cut off from its stylet when it has started to feed. The stylet is left as a very narrow tube, through which sap continues to emerge.

When radioactively labelled isotopes became available from the 1940s onwards, it became possible to do more sophisticated phloem experiments using aphids. If radioactively labelled carbon dioxide ($^{14}CO_2$) is supplied to the leaf of a photosynthesizing plant, radioactive sucrose is made in the leaf and loaded into the phloem. The time taken for this radioactive sucrose to emerge from severed aphid stylets at different distances from the leaf can be used to give a measure of the rate of movement of phloem sap.

An example of apparatus used for this research is shown (diagram below) and results obtained using it (table below).

Experiment number	1	2	3
Distance between aphid colonies (mm)	650	340	630
Time for radioactivity to travel between colonies (hours)	2.00	1.25	2.50
Rate of movement (mm hours^{-1})	32.5	27.2	25.2

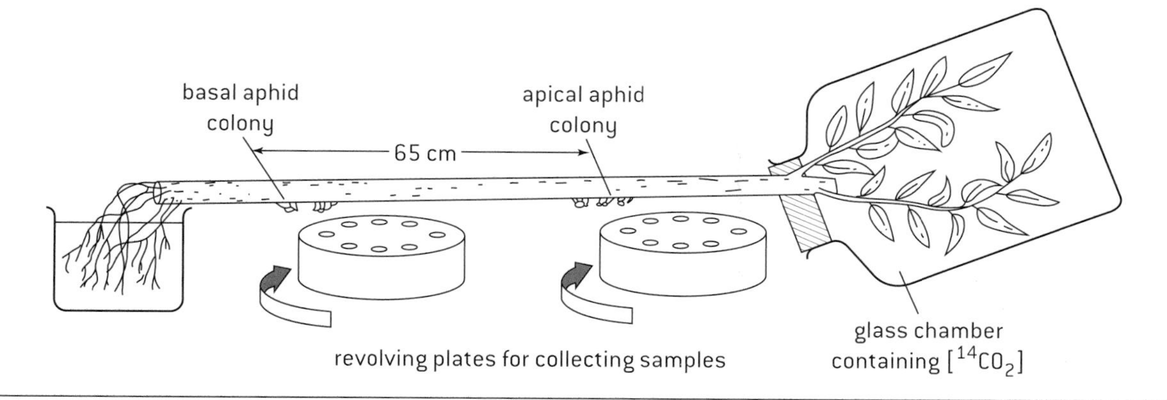

DETECTING TRACES OF PLANT HORMONES

Plant hormones were discovered in the 20th century but research into their effects was hampered by the very low concentrations in plant tissues. Even traces of plant hormones have significant effects on plant physiology, because in most cases they act as regulators of gene transcription. The concentrations at which plant hormones are active can be as low as picograms of hormone per gram of plant tissue. One picogram is a million millionth of a gram. Another problem is that there are five groups of plant hormones that are chemically very diverse, so different extraction methods are needed.

Analytical techniques have improved greatly. A variety of techniques has been used:

- ELISA (enzyme linked immunosorbent assays)
- gas chromatography–mass spectrophotometry
- liquid chromatography–mass spectrophotometry (right).

Very low concentrations of plant hormones are now detectable and previously unknown hormones have been discovered.

Recently developed techniques of molecular biology have been employed in research into plant hormones. Changes in the pattern of gene expression due to a hormone can be detected using microarrays. Proteins have been discovered to which specific hormones bind. This activates the protein, allowing it to bind to promoters of specific genes and cause their transcription. For example, five genes have been shown to be expressed on the shadier side of a shoot tip, where the auxin concentration is higher.

Plant hormones and growth of the shoot

INDETERMINATE GROWTH IN PLANTS

Plants have regions where small undifferentiated cells continue to divide and grow, often throughout the life of the plant. These regions are called **meristems**.

Flowering plants have meristems at the tip of the root and the tip of the stem. They are apical meristems as they are at the apex of the root and stem.

Growth in apical meristems allows roots and stems to elongate. The shoot apical meristem also produces new leaves and flowers.

In animal embryos a fixed number of parts develop, such as two legs and two arms in humans. This is called determinate growth.

The growth of plants by contrast is **indeterminate**, because apical meristems can continue to increase the lengths of stem and root throughout the life of a plant and can produce any number of extra branches of the stem or root. They can also produce any number of extra leaves or flowers.

GROWTH OF THE SHOOT

The leaves of a plant are attached to the stem.

The **shoot** of the plant is the stem together with the leaves. At the tip of the shoot there is a meristem, called the **shoot apical meristem**.

The cells in this meristem carry out **mitosis** and **cell division** repeatedly, to generate the cells needed for extension of the stem and development of leaves.

- Some of the cells always remain in the meristem and continue to go through the cell cycle, producing more cells.
- This production of new cells causes other cells to be displaced to the edge of the meristem.
- Cells at the edge stop dividing and undergo rapid growth and differentiation to become either stem or leaf tissue.
- Leaves are initiated as small bumps at the side of the apical dome. These bumps are called leaf primordia and through continued cell division and rapid growth they develop into mature leaves.

The diagram below shows the structure of the shoot apex of a plant.

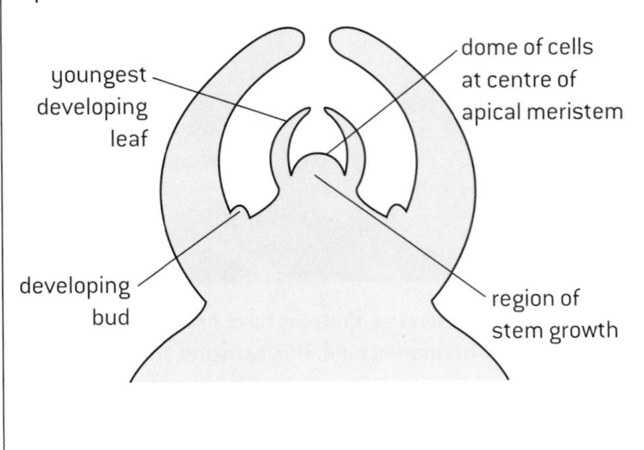

AUXIN AND PHOTOTROPISM

Plant hormones are used to control growth at the shoot tip. The main hormone is **auxin**, which acts as a growth promoter. One of the processes that auxin controls is **phototropism**.

Tropisms are directional growth responses to directional stimuli. Shoots are positively phototropic – they grow towards the brightest source of light. Charles Darwin observed this response in canary grass and made these drawings:

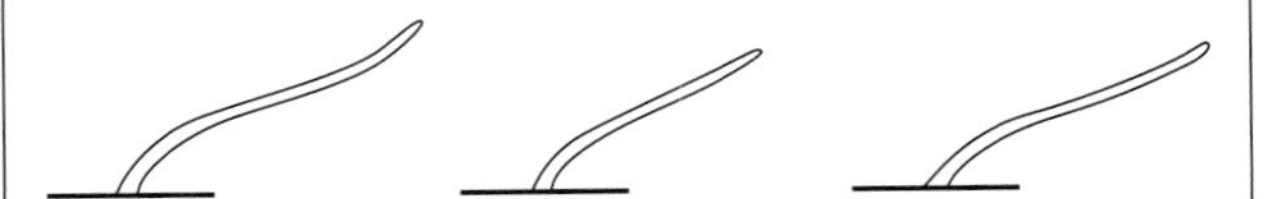

Phalaris canariensis: cotyledons after exposure in a box open on one side in front of a south-west window during 8 h. Curvature towards the light accurately traced. The short horizontal lines show the level of the ground.

Shoot tips can detect the source of the brightest light and also produce auxin. According to a long-standing theory, auxin is redistributed in the shoot tip from the lighter side to the shadier side. It then promotes more growth on the shadier side, causing the shoot to bend towards the light.

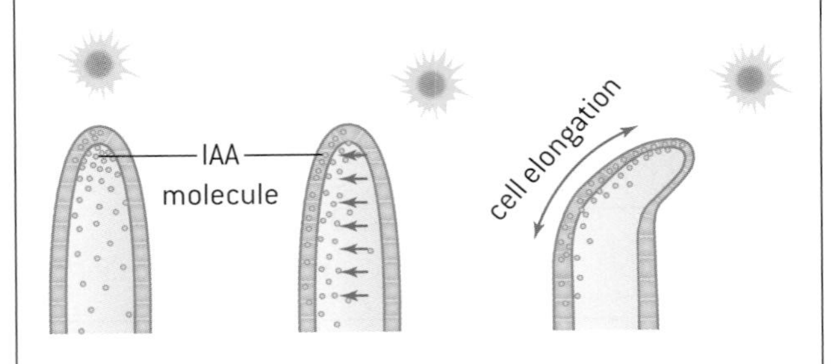

Light is detected using several types of pigment, but the most important are a group of proteins called **phototropins**. When these detect differences in the intensity of blue light in the shoot tip they trigger off movements of auxin by active transport. This is carried out by auxin pumps in the plasma membranes. They are efflux pumps as they move auxin from the cytoplasm out into the cell wall. Auxin molecules in the cytoplasm carry a negative charge and it is these that are moved by the efflux pumps.

In the cell wall a proton binds to the auxin and it can then diffuse into a cell through the plasma membrane. Once in a cell the auxin loses its proton again and is trapped in the cytoplasm until an efflux pump ejects it.

Auxin efflux pumps are moved in response to the differences in light intensity so they set up a concentration gradient of auxin from lower on the lighter side to a higher concentration on the shadier side.

Plant cells contain an **auxin receptor**. When auxin binds to it, transcription of specific genes is promoted. The expression of these genes causes secretion of hydrogen ions into cell walls. This loosens connections between cellulose fibres, allowing cell expansion.

Reproduction in flowering plants

STRUCTURE OF FLOWERS

A half view is a drawing of the structure of a flower with either the left or right half dissected away. As with all biological drawings, a sharp pencil should be used, so that single narrow lines can be drawn to show structures. The drawing below shows the structure of a *Lamium album* flower. The large petals, nectaries and the position of the anthers and stigma shows that the flowers of *Lamium album* are adapted for insect pollination, usually by bees.

Structure of *Lamium album* flower

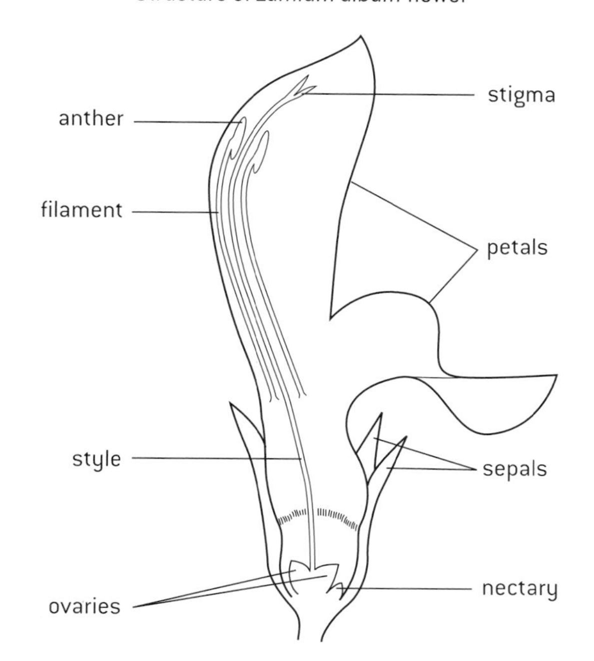

DAY LENGTH AND FLOWERING

The shoot apex produces more stem and leaves until it receives a stimulus that makes it change to producing flowers. This switch involves a change in gene expression in the cells of the shoot apex.

In many plants the stimulus is a change in the length of light and dark periods. Some plants only flower at the time of year when days are short and other plants only flower when the days are long. They are called short-day plants and long-day plants. Experiments have shown that it is not the length of day but the length of night that is significant. For example, chrysanthemums are short-day plants and only flower when they receive a long continuous period of darkness (14.5 hours or more). They therefore naturally flower in the autumn (fall).

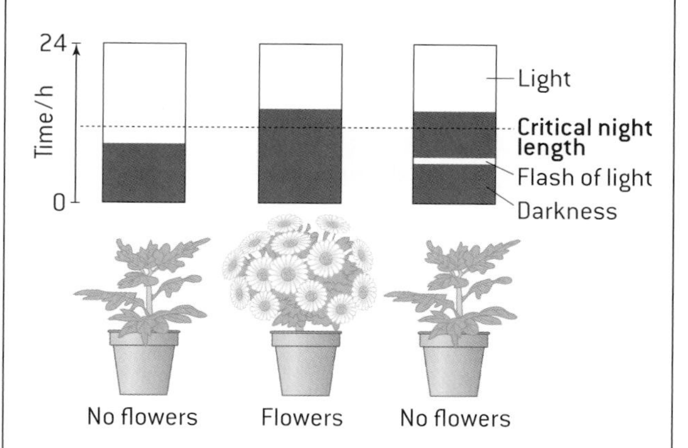

Growers can produce pots of flowering chrysanthemums out of the normal flowering season by keeping them in greenhouses with blinds. When the nights are not long enough to induce flowering, the blinds are closed to extend the nights artificially.

POLLINATION, FERTILIZATION AND SEED DISPERSAL

Flowers are the structures used by flowering plants for sexual reproduction. Female gametes are contained in ovules in the ovaries of the flower. Pollen grains, produced by the anthers, contain the male gametes. A zygote is formed by the fusion of a male gamete with a female gamete inside the ovule. This process is called **fertilization**.

Before fertilization, another process called **pollination** must occur. Pollination is the transfer of pollen from an anther to a stigma. Pollen grains containing male gametes cannot move without help from an external agent. Most plants use either wind or an animal for pollination.

Pollen grains germinate on the stigma of the flower and a pollen tube containing the male gametes grows down the style to the ovary. The pollen tube delivers the male gametes to an ovule, which they fertilize.

Fertilized ovules develop into seeds. Ovaries containing fertilized ovules develop into fruits. The function of the fruit is **seed dispersal**. This is the spreading of seeds away from the parent plant to sites where they can germinate and grow without competing with their parent.

So, success in plant reproduction depends on three different processes: pollination, fertilization and seed dispersal.

MUTUALISM IN POLLINATION

More than 85% of the world's 250,000 species of flowering plant depend on insects or other animal pollinators for reproduction. Both species in a mutualistic relationship benefit. In a plant–pollinator relationship, the plant benefits by its flowers being pollinated and the pollinator benefits by obtaining nectar (a source of energy) and pollen (a source of protein).

There is a trend for plant species to develop mutualistic relationships for pollination with one specific species of insect. For example, the vanilla orchid (right) is pollinated by a species of *Melipona* bee. The advantage of this is that the insect will transfer pollen from flower to flower of the species and not to other species. This is an example of why it is essential to protect entire ecosystems, not individual species – species cannot live in isolation as they depend on each other.

Propagating plants

SEED STRUCTURE

Flowering plants can be propagated by sowing seeds. A seed contains an embryo plant and food reserves for the embryo to use during germination.

Beans are large seeds with a structure that is easy to observe. The seed coat (testa) must be removed. The food reserve consists of two large modified embryo leaves, called **cotyledons**. If one of these is removed, the internal structure of the seed can be seen. The drawing below shows this structure.

External structure

seed coat (testa)
scar where seed was attached to the ovary

Internal structure

embryo root (radicle)
seed coat
embryo shoot (plumule)
cotyledon – one of two in the seed

FACTORS NEEDED FOR SEED GERMINATION

Seeds will not germinate unless external conditions are suitable. Three hypotheses about factors needed for germination are suggested here:

- Water must be available (to rehydrate the seed).
- Oxygen must be available (for cell respiration).
- Warmth is needed (for enzyme activity).

The basic design of an experiment to test one of these hypotheses requires at least two treatments:

1. a control treatment giving seeds all factors needed
2. a treatment giving seeds all factors except one.

If seeds given the control treatment germinate but those denied one factor do not, that factor must be needed for germination.

Apparatus for testing whether oxygen is needed is shown below:

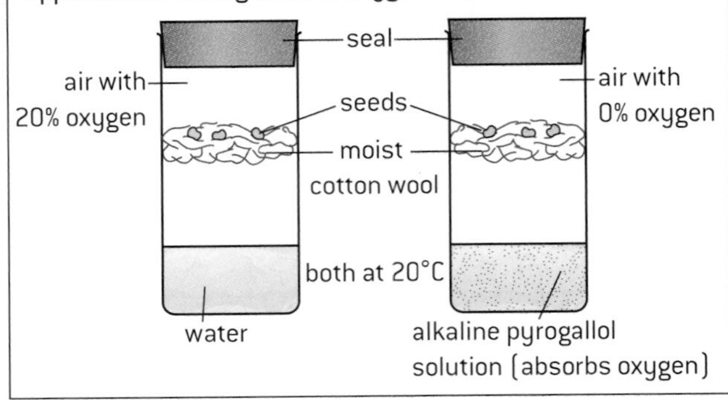

MICROPROPAGATION OF PLANTS

Desirable varieties of plants are propagated by asexual reproduction, so that all the plants produced have the desirable characteristics. There are many traditional methods for this, but more recently a technique called micropropagation has been developed. This name is used because the propagation can be done with very small pieces of tissue taken from the shoot apex of a plant. Stages in micropropagation are shown in the diagrams below. Three advantages of the procedure explain why micropropagation is now very widely used.

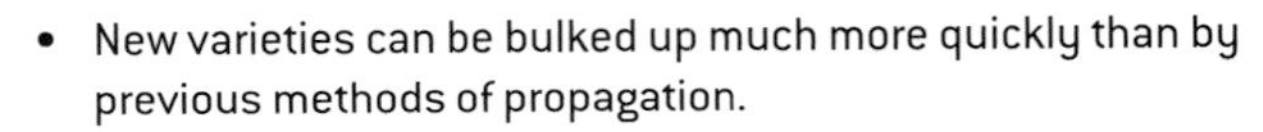

- New varieties can be bulked up much more quickly than by previous methods of propagation.
- Virus-free strains of existing varieties can be produced because cells in the shoot apex normally do not contain viruses that reduce plant growth even if other cells in a plant do contain them.
- Large numbers of rare plants such as orchids can be produced, reducing the cost to people who want to buy them and making it unnecessary to take them from wild habitats.

A small piece of tissue is removed from the plant that is being cloned. Often the tissue comes from a shoot tip. The tissue is sterilized. All apparatus and growth media must be sterilized to prevent infections. This is called aseptic technique.

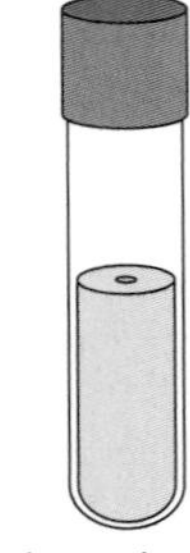

The tissue is placed on sterile nutrient agar gel, containing a high auxin concentration. This stimulates cell growth and division.

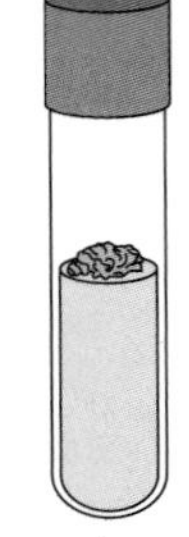

An amorphous lump of tissue called a callus grows, which can be cut up and made to grow more using the same type of nutrient agar containing auxin.

Eventually the callus is transferred to nutrient agar gel containing less auxin but high concentrations of cytokinin which stimulates plantlets with roots and shoots to develop. Gibberellin is sometimes added to increase shoot growth and prevent dormancy.

The plantlets are separated and transferred to soil, where they should grow strongly.

Questions – plant biology

1. What conditions cause most rapid transpiration?

   A. still, hot and humid B. windy, warm and dry

   C. windy, cold and humid D. still, cool and dry

2. What is an advantage of micropropagation of plants?

   A. produces smaller plants

   B. avoids using hormones

   C. small companies do it

   D. produces virus-free plants

3. The micrograph is a transverse section through the stem of *Salicornia europaea*, which is adapted to saline soils in salt marshes. The leaves of this plant are very small so the stem is the main organ of photosynthesis.

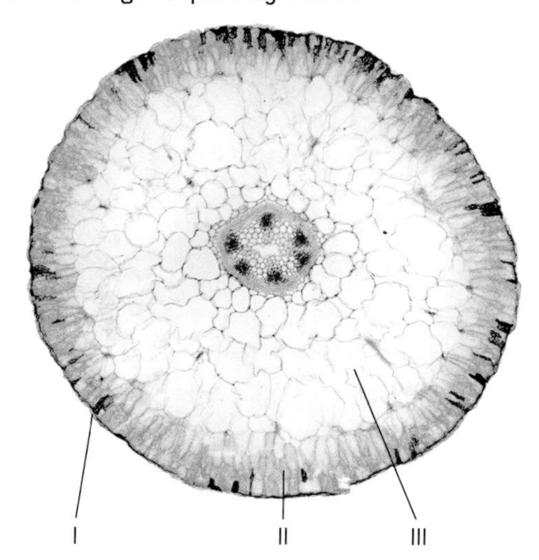

   a) Explain how structure I helps conserve water. [2]

   b) Structure II is similar to one of the tissues in a typical leaf. Deduce which leaf tissue this is. [1]

   c) Suggest a role for tissue III. [2]

   d) There are six vascular bundles in the centre of the stem, with relatively small amounts of xylem. Suggest reasons for the stem having little xylem. [2]

   e) *Salicornia* has C_4 metabolism that allows CO_2 fixation by rubisco at lower CO_2 concentrations than in plants with standard C_3 metabolism. Suggest how this might help to reduce water loss by transpiration. [2]

   The density of stomata was measured in the leaves of *Salicornia persica* when grown in soils with different salt (NaCl) concentrations. The results are below.

NaCl salinity (mmol dm^{-3})	0	100	200	500
Stomatal density (mm^{-2})	194	101	74	68

   f) (i) Outline the relationship between stomatal density and salinity. [1]

   (ii) Suggest reasons for the relationship. [2]

4. a) State two processes occurring in the shoot apex that are needed for stem growth. [2]

   b) (i) Outline how concentration gradients of auxin are established in a shoot apex. [4]

   (ii) Explain the role of these auxin gradients. [4]

5. The apparatus below was set up and left for three hours. The mass on the left dropped from 158.47 g to 155.77 g and the mass on the right increased from 158.80 g to 161.41 g.

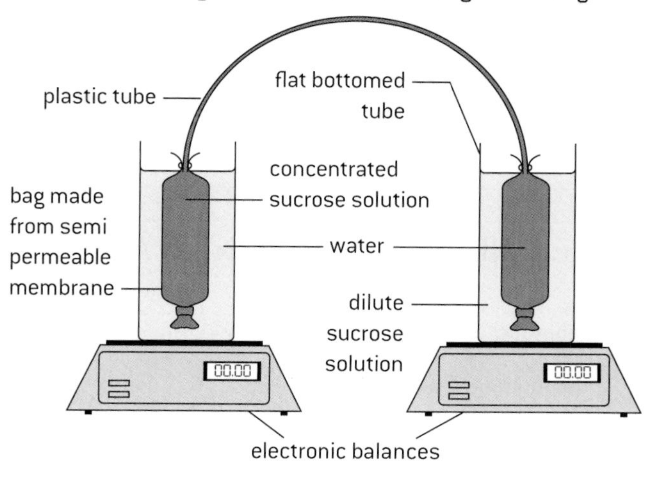

   a) Calculate the rate of mass change on each side. [2]

   b) Explain the mass changes. [4]

   c) (i) Identify a plant transport system that is similar to the apparatus. [1]

   (ii) Describe the similarities. [3]

6. The diagram is a half view of a flower of *Anticlea elegans*.

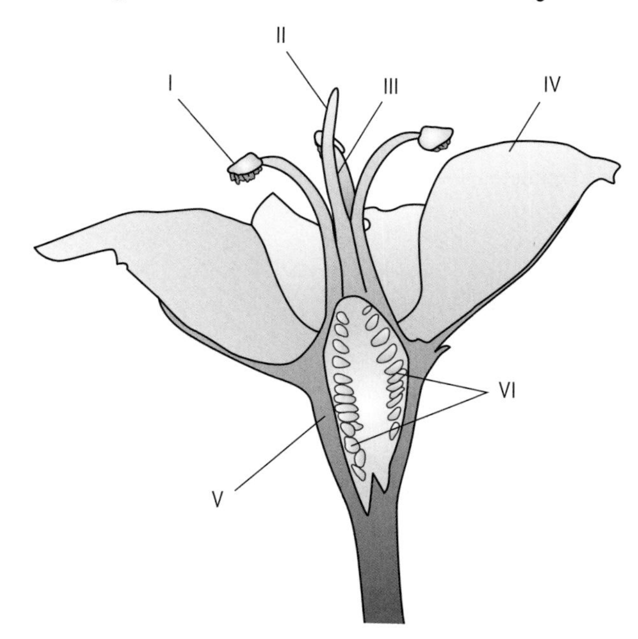

   a) State the names of structures I–VI. [6]

   b) Explain the mutualistic relationship between the plant and the animals that visit its flowers. [4]

   c) Suggest reasons for seeds not developing if no insects visit the flower. [3]

   d) The flower develops from a shoot apex that was previously developing leaves and stem. Outline how the change to developing a flower occurs. [5]

   e) State two parts of the embryo in a seed. [2]

Mendel's law of independent assortment

Gregor Mendel's monohybrid crosses with pea plants showed that the two alleles of a gene separate into different haploid gametes during meiosis. This is called the **law of segregation**.

Mendel discovered the law of independent assortment by doing crosses in which the parents differed in two characteristics that are controlled by two different genes. These are called **dihybrid crosses**. An example of one of his crosses is shown below. The parents in this cross differ in seed shape, controlled by one gene, and in seed colour, controlled by a different gene.

RATIOS IN GENETIC CROSSES

The **genotypic ratio** is the proportions of the various genotypes produced by the cross.

The **phenotypic ratio** is the proportions of the various phenotypes.

INDEPENDENT ASSORTMENT IN A DIHYBRID CROSS

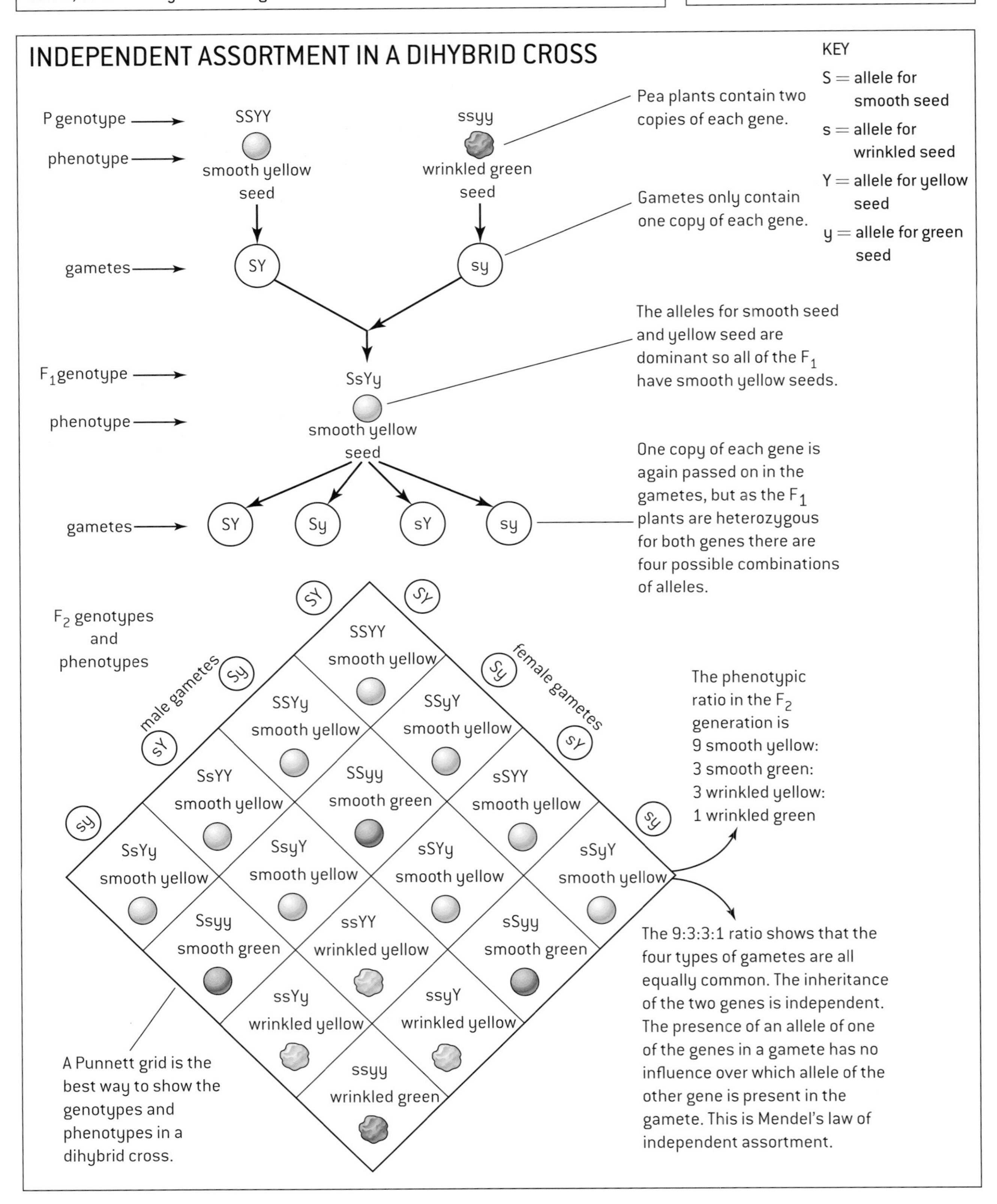

Dihybrid crosses

PREDICTING RATIOS IN DIHYBRID CROSSES

The 9:3:3:1 ratio is often found when parents that are heterozygous for two genes are crossed together. The ratio is the product of two 3:1 ratios – each of the two genes would give a 3:1 ratio in a monohybrid cross between two heterozygous parents. In a dihybrid cross they follow Mendel's law of independent assortment because they are unlinked.

Dihybrid crosses can give other ratios if:

- either of the genes has co-dominant alleles;
- either of the parents is homozygous for one or both of the genes; or
- either of the genes is not autosomal – in other words if it is sex-linked.

The figure (right) shows ratios that these types of gene could give.

Another cause of unusual ratios is interaction between genes (epistasis). The figure (below) shows an example of a dihybrid cross where there is interaction between genes.

POSSIBLE RATIOS IN DIHYBRID CROSSES

	3	1
3	9	3
1	3	1

	1	2	1
3	3	6	3
1	1	2	1

	1	1
3	3	3
1	1	1

	1	1
1	1	1
2	2	2
1	1	1

GENOTYPIC AND PHENOTYPIC RATIOS IN A CROSS WITH INTERACTION BETWEEN GENES

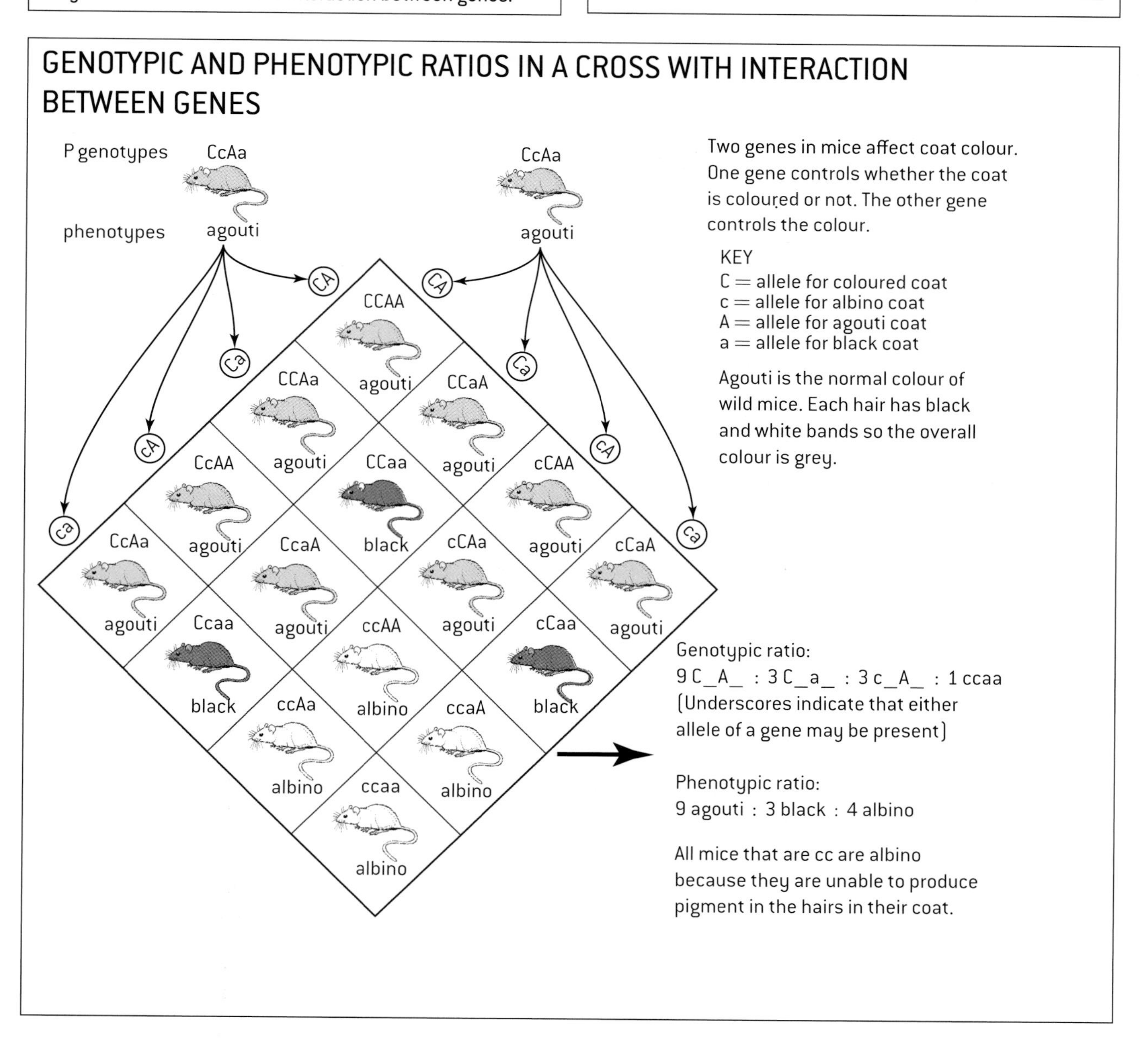

Two genes in mice affect coat colour. One gene controls whether the coat is coloured or not. The other gene controls the colour.

KEY
C = allele for coloured coat
c = allele for albino coat
A = allele for agouti coat
a = allele for black coat

Agouti is the normal colour of wild mice. Each hair has black and white bands so the overall colour is grey.

Genotypic ratio:
9 C_A_ : 3 C_a_ : 3 c_A_ : 1 ccaa
(Underscores indicate that either allele of a gene may be present)

Phenotypic ratio:
9 agouti : 3 black : 4 albino

All mice that are cc are albino because they are unable to produce pigment in the hairs in their coat.

Genes – linked and unlinked

UNLINKED GENES

Genes that assort independently are **unlinked genes**. Genes assort independently if they are located on different chromosomes.

Independent assortment of unlinked genes can be explained in terms of chromosome movements during meiosis. When pairing of homologous chromosomes occurs during prophase I of meiosis, the alleles of unlinked genes are on different pairs of homologous chromosomes. A pair of homologous chromosomes is called a **bivalent**.

Bivalents are orientated randomly on the equator during metaphase I of meiosis. The orientation of one bivalent does not affect the orientation of other bivalents, so the pole to which an allele on one bivalent moves when homologous chromosomes separate in anaphase I of meiosis does not affect the pole to which alleles on another bivalent move. For example, when a parent with the genotype AaBb produces gametes, AB, Ab, aB and ab are all equally probable if genes A and B are located on different chromosomes. This is shown more fully by means of a diagram in Topic 3.

LINKED GENES

Some pairs of genes do not follow the law of independent assortment and expected ratios for unlinked genes are not found. Combinations of genes tend to be inherited together. This is called **gene linkage**. It is caused by pairs of genes being located on the same type of chromosome. The scientific name for the location of a gene on a chromosome is a **locus**, so linked genes have loci on the same chromosomes.

New combinations of the alleles of linked genes can only be produced if DNA is swapped between chromatids. This is called **recombination** and involves a process called **crossing-over**. Individuals that have a different combination of characters from parents, due to crossing over, are **recombinants**.

The figure (below) shows the first example of gene linkage to be discovered. The results show that there were more offspring than expected with the parental character combinations – purple long and red round. There were fewer than expected with the new combinations – purple round and red long.

MENDEL AND MORGAN

Mendel (below left) performed careful dihybrid crosses, with meticulous recording of results. He developed a theory that explained all his results – the law of independent assortment. In the 20th century anomalous results were obtained that did not fit the theory.

An American geneticist, Thomas Hunt Morgan (above right), developed the idea of linked genes to account for the anomalies. He did this while investigating examples in *Drosophila* where the inheritance pattern is different in males and females – sex-linkage.

His explanation for sex-linkage was that genes were located on the sex chromosomes.

Other anomalies, where the pattern of inheritance was the same in both males and females, but the ratios were non-Mendelian, could be explained by two genes being located together on the same non-sex chromosome (autosome).

Mendel's law of independent assortment still works for most pairs of genes because they are on different chromosomes so are unlinked, but Morgan's idea of gene linkage is an essential refinement of the theory for the groups of genes that occur together on a chromosome.

LINKAGE AND RECOMBINANTS

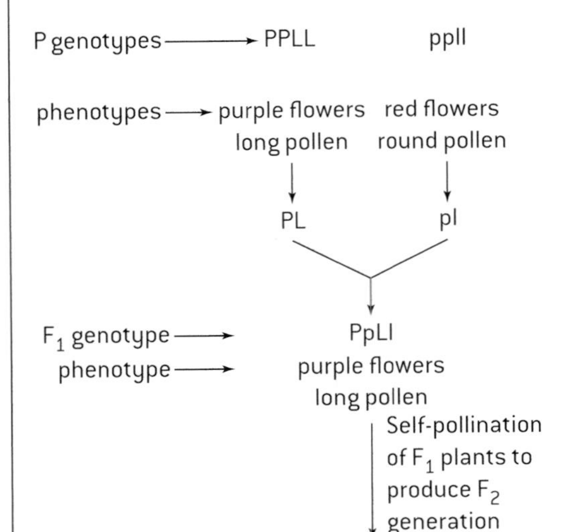

Expected F_2 ratio	9 purple long	3 purple round	3 red long	1 red round
Expected results (6952 plants in total)	3910.5	1303.5	1303.5	434.5
Observed results	4831	390	393	1338

Chi-squared = 372 at 3 degrees of freedom
Significance level is less than 0.001
So there is 99.9% confidence of a significant difference between the observed and expected results.

The purple–round and red–long offspring are recombinants because they are different from the parental phenotypic combinations.

This is clearly a case of gene linkage because the expected ratio for unlinked genes is 9:3:3:1 but there are more of the parental combinations and fewer recombinants than this.

Crossing-over

PROPHASE I OF MEIOSIS

Homologous chromosomes pair up in prophase I of meiosis. Each homologous chromosome consists of two **sister chromatids**, because all DNA has been replicated in interphase before the start of meiosis. Chromatids of the two different chromosomes in a pair are **non-sister chromatids**. While the chromosomes are paired, sections of chromatid are exchanged in a process called **crossing-over**. The figure (right) shows how it occurs.

RECOMBINATION OF LINKED GENES

Without crossing-over it would be impossible to produce new combinations of linked genes – parental combinations would always be passed unaltered to offspring.

Homologous chromosomes separate in meiosis I and sister chromatids separate in meiosis II, so each of the four haploid cells produced by meiosis receives one chromatid from each bivalent. There is always at least one cross-over per bivalent so most of these chromatids will have a new combination of alleles.

The point where crossing-over occurs along chromosomes is random – it can occur at a vast number of different points so meiosis produces an almost infinite amount of genetic variety.

The figure (below) shows how crossing-over can cause recombination of linked genes.

The figure (right) shows an example of a cross involving gene linkage.

Parental gene combinations are AB and ab

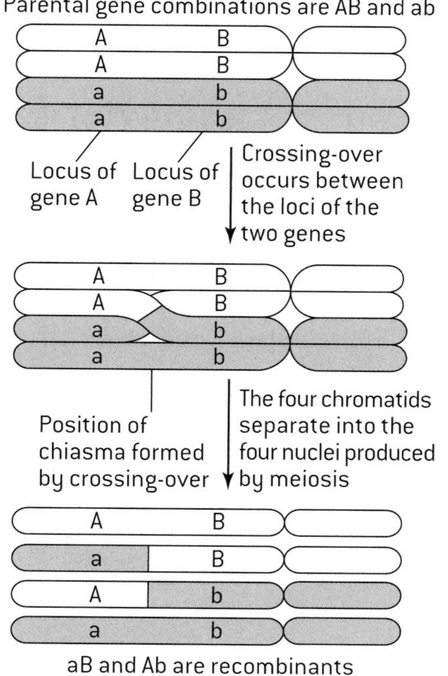

aB and Ab are recombinants

THE PROCESS OF CROSSING-OVER

Crossing-over is the exchange of DNA material between non-sister homologous chromatids.

At one stage in prophase I all of the chromatids of two homologous chromosomes become tightly paired up together. This is called **synapsis**.

four chromatids in total, long and thin at this stage

centromeres

The DNA molecule of one of the chromatids is cut. A second cut is made at exactly the same point in the DNA of a non-sister chromatid.

DNA is cut at the same point in two non-sister chromatids

The DNA of each chromatid is joined up to the DNA of the non-sister chromatid. This has the effect of swapping sections of DNA between the chromatids.

In the later stages of prophase I the tight pairing of the homologous chromosomes ends, but the sister chromatids remain tightly connected. Where each cross-over has occurred there is an X-shaped structure called a chiasma.

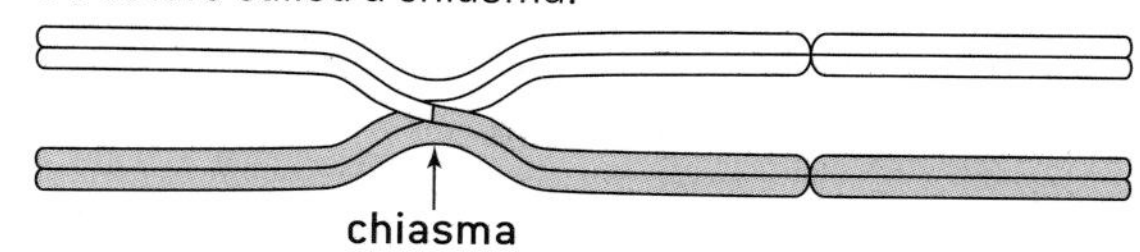

AN EXAMPLE OF GENE LINKAGE AND TEST CROSSING

Bars are used to represent chromosomes on which genes are linked.

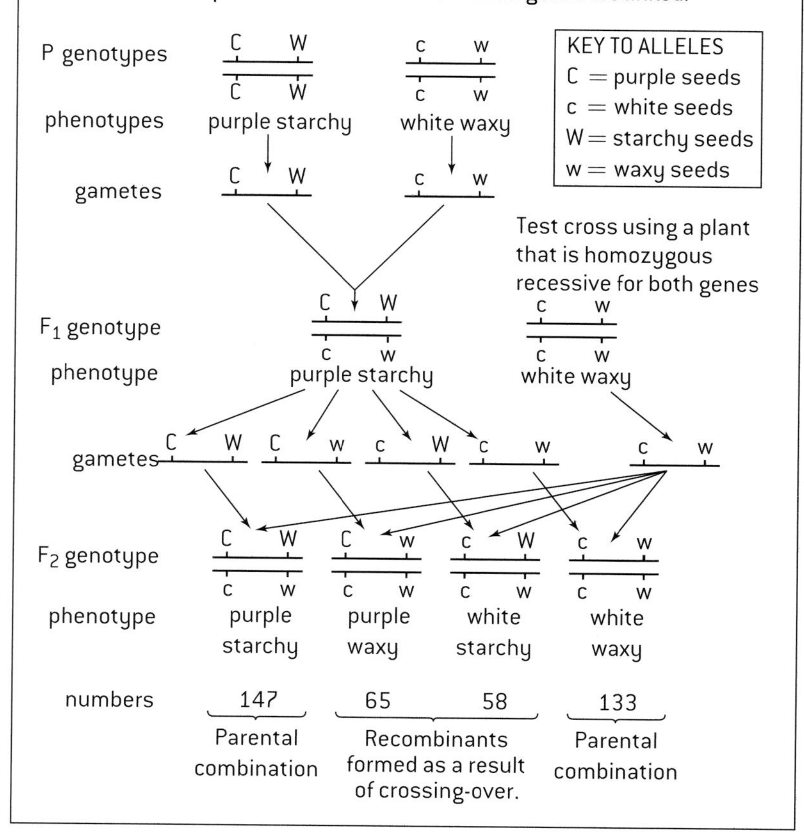

Chi-squared and continuous variation

CHI-SQUARED TESTS IN GENETICS

The use of this statistical test in ecology was described in Topic 4. It can also be used in genetics to find out whether there is a significant difference between observed and and expected results.

Method for chi-squared test

1. Draw up a contingency table of observed frequencies, which are the numbers of individuals of each phenotype resulting from the cross.
2. Calculate the expected frequencies, based on a Mendelian ratio and the total number of offspring.
3. Determine the number of degrees of freedom, which is one less than the total number of possible phenotypes. In a dihybrid cross there are four phenotypes so there are 3 degrees of freedom.
4. Find the critical region for chi-squared from a table of chi-squared values, using the degrees of freedom that you have calculated and a significance level (p) of 0.05 (5%). The critical region is any value of chi-squared larger than the value in the table.
5. Calculate chi-squared using this equation:

$$\chi^2 = \sum \frac{(\text{obs} - \text{exp})^2}{\text{exp}}$$

6. Compare the calculated value of chi-squared with the critical region. If the calculated value is in the critical region, the differences between the observed and the expected results are statistically significant – the results do not fit the Mendelian ratio closely, perhaps because the two genes in the cross are linked. If the calculated value is outside the critical region the differences between the observed and the expected results are not statistically significant – the results fit the Mendelian ratio, suggesting that the genes are unlinked and assort independently.

Example:

White leghorn chickens with large single combs were crossed with Indian game fowl with dark feathers and small pea combs. All of the F_1 crosses were white with pea combs. They were crossed with each other so the expected ratio in the F_2 generation was 9:3:3:1.

	white pea	white single	dark pea	dark single	total
observed	111	37	34	8	190
expected	$\frac{9}{16} \times 190 = 106.9$	$\frac{3}{16} \times 190 = 35.6$	$\frac{3}{16} \times 190 = 35.6$	$\frac{1}{16} \times 190 = 11.9$	190

Degrees of freedom = 4 − 1 = 3

Critical values of the χ^2 distribution

df	p 0.995	0.975	0.9	0.5	0.1	0.05	0.025	0.01	0.005	df
1	0.000	0.000	0.016	0.455	2.706	3.841	5.024	6.635	7.879	1
2	0.010	0.051	0.211	1.386	4.605	5.991	7.378	9.210	10.597	2
3	0.072	0.216	0.584	2.366	6.251	7.815	9.348	11.345	12.838	3

At the 0.05 level of significance, the critical value is 7.815.

$$\text{Chi-squared} = \frac{(111 - 106.9)^2}{106.9} + \frac{(37 - 35.6)^2}{35.6} + \frac{(34 - 35.6)^2}{35.6} + \frac{(8 - 11.9)^2}{11.9}$$

$$= 1.56$$

The calculated value for chi-squared is outside of the critical region so the differences between the observed and expected results are not statistically significant. The results fit the 9:3:3:1 ratio so we conclude that the genes for comb shape and colour of feathers are unlinked and assort independently.

CONTINUOUS VARIATION

Variation can be **discrete** or **continuous**.

With discrete variation every individual fits into one of a number of non-overlapping classes. For example, all humans are in blood group A, B, AB or O.

With continuous variation any level of the characteristic is possible, between the two extremes. For example, any height is possible in humans between the smallest and the largest height.

Discrete variation is usually due to one gene. If continuous variation is genetically determined it is due to the combined effects of two or more genes. This is known as **polygenic inheritance**.

Example: wheat grains vary in colour from white to dark red, depending on the amount of a red pigment they contain. Three genes control the colour. Each gene has two alleles, one that causes pigment production and one that does not. Wheat grains can therefore have between 0 and 6 alleles for pigment production.

The figure (below) shows the expected distribution of grain colour from a cross between two plants that are heterozygous for each of the three genes.

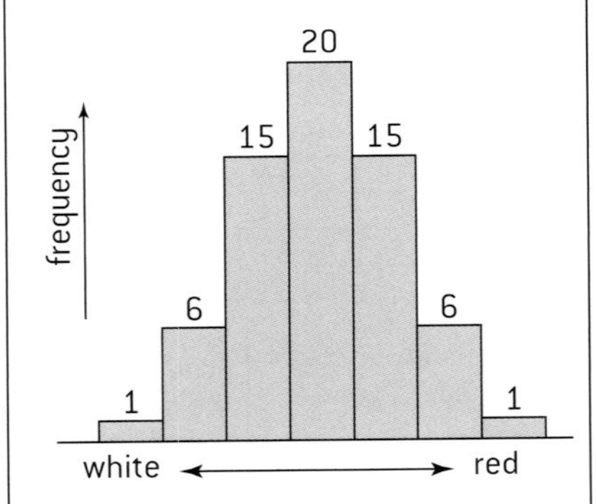

Many polygenic traits are also influenced by environmental factors. Human height is influenced both by many genes and also by environmental factors such as the quality of nutrition given to a child when he/she is growing. Skin colour is influenced by genes and also by the amount of light the skin receives.

Speciation

GENE POOLS

The concept of the gene pool is used in a branch of biology called population genetics. A gene pool consists of all the genes and their different alleles in an interbreeding population.

A new individual, produced by sexual reproduction, inherits genes from its two parents. Assuming there is random mating, any two individuals in an interbreeding population could be the parents, so the individual could inherit any of the genes in the gene pool.

Many genes have different alleles. In a typical interbreeding population, some alleles are more common than others. Evolution always involves a change over time in allele frequency in a population's gene pool.

DIFFERENCES IN ALLELE FREQUENCY

The frequency of an allele is the number of that allele there is in a population divided by the total number of alleles of the gene. Allele frequency can range from 0.0 to 1.0 and the total frequency of all alleles is 1.0.

Geographically isolated populations often have different allele frequencies from the rest of a species, For example, the frequency of delta F508, an allele that causes cystic fibrosis, is 0.04 on the Faroe Islands but only 0.03 in northern Europe and is below 0.01 in most other parts of the world. Differences in allele frequency may be due either to differences in natural selection or to random drift.

TYPES OF NATURAL SELECTION

There are three patterns of natural selection:

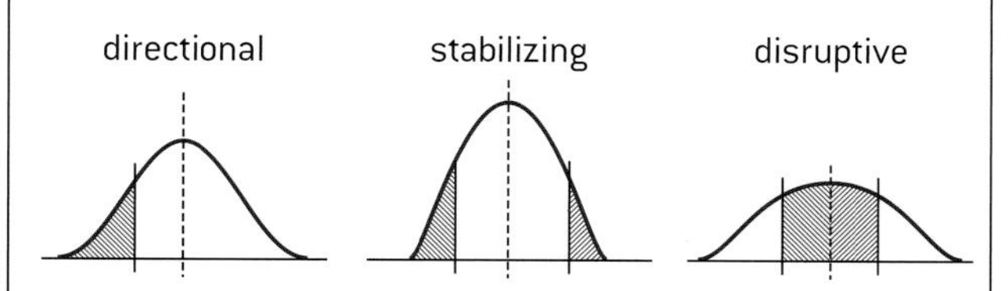

Directional – one extreme in the range of variation is selected for; the other extreme is selected against.

Example: in the bird species *Parus major* (great tit), breeding success has been greater with birds that breed early than with those that breed later because the peak availability of prey is now earlier in the year as a consequence of climate change.

Stabilizing – intermediates are selected for and extremes are selected against.

Example: in the bird species *Parus major* (great tit), breeding success is greatest with intermediate clutch sizes (number of eggs) because in large clutches the offspring have lower survival rates and in small clutches there are fewer offspring with no greater chance of survival than in intermediate clutches.

Disruptive – extreme types are selected for and intermediates are selected against.

Example: in the bird species *Passerina amoena* (lazuli bunting), year-old males with the dullest and brightest plumage are more successful than males with intermediate plumage at obtaining high-quality territories, pairing with females and siring offspring.

SPECIATION AND REPRODUCTIVE ISOLATION

The formation of new species is called **speciation**. New species are formed when a pre-existing species splits. This usually involves one population not interbreeding with any other populations of its species – **reproductive isolation**. The isolated population's gene pool is therefore separate. If natural selection acts differently on this population, it will gradually diverge from the other populations. Eventually the isolated population will be incapable of interbreeding with the rest of the species – it has become a new species.

Speciation can occur by **gradual** divergence over thousands of years or it can be **abrupt** and happen suddenly. The former type of evolution is called **gradualism** and the latter type is **punctuated equilibrium** – long periods without appreciable change and short periods of rapid evolution.

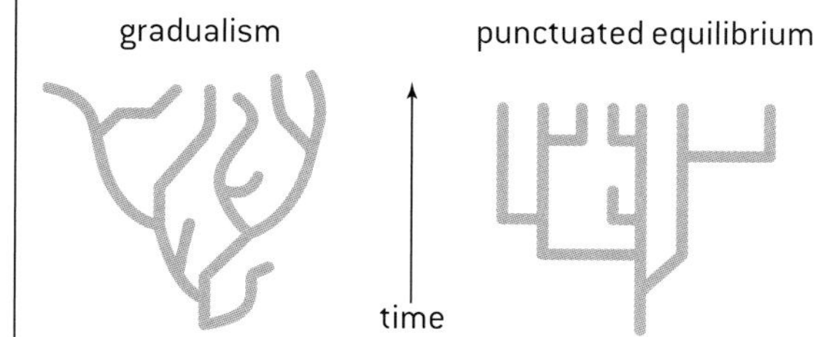

How fast speciation occurs depends on the type of reproductive isolation. There are three main types.

1. **Temporal** – when populations of a species breed at different times. For example, some species of cicada only breed every 13th year. If some individuals breed with each other in a different year from the rest of the species, they will be isolated.
2. **Behavioural** – when populations of a species have behaviour that prevents interbreeding. For example, if mating calls used by male tree frogs to attract females vary and different females favour different calls, the species may split into non-interbreeding populations with distinguishable mating calls.
3. **Geographical** – when populations of a species live in different areas and therefore do not interbreed. For example, both lava lizards and finches on the Galápagos archipelago migrated from island to island, becoming reproductively isolated and splitting into different species.

SPECIATION BY POLYPLOIDY IN ALLIUM

In some plant groups there is a trend for the species to have chromosome numbers that are all multiples of one basic number. For example, most *Allium* species have a diploid number that is a multiple of 16. The ancestral *Allium* probably had this number. *Allium* species with 32 chromosomes evolved by polyploidy. This is when an error leads to an individual having more than two sets of chromosomes. In a species with a diploid number of 16, an individual with 32 chromosomes is tetraploid. If it crosses with a diploid individual, all the offspring are infertile triploids. Because of this, a tetraploid is reproductively isolated from diploids. Polyploidy is therefore instant speciation. Many plant species have been produced by polyploidy.

Questions – genetics and evolution

1. Two stages of meiosis are shown below.

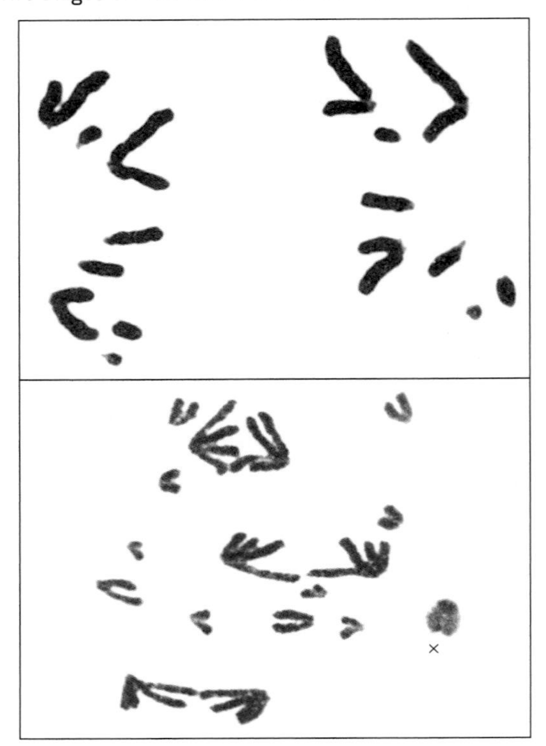

   a) Identify the stage of meiosis in (i) the upper and (ii) the lower micrograph. [2]

   b) Deduce whether the nuclei being produced are haploid or diploid in (i) the upper and (ii) the lower micrograph. [2]

2. The micrograph below shows a pair of homologous chromosomes in a cell carrying out meiosis in the grasshopper *Chorthippus parallelus*.

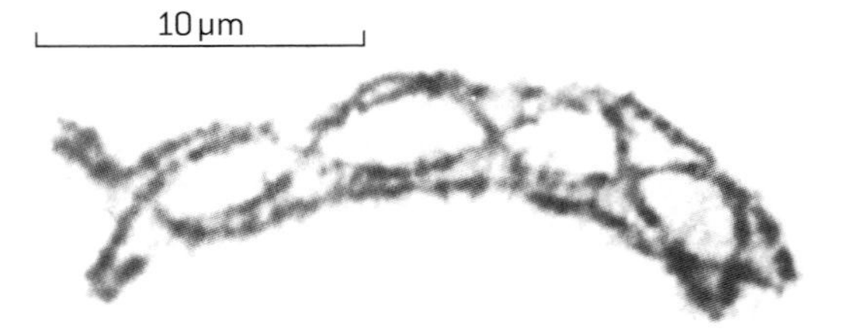

   a) Identify the stage of meiosis of the cell that contained the pair of chromosomes. [2]

   b) In the pair of chromosomes in the micrograph deduce the number of

      (i) chromatids [1]

      (ii) chiasmata [1]

   c) Outline how chiasmata are produced. [3]

3. In some plants two genes control flower colour.

   Plants with the genotype A_B_ have blue flowers.

   Plants with the genotype A_bb have red flowers.

   Plants with the genotype aa_have white flowers.

   The underscore symbol (_) represents any allele.

   a) State the name given to the type of inheritance where more than one gene controls a single phenotypic characteristic. [1]

   A homozygous blue-flowered plant (AABB) is crossed with a homozygous white-flowered plant (aabb).

   b) State genotypes and phenotypes of F_1 offspring. [2]

   c) The F_1 plants are allowed to pollinate each other. Deduce, using a Punnett grid, the genotypes of the gametes produced by the F_1 plants and the genotypes and phenotypes of all the possible F_2 offspring. [5]

   d) State the expected colour ratio in F_2 offspring. [1]

   e) The two genes code for enzymes used to convert a white substance into a red pigment and the red pigment into a blue pigment. Deduce the effect of the enzymes produced from gene A and gene B. [1]

4. When grey-bodied, long-winged *Drosophila* flies were test-crossed with black-bodied, vestigial-wing flies the F_1 generation was found to contain:

   407 grey-bodied, long-winged flies

   396 black-bodied, vestigial-winged flies

   75 black-bodied, long-winged flies

   69 grey-bodied, vestigial-winged flies

   a) State the name for a cross involving two genes. [1]

   b) Identify which of the flies were recombinants. [2]

   The F_1 generation does not follow Mendel's law of independent assortment.

   c) (i) State the expected ratio for a test cross that follows Mendel's law of independent assortment. [2]

      (ii) Explain how the observed ratio could have arisen, including a key to the symbols used for the alleles. [5]

5. The table shows frequencies of ABO blood groups in three populations that do not breed with each other.

Population	Frequency (%)			
	O	A	B	AB
Andamanese (India)	9	60	23	9
Navajo (N. America)	73	27	0	0
Kalmyk (Mongolia)	26	23	41	11

   a) Compare the frequencies of the I^A, I^B and i alleles in the three populations. [5]

   b) Suggest two reasons for the differences. [2]

   c) Suggest two processes that could cause the blood group frequencies to change in a population. [2]

   d) State the term used for all the alleles in an interbreeding population. [1]

6. In the species *Allium schoenoprasum* (chives) many adult plants have 16 chromosomes and many have 32. Smaller numbers of plants have 24 chromosomes.

   a) The plants with 16 and 24 chromosomes are reproductively isolated, even when growing close together. Explain the causes of this. [4]

   b) State two other causes of reproductive isolation. [2]

   c) Discuss whether the plants with 16 and 32 chromosomes are separate species. [4]

Antigens and allergy

SUBSTANCES ON CELL SURFACES

All living organisms have proteins and other substances in the plasma membranes on the surface of their cells, especially proteins. Some organisms have a cell wall outside their plasma membranes made of polysaccharides or other substances. There is so much variety in the types of substance on the surfaces of cells that every species has unique molecules.

Viruses are not considered living organisms and are not composed of cells, but they also have unique molecules on their surface. The surface of most viruses is a protein coat (capsid). The capsid of some viruses is enveloped in a membrane taken from the plasma membrane of the host cell. The image below shows the capsid of an adenovirus.

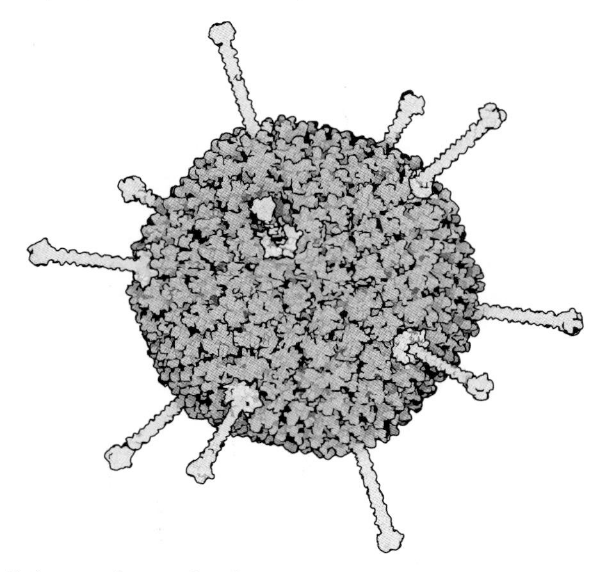

Unique surface molecules are used in several ways:

- viruses recognize and bind to their host using molecules on the surface of the host's cells
- living organisms recognize their own cells and cell types using surface molecules
- living organisms recognize cells that are not part of the organism and also viruses by surface molecules that are not present in that organism (foreign). These molecules trigger the production of antibodies, so they are **antigens**.

HOST SPECIFICITY OF PATHOGENS

Some pathogens are species-specific and only infect members of a single species.

Examples:
Polio, measles and syphilis only affect humans.

Other pathogens can cross species barriers, so can be transmitted from infected members of one species to uninfected members of another species.

Examples:
Tuberculosis can infect both cattle and badgers and can pass in milk from cattle to infect humans; rabies can pass from infected dogs to humans. A disease that can be passed to humans from other animals is called a **zoonosis**.

ANTIGENS ON RED BLOOD CELLS

The ABO blood groups system is based on the presence or absence of a group of glycoproteins in the membranes of red blood cells. Glycoproteins in this group cause antibody production if a person does not naturally possess them, so they are known as antigens. O, A and B antigens are three different versions of the glycoprotein. The O antigen is always present. The A antigen is made by adding an N-acetyl-galactosamine molecule to the O antigen, and the B antigen is made by adding galactose.

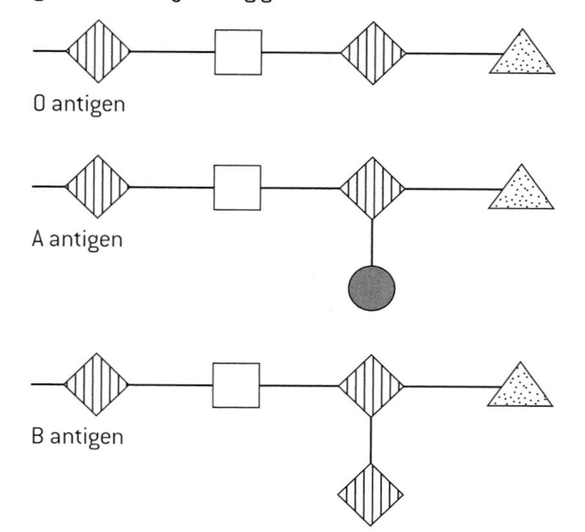

Blood group	Antigens present	Antigens that cause antibody production
O	O	A or B (A, B or AB blood)
A	O and A	B (B or AB blood)
B	O and B	A (A or AB blood)
AB	O, A and B	None

HISTAMINE AND ALLERGIES

Two types of cell in the body secrete histamine:

- **basophils**, which are a type of white blood cell
- **mast cells**, which are similar to basophils but are found in connective tissue.

Histamine is secreted in response to local infection and causes the dilation of the small blood vessels in the infected area. The vessels become leaky, increasing the flow of fluid containing immune components to the infected area and allowing these components to leave the blood vessel, resulting in both specific and non-specific immune responses.

Allergies are reactions by the immune system to substances in the environment that are normally harmless, such as pollen, bee stings or specific foods, for example peanuts. Substances in these allergens cause over-activation of basophils and mast cells and therefore excessive secretion of histamine. This causes the symptoms associated with allergies: inflammation of tissues, itching, mucus secretion and sneezing. Histamine is also implicated in the formation of allergic rashes and in the dangerous swelling known as anaphylaxis. To lessen the effects of allergic responses, anti-histamine drugs can be used.

Antibody production

STAGES IN ANTIBODY PRODUCTION

The production of antibodies by the immune system is one of the most remarkable biological processes. When a pathogen invades the body, the immune system gears up to produce large amounts of the specific antibodies needed to combat the pathogen. This process only takes a few days. The production of antibodies by B-cells is shown in a simplified form in Topic 6 and is explained more fully here.

1. **Activation of helper T-cells**

   Helper T-cells have antibody-like receptor proteins in their plasma membrane to which one specific antigen can bind. When the antigen binds, the helper T-cell is activated. The antigen is brought to the helper T-cell by a macrophage – a type of phagocytic white blood.

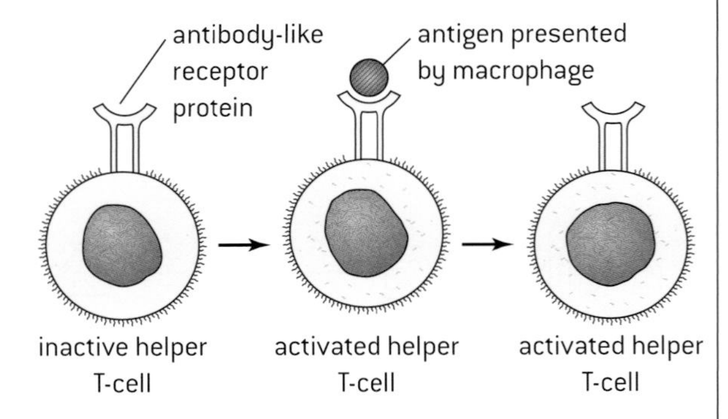

2. **Activation of B-cells**

   Inactive B-cells have antibodies in their plasma membrane. If these antibodies match an antigen, the antigen binds to the antibody. An activated helper T-cell with receptors for the same antigen can then bind to the B-cell. The activated helper T-cell sends a signal to the B-cell, activating it.

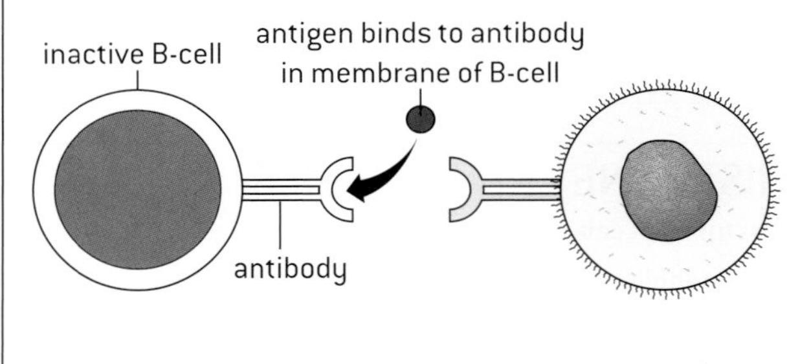

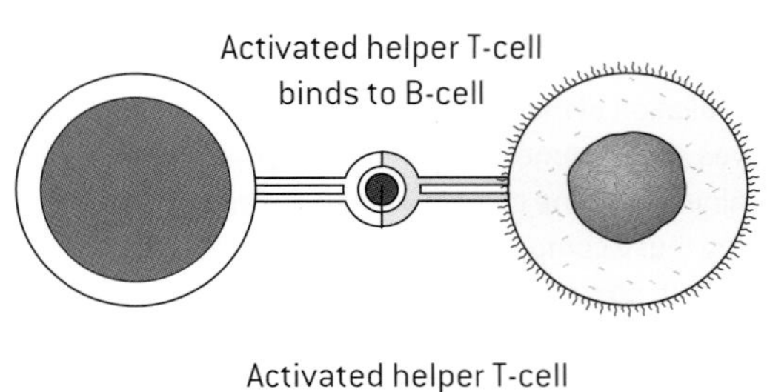

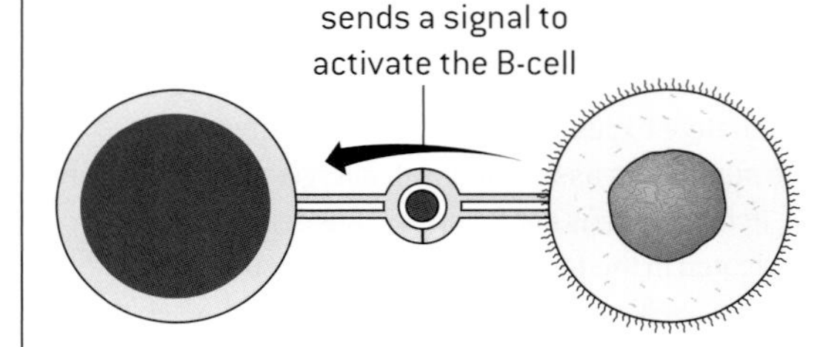

3. **Production of plasma cells**

   Activated B-cells start to divide by mitosis to form a clone of cells. These cells become active, with a much greater volume of cytoplasm. They are then known as plasma cells. They have a very extensive network of rough endoplasmic reticulum. This is used for synthesis of large amounts of antibody, which is then secreted by exocytosis.

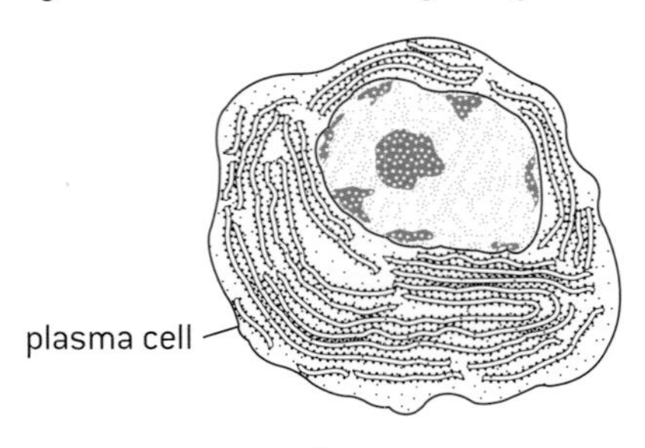

4. **Production of memory cells**

   Memory cells are B-cells and T-cells that are formed at the same time as activated helper T-cells and B-cells, when a disease challenges the immune system. After the activated cells and the antibodies produced to fight the disease have disappeared, the memory cells persist and allow a rapid response if the disease is encountered again. Memory cells give long-term immunity to a disease.

THE ROLE OF ANTIBODIES

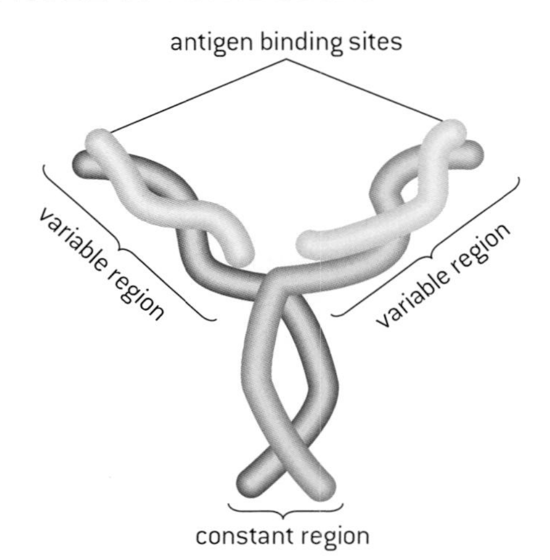

The diagram (above) shows the structure of an antibody molecule (an immunoglobulin). The tips of the variable region are the antigen binding sites. The constant region is the part of the molecule that aids the destruction of the pathogen.

There are different versions of the constant region, which use different tactics to destroy the pathogen. Five are outlined here:

- making a pathogen more recognizable to phagocytes so they are more readily engulfed.
- preventing viruses from docking to host cells
- neutralizing toxins produced by pathogens
- binding to the surface of a pathogen cell and bursting it by causing the formation of pores
- sticking pathogens together (agglutination) so they cannot enter host cells and phagocytes can ingest them more easily.

Vaccination and monoclonal antibodies

VACCINATION

Vaccines contain antigens that trigger immunity to a disease without actually causing the disease in the person who is vaccinated. Most vaccines contain weakened or killed forms of the pathogens. Some vaccines just contain the chemical that acts as the antigen. The vaccine is either injected into the body or sometimes swallowed. The principle of vaccination is that antigens in the vaccine cause the production of the antibodies needed to control the disease. Sometimes two or more vaccinations are needed to stimulate the production of enough antibodies. The figure (right) shows a typical response to a first and second vaccination against a disease. The first vaccination causes a little antibody production and the production of some memory cells. The second vaccination, sometimes called a booster shot, causes a response from the memory cells and therefore faster and greater production of antibodies. Memory cells produced as a result of vaccination should persist to give long-term immunity.

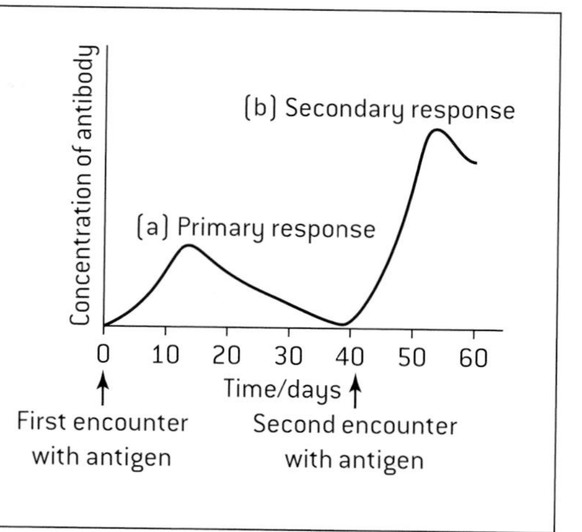

JENNER AND SMALLPOX VACCINATION

Smallpox was the first infectious disease of humans to have been eradicated by vaccination. This was done by a worldwide vaccination programme in the 1960s and 70s, with the last ever case of the disease in 1977. Smallpox was also the first disease for which a vaccine was tested on a human. In 1796 Edward Jenner deliberately infected an 8-year-old boy with cowpox using pus from a blister of a milkmaid with this disease. He then tried to infect the boy with smallpox, but found that he was immune. Cowpox is a less virulent disease caused by viruses similar enough to the smallpox virus for antibodies produced in response to cowpox to give immunity to smallpox. Jenner then tested his procedure on 23 other people including himself.

Today Jenner's tests would be considered ethically unacceptable as they involved a child too young to understand the dangers who could not therefore give informed consent, and he had not first done tests to find out if the vaccine had harmful side-effects.

ANALYSING EPIDEMIOLOGICAL DATA

Epidemiology is the study of the distribution, patterns and causes of disease in a population. Epidemiological data can be used to help plan vaccination programmes, such as the programme aimed at eliminating polio. Cases are monitored carefully to find out where further vaccination is required to give the population immunity and prevent further spread of the disease.

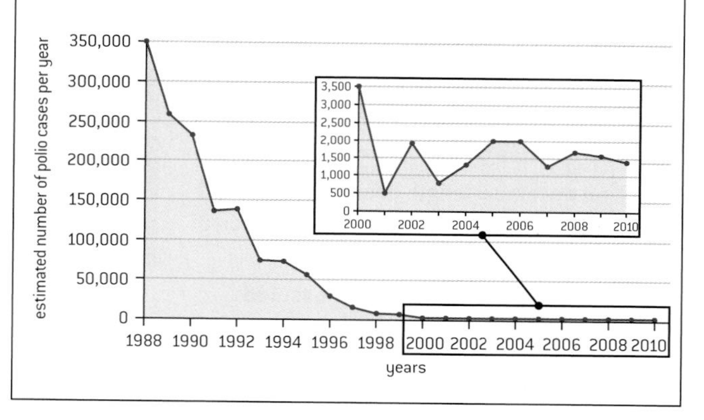

PRODUCTION OF MONOCLONAL ANTIBODIES

Large quantities of a single type of antibody can be made using an ingenious technique. Antigens that correspond to a desired antibody are injected into an animal. Plasma cells producing the desired antibody are extracted from the animal. Tumour cells that grow and divide endlessly are obtained from a culture. The **plasma cells** are fused with the **tumour cells** to produce **hybridoma cells**, which divide endlessly to produce a clone of one specific type of hybridoma cell. The hybridoma cells are cultured and the antibodies that they produce are extracted and purified. The antibodies produced by this method are called **monoclonal antibodies** because they are all produced from one clone of hybridoma cells, so are identical.

Monoclonal antibodies are used in many different ways. One use is in pregnancy test kits. The urine of pregnant women contains hCG, a protein secreted by the developing embryo and later by the placenta. Pregnancy test kits contain monoclonal antibodies to which hCG binds. This causes a coloured band to appear, indicating that the hCG was present in the urine sample and the woman who produced the urine is pregnant.

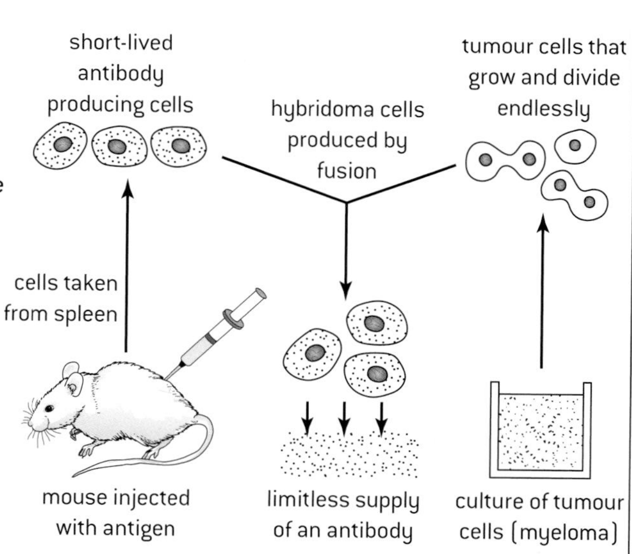

Muscle

STRUCTURE OF SKELETAL MUSCLE

Skeletal muscle is attached to bone and causes movement of animal bodies. It consists of large multinucleate cells called muscle fibres. Within each muscle fibre are cylindrical structures called **myofibrils** and around these is a specialized type of endoplasmic reticulum – the **sarcoplasmic reticulum**. There are also mitochondria between the myofibrils.

Myofibrils consist of repeating units called sarcomeres, which have light and dark bands. The light and dark bands extend across all the myofibrils in a muscle fibre, giving it a striated (striped) appearance. Each sarcomere is able to contract and exert force.

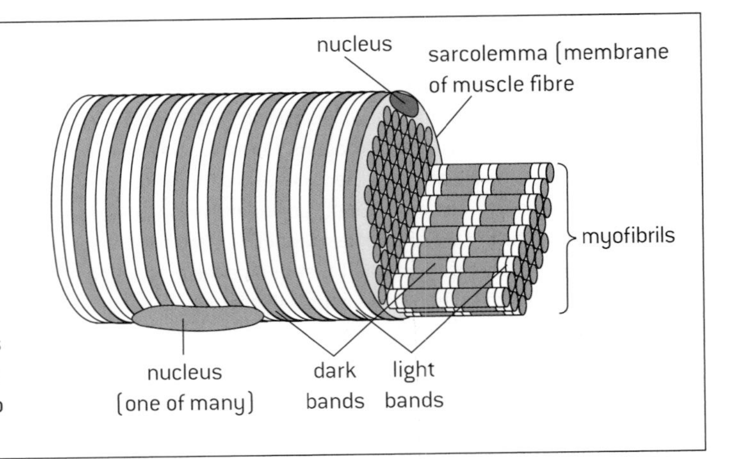

MEASURING SARCOMERE LENGTHS WITH LIGHT MICROSCOPES

The image on the right is a light micrograph of skeletal muscle fibres showing more than one nucleus per fibre and also light and dark bands. Use these instructions to measure the length of one sarcomere:

1. Measure the distance in millimetres from the start of one dark band to the start of a dark band ten bands away.
2. Divide by ten to find the length of one sarcomere in the micrograph.
3. Convert this length in millimetres to micrometres by multiplying by a thousand.
4. Find the actual length of a sarcomere by dividing this length by the magnification of the micrograph, which is 200×.

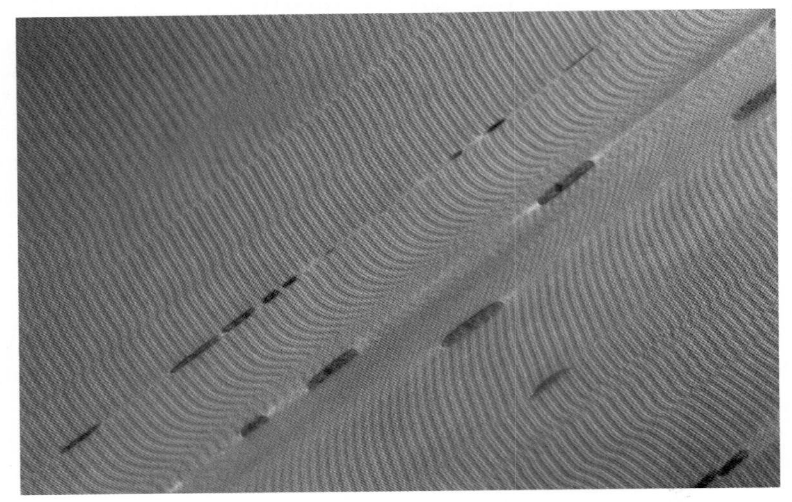

Sarcomere lengths can also be measured using a slide of skeletal muscle and a light microscope with an **eyepiece scale**. As in the method above, it is best to measure the length of ten or more sarcomeres and divide to find the length of one.

The eyepiece scale does not have units on it and must be **calibrated** using a slide that has an accurate scale of known lengths marked on it. This type of slide is called a **stage micrometer**.

Calibration shows how many micrometres each division of the eyepiece scale represents.

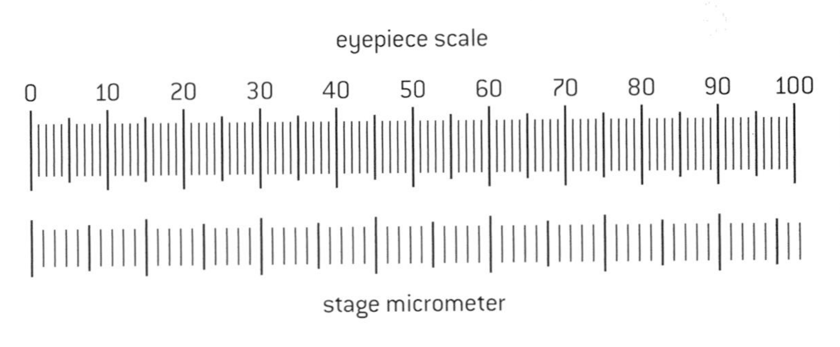

STRUCTURE OF A SARCOMERE

A sarcomere is a subunit of a myofibril. At either end is a Z line to which narrow actin filaments are attached. The actin filaments stretch inwards towards the centre of the sarcomere. Between them, there are thicker myosin filaments, which have heads that form cross-bridges by binding to the actin. The part of the sarcomere containing myosin is the dark band and the part containing only actin filaments is the light band. The figure (right) shows the structure of a sarcomere.

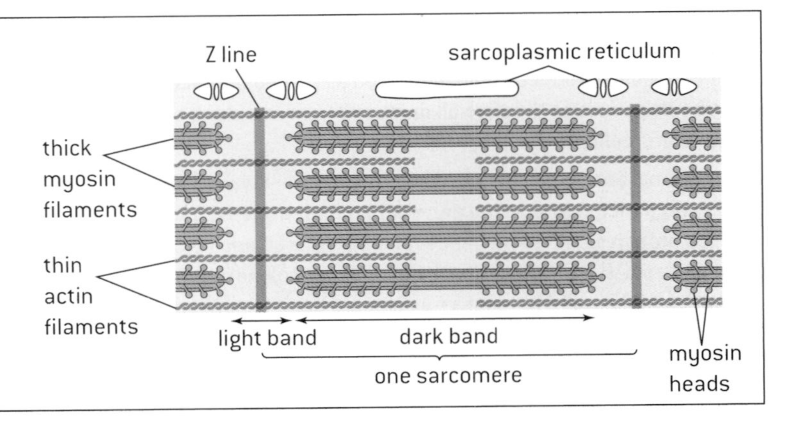

Muscle contraction

SLIDING FILAMENTS AND CONTRACTION

The contraction of the skeletal muscle is achieved by the **sliding of actin and myosin filaments** over each other. This pulls the ends of the sarcomeres together, making the muscle shorter.

The sliding of the filaments is an active process and requires the use of energy from **ATP**. The hydrolysis of one molecule of ATP provides enough energy for a myosin filament to slide a small distance along an actin filament. A repeated cycle of events is used to contract muscle sufficiently to move part of an animal body in the desired way.

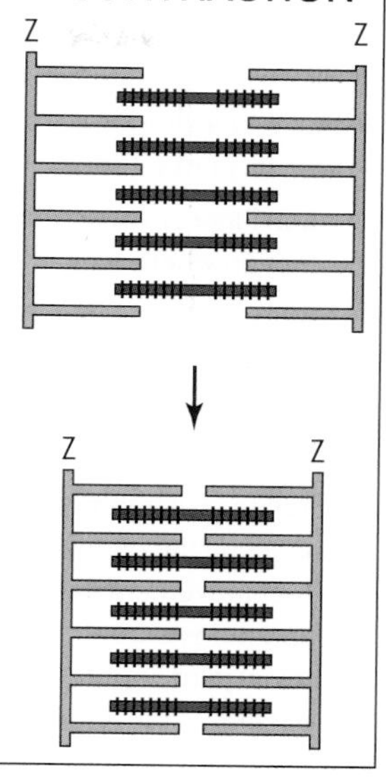

CONTROL OF MUSCLE CONTRACTION

When a motor neuron stimulates a striated muscle fibre, **calcium ions** are released from the sarcoplasmic reticulum inside the fibre. The calcium binds to **troponin**, a protein that is associated with the actin filaments in muscle. The calcium causes the shape of troponin to change and this causes the movement of **tropomyosin**, another protein associated with actin, exposing binding sites on actin. This allows myosin heads to form cross-bridges by binding to actin.

Radioactive calcium (^{45}Ca) has been used to investigate the control of muscle contraction. For example, using autoradiography it was shown that radioactive calcium is concentrated in the region of overlap between actin and myosin filaments in contracted muscle, but not in relaxed muscle. This is because calcium ions are bound to troponin, allowing cross-bridge formation and sliding of filaments.

THE MECHANISM OF MUSCLE CONTRACTION

The sliding of actin filaments over myosin filaments towards the centre of the sarcomere is achieved by a repeated cycle of stages, in which cross-bridges are formed and broken and energy is released by the hydrolysis of ATP.

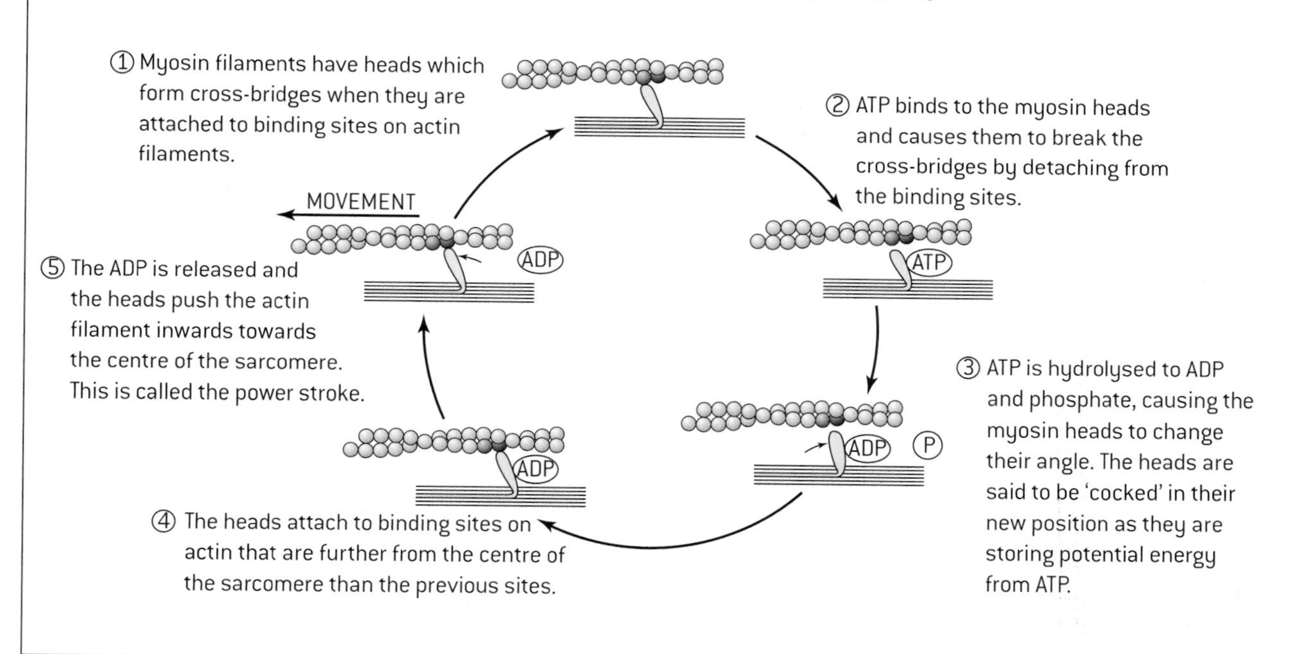

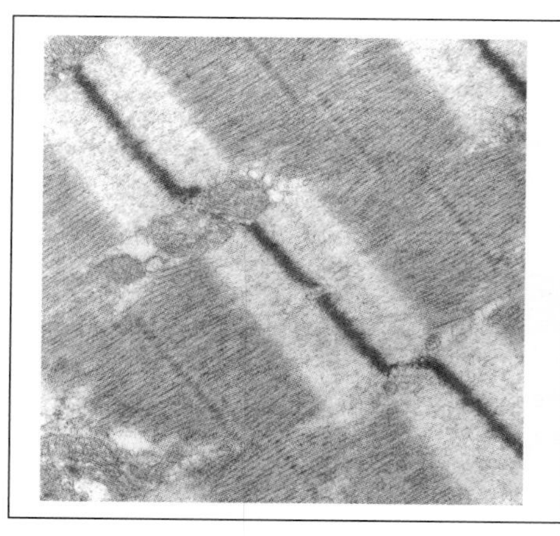

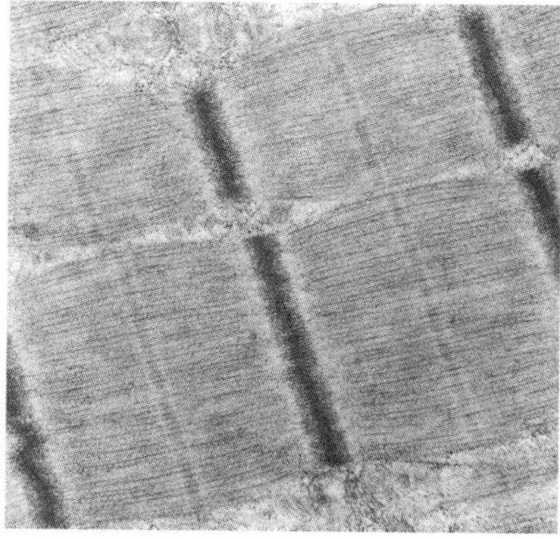

CONTRACTED AND RELAXED MUSCLE FIBRES IN ELECTRON MICROGRAPHS

Contraction of striated (skeletal) muscle makes the light bands narrower and the sarcomeres shorter. The electron micrographs show relaxed muscle with wide light bands (far left) and fully contracted muscle with very narrow light bands (near left).

Movement

MUSCLES AND MOVEMENT

Muscles provide the forces that move animal bodies. As muscles only exert force when they contract and not when they relax and lengthen, a muscle can only cause a movement in one direction. For opposite movements there has to be a pair of muscles that exert force in opposite directions – an **antagonistic pair** of muscles.

Muscles are typically elongated structures, with tendons forming attachments at both ends. One end of the muscle is the anchorage, which is a firm point of attachment that does not move when the muscle contracts. Bones are used as **anchorages** in humans and other vertebrates. In insects and other arthropods the exoskeleton provides the anchorage. The opposite end of the muscle from the anchorage is the **insertion**. Bones and exoskeletons are again used for muscle insertions. Muscle contraction causes the bone or section of exoskeleton forming the insertion to move, together with surrounding tissues. Bones and exoskeleton can change the size and direction of the force exerted by a muscle, so they act as **levers**.

SYNOVIAL JOINTS

Junctions between bones are called **joints**. Some joints are fixed, such as joints between the plates of bone in the skull. Other joints allow movement (articulation). Most of these are **synovial joints**. They have three main parts:

- **Cartilage** covering the surface of the bones to reduce friction where they could rub against each other.
- **Synovial fluid** between the cartilage-covered surfaces, to lubricate the joint and further reduce friction.
- **Joint capsule** that seals the joint and holds in the synovial fluid.

There are also **ligaments** which are tough cords of tissue connecting the bones on opposite sides of a joint. They restrict movement and help to prevent dislocation. Ligaments ensure that certain movements can occur at a synovial joint but not others. For example the elbow allows considerable movement in one plane: bending (flexion) or straightening (extension), but little movement in the other two planes.

THE ELBOW JOINT

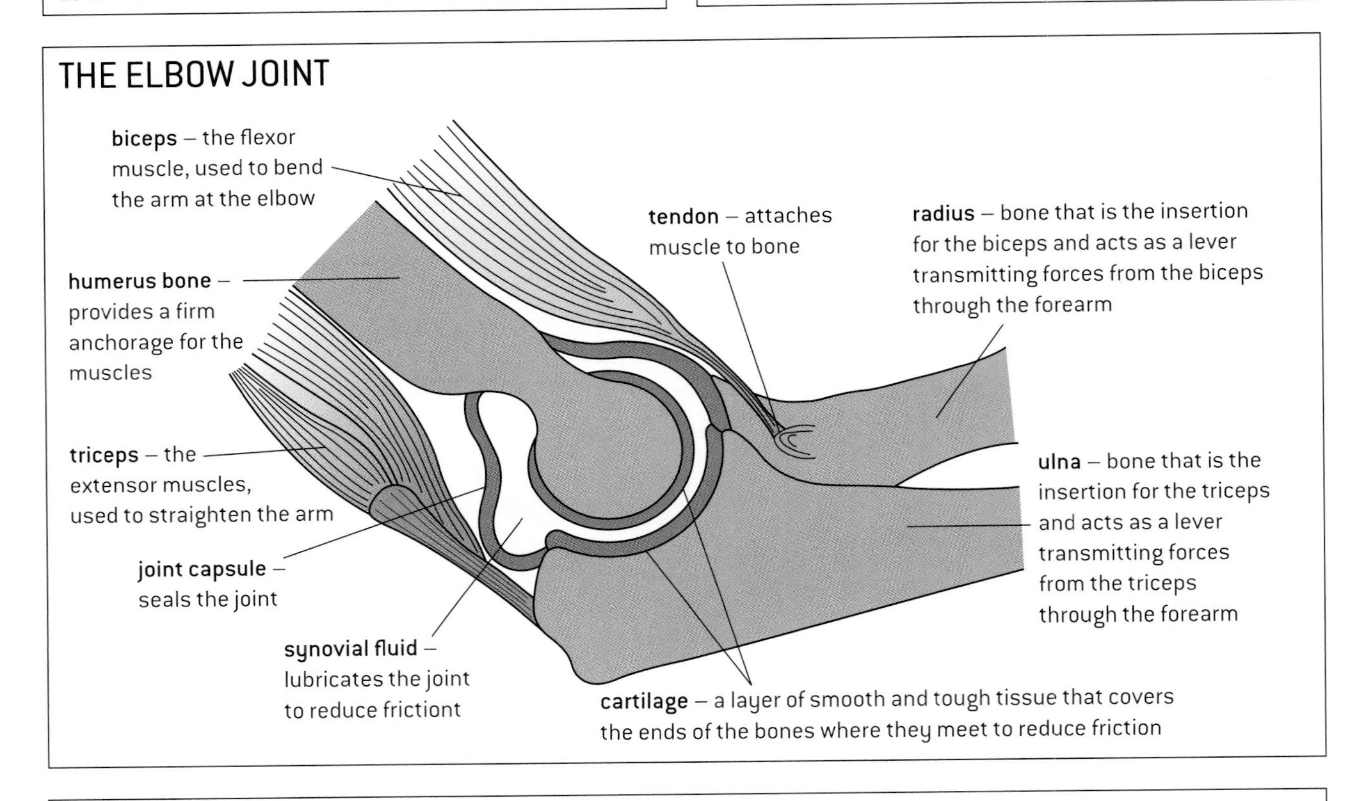

ANTAGONISTIC MUSCLES IN AN INSECT LEG

Insects have many joints in their legs, most of which move on one plane and can either flex (bend) or extend (straighten). A pair of antagonistic muscles causes these opposite movements.

For example in the legs of crickets there are two large muscles inside the femur. The tendons at the distal ends of these muscles are attached to opposite sides of the exoskeleton of the tibia, so one of them is a flexor of the joint between the femur and tibia and the other is an extensor.

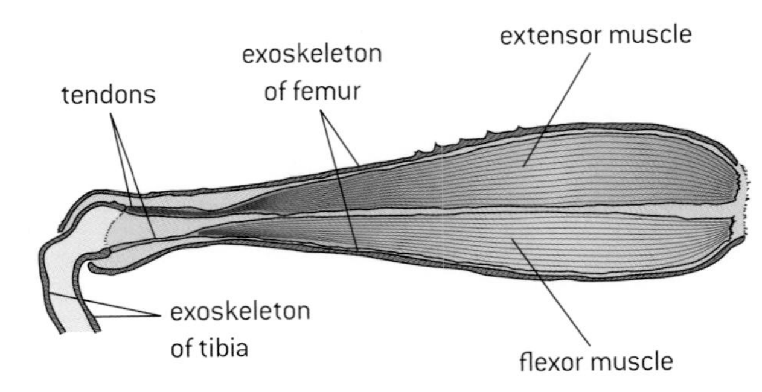

Excretion and osmoregulation

EXCRETION

Metabolic pathways are chains and cycles of reactions in living cells used to build up and break down biochemicals.

In all organisms the metabolic pathways produce waste products that would be toxic if they were allowed to accumulate in cells, so they must be removed.

The removal from the body of potentially toxic waste products of metabolic pathways is **excretion**.

OSMOREGULATION

Water moves into and out of cells by osmosis. The direction in which water moves is determined by hydrostatic pressure and solute concentration. If the pressures are equal, water moves from a lower to a higher solute concentration by osmosis. Living organisms can control the movement of water by adjusting the solute concentrations of their cells and body fluids. This is **osmoregulation** – control of the internal solute concentration of a living organism.

NITROGENOUS WASTE PRODUCTS

Three nitrogenous compounds are excreted by animals: ammonia, urea and uric acid. The table below shows which of these compounds are excreted by the main groups of animal:

Waste product	Groups in which this is the main nitrogenous waste product
Ammonia	freshwater fish amphibian larvae
Urea	marine mammals terrestrial mammals marine fish adult amphibians
Uric acid	birds insects

Two trends can be seen in this table.

1. The type of type of nitrogenous waste in animals is correlated with **habitat**.
   - **Ammonia** is toxic and has to be excreted as a very dilute solution, so a large volume of water is required. It is therefore only excreted by animals that live in water, where abundant supplies of water are always available.
   - **Urea** is less toxic, so can be excreted as a more concentrated solution, with less loss of water. Conversion of ammonia to urea requires energy but it is worthwhile if an animal needs to conserve water.
   - **Uric acid** is not toxic even when concentrated so much that it precipitates to form a semi-solid paste. Conversion of ammonia to uric acid requires much energy, but it is worthwhile for animals that live in arid habitats so need to conserve as much water as possible. It also benefits animals that fly, as a concentrated paste of uric acid contains less water than dilute urine, reducing body mass during flight.
2. The type of nitrogenous waste in animals is correlated with **evolutionary history**. For example, mammals excrete urea, even though some mammals such as beavers and otters live in aquatic habitats and do not need to conserve water and presumably could excrete ammonia in a large volume of dilute urine, but instead they excrete urea, like terrestrial mammals.

OSMOCONFORMERS AND OSMOREGULATORS

Many marine organisms allow their internal solute concentration to fluctuate with that of the water around them – they do not attempt to maintain constant internal solution concentrations. These organisms are **osmoconformers**.

Examples: squids and sea squirts.

A disadvantage of being an osmoconformer is that cells inside the body may not contain the ideal solute concentration for body processes.

Most terrestrial organisms are **osmoregulators** because they maintain a constant internal solute concentration, whatever the external solute concentration.

Example: humans.

A disadvantage of being an osmoregulator is that energy has to be used to keep solute concentrations in the body constant.

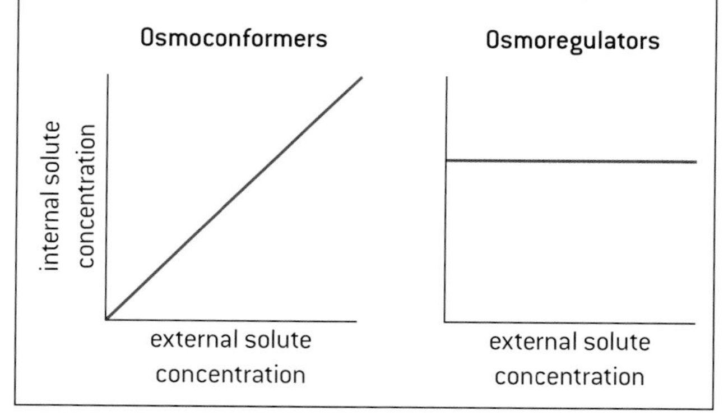

DEHYDRATION AND OVER-HYDRATION

To explain dehydration and over-hydration in osmoregulators, three terms are required:

isotonic – a solute concentration equal to that of normal body fluids; **hypotonic** – a lower solute concentration than normal body fluids; **hypertonic** – a higher solute concentration than normal body fluids.

Dehydration is due to loss of water from the body, but not an equivalent quantity of solutes, so body fluids become hypertonic. The consequences are thirst, small quantities of dark coloured urine, lethargy, a raised heart rate, low blood pressure and in severe cases seizures, brain damage and death. **Over-hydration** is due to excessive intake of water, so the body fluids become hypotonic. The consequences are behaviour changes, confusion, drowsiness, delirium, blurred vision, muscle cramps, nausea and in acute cases seizures, coma and death.

Kidney structure and ultrafiltration

STRUCTURE AND FUNCTIONS OF THE KIDNEY

The kidney has two functions, excretion and osmoregulation.

The diagram (below) shows the structure of the kidney. The **cortex** and **medulla** contain many narrow tubes called **nephrons**. The renal pelvis consists of spongy tissue into which urine drains from collecting ducts.

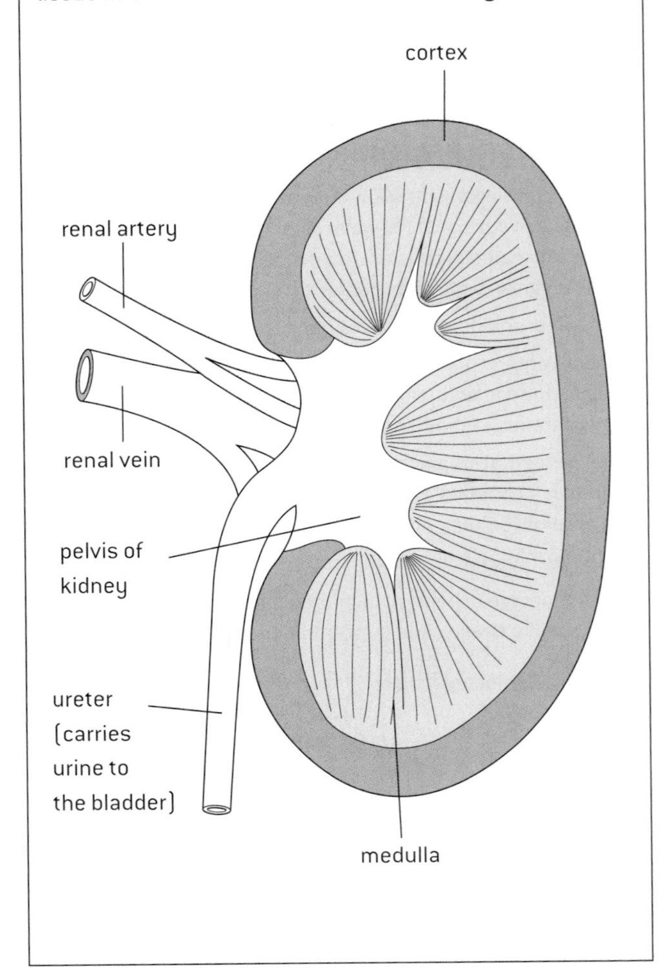

STRUCTURE OF THE NEPHRON

The figure (below) shows the structure of a nephron, together with the associated **glomerulus**. A group of nephrons join up to form one **collecting duct**.

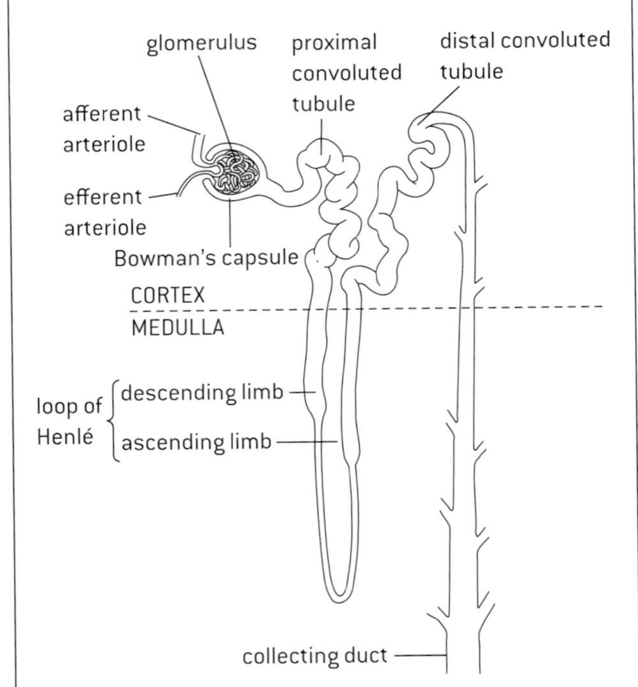

- The glomerulus and Bowman's capsule produce a filtrate from the blood by **ultrafiltration**.
- The proximal convoluted tubule transfers useful substances from the filtrate back into the blood by **selective reabsorption**.
- The loop of Henlé establishes high solute concentrations in the medulla, so hypertonic urine can be produced.
- The distal convoluted tubule adjusts individual solute concentrations and the pH of the blood.
- The collecting duct carries out **osmoregulation** by varying the amount of water reabsorbed.

ULTRAFILTRATION IN THE GLOMERULUS

The glomerulus is a knot-like ball of blood capillaries. All capillaries let some fluid leak out but 20% of the plasma escapes from glomerulus capillaries which is a very large amount. There are two reasons:

- very high blood pressure, because the vessel taking blood away from the glomerulus is narrower than the vessel bringing blood
- many large pores (fenestrations) in the capillary walls.

These pores would allow any molecules through, but there are two filters beyond the pores that only small to medium sized particles can pass through (68,000 molecular mass or less):

- **basement membrane** – a gel on the outside of the capillary, with small gaps through a mesh of protein fibres
- **filtration slits** – narrow gaps between the foot process of **podocyte** cells where they wrap around the capillaries.

The filtrate that enters the Bowman's capsule contains all substances in blood plasma except plasma proteins.

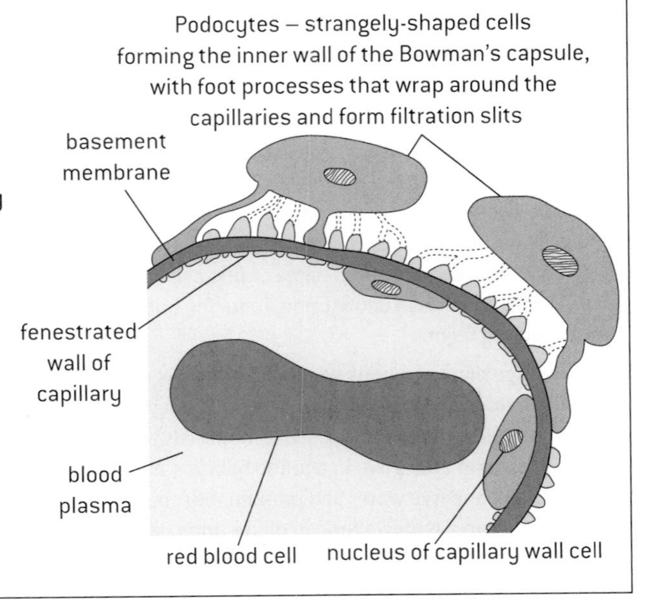

Urine production and osmoregulation

SELECTIVE REABSORPTION

Large volumes of glomerular filtrate are produced – about one litre every 10 minutes by the two kidneys. As well as waste products, the filtrate contains substances that the body needs, which must be reabsorbed into the blood. Most of this selective reabsorption happens in the **proximal convoluted tubule**. The wall of the nephron consists of a single layer of cells. In the proximal convoluted tubule the cells have **microvilli** projecting into the lumen (right), giving a large surface area for absorption. Pumps in the membrane reabsorb useful substances by **active transport**, using ATP produced by **mitochondria** in the cells. All of the glucose in the filtrate is reabsorbed. About 80% of the mineral ions, including sodium, are reabsorbed. Active transport of solutes makes the total solute concentration higher in the cells of the wall than in the filtrate in the tubule. Water therefore moves by osmosis from the filtrate to the cells and on into the adjacent blood.

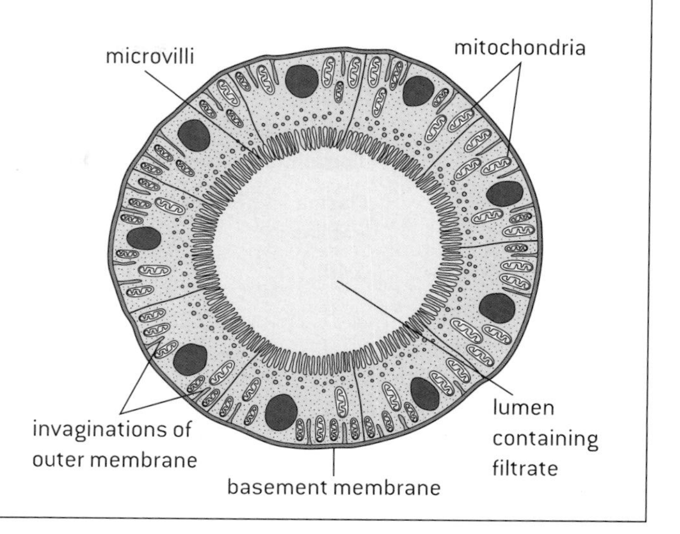

ROLE OF THE LOOP OF HENLÉ

Glomerular filtrate flows deep into the medulla in descending limbs of the loops of Henlé and then back out to the cortex in ascending limbs. Descending limbs and ascending limbs are opposite in terms of permeability. Descending limbs are very permeable to water but are relatively impermeable to sodium ions. Ascending limbs are very permeable to sodium ions but are relatively impermeable to water.

Ascending limbs pump sodium ions from the filtrate into the medulla by active transport, creating a high solute concentration in the medulla. As the filtrate flows down the descending limb into this region of high solute concentration, some water is drawn out by osmosis. This dilutes the fluids in the medulla slightly. However the filtrate that leaves the loop of Henlé is more dilute than the fluid entering it, showing that the overall effect of the loop of Henlé is to increase the solute concentration of the medulla. This is the role of the loop of Henlé – to create an area of higher solute concentrations in the medulla than in normal body fluids (hypertonic).

After the loop of Henlé, the filtrate passes through the distal convoluted tubule, where the ions can be exchanged between the filtrate and the blood to adjust blood levels. It then passes into the collecting duct.

The diagram shows movements of water and sodium ions in the loop of Henlé and the collecting duct. Concentrations of solutes inside and outside the nephron are shown as a percentage of normal blood solute concentration.

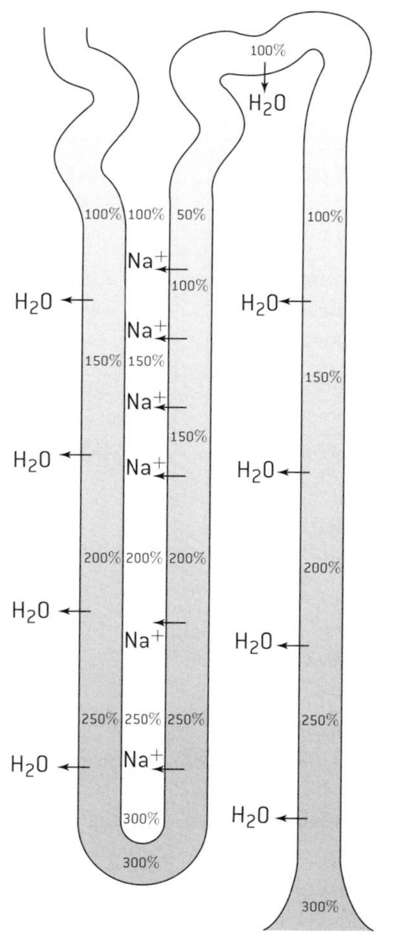

Movements of water and sodium ions in the loop of Henlé and the collecting duct. Solute concentrations inside and outside the nephron are shown as a percentage of normal blood solute concentration

ADH AND OSMOREGULATION

Osmoregulation is the control of solute concentrations in the body fluids, especially the blood plasma. The collecting duct has an important role in osmoregulation. If the water content of the blood is too low, the pituitary gland secretes ADH (anti-diuretic hormone), which is also sometimes called vasopressin. This hormone makes the cells of the collecting duct increase the permeability of their plasma membranes to water. The cells do this by putting water channels, called aquaporins, into their membranes. As the filtrate passes down the collecting duct through the medulla, the high solute concentration of the medulla causes much of the water in the filtrate to be reabsorbed by osmosis. ADH is secreted when the internal solute concentration of body fluids is too high and, as it causes a small volume of concentrated urine to be produced, the result is that the blood plasma becomes more dilute.

If the solute concentration of body fluids is too low, ADH is not secreted and the collecting duct becomes much less permeable to water by removal of aquaporins from its membranes. Only a small amount of water is reabsorbed as the filtrate passes down the collecting duct and a large volume of dilute urine is produced, making the solute concentration of the blood higher.

Kidney function and kidney failure

FILTRATE AND URINE CONCENTRATIONS

The table below shows differences in composition between blood in the glomerulus, filtrate at various points in the nephron and urine.

	Concentration (mg per 100ml)		
	Plasma proteins	Glucose	Urea
Blood in glomerulus	740	90	30
Glomerular filtrate	0	90	30
Filtrate at start of loop of Henlé	0	0	90
Filtrate at end of loop of Henlé	0	0	200
Urine with ADH	0	0	1800
Urine without ADH	0	0	180

BLOOD IN THE RENAL ARTERY AND VEIN

The composition of blood is altered as it flows through the kidney, so there are differences between blood in the renal artery and vein.

	Comparison of concentrations		Reason for difference between renal artery and vein
	Renal artery	Renal vein	
oxygen	higher	lower	aerobic respiration to provide ATP for kidney function
carbon dioxide	lower	higher	
glucose	slightly higher	slightly lower	use of glucose in aerobic respiration
urea	higher	about 20% lower	excretion of urea in urine
plasma proteins	equal		not added or removed
sodium and chloride ions	variable	always at normal levels	kidney raises or lowers concentrations to normalize them

URINE TESTS

Samples of urine are easily obtained and can be tested for the presence of abnormalities that are indicators of disease:

Blood cells – their presence is caused by a variety of diseases including infections and some cancers.

Glucose – almost always indicates diabetes.

Proteins – very small amounts of protein in the urine are normal because some proteins such as the hormones hCG and insulin are small enough to be filtered out of the blood, but larger amounts of proteins in urine are a sign of kidney disease.

Drugs – many drugs pass out of the body in the urine so tests can show if a person is a drug abuser, either for recreational reasons or to gain unfair advantage in sports competitions.

TREATMENT OF KIDNEY FAILURE

Kidney failure is a serious condition because toxins build up in the body and solute concentrations are not maintained at the normal level. Untreated kidney failure makes the patient feel increasingly ill and is eventually fatal. There are two approaches to the treatment of kidney failure.

1. **Hemodialysis**

   Blood is drawn out of a vein in the arm and passed though a kidney machine for 3 to 4 hours, 3 times per week. The blood flows next to a semi-permeable dialysis membrane with dialysate (dialysis fluid) on the other side. Pores through the membrane allow small particles to diffuse in either direction, but plasma proteins and cells are retained in the blood. Dialysate has these features:

   - no urea or other waste products so they diffuse from the blood to the fluid
   - ideal concentrations of glucose and other metabolites so ideal concentrations are achieved in the blood by diffusion to or from the fluid
   - high calcium and low potassium concentrations to extract potassium and add calcium to the blood
   - hydrogencarbonate ions (HCO_3^-) to reduce the acidity of the blood
   - a total solute concentration that will cause excess water to be removed from the blood by osmosis across the dialysis membrane.

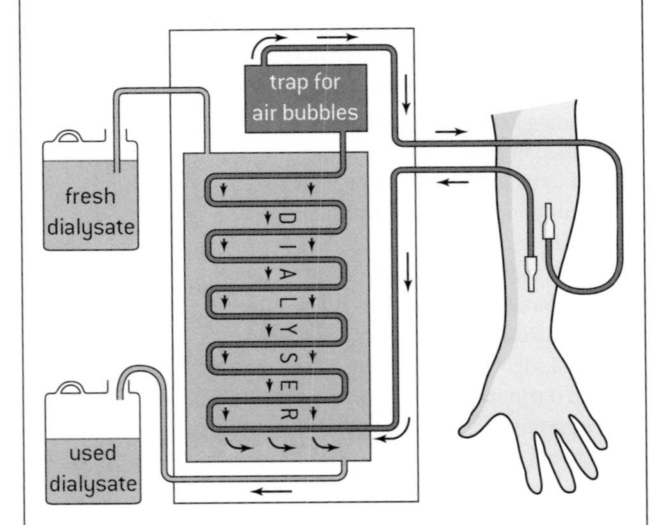

2. **Kidney transplants**

   Dialysis can keep patients alive for years, but a better long-term treatment is a kidney transplant. Sometimes a living donor provides one of their two kidneys for transplant and in other cases the kidneys of a person who has recently died are donated to two patients needing a transplant. It is essential that the donor and recipient are in the same blood group and their tissues match as closely as possible to minimize the chance of rejection of the kidney by the recipient's immune system. The new kidney is grafted in to the lower abdomen with the renal artery, renal vein and ureter connected to the recipient's blood vessels and bladder.

Excretion and osmoregulation in animals

EXCRETION AND WATER CONSERVATION

The maximum solute concentration of urine varies considerably between species. This observation led to research into the kidney physiology of different species and in particular how desert animals are able to conserve water by producing very concentrated urine. This research revealed some of the basic physiology of all mammalian kidneys so is a good example of how curiosity about a particular phenomenon can lead to progress in science.

The table below shows the maximum solute concentration (MSC) and concentration factor of urine (CF) and the habitat:

Species	MSC (mOsm dm^{-3})	CF	Habitat
beaver	520	×2	aquatic
human	1200	×4	intermediate
brown rat	2900	×9	intermediate
kangaroo rat	5500	×18	desert
hopping mouse	9400	×25	desert

Longitudinal sections through the kidneys of mammals of aquatic, desert and intermediate habitats show significant differences. One example of each is shown (below). These species show a general trend in mammals: there is a positive correlation between the thickness of the medulla compared to the overall size of the kidney, and the need for water conservation. This is because a thicker medulla allows the loops of Henlé and collecting ducts to be longer, so more water can be reabsorbed and the urine can be made more concentrated.

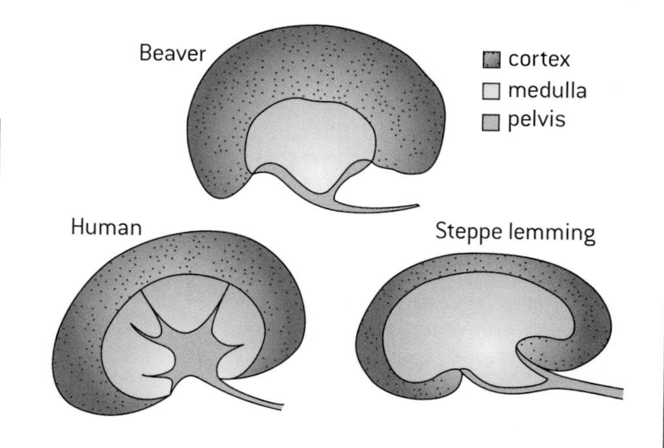

THE MALPIGHIAN TUBULE SYSTEM

The circulatory system of insects uses hemolymph rather than blood. Hemolymph is pumped by a vessel that runs from the abdomen forwards through the thorax to the head. Branches of this vessel carry the hemolymph to different parts of the body and it is then released and is free to flow gradually through tissues until being drawn back into the vessel for re-pumping. Body cells are therefore bathed in hemolymph and release waste products into it.

Between the midgut and hindgut of insects there is a ring of narrow blind-ended ducts, called **Malpighian tubules**, which extend through the body cavity of the insect. Cells in the tubule walls extract waste products from the hemolymph and pass them into the lumen of the tubule. Ammonia is extracted and converted by Malpighian tubule cells into uric acid.

To create a flow of fluid that will carry uric acid and other waste products along the Malpighian tubules to the hindgut, cells in the tubule wall transfer mineral ions by active transport from the hemolymph to the lumen of the tubule and water follows passively by osmosis. The solution that is produced in this way drains into the lumen of the hindgut where it mixes with the semi-digested food. The mixture is carried on to the last section of the gut – the rectum. Mineral ions are pumped by cells in the wall of the rectum from the feces in the rectum to the hemolymph and again water follows passively by osmosis. By moving solutes and water into and out of the hemolymph, the Malpighian tubules and rectum together prevent dehydration and achieve osmoregulation.

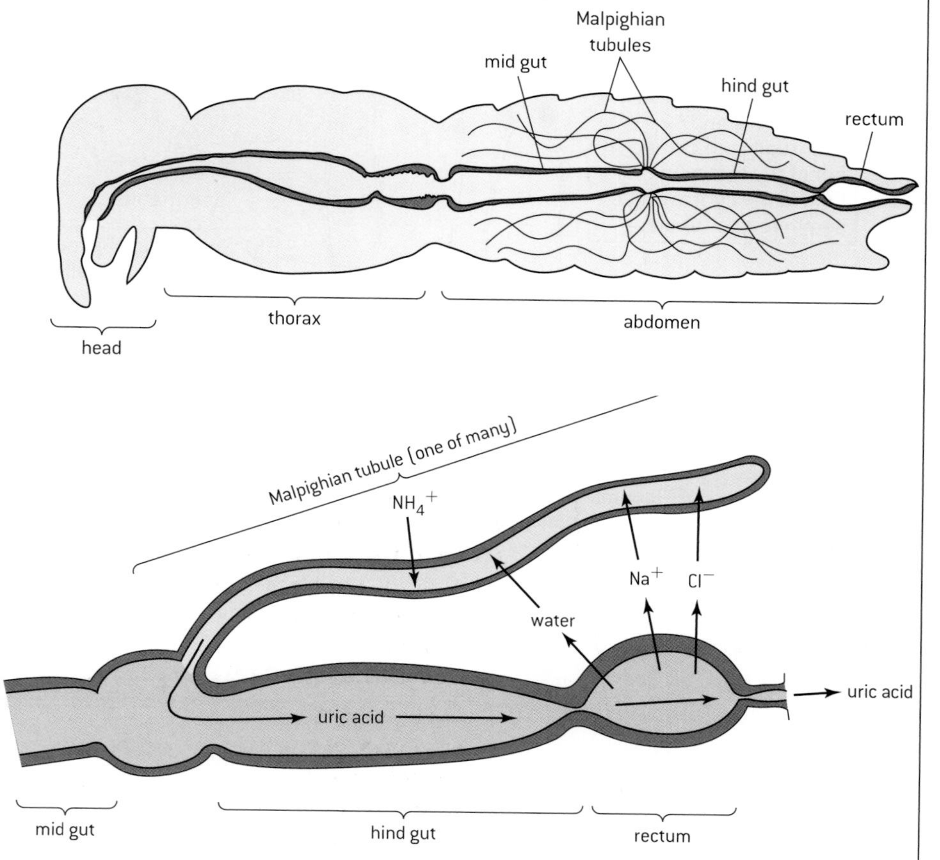

Spermatogenesis

STAGES IN GAMETOGENESIS

Spermatogenesis is the production of male gametes in the testes. Oogenesis is production of female gametes in the ovaries. Both processes have the same basic stages:

- mitosis to generate large numbers of diploid cells
- cell growth so the cells have enough resources to undergo two divisions of meiosis
- meiosis to produce haploid cells
- differentiation so the haploid cells develop into gametes with structures needed for fertilization.

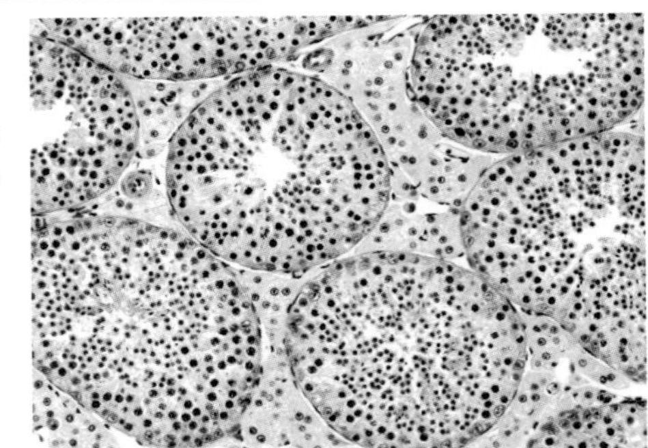

The micrograph (left) shows the testis tissue. Most of it is seminiferous tubules. The tubule walls produce sperm.

STAGES IN SPERMATOGENESIS

The five stages of spermatogenesis are shown in this diagram of cells in the wall of the seminiferous tubule.

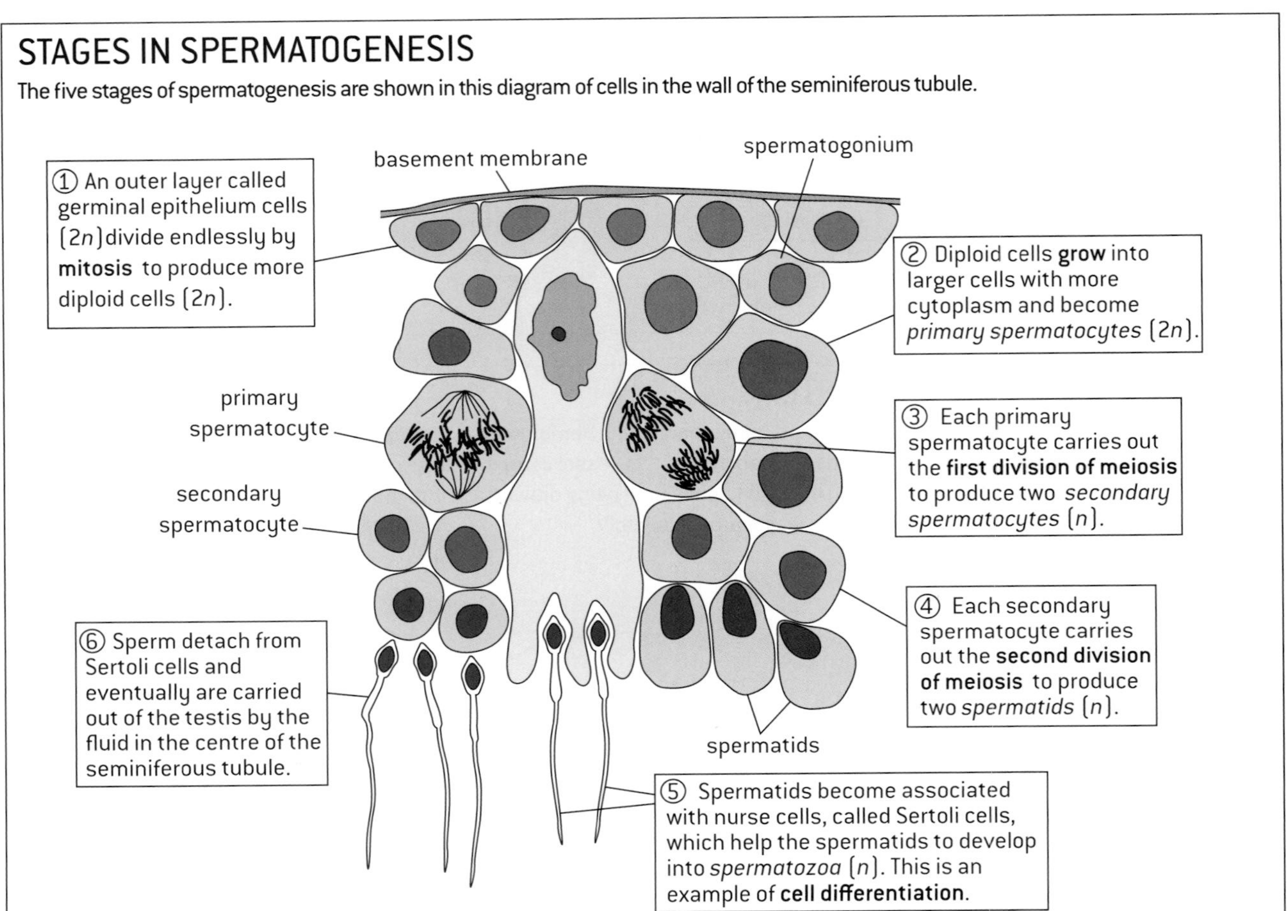

STRUCTURE OF HUMAN SPERM

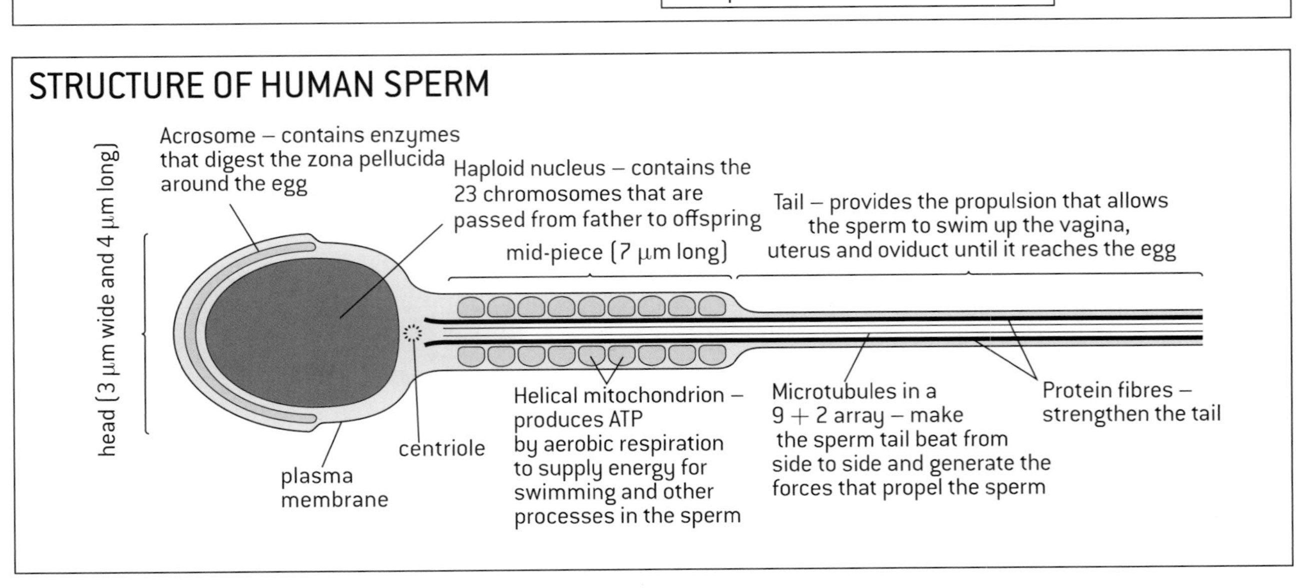

Oogenesis

STAGES IN OOGENESIS

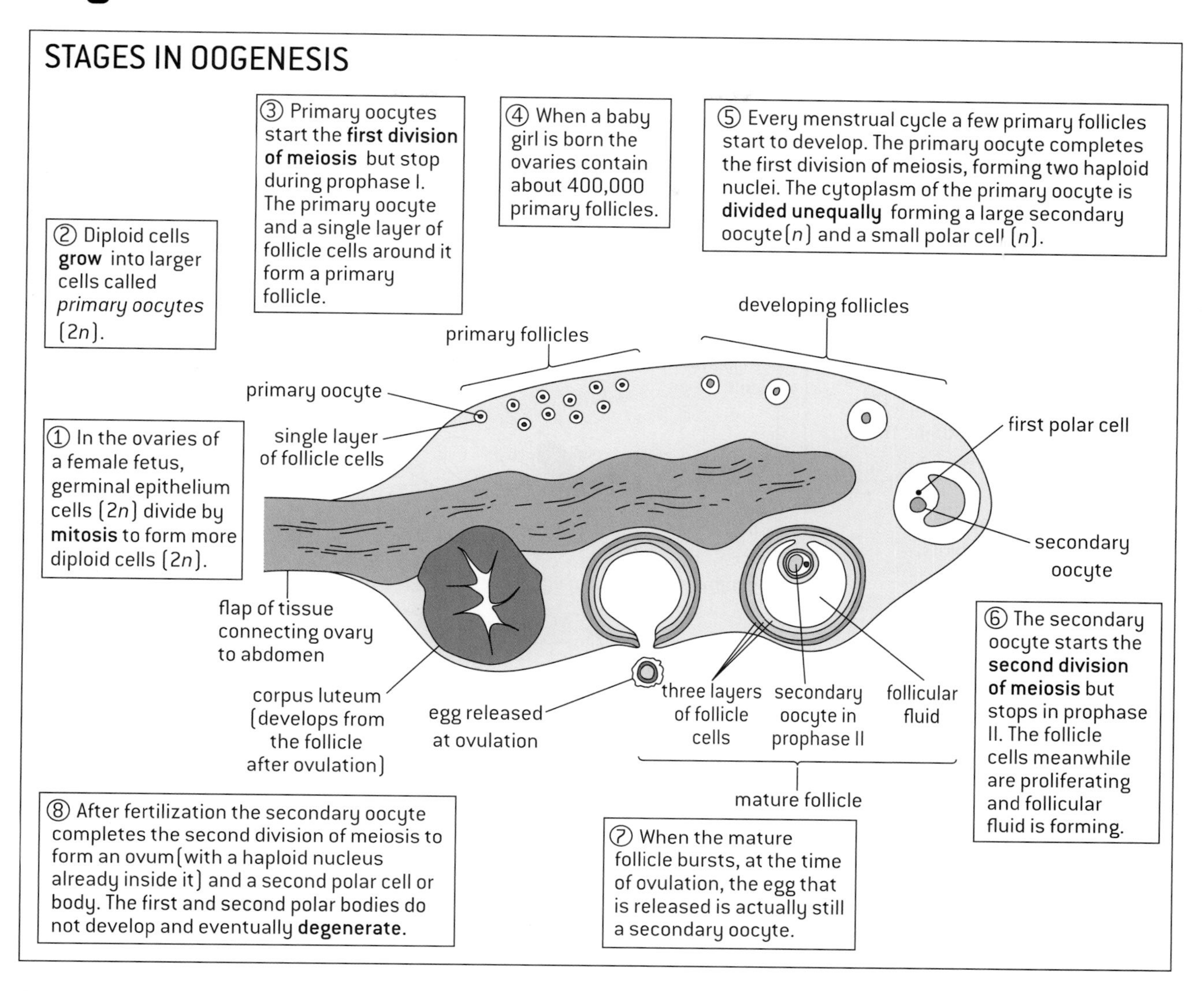

STRUCTURE OF A MATURE HUMAN EGG

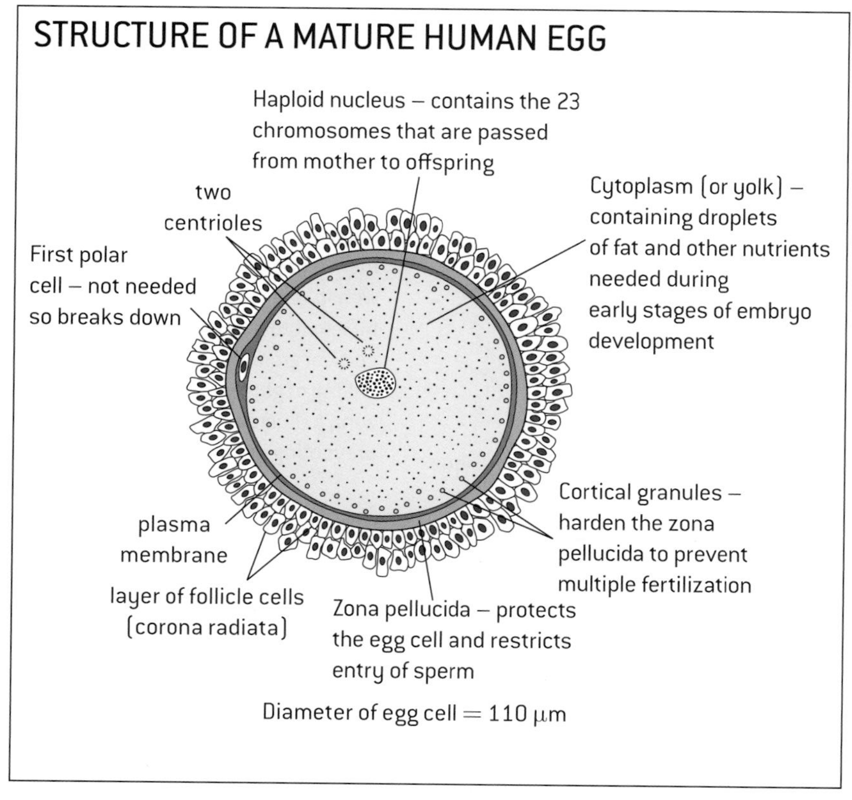

Diameter of egg cell = 110 μm

COMPARING OOGENESIS AND SPERMATOGENESIS

There are some significant differences between spermatogenesis and oogenesis:

1. Millions of sperm are produced by men each day from puberty onwards and they can be released frequently by ejaculation. From puberty until menopause women who are not pregnant produce and release just one egg every 28 days.
2. Nearly all the cytoplasm is removed during the latter stages of spermatogenesis so sperm contain very little. Egg cells have more cytoplasm than any other human cell. The mitochondria of the zygote are all derived from the cytoplasm of the egg cell. The egg cell destroys the helical mitochondria of the sperm after fertilization.

Fertilization

INTERNAL AND EXTERNAL FERTILIZATION

In some species females release unfertilized eggs and males put their sperm over the eggs, so fertilization takes place outside the body. This is **external fertilization**.

Examples:
salmon and other fish, frogs and other amphibians.

In other species the male passes his sperm into the female's body and fertilization takes place there. This is **internal fertilization**.

Examples:
pythons and other reptiles,
albatrosses and other birds,
humans and other mammals.

AVOIDING POLYSPERMY

A diploid zygote is produced when one haploid sperm fuses with a haploid egg – this is fertilization. Fusion of two or more sperm with an egg cell results in a cell that has three of each chromosome type (triploid), or more. This is called polyspermy. Cells produced in this way often die and those that survive are almost always sterile. There are therefore mechanisms in fertilization that normally prevent polyspermy.

DECLINING MALE FERTILITY

During the last fifty years the average number of sperm per unit volume of human semen has fallen by 50% and it continues to drop by about 2% per year. Various factors may be contributing to this, but one is the presence in the environment of estrogen and progesterone since the introduction of the female contraceptive pill. The effects of these chemicals on male fertility were not tested before the contraceptive pill started to be used by millions of women. There are also steroids that are chemically related to these female sex hormones in a wide range of products including plastics, food packaging and furniture. Again, adequate testing has not been done. The enormous drop in male fertility shows how essential it is to test for harmful side effects before scientific or technological developments are introduced.

STAGES IN THE FERTILIZATION OF A HUMAN EGG

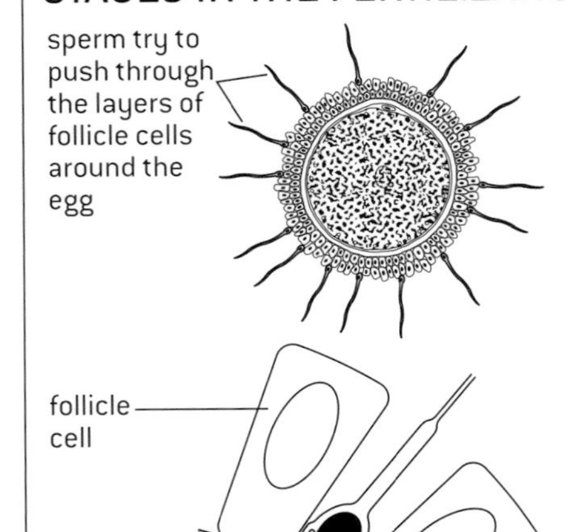

1. **Arrival of sperm**

   Sperm are attracted by a chemical signal and swim up the oviduct to reach the egg. Fertilization is only successful if many sperm reach the egg.

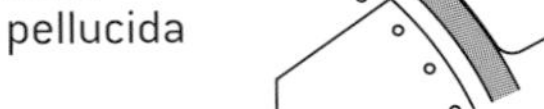

2. **Binding**

   The first sperm to break through the layers of follicle cells binds to the zona pellucida. This triggers the acrosome reaction.

acrosomal cap

3. **Acrosome reaction**

   The contents of the acrosome are released, by the separation of the acrosomal cap from the sperm. Enzymes from the acrosome digest a route for the sperm through the zona pellucida, allowing the sperm to reach the plasma membrane of the egg.

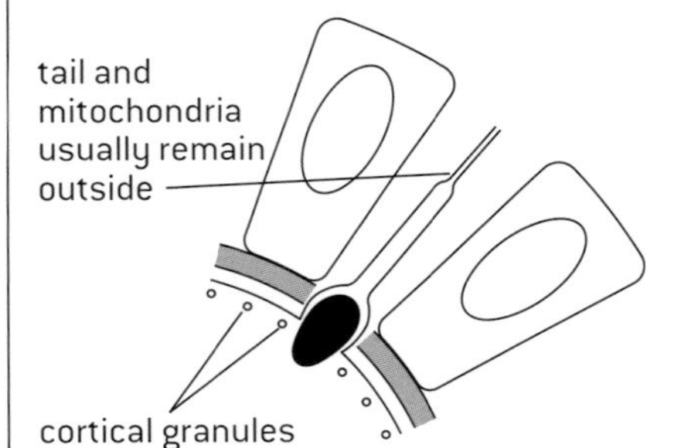

4. **Fusion**

   The plasma membranes of the sperm and egg fuse and the sperm nucleus enters the egg and joins the egg nucleus. Fusion causes the cortical reaction.

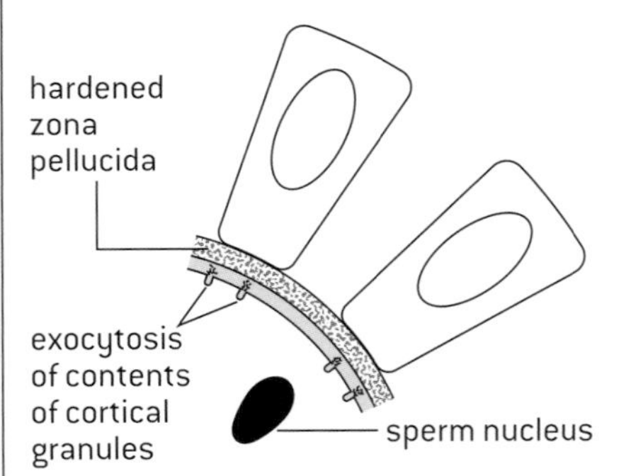

5. **Cortical reaction**

   Small vesicles called cortical granules move to the plasma membrane of the egg and fuse with it, releasing their contents by exocytosis. Enzymes from the cortical granules cause cross-linking of glycoproteins in the zona pellucida, making it hard and preventing polyspermy.

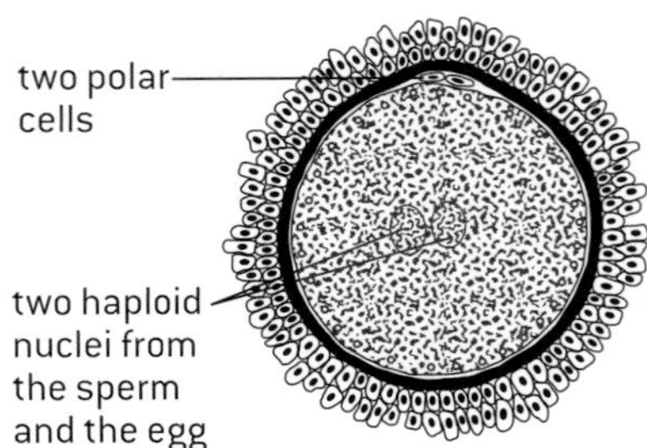

6. **Mitosis**

   The nuclei from the sperm and egg do not fuse together. Instead, both nuclei carry out mitosis, using the same centrioles and spindle of microtubules. A two-cell embryo is produced.

Pregnancy and childbirth

EARLY EMBRYO DEVELOPMENT AND IMPLANTATION

If a couple want to have a child, they have sexual intercourse without using any method of contraception. Semen is ejaculated into the vagina and sperm that it contains swim through the cervix, up the uterus and into the oviducts. If there is an egg in the oviducts, a sperm can fuse with it to produce a zygote.

The zygote produced by fertilization in the oviduct is a new human individual. It starts to divide by mitosis to form a 2-cell embryo, then a 4-cell embryo (right) and so on until a hollow ball of cells called a **blastocyst** is formed. While these early stages in the development of the embryo are happening, the embryo is transported down the oviduct to the uterus. When it is about 7 days old, the embryo implants itself into the **endometrium** (the lining of the wall of the uterus), where it continues to grow and develop. If implantation does not occur then the embryo is not supplied with enough food and the pregnancy does not continue.

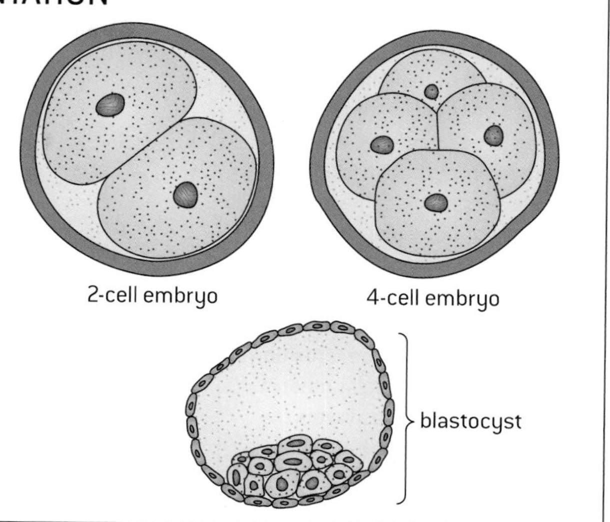

ANIMAL SIZE AND DURATION OF GESTATION

The graph below shows the relationship between body mass and duration of gestation (pregnancy) in a wide range of species of mammal. Both scales are logarithmic. The cross is the data point for humans (266 day gestation and 60kg body mass). Although there is a positive correlation overall between body mass and duration of gestation, there are examples of species that have the same length of gestation but body masses differing by more than two orders of magnitude. In animals with a relatively long gestation the offspring are more advanced in their development when they are born than animals with a short gestation time in relation to adult body mass.

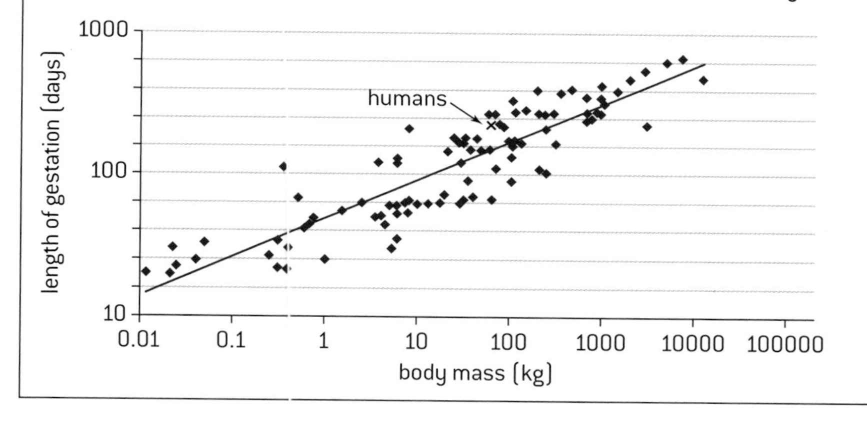

HORMONAL CONTROL OF PREGNANCY

Human embryos secrete the hormone **hCG** (human chorionic gonadotrophin) from a very early stage. hCG stimulates the ovary to maintain the secretion of **progesterone** during the first three months of pregnancy. Progesterone causes the uterus lining to continue to thicken so it can support the embryo after implantation.

By about the 12th week of pregnancy the ovary stops secreting progesterone, but by this time the **placenta** has developed and takes over the task of secreting the progesterone that is needed to sustain the pregnancy until the time of childbirth (labour). The placenta also secretes **estrogen**.

HORMONAL CONTROL OF CHILDBIRTH

Through the 9 months of pregnancy, rising levels of the hormone **progesterone** ensure that the uterus develops and sustains the growing fetus. It also prevents uterine contractions and so prevents spontaneous abortions. The level of progesterone starts to fall in the last third of the pregnancy and more steeply shortly before the end. This allows the mother's body to secrete another hormone – **oxytocin**. There is also a rise in **estrogen**, which causes an increase in the number of oxytocin receptors on the muscle in the uterus wall. When oxytocin binds to these receptors it causes the muscle to contract. Uterine contractions stimulate the secretion of more oxytocin. The uterine contractions therefore become stronger and stronger. This is an example of **positive feedback**.

uterus wall contracting

vagina – the birth canal

While the muscle in the wall of the uterus is contracting, the cervix relaxes and becomes wider. The amniotic sac bursts and the amniotic fluid is released. Finally, often after many hours of contractions, the baby is pushed out through the cervix and the vagina. The umbilical cord is cut and the baby begins its independent life. Contractions continue for a time until the placenta is expelled as the afterbirth. The diagram shows the baby's head emerging during childbirth.

Structure and function of the placenta

FUNCTION OF THE PLACENTA

By the time that the embryo is about 8 weeks old, it starts to develop bone tissue and is known from then onwards as a **fetus**. The fetus develops a placenta and an umbilical cord. The placenta is a disc-shaped structure, with many projections called placental villi embedded in the uterus wall. In the placenta the blood of the fetus flows close to the blood of the mother in the uterus wall. This facilitates the exchange of materials between maternal and fetal blood.

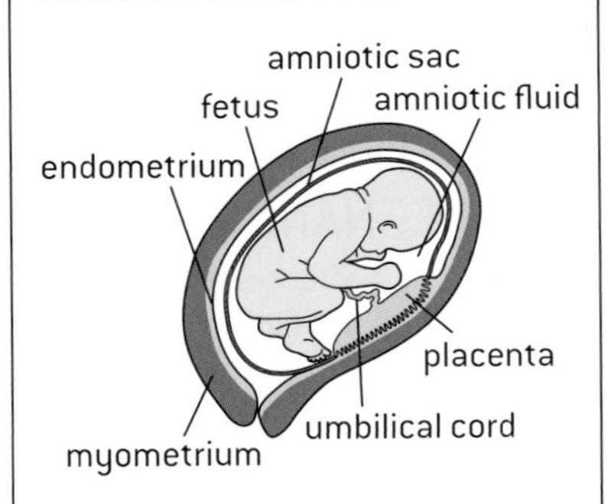

STRUCTURE OF THE PLACENTA

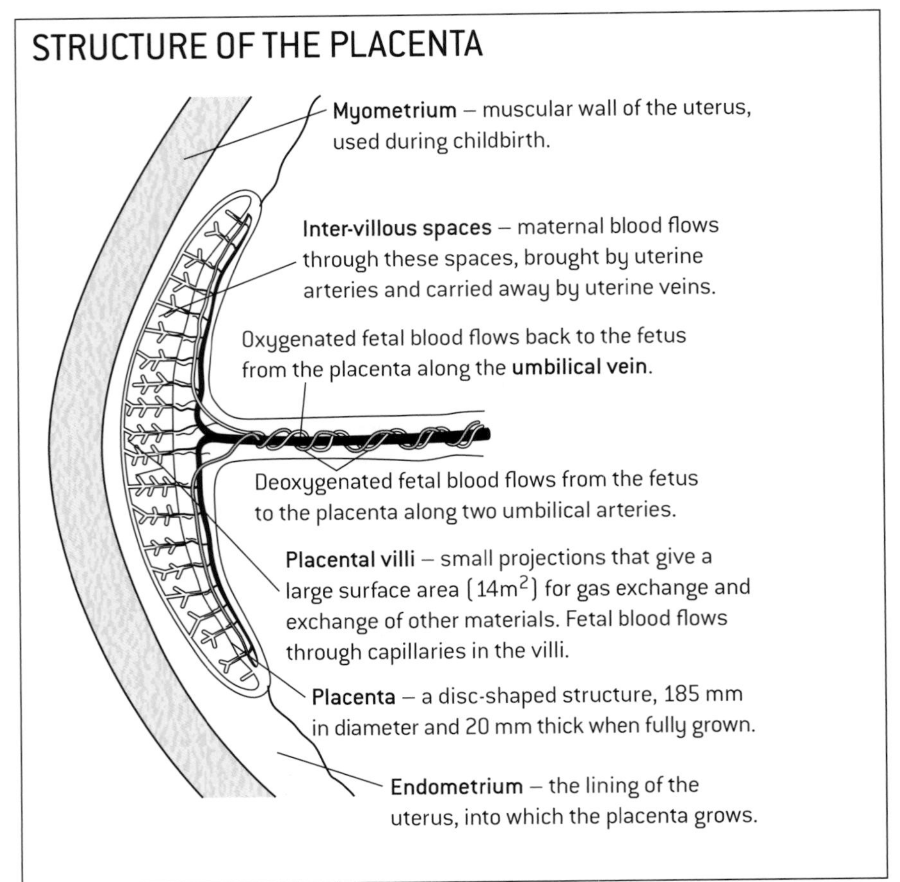

EXCHANGE OF MATERIALS ACROSS THE PLACENTA

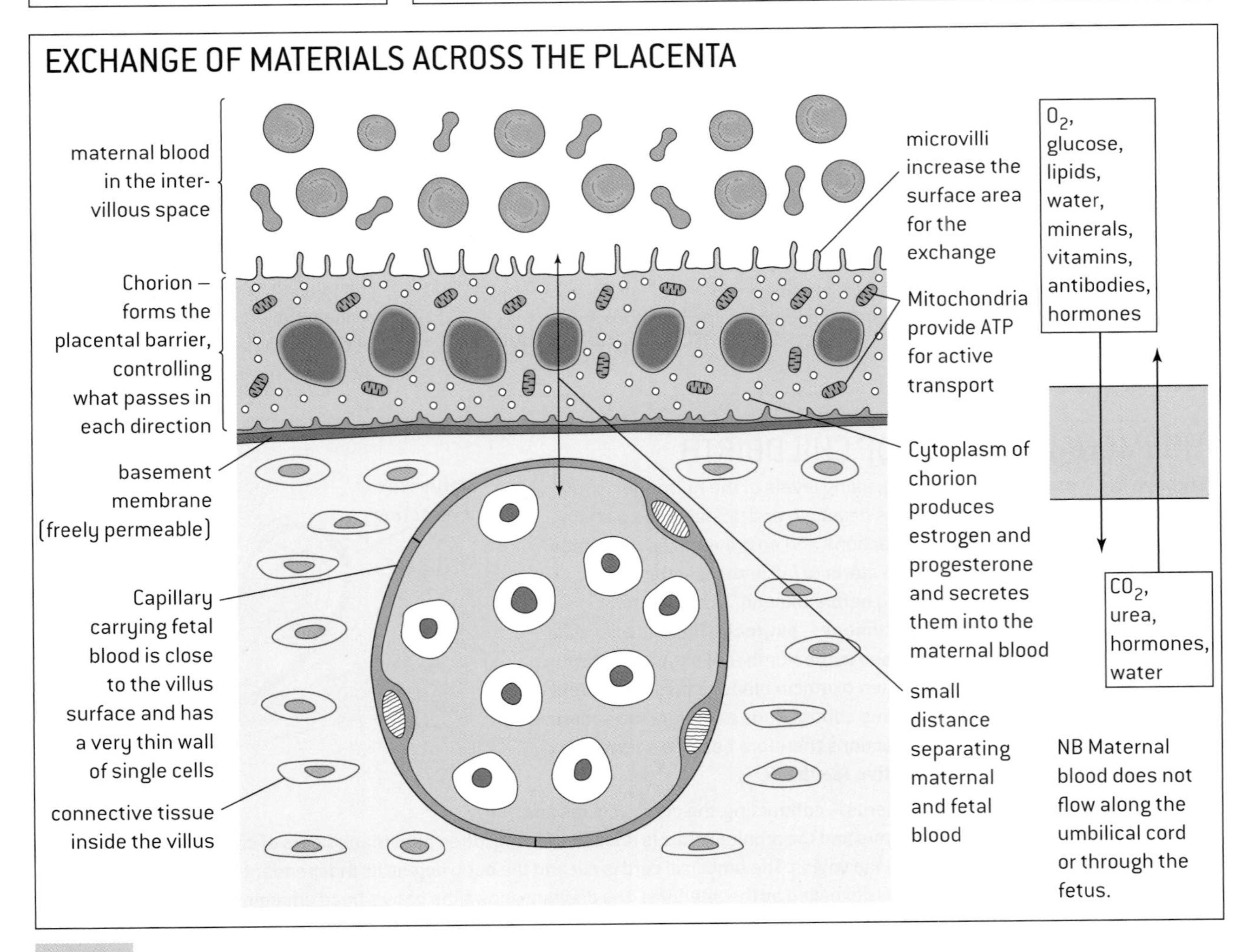

NB Maternal blood does not flow along the umbilical cord or through the fetus.

Questions – animal physiology

1. The graph shows cases of polio in Brazil. On dates with arrows all 0–4 year old children were vaccinated.

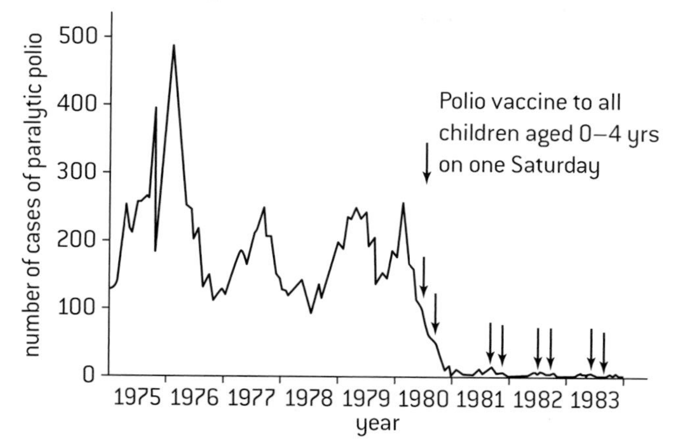

   a) State the maximum number of cases of polio. [1]
   b) Evaluate the success of the vaccination programme using the data in the graph. [3]
   c) Describe the response of the immune system to a polio vaccination. [5]
   d) Suggest reasons for
      (i) a second vaccination a few weeks after the first [2]
      (ii) vaccinations being repeated each year in Brazil. [2]
   e) Explain the reasons for polio vaccinations still being given to children in Brazil despite no cases in North or South America since 1991. [2]

2. The electron micrograph below shows part of a myofibril.

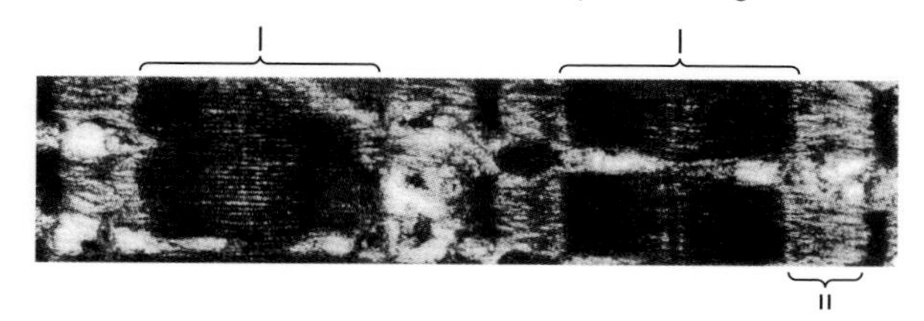

   a) State the type of filament present only in part I. [1]
   b) State one other type of filament in myofibrils. [1]
   c) The myofibril is partly contracted. Deduce any changes in length of regions I and II if
      (i) the myofibril contracted more [2]
      (ii) the myofibril relaxed and the antagonistic muscle contracted. [2]
   d) Draw and label a diagram to show the structure of one sarcomere from the partly contracted myofibril. [4]

3. The X-ray shows an elbow joint with osteoarthritis.

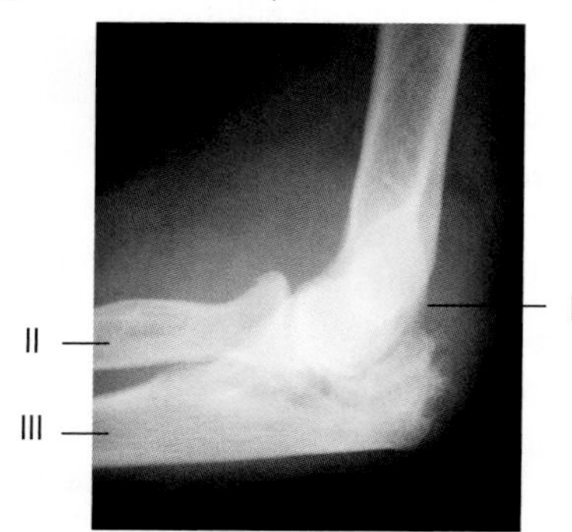

   a) State the names of bones I, II and III. [3]
   b) (i) State the muscles attached to each bone in the region shown. [4]
      (ii) Deduce with reasons whether these muscles were contracted or relaxed when the X-ray was taken. [3]

4. The photograph shows the underside of a beetle.

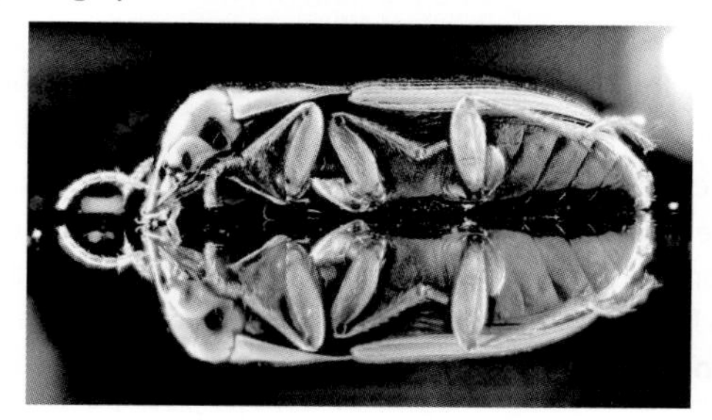

   a) Deduce with reasons
      (i) the phylum in which the beetle is classified [3]
      (ii) the kingdom in which the beetle is classified. [2]
   b) Outline the structure of the beetle's legs. [4]
   c) Explain how the beetle carries out osmoregulation and excretion. [6]

5. The diagram shows two nephrons and a collecting duct in the kidney.

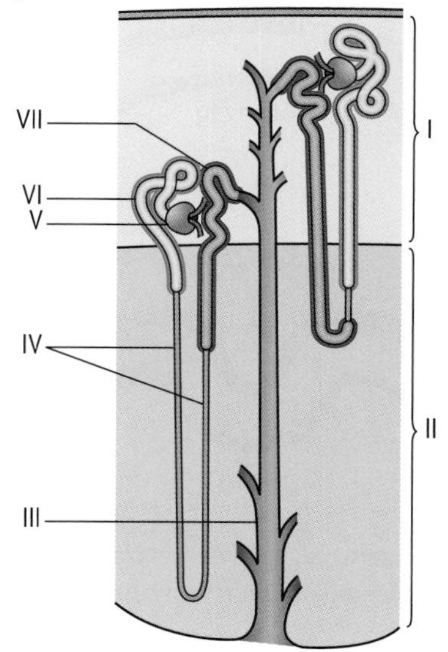

   a) State the names of tissues I and II [2]
   b) (i) List structures III to VII in the sequence that glomerular filtrate flows through them. [2]
      (ii) State the name and role of each structure in this sequence in the production of urine. [10]
   c) State the effect of ADH on part III. [1]

6. a) Compare spermatogenesis and oogenesis in humans. [4]
   b) Distinguish between spermatogenesis and oogenesis in humans. [3]
   c) Explain how polyspermy is prevented in humans. [2]
   d) Describe the different methods of nutrition that are used from conception until birth in humans. [5]
   e) Explain how progesterone levels are maintained from conception until shortly before birth. [3]

Neurulation

NEURULATION

Humans are in the phylum Chordata. All animals in this phylum develop a dorsal **nerve cord** at an early stage in their development. The process is called **neurulation** and in humans it occurs during the first month of life. The dorsal nerve cord develops from **ectoderm**, which is the outer tissue layer. An area of ectoderm cells on the dorsal surface of the embryo develops differently from the rest of the ectoderm and becomes the neural plate.

The cells in the neural plate change shape and this causes the plate to fold inwards forming a groove along the back of the embryo and then separating from the rest of the ectoderm. This forms the **neural tube**, which elongates and develops into the **nerve cord** and later into the brain and spinal cord.

NEURULATION IN XENOPUS

The diagrams below show how neurulation takes place in *Xenopus* (African clawed frog). This species is an ideal model for research into neurulation because the embryo is transparent.

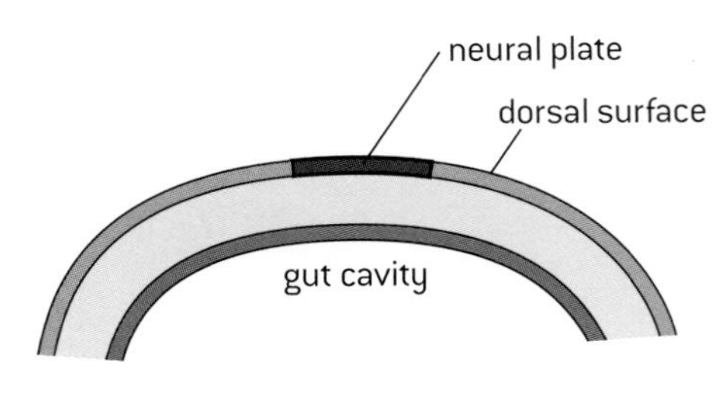

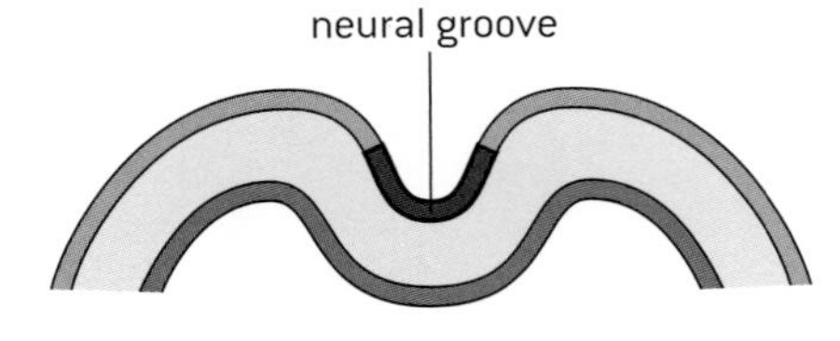

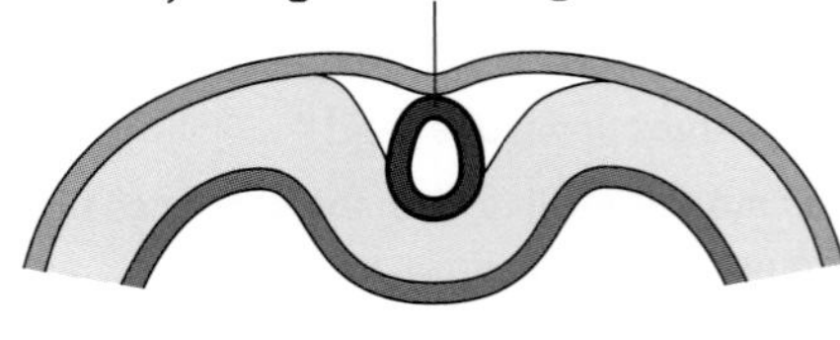

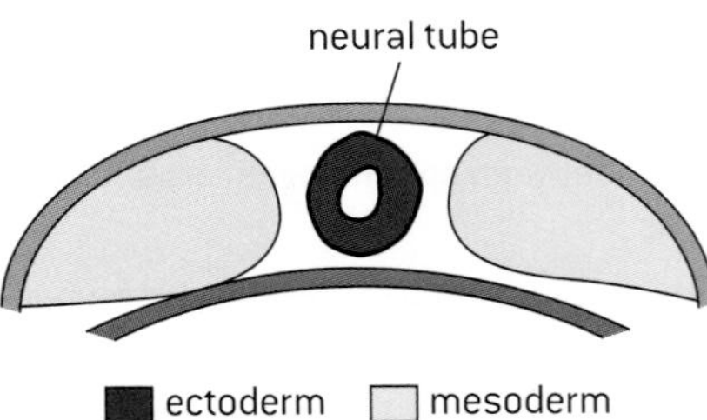

SPINA BIFIDA

In vertebrates there are a series of bones called vertebrae. Each of these has a strong centrum that provides support and a thinner vertebral arch, which encloses and protects the spinal cord. The centrum develops on the ventral side of the neural tube at an early stage in embryonic development. Tissue migrates from both sides of the centrum around the neural tube and normally meets up to form the vertebral arch.

In some cases the two parts of the arch never become properly fused together, leaving a gap. This condition is called **spina bifida**. It is probably caused by the embryonic neural tube not closing up completely when it is formed from the neural groove. Spina bifida is commonest in the lower back. It varies in severity from very mild with no symptoms, to severe and debilitating.

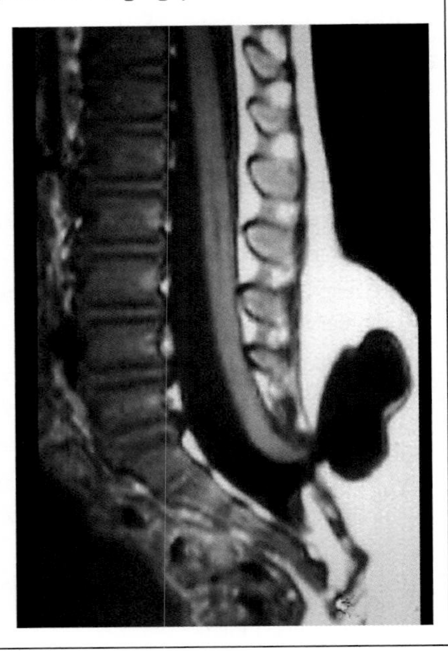

USING ANIMAL MODELS

Neuroscience is the branch of biology concerned with neurons and nervous systems. The aim of research in developmental neuroscience is to discover how nervous systems are formed as animals grow from an embryo into an adult. The aim of many neuroscientists is to understand and develop treatments for diseases of the nervous system, but most experiments are impossible to perform in humans for ethical reasons. Also, research into other animal species is usually easier because development of the nervous system is more rapid, less complex and is easier to observe because the embryo develops externally rather than in a uterus.

For these reasons, even when researchers are trying to make discoveries about humans, they work with other species. A relatively small number of species is used for most of this research and these species are known as animal models:

- *Caenorhabditis elegans* (flatworm)
- *Drosophila melanogaster* (fruit fly)
- *Danio rerio* (zebrafish)
- *Xenopus laevis* (African clawed frog)
- *Mus musculus* (mouse).

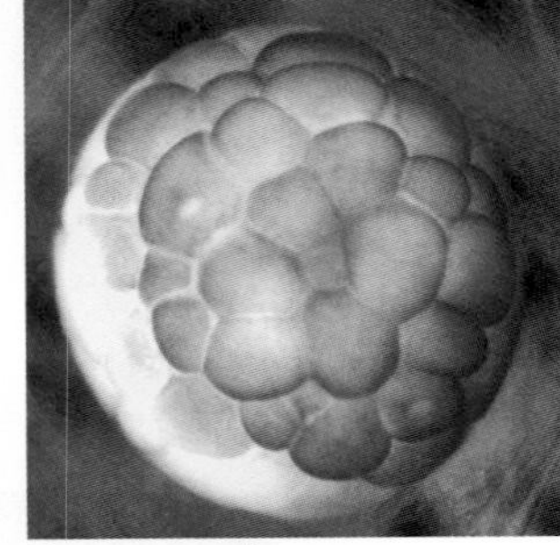

Xenopus embryos

Development of the nervous system

DEVELOPMENT OF NEURONS

Cell division in the neural tube produces large numbers of cells that gradually differentiate into neurons. Some immature neurons migrate from where they are produced in the neural tube to a final location.

Axons grow out from each immature neuron. They are stimulated to do this by chemical stimuli. In some cases the axon grows out of the neural tube to other parts of the embryo and the neuron develops into a sensory or a motor neuron.

Developing neurons produce connections with many other neurons, called **multiple synapses**, but not all of them persist. Synapses that are not used are removed, following the principle 'use it or lose it'.

There is also a process of removing entire neurons that are not being used. This is called **neural pruning**. This is an example of the plasticity of the nervous system – throughout life it can change with experience.

DEVELOPMENT OF THE CNS

The nervous system has two main parts:

- the **peripheral nervous system** consisting of nerves and sensory receptors,
- the **central nervous system** (CNS) consisting of the spinal cord and brain.

Both the brain and spinal cord develop from the neural tube. As the embryo grows, the neural tube elongates. The anterior part of the neural tube develops into the brain and the rest thickens to form the spinal cord. The channel at the centre of the neural tube persists as the very small neural canal in the middle of the spinal cord.

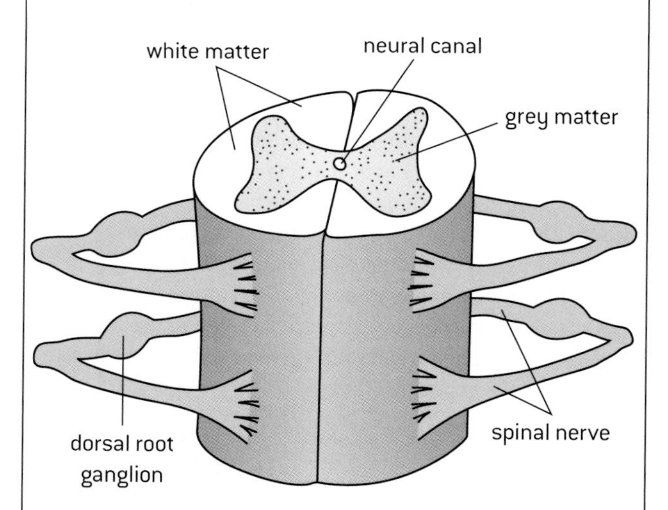

Far more neurons are needed than are initially present in the embryonic neural tube, so cell proliferation continues in both the developing spinal cord and brain. Although this ceases before birth in most parts of the nervous system, there are parts of the brain where extra neurons are produced during adulthood.

STRUCTURE OF THE BRAIN

The brain has a complex structure with distinctive parts, which are shown in the vertical section below.

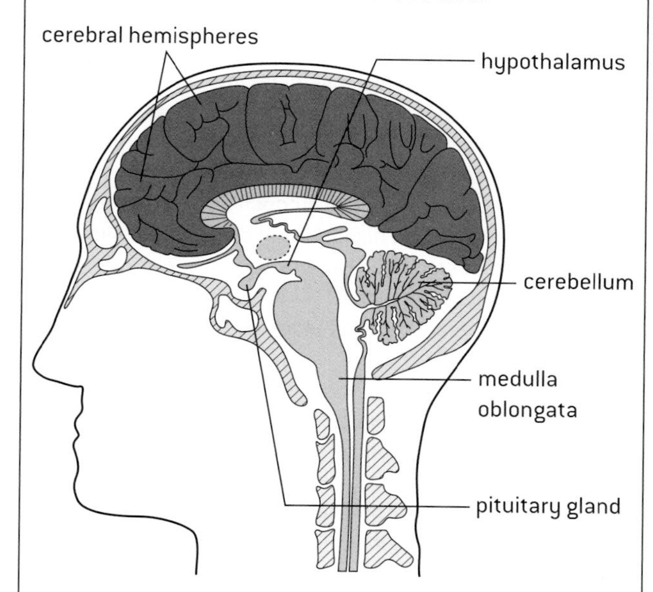

Photographs of the exterior of the brain show the cerebral hemispheres and cerebellum, with the spinal cord connected to the hindbrain (lower part of brain).

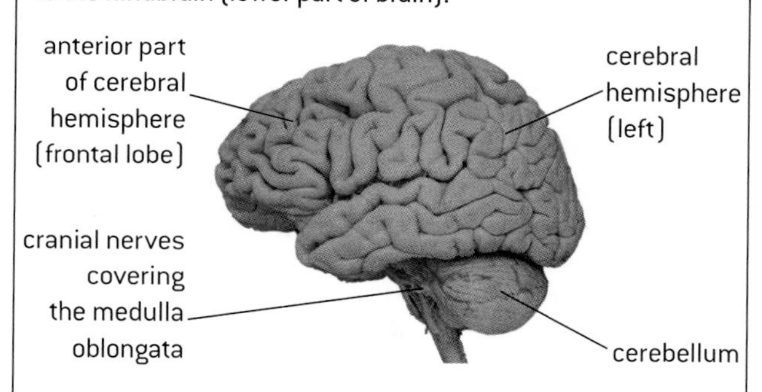

MRI and CAT scans reveal the internal structure of the brain and are widely used to investigate health problems. The image below is a CAT scan.

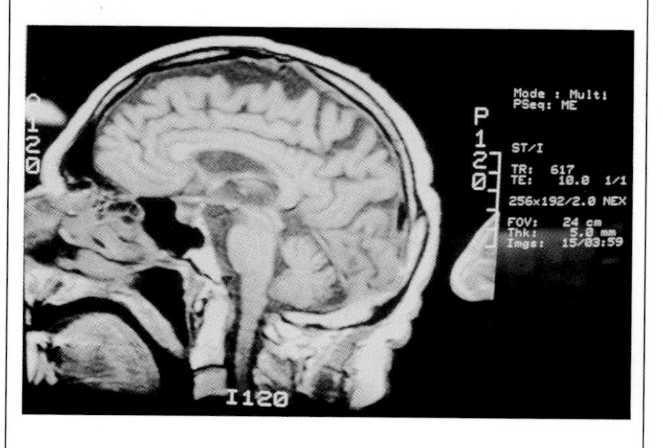

Functions of the brain

METHODS OF BRAIN RESEARCH

Various approaches have been used to identify the roles of parts of the brain.

Lesions and autopsy

A **lesion** is a region of damage or injury to an organ. An **autopsy** is dissection after death of an animal or human body. Brain lesions can be caused by tumours, strokes or accidental damage. Many lesions have been investigated by carrying out an autopsy and relating its position to changes that had been observed to behaviour or capabilities. There are some famous cases from the 19th century including that of Phineas Gage and those investigated by Jean-Martin Charcot.

Animal experiments

Although lesions due to natural causes have revealed much about the brain, more can be learned using animals. Removal of parts of the skull gives access to the brain and allows experimental procedures to be performed. The effects of local stimulation or surgery to a specific part of an animal's brain can be observed.

There are widespread objections to such research, because of the suffering caused to the animal and because at the end the animal is often sacrificed. Neuroscientists have argued that these experiments can increase our understanding of conditions such as epilepsy, Parkinson's disease and multiple sclerosis. There are now strict regulations in most countries to ensure that the benefits of the research justify any harm caused to the animals used.

Functional MRI (fMRI)

Magnetic resonance imaging is used to investigate the internal structure of the body, including looking for tumours or other abnormalities in patients. A specialized version of MRI, called **functional MRI (fMRI)** allows parts of the brain that have been activated by specific thought processes to be identified. Active parts of the brain receive increased blood flow, often made visible by injecting a harmless dye, which fMRI records.

The subject is placed in the scanner and a high-resolution scan of the brain is taken. A series of low-resolution scans is then taken while the subject is being given a stimulus. These scans show which parts of the brain are activated in the response to the stimulus.

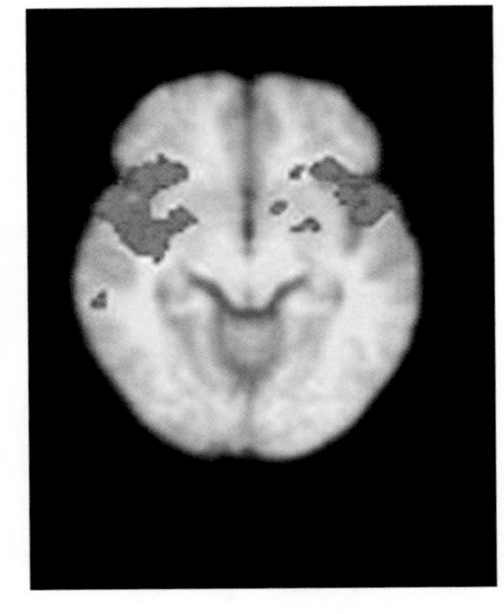

fMRI scan of endometriosis pain

FUNCTIONS OF PARTS OF THE BRAIN

The brain has easily distinguishable parts which have different roles.

The **medulla oblongata** controls automatic and homeostatic activities such as swallowing, digestion, vomiting, breathing and heart rate. Three examples are explained below.

The **cerebellum** coordinates unconscious functions, such as movement and balance.

The **hypothalamus** is the interface between the brain and the pituitary gland, controlling the secretion of pituitary hormones.

The **pituitary gland** secretes at least ten hormones that regulate many body functions.

The **cerebral hemispheres** have many different functions. They are explained on the next page.

THE AUTONOMIC NERVOUS SYSTEM

The peripheral nervous system has two parts: the voluntary and autonomic nervous systems. The autonomic nervous system controls unconscious processes using centres in the **medulla oblongata**.

Swallowing

In the first phase of swallowing food is passed from the mouth cavity to the pharynx. This is voluntary and controlled by the cerebral cortex. The food then passes down the esophagus to the stomach by involuntary muscle contraction, coordinated by the swallowing centre of the medulla oblongata.

Breathing

Two centres in the medulla control the rate and depth of ventilation in response to changes in blood pH, which is monitored by **chemoreceptors** in blood vessels and in the medulla. The depth and rate of inspiration are increased if blood pH falls, as this indicates an increase in CO_2 concentration. They are decreased if blood pH rises.

Heart rate

The cardiovascular centre of the medulla regulates the rate at which the heart beats. It increases or decreases the heart rate by sending impulses to the heart's pacemaker (SAN). Impulses carried by **sympathetic nerve fibres** cause the heart rate to speed up; impulses carried by **parasympathetic nerve fibres** cause the rate to slow down. The sympathetic and parasympathetic systems are the two parts of the **autonomic nervous system**. In many cases, like this, they have opposite effects.

STROKES

A **stroke** is a disruption of the blood supply to part of the brain, caused either by a blockage or by bleeding. Brain tissue is deprived of oxygen for a time and is often damaged. Patients frequently recover from minor strokes, even though a part of the brain is no longer able to function as it did before. This shows that there can be reorganization of certain functions and that not all functions are invariably carried out by one part of the brain. Scans of the brain show that some activities involve many different areas and there may be alternative ways to carry them out.

Cerebral hemispheres

EVOLUTION OF THE CEREBRAL CORTEX

The **cerebral cortex** is the outer layer of the cerebral hemispheres. Although it is only between two and four millimetres thick, it consists of up to six layers of neurons with a complex architecture. It forms a larger proportion of the brain and is more highly developed in humans than other mammals.

Over millions of years of evolution the human cerebral cortex has become immensely enlarged, principally by an increase in total area. There is extensive folding, without which the cerebral cortex could not be accommodated within the cranium.

FUNCTIONS OF THE CEREBRAL HEMISPHERES

The cerebral hemispheres act as the integrating centres for higher order functions such as learning, memory and emotions. As with other parts of the brain, specific functions are carried out by specific parts of the left and right cerebral hemispheres.

The **somatosensory cortex** receives sensory inputs. The left hemisphere receives sensory inputs from the right side of the body and vice versa for the right cerebral hemisphere.

The **motor cortex** controls voluntary muscle contractions by skeletal (striated) muscles. The left cerebral hemisphere controls muscle contraction in the right side of the body, and vice versa for the right cerebral hemisphere.

The **visual cortex** processes visual stimuli detected by light-sensitive rod and cone cells in the retina. In the eye the visual field (the area of vision) is divided into left and right halves. Impulses generated by the right half of the visual field in both eyes are passed to the left cerebral hemisphere and impulses from the left half of the visual field of both eyes are passed to the right cerebral hemisphere. This allows stimuli from the two eyes to be combined, so distance and relative size of objects can be judged. Analysis in the visual cortex also includes pattern recognition and judging the speed and direction of moving objects.

Broca's area is a part of the left cerebral hemisphere that controls the production of speech. If there is damage to this area an individual knows what they want to say and can produce sounds, but they cannot put sounds together into words that have meaning. For example, if we see a horse-like animal with black and white stripes, Broca's area allows us to say 'zebra', but a person with a damaged Broca's area knows that it is a zebra yet cannot say the word.

The **nucleus accumbens** in each of the cerebral hemispheres act as the pleasure or reward centres of the brain. A variety of stimuli including food and sex cause the release of the neurotransmitter dopamine in the nucleus accumbens, which causes feelings of well-being, pleasure and satisfaction. Cocaine, heroin and nicotine are addictive because they also cause release of dopamine in the nucleus accumbens even when nothing has happened in a person's life to justify these feelings.

SENSORY AND MOTOR HOMUNCULI

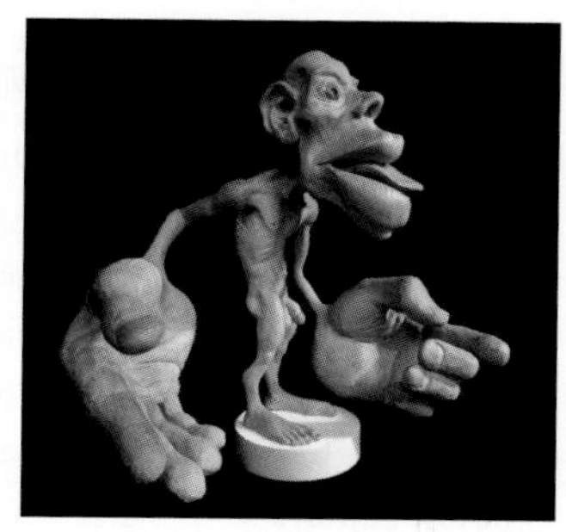

Models of the human body can be made with the size of each part of the body proportional either to the area of the somatosensory cortex that receives inputs from that part (right), or to the area of the motor cortex that controls muscles in that part of the body (below) These types of model are called sensory and motor homunculi.

There are striking differences between the proportions of both homunculi and actual human bodies. Hands appear disproportionately large for example, as they contain a relatively large number of sensory receptor cells and many small muscles.

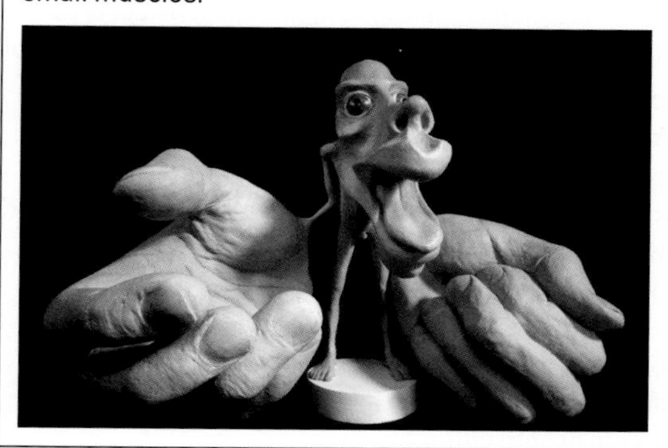

BRAIN AND BODY SIZE

The graph below shows the relationship between brain mass and body mass in animal species.

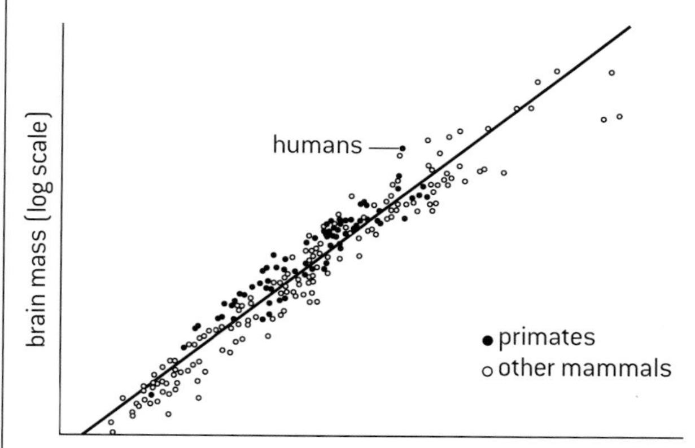

The correlation coefficient for the data in the graph is 0.75, so there is quite a strong positive correlation between brain and body mass. Humans do not have the largest brain size of any animal – species with a larger body mass such as blue whales and elephants have larger brains. However, the data point for humans is further above the correlation curve than any other species, indicating that humans have a larger brain in relation to their body mass than other animals. The graph also shows that most but not all primates have relatively large brains in relation to their body mass.

Perception of stimuli

DIVERSITY OF SENSORY RECEPTORS

Sensory receptors detect changes in the environment. The change detected by a receptor is a **stimulus**. In humans there are four types of sensory receptor, which together can detect a wide range of stimuli.

Type	Stimulus	Example
Mechanoreceptors	Mechanical energy in the form of sound waves Movements due to pressure or gravity	Hair cells in the cochlea of the ear Pressure receptor cells in the skin
Chemoreceptors	Chemical substances dissolved in water (tongue) Chemical substances as vapours in the air (nose)	Receptor cells in the tongue Nerve endings in the nose
Thermoreceptors	Temperature	Nerve endings in skin detect warm or cold
Photoreceptors	Electromagnetic radiation, usually in the form of light	Rod and cone cells in the eye

PHOTORECEPTORS

The photoreceptors of the eye are contained in the retina. There are two types of photoreceptor cell – **rod cells** and **cone cells**. The diagram of the retina on the next page shows the structure of rod and cone cells. These cell types both absorb light and then transmit messages to the brain, via the optic nerve. They are different in these ways:

1. Rod cells are more sensitive to light than cone cells, so they function better in dim light, for example at night. Rod cells become bleached in bright light, for example in daylight, but cone cells function well in high light intensities.
2. All rod cells contain the same pigment, which absorbs a wide range of wavelengths of light, so they do not distinguish between different colours and only give monochrome vision. There are three types of cone cell, each of which contains a different pigment. These pigments absorb different ranges of wavelength, with peaks of absorbance in blue, green and red light. Cone cells can therefore distinguish between light of different wavelengths and so give colour vision.

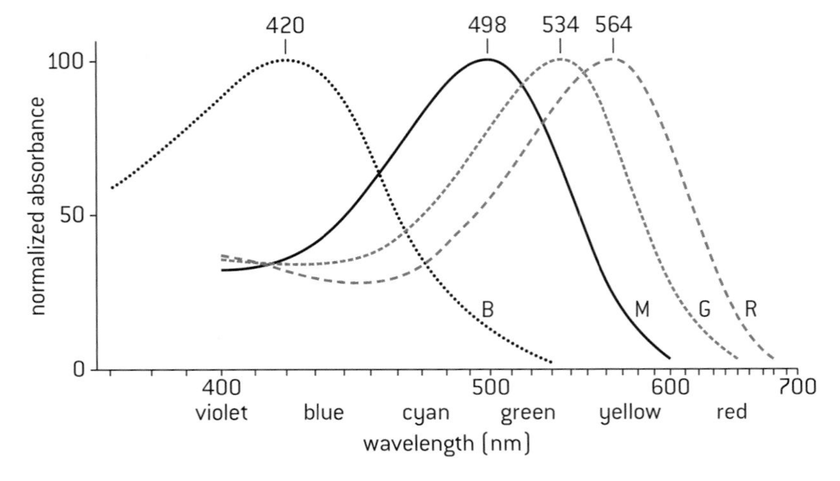

The graph shows the absorbance of wavelengths of light by the three pigments in cone cells (B, G and R) and the pigment in rod cells (M).

OLFACTORY RECEPTORS

The sense of smell (olfaction) is due to **olfactory receptor cells** located in the epithelium inside the upper part of the nose. These cells have cilia which project into the air in the nose. In the membrane of these cilia are the **receptors**, which are proteins that can detect specific chemicals in the air.

Only chemicals that are volatile and can pass through the air can be detected (smelled). Odorants from food in the mouth can pass through the air in the mouth and nasal cavities to be detected in the nose.

Each olfactory receptor cell has just one type of odorant receptor in its membrane, but there are many different types of receptor, each of which is encoded by a different gene and detects a different group of odorants.

In some mammals such as mice there are over a thousand different receptors, but humans have fewer. Using these olfactory receptors a large number of chemicals in the air can be distinguished.

RED–GREEN COLOUR-BLINDNESS

The photoreceptor pigments in blue, green and red cone cells are all members of a group of proteins called **opsins**. There is a separate gene coding for each of the three pigments. The genes for the pigments in both red and green cones are located on the X chromosome. Red–green colour-blindness is a common inherited condition. It is due to a lack of functioning pigment in either red or green cone cells. Whichever pigment is missing, light with the wavelengths in the green to red part of the spectrum cannot be distinguished. As genes for both pigments are on the X chromosome, red–green colour-blindness is sex-linked, whether it is the green- or red-detecting pigment that is missing. The normal alleles of both genes are dominant and the alleles that cause red–green colour-blindness are recessive. Red–green colour-blindness is therefore much commoner among males than females and males inherit the allele that causes the condition from their mother.

Vision in humans

STRUCTURE AND FUNCTION OF THE RETINA

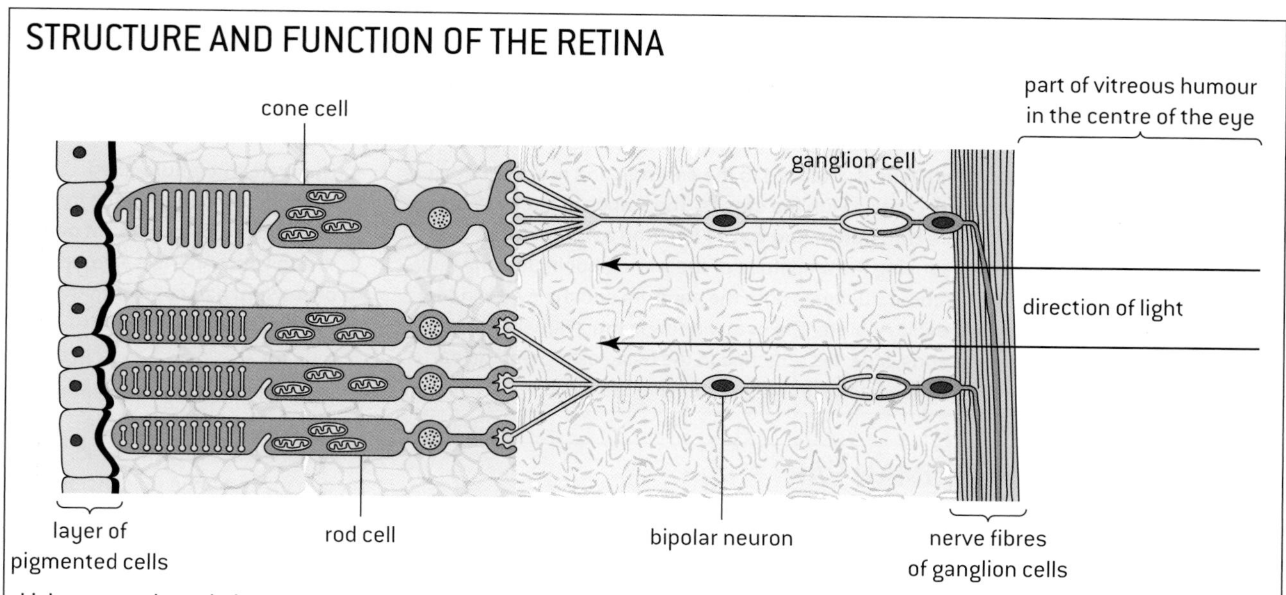

Light passes through the nerve fibres of **ganglion cells** and the layer of **bipolar cells** in the outer part of the retina until it reaches the rod and cone cells. When rod or cone cells absorb light they pass impulses to bipolar cells, which pass them on to ganglion cells. Groups of up to two hundred rod cells pass impulses to the same bipolar cell, whereas as few as one cone cell may pass impulses to a single bipolar cell, so cone cells give greater visual acuity. Impulses from rod and cone cells are processed in bipolar and ganglion cells and are then transmitted to the brain in the nerve fibres of ganglion cells, which are located in the **optic nerve**.

ASSESSING BRAIN DAMAGE USING THE PUPIL REFLEX

Muscle cells in the iris control the size of the pupil of the eye. Impulses carried to radial muscle by neurons of the sympathetic system cause them to contract and dilate the pupil; impulses carried to circular muscle by neurons of the parasympathetic system cause the pupil to constrict. The pupil reflex occurs when bright light suddenly shines into the eye. Photoreceptive ganglion cells in the retina perceive the bright light, sending signals through the optic nerve to the mid-brain, immediately activating the parasympathetic system, which stimulates circular muscle in the iris to constrict the pupil, reducing the amount of light entering the eye and protecting the delicate retina from damage. The mid-brain is part of the brain stem – the region of the brain that is adjacent to the spinal cord.

Doctors sometimes use the pupil reflex to test a patient's brain function. A light is shone into each eye. If the pupils do not constrict at once, the brain stem is probably damaged. If this and other tests of brain stem function repeatedly fail, the patient is said to have suffered brain death. It may be possible to sustain other parts of the patient's body on a life support machine, but full recovery is extremely unlikely.

OPTIC NERVES

The diagram below shows how information from the left field of vision reaches the right visual cortex and vice versa.

Nerve fibres cross over at the optic chiasma so that impulses from the left field of vision in both eyes go to the right side of the visual cortex and vice versa for the right field of vision. This is called contralateral processing and gives us stereoscopic vision so we can judge distances.

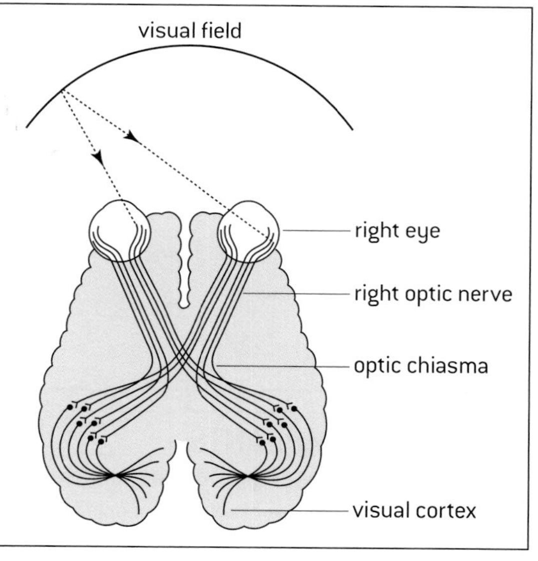

STRUCTURE OF THE HUMAN EYE

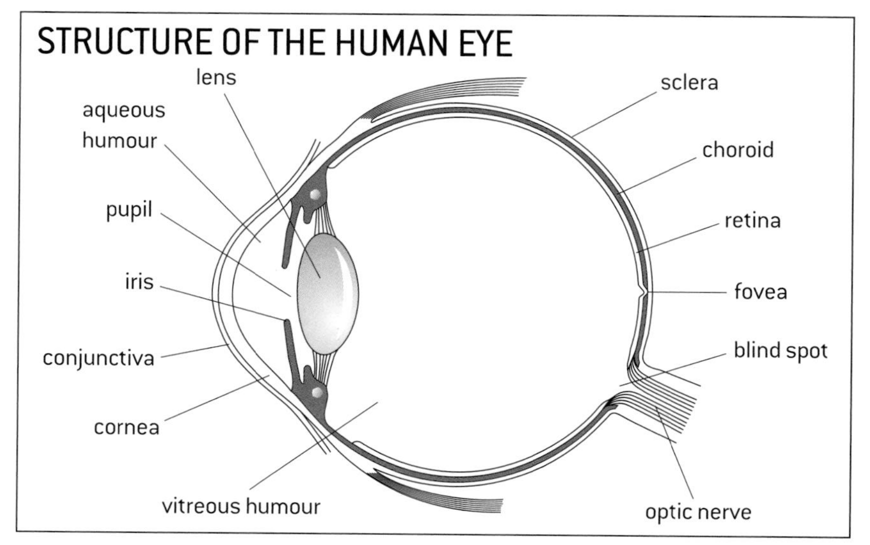

Hearing in humans

FUNCTIONS OF THE MIDDLE EAR

When sound waves reach the eardrum at the end of the outer ear, they make it vibrate. The vibration consists of rapid movements of the eardrum, towards and away from the middle ear. In the middle ear is a series of very small bones called **ossicles**, which are shown in the diagram below. The **malleus** is attached to the **eardrum** and makes contact with the **incus**, which in turn makes contact with the **stapes**. The stapes is attached to the **oval window**. The ossicles therefore **transmit** sound waves from the eardrum to the oval window. They also act as levers, reducing the amplitude of the waves, but increasing their force, which **amplifies** sounds by about 20 times. Both the eardrum and the oval window are thin layers of tissue that can readily vibrate. The oval window is much smaller than the eardrum. This helps to amplify sounds. Muscles attached to the ossicles protect the ear from loud sounds by contracting, which damps down vibrations in the ossicles.

STRUCTURE OF THE HUMAN EAR

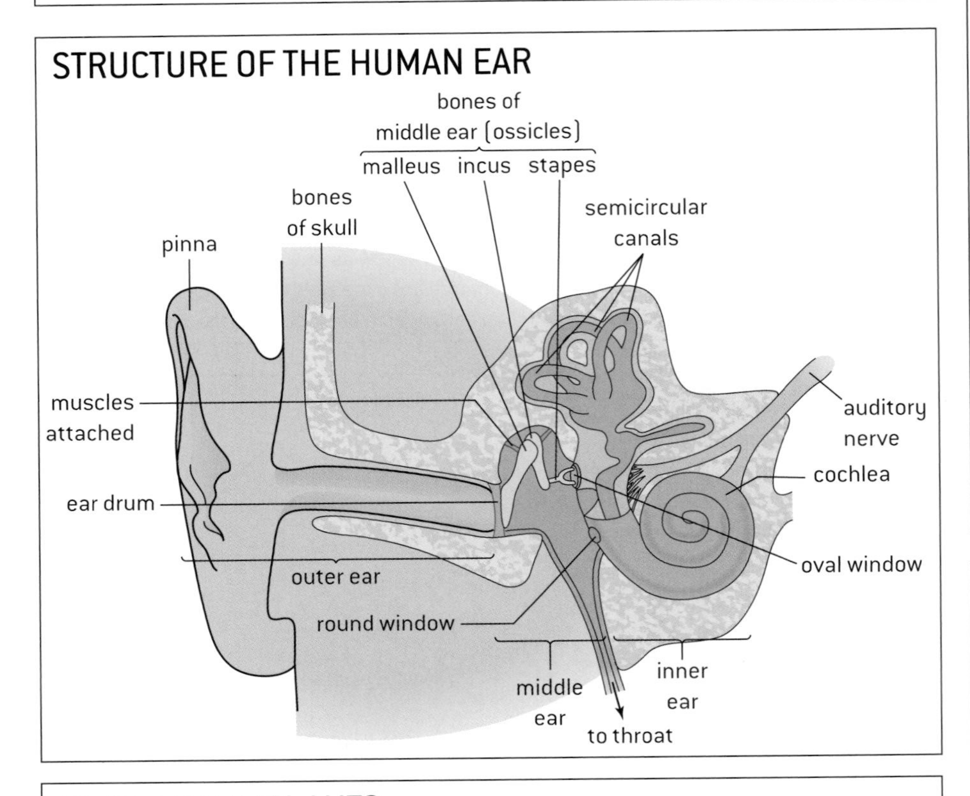

COCHLEAR IMPLANTS

Research from the 1950s onwards led to the development of **cochlear implants**, which can help give some sense of sound to people with non-functional cochlear hair cells. The external parts are a microphone to detect sounds, a speech processor to filter out frequencies apart from those used in speech and a transmitter.

The internal parts are implanted in bone behind the ear. They consist of a receiver that picks up sound signals from the transmitter, a stimulator to convert the signals into electrical impulses and an array of electrodes to carry the impulses to the cochlea. The electrodes stimulate the auditory nerve directly and so bypass the non-functional hair cells.

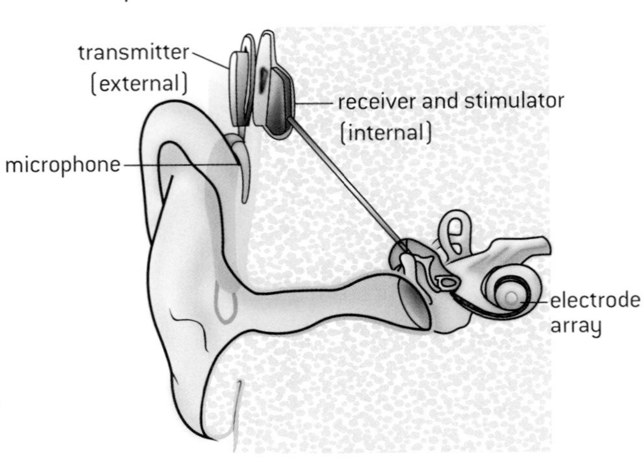

FUNCTION OF THE SEMICIRCULAR CANALS

There are three fluid-filled semicircular canals in the inner ear. Each has a swelling at one end in which there is a group of sensory hair cells, with their hairs embedded in gel.

When the head moves in the plane of one of the semicircular canals, the stiff wall of the canal moves with the head, but due to inertia the fluid inside lags behind. There is therefore a flow of fluid past the hairs, stimulating the hair cells to send impulses to the brain.

The three semicircular canals are at right angles to each other, so each is in a different plane. They can therefore detect movements of the head in any direction. The brain can deduce the direction of movement by the relative amount of stimulation of the hair cells in each of the semicircular canals.

FUNCTION OF THE COCHLEA

The cochlea consists of a spiral fluid-filled tube. Within the tube are membranes with receptors called **hair cells** attached. These cells have hair bundles, which stretch from one of the membranes to another. When sound waves transmitted by the oval window pass through the fluid in the cochlea, the hair bundles vibrate.

Gradual variations in the width and thickness of the membranes allow different **frequencies** of sound to be distinguished, because each hair bundle only resonates with particular frequencies. When the hair bundles vibrate, the hair cells send messages to the brain via the **auditory nerve**.

Innate behaviour (HL only)

ORIGINS OF INNATE BEHAVIOUR

Innate behaviour is inherited from parents and is not influenced by an organism's environment, including experiences that the organism has during its life. Innate behaviour is therefore due to the organism's genes, so develops by natural selection, as with other heritable features. If one allele of a gene that affects behaviour gives a greater chance of survival and reproduction than other alleles of the gene, both the allele and the behaviour pattern will increase in frequency in the species.

REFLEXES

A **stimulus** is a change in the environment, either internal or external, that is detected by a receptor and elicits a **response**. A response is a change in an organism, often carried out by a muscle or a gland.

Some responses happen without conscious thought and are therefore called **involuntary responses**. Many of these are controlled by the **autonomic** nervous system. These autonomic and involuntary responses are known as **reflexes**. A reflex is a rapid unconscious response to a stimulus.

COMPONENTS OF A REFLEX ARC

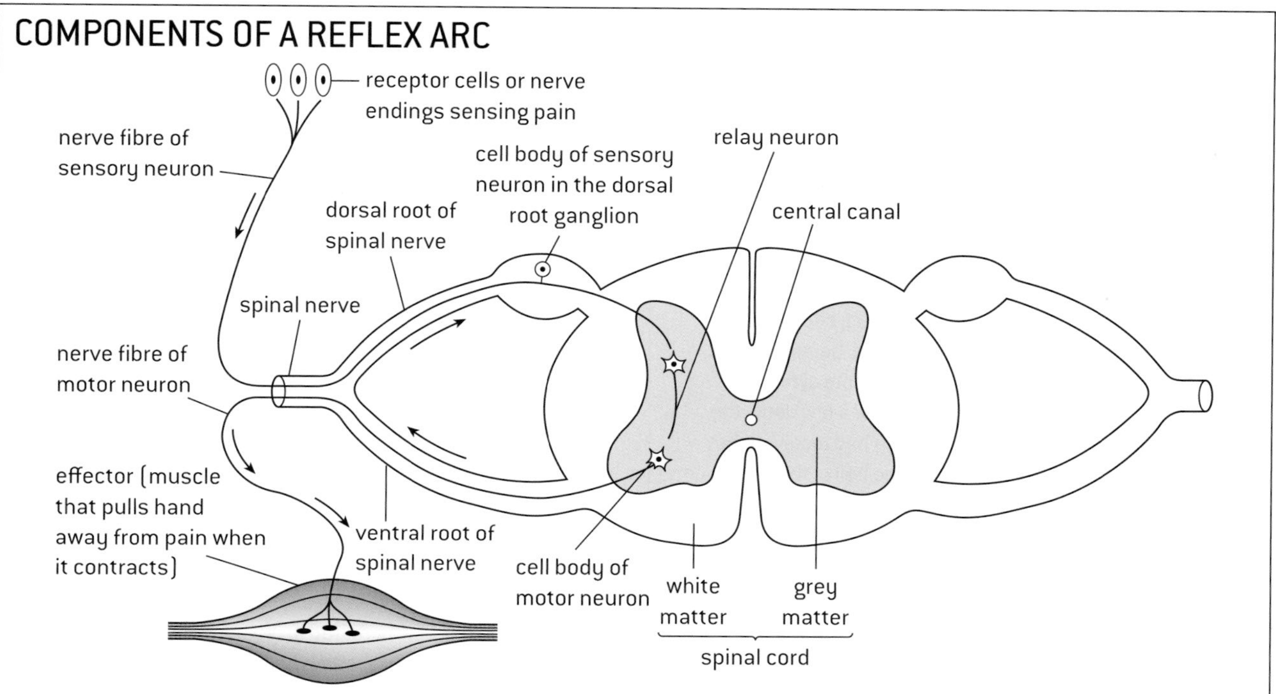

Reflexes are mediated by a series of neurons, called a reflex arc. The diagram above shows a reflex arc used to carry out a **withdrawal reflex**.

The withdrawal reflex is carried out when the hand receives a pain stimulus, for example when touching a hot object. The response is to pull away the hand (withdrawal) by contracting muscles in the arm.

ANALYSING INNATE BEHAVIOUR IN INVERTEBRATES

Most behaviour in invertebrates is innate, not learned. It can be investigated by simple experiments, for example chemotaxis in *Planaria* (flatworms). A taxis is a movement towards or away from a directional stimulus. If *Planaria* are placed in a shallow dish with small pieces of food in part of the dish, they usually move towards the food. This response has clear benefits in terms of **survival** and **reproduction**. Other variables need to be kept constant, for example the amount of light in different parts of the dish. Also, in behaviour experiments like this, results should be quantitative, not merely descriptive, so they can be analysed using statistical tests.

An example of the analysis of the results of an experiment into colour preferences of slaters (woodlice) in a choice chamber is given (right).

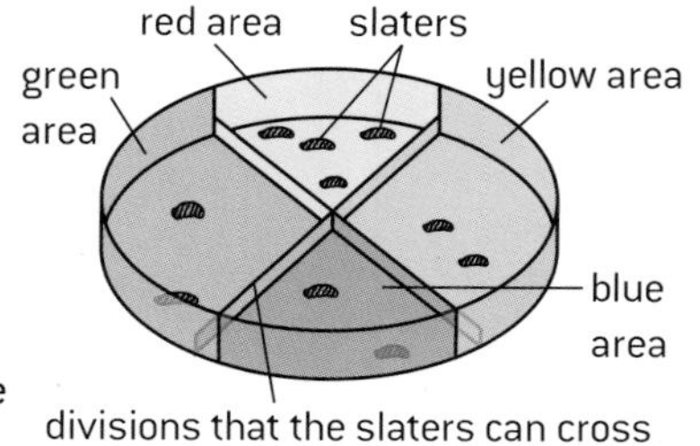

Results (4 trials with 10 slaters)					
	1	2	3	4	
Colour	Number per trial				Total
Blue	1	1	0	2	4
Green	1	1	1	1	4
Red	6	4	9	6	25
Yellow	2	4	0	1	7

Analysis using the chi² test (explained in Topic 4).

Expected results if slaters have no colour preference: 10 per colour

$$\text{Chi-squared} = \frac{(10-4)^2}{10} + \frac{(10-4)^2}{10} + \frac{(10-25)^2}{10} + \frac{(10-7)^2}{10} = 30.6$$

This value of chi-squared lies in the critical region whether significance levels of 5%, 1% or 0.1% are used, so the hypothesis that the slaters have no colour preference is rejected at all these levels. Slaters inhabit damp, dark habitats and show negative phototaxis. These results suggest that it is shorter wavelength light that they perceive and avoid.

Learned behaviour (HL)

LEARNING AND MEMORY

In biology the words **learning** and **memory** have specific meanings:

Learning is the acquisition of skill or knowledge.

Memory is the process of encoding, storing and accessing information.

Learned behaviour develops as a result of the experiences that an animal has during its life. The animal's genes give it the capacity to develop learned behaviour, but without experiences it does not develop. Research into animal behaviour has revealed several different types of learned behaviour: reflex conditioning, imprinting and operant conditioning.

REFLEX CONDITIONING

Ivan Pavlov investigated the salivation reflex in dogs. He observed that his dogs secreted saliva when they saw or tasted food. The sight or taste of meat is called the **unconditioned stimulus** and the secretion of saliva is called the **unconditioned response**.

Pavlov then gave the dogs a neutral stimulus, such as the sound of a ringing bell or ticking metronome, before he gave the unconditioned stimulus – the sight or taste of food. He found that after repeating this procedure for a few days, the dogs started to secrete saliva before they had received the unconditioned stimulus. The sound of the bell or the metronome is called the **conditioned stimulus** and the secretion of saliva before the unconditioned stimulus is the **conditioned response**.

The dogs had learned to associate two external stimuli – the sound of a bell or metronome and the arrival of food. This is called **reflex conditioning** – an alteration in behaviour as a result of an animal forming new associations.

IMPRINTING

Konrad Lorenz investigated learning in greylag geese. He removed half of the eggs laid by a female and kept them in an incubator. Lorenz was with the goslings when they hatched and stayed with them for a few hours. He was therefore the first moving object that they saw. The goslings followed him around instead of their mother and some of them even tried to mate with humans when they became adults. The goslings that hatched from the eggs left with their mother showed normal behaviour. They followed their mother around while young and mated with other geese. Lorenz deduced that there is a sensitive period after hatching during which goslings normally learn to identify and become attached to their mother. This and other cases where an animal learns a response to a stimulus during a sensitive period is **imprinting**. It is independent of the consequences of the behaviour.

Further experiments showed that the newly hatched goslings only became imprinted on moving objects, which in nature would almost always be the mother. Imprinting involves an **innate releasing mechanism** which the animal uses to filter stimuli that they receive and only respond to a stimulus that is significant – called a **sign stimulus**.

OPERANT CONDITIONING

Burrhus Frederic Skinner designed a piece of apparatus called a Skinner box, to investigate learned behaviour in animals. When an animal such as a rat pressed a lever inside the box, a small pellet of food dropped into the box, which the rat could then eat. When a hungry rat is placed into the box it moves around looking and sniffing at everything within the box. It eventually presses the lever by accident but soon learns to associate pressing the lever with the reward of food. The food reward is called the **reinforcement**. Pressing the lever is called the **operant response**. This form of learning is due to **trial and error** and is called **operant conditioning**. The more quickly the reinforcement is given, the more quickly the operant response develops. Surprisingly, Skinner found that if the reinforcement is only given sometimes after the operant response, the operant response develops more quickly.

DEVELOPMENT OF BIRDSONG

Birdsong has been investigated intensively in some species and evidence has been found for it being partly innate and partly learned. All members of a bird species share innate aspects of song, allowing each individual to recognize other members of the species. In many species, including all passerines, males learn mating calls from their father. The learned aspects introduce differences, allowing males to be recognized by their song and in some species mates to be chosen by the quality of their singing.

The chaffinch (*Fringilla coelebs*) is an example.

The figures (below) show the normal song of a male, reared where he could hear the song of adult chaffinches, and the song of a male that was reared in isolation in a soundproof box. The song of the bird reared in isolation had some features of the normal song, including the correct length and number of notes, which must have been innate. However, there is a narrower range of frequencies, and fewer distinctive phases. These must normally be learned from other chaffinches.

(a) a normal chaffinch song

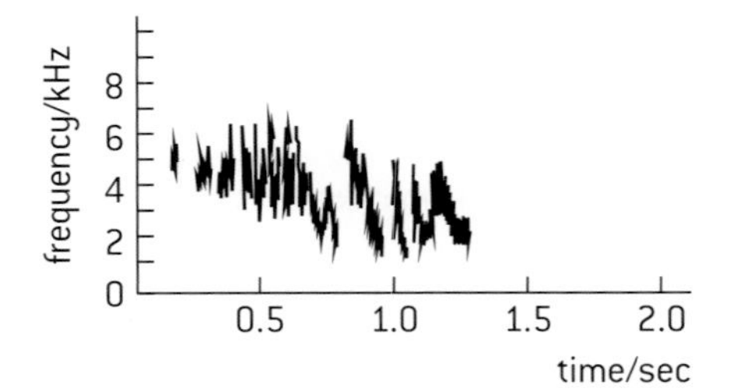

(b) a song from a bird reared in isolation

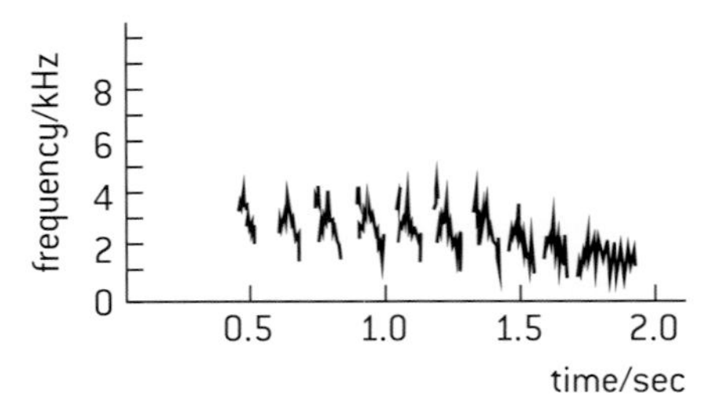

Neurotransmitters and synapses (HL only)

EXCITATION AND INHIBITION

In Topic 6 the effects of excitatory neurotransmitters at a synapse were described. Inhibitory neurotransmitters are released into synapses by some pre-synaptic neurons.

	Excitatory	Inhibitory
Ion that is caused to enter the post-synaptic neuron	Positively charged e.g. sodium Na^+	Negatively charged e.g. chloride Cl^-
Typical effect on the post-synaptic membrane potential	Rises (−70 to −50 mV)	Drops (−70 to −80 mV or lower)
Typical effect on propagation of a nerve impulse in the post-synaptic neuron	Excites an impulse as threshold potential is reached	Inhibits an impulse as post-synaptic membrane is hyperpolarized

One post-synaptic neuron may have synapses with many excitatory and inhibitory pre-synaptic neurons. Whether or not a nerve impulse is stimulated in the post-synaptic neuron depends on summation of all excitatory and inhibitory neurotransmitters received.

SLOW-ACTING NEUROTRANSMITTERS

The neurotransmitters above are fast-acting, affecting post-synaptic neurons a few milliseconds after release. Other neurotransmitters take hundreds of milliseconds to have effects so they are slow-acting. Rather than having an effect on a single post-synaptic neuron they may diffuse through surrounding fluid and affect groups of neurons. Norepinephrine, dopamine and serotonin are examples. Slow-acting neurotransmitters do not affect ion movement across post-synaptic membranes directly, but instead cause release of secondary messengers inside post-synaptic neurons. These can modulate fast synaptic transmission for days and may be central to the synaptic plasticity that is necessary for early stages of memory and learning. Long-term memories probably require a remodelling of the synaptic connections between neurons.

DRUG ADDICTION

Three factors increase addiction rates:

Dopamine secretion: Many addictive drugs increase secretion of dopamine. Synapses where dopamine is the neurotransmitter are involved in the reward pathway, so users of addictive drugs find it very difficult to stop, because they have become dependent on the feelings that dopamine promotes.

Genetic predisposition: Even with many drugs that are potentially addictive, not everyone becomes an addict. Addictions are much commoner in some families than others. This suggests that certain genes are implicated.

Social factors: Cultural traditions, peer pressure, poverty and social deprivation, traumatic life experiences and mental health problems all increase the chances of an addiction developing.

PSYCHOACTIVE DRUGS

Stimulants are psychoactive drugs that increase transmission at synapses in the brain. **Nicotine** has many effects on the nervous system but is strongly addictive due to its activity in synapses where dopamine is the neurotransmitter. Nicotine binds to receptors in the pre-synaptic membrane, leading to depolarization and increased dopamine release. **Cocaine** also acts at synapses that use dopamine as a neurotransmitter. It binds to and blocks dopamine reuptake transporters, which pump dopamine back into the pre-synaptic neuron, causing high concentrations in the synapse and continual excitation of the post-synaptic membrane.

Sedatives are psychoactive drugs that decrease transmission at synapses in the brain. **Diazepam** (a benzodiazepine drug) binds to an allosteric site on GABA receptors in post-synaptic membranes. GABA is an inhibitory neurotransmitter and when it binds to its receptor a chloride channel opens, causing hyperpolarization of the post-synaptic neuron. When diazepam is bound to the receptor the chloride ions enter at a greater rate, inhibiting nerve impulses in the post-synaptic neuron. **Tetrahydrocannabinol** (THC) is present in cannabis. It binds to cannabinoid receptors in pre-synaptic membranes, inhibiting the release of neurotransmitters that cause excitation of post-synaptic neurons. Cannabinoid receptors are found in synapses in various parts of the brain, including the cerebellum and cerebral hemispheres.

EFFECTS OF ECSTASY

MDMA (ecstasy) is an amphetamine. There is strong evidence for MDMA promoting the release of the slow-acting neurotransmitters serotonin and dopamine in the brain. An example of this evidence is given in the questions at the end of this topic.

ENDORPHINS AND ANAESTHETICS

Pain receptors are endings of sensory neurons that convey impulses to the cerebral cortex, where they cause the sensation of pain. **Endorphins** secreted by the pituitary gland bind to and block receptors at synapses in pathways used in this perception of pain. Endorphins therefore act as natural painkillers.

Anaesthetics act by interfering with neural transmission between areas of sensory perception and the CNS. They cause a reversible loss of sensation in part of the body (local anaesthetics) or all (general anaesthetics). General anaesthetics cause unconsciousness, so the patient is not aware of what is happening. This is useful in many surgical procedures.

DRUG APPROVAL PROCEDURES

Modern drug testing procedures are rigorous and take many years. This can be frustrating for patients who might benefit. They often act as advocates for new drugs, pressing for shorter approval processes and encouraging more tolerance of risk. The role of drug testing agencies is to minimize risk, so only in exceptional cases is testing abbreviated or are placebo groups given the drug immediately.

Ethology (HL only)

EVOLUTION OF ANIMAL BEHAVIOUR

Ethology is the study of animal behaviour in natural conditions. Natural selection can change the frequency of a particular type of animal behaviour: if the behaviour increases the chances of survival and reproduction, it will become more prevalent in the population. Changes in innate behaviour depend on a change in frequencies of the alleles that cause the behaviour. Learned behaviour can spread through a population or be lost from it more rapidly than innate behaviour because it is passed from individual to individual without allele frequencies having to change.

EXAMPLES OF ANIMAL BEHAVIOUR

1. Migration in blackcaps

Blackcaps breed in the early summer across much of central and northern Europe. They then migrate to warmer areas before the winter. Until recently, populations in Germany migrated to Spain or other Mediterranean areas. Recent studies have shown a change in migration pattern, with 10% of the birds migrating to the UK. Experiments with eggs have shown that the direction of migration has a **genetic basis** so can be inherited. The blackcaps that migrate from Germany to the UK for the winter instinctively tend to fly west, whereas those still migrating to Spain tend to fly southwest. Migration to the UK has increased by **natural selection** recently as winters have become warmer and many people there provide food for wild birds in winter.

2. Blood sharing in vampire bats

Blood sharing was investigated in a population of vampire bats in Costa Rica. They live in groups and feed at night by sucking blood from larger animals. If one of the bats in the group fails to feed for more than two consecutive nights it may die of starvation. However, bats that have fed successfully regurgitate blood for a bat that has failed to feed. Tests have shown that this is done whether the two bats are genetically related or not. This is called **reciprocal altruism** because the bat that donates food to a hungry bat might in the future receive blood when it is hungry. There is an advantage for the whole group, because the benefit of receiving blood when starving is greater than the cost of donating blood after feeding well.

3. Foraging in shore crabs

Foraging is searching for food. Animals must decide what type of prey to search for and how to find it. Studies have shown that animals increase their chance of survival by **optimal prey choice**. For example, the shore crab prefers to eat mussels of intermediate size when presented in an aquarium with different sizes of mussel. It is the mussels of intermediate size that are the most profitable in terms of the energy yield per second of time spent breaking open the shells.

4. Courtship in birds of paradise

Male birds of paradise have showy plumage and gather at a site called a lek, to perform a distinctive **courtship dance**. Females watch the males before selecting one of them for mating. If a male bird of paradise has survived in the rainforest with the encumbrance of long tail feathers and bright plumage that makes it visible to predators, and if it has enough energy to carry out very vigorous courtship displays, it must have a high level of overall fitness. If females use showy plumage and particularly vigorous and spectacular courtship dances for **mate selection** they will tend to have offspring fathered by males with greater overall fitness. Natural selection therefore causes these traits to become more and more exaggerated.

5. Breeding strategies in salmon

Coho salmon breed in rivers that discharge into the North Pacific Ocean. About a year after spawning, young fish migrate to the ocean where they remain for several years before returning to breed. Males that grow rapidly and are able to return two years after they were spawned are called **jacks**. Males that grow less rapidly remain in the ocean for one year longer, but are then significantly larger and are **hooknoses**. Jacks and hooknoses adopt different breeding strategies to maximize their **chances of survival and reproduction**. Hooknoses fight each other for access to females laying eggs, with the winner shedding sperm over the eggs to fertilize them. Jacks are unlikely to win fights, so avoid them and instead sneak up on females and attempt to shed sperm over their eggs before being noticed. The larger hooknoses are unlikely to sneak up on a female without being noticed, so must fight other hooknoses and fend off jacks if they are to be successful in breeding.

6. Synchronized oestrus in lionesses

Female lions remain in the pride (group) into which they were born, but males are expelled when 3 or so years old. Adult males can only breed if they overcome the dominant male in another pride by fighting. Sometimes two or more closely related young males together fight for dominance of a pride. This increases their chance of success, especially if they are fighting a single dominant male. All females in a pride tend to come into oestrus and mate at the same time. This has several benefits: females have their cubs at the same time so are all lactating while the cubs are suckling and can suckle each other's cubs when they are hunting, increasing the cubs' chance of **survival**. Also a group of male cubs of the same age are ready to leave the pride at the same time so can compete for dominance of another pride more effectively, increasing their chance of **reproduction**.

7. Feeding on cream by blue tits

Blue tits were first observed in the 1920s pecking through the aluminium caps of milk bottles left outside houses. Amateur birdwatchers followed the spread of this behaviour, in both blue tits and great tits, across Europe from England to the Netherlands, Sweden and Denmark. The rapid spread of the behaviour pattern shows that it must be due to **learned** rather than innate behaviour. Newspaper articles recently reported that blue tits had stopped feeding on cream from milk bottles. Much less milk is now delivered to doorsteps because milk in supermarkets is cheaper. Also skimmed milk, without cream at the top, has become popular with humans. This may explain why blue tits have lost this learned behaviour pattern.

Questions – neurobiology and behaviour

1. The diagram shows part of the CNS.

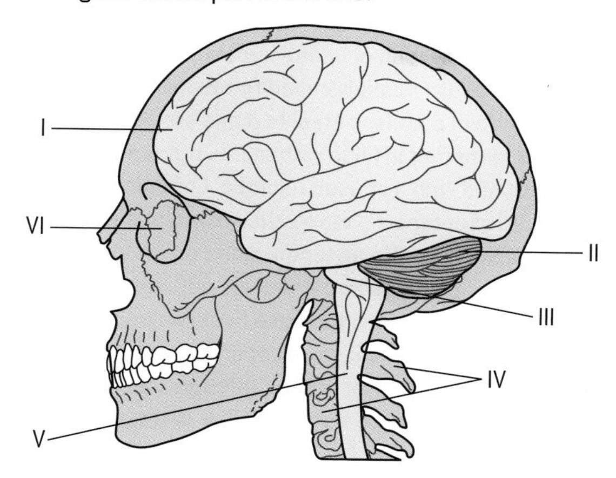

   a) State the name of structures I to V. [5]

   b) Outline the functions of structures I to III. [6]

   c) Explain the development of structure V in an embryo. [3]

   d) Outline the types of receptor in the organ occupying the position marked VI. [3]

   e) State three types of procedure that are used to investigate the functions of parts of structure I. [3]

2. The graph below shows the density of rods and cones across the retina.

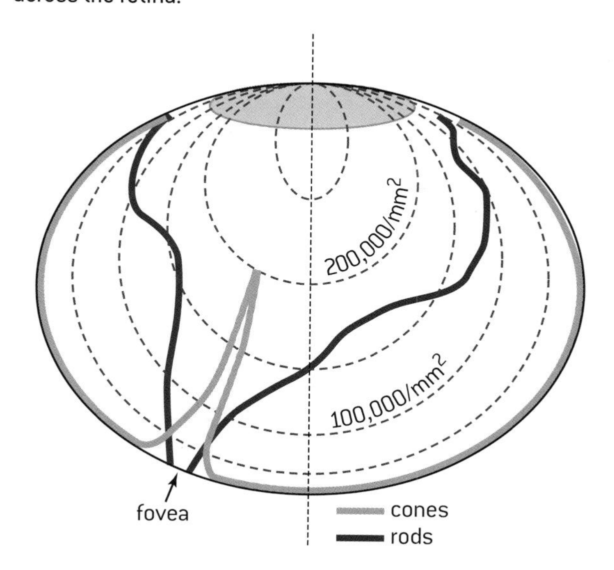

   a) Distinguish between the distribution of rod and cone cells across the retina. [3]

   b) Distinguish between the roles of rods and cones. [4]

   c) Compare and contrast processing in the retina of visual stimuli in rods and cones. [3]

3. A pedestrian who uses cochlear implants to help with deafness is waiting to cross a road. Vehicles are approaching from the right.

   a) Explain which side of the brain is responsible for

      (i) the pedestrian seeing approaching vehicles [2]

      (ii) the pedestrian hearing approaching vehicles. [2]

   b) The pedestrian turns their head to the right. Explain how this movement is detected by sensory receptors, apart from those in the eye. [2]

   c) A garbage truck drives past the pedestrian. Explain how the smell of the garbage is detected. [2]

   d) Explain how the cochlear implants help the pedestrian to hear. [4]

4. (HL) The diagram shows a rat in a Skinner box.

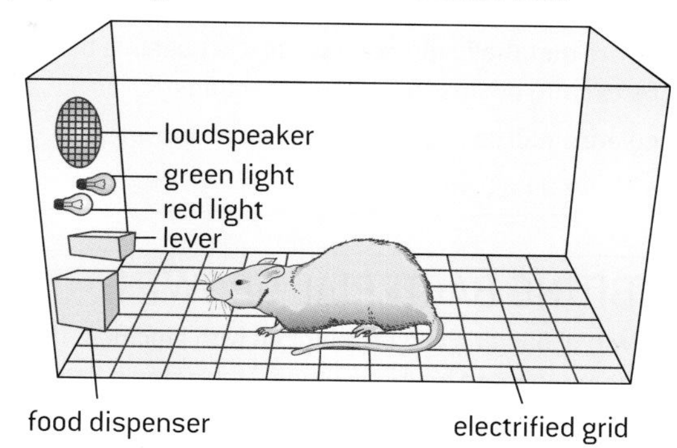

   a) State two names for the type of learning that was investigated using Skinner boxes. [2]

   b) Outline the role in this type of learning of

      (i) the food reward [2]

      (ii) pressing the lever. [2]

   c) Design an experiment into operant conditioning involving three labelled structures within the Skinner box. [6]

5. (HL) To investigate the effects of MDMA (ecstasy), healthy human volunteers were given one of three different drugs:

   - citalopram (which inhibits reuptake of serotonin from synapses into pre-synaptic neurons and reduces secretion of serotonin from them),
   - ketanserin (which binds to serotonin receptors and blocks them), or
   - haloperidol (which binds to dopamine receptors and has opposite effects to dopamine).

   The volunteers were then given 1.5 mg/kg of MDMA.

   Citalopram markedly reduced most of the subjective effects of MDMA, including positive mood, increased extroversion and self-confidence. Haloperidol selectively reduced MDMA-induced positive mood. Ketanserin selectively reduced MDMA-induced perceptual changes and emotional excitation.

   a) Subjective effects can only be perceived by the person who takes a drug. Suggest difficulties of investigating the subjective effects of drugs. [2]

   b) List three subjective effects of MDMA. [3]

   c) State the evidence from this research for the subjective effects of MDMA being mediated through serotonin and dopamine metabolism. [5]

6. (HL) a) Outline what is studied by ethologists. [2]

   b) Using the examples of migration in blackcaps and feeding on cream from milk bottles by blue tits,

      (i) discuss whether patterns of behaviour are learned or genetic. [3]

      (ii) deduce whether behaviour patterns spread through populations faster if learned or genetic. [2]

   c) Explain the advantages of synchronized oestrus in female lions. [4]

Microorganisms and fermenters

REASONS FOR USING MICROORGANISMS

Microorganisms are organisms that are too small to see without magnification: bacteria, fungi and some protoctista. Microorganisms are often referred to simply as **microbes**. They are very widely used in industry for these reasons:

- they are **metabolically diverse**, so it is possible to find a type to carry out many different reactions
- they are small so large numbers can be grown
- they have a fast growth rate.

INHIBITING BACTERIAL GROWTH

The growth of bacteria can be inhibited with **biocides**. This can be demonstrated by this method: Make sterile Petri dishes containing nutrient agar. Spread a pure culture of a bacterium over the surface of the agar. Place paper discs soaked in biocide on the agar surface, or cut wells in the agar and fill with biocide. Incubate the dishes at the optimum temperature for the bacterium and examine after 36 hours.

This technique is used to test which antibiotics kill the bacteria causing a patient's disease. Clear areas are zones of inhibition of bacterial growth (see page 66).

GRAM STAINING

Differences in bacterial resistance to biocides are due to differences in metabolism or to the structure of the cell wall. There are two main types of wall structure, called Gram-positive and Gram-negative.

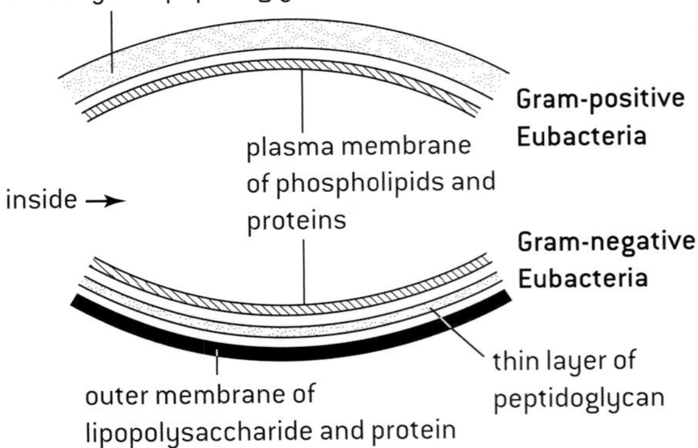

Procedure for Gram staining:

- smear a small sample of a pure bacterial culture on a microscope slide with an inoculating loop
- pass through a flame to fix the bacteria to the slide
- stain with crystal violet for 30 seconds
- treat with Gram's iodine for 30 seconds (to bind crystal violet to the outer surface of the bacteria)
- decolorize with alcohol for 20 seconds (to dissolve the outer membrane of Gram-negative bacteria and remove the crystal violet staining)
- counterstain with safranin (which is red) for 30 seconds, then rinse and blot dry.

Under the microscope Gram-negative bacteria will be red or pink. Gram-positive bacteria will be violet.

FERMENTERS

Large-scale production of useful substances by microbes requires the use of vessels called **fermenters**. They are usually made of stainless steel to make sterilization easy. The fermenter is filled with sterile nutrient medium and inoculated with a chosen microbe. Conditions are maintained at optimal levels for the growth of the microbe. Conditions such as pH and temperature are monitored in the fermenter using **probes** and levels are adjusted if they move too far from the optimum. Because heat can build up as a waste product of metabolism, a cooling jacket surrounds the vessel with cool water flowing through. Sedimentation of microbes is prevented by an impeller (a rotating set of paddles). Sterile air is bubbled through if the desired metabolic process is aerobic. A pressure gauge detects gas build-up and allows waste gases to escape. Other waste products may build up in the medium and eventually limit the fermentation.

There are two main types of fermentation. Nutrients are only added at the start with **batch** fermentation and when the yield has reached a maximum the fermenter is drained to harvest the product. In **continuous culture** nutrients are added during the fermentation, so they do not run out and the product is harvested during the fermentation.

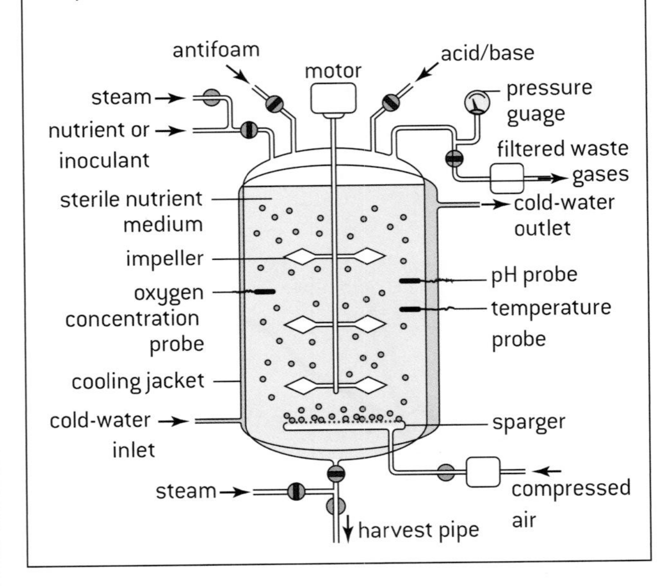

PATHWAY ENGINEERING

The useful product of microbes grown in fermenters is the **metabolite of interest**. Detailed knowledge and analysis of metabolic pathways allows scientists to change conditions at multiple points to improve the yield of the metabolite of interest. This is called **pathway engineering**.

Extra substrates may be added, by-products that slow down the pathway may be removed and the range of products may be extended.

In many cases genetic engineering is used to introduce extra genes or change how the expression of existing microbial genes is regulated.

Microorganisms in industry

PENICILLIN

Alexander Fleming discovered penicillin in the 1920s by **chance**. A Petri dish on which he was growing bacteria became contaminated with *Penicillium* and bacterial colonies near the fungus died. Fleming realized that a chemical produced by the fungus was acting as a biocide. This is one of the best-known examples of **serendipity** in biology.

In the 1940s methods for mass producing penicillin by **deep-tank fermentation** were developed. The fungus *Penicillium* is an obligate aerobe so oxygen is bubbled through the fermenter, with paddles to distribute it evenly. Optimum conditions are 24 °C and slightly alkaline pH. The nutrient source is corn steep liquor. Penicillin is a secondary metabolite that is only produced if nutrient concentrations are low, so **batch culture** is used. Initially high nutrient concentrations stimulate the fungus to grow. About 30 hours after the start of the batch culture nutrient concentrations have dropped so penicillin production starts and continues for about 6 days, after which the fermenter is drained and the liquid filtered. Solvents are used to precipitate the penicillin.

CITRIC ACID

Citric acid is a food additive, used as a **flavour enhancer** and **preservative**. Industrial production of citric acid relies on the fungus *Aspergillus niger*. While most industrially produced citric acid is made by batch fermentation, **continuous fermentation** is also sometimes used. The optimal conditions are high dissolved oxygen and sugar concentrations, an acidic pH and a temperature of about 30 °C. Citric acid is produced in the Krebs cycle, so is a primary metabolite. If the culture medium is under-supplied with minerals such as iron, citric acid builds up in the fermenter and can be harvested by draining off fluid, filtration, then precipitation by adding calcium hydroxide.

BIOGAS

Biogas is the combustible gas produced in a fermenter by the anaerobic breakdown of organic matter such as manure, waste plant matter from crops and household organic waste. Depending on the construction of the fermenter, biogas is mostly methane with some carbon dioxide. A series of processes is carried out by different bacteria:

- conversion of raw organic waste into organic acids, alcohol, hydrogen and carbon dioxide;
- conversion of organic acids and alcohol to ethanoic acid, carbon dioxide and hydrogen;
- production of methane by reducing carbon dioxide with hydrogen or splitting ethanoic acid:

$$CO_2 + 4H_2 \rightarrow CH_4 + 2H_2O$$
$$\text{or } CH_3COOH \rightarrow CH_4 + CO_2$$

Bacteria that produce methane are **methanogens**.

The figure below shows a simple biogas generator. Mylar balloons are the type that are filled with helium as party balloons. The feedstock bottle should be plastic rather than glass due to the risk of explosion. The tube clamps can be used to prevent gas leakage when the balloon is disconnected. Insulating tape seals the balloon to the tube junction.

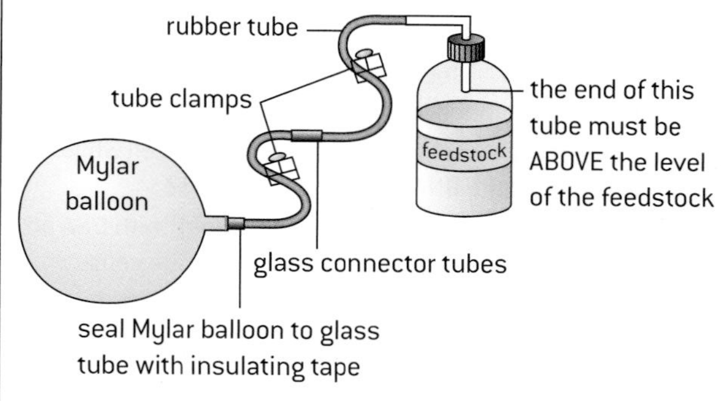

TRANSGENIC ORGANISMS

The **proteome** of a species is the complete set of proteins that it can produce. It depends on the genes that are in the **genome**. The proteome can be altered by genetic engineering. A new gene is added to the genome, which is expressed to produce a protein not previously in the proteome. An organism is **transgenic** if it has been genetically modified with a gene from another organism.

The new gene that is expressed is the **target gene**. If the transgenic organism is to be useful it is important that the target gene is expressed at an appropriate rate when required and not at other times. This is achieved by linking it to other base sequences that control its expression. Marker genes are also linked to the target gene to indicate whether it has been successfully taken up by the intended transgenic organism. For example, a gene for antibiotic resistance is used to indicate whether cells have taken up a target gene. Cells that have not are killed when treated with the antibiotic.

Bioinformatics makes it relatively easy to find target genes in other species, as explained on this page (right, open reading frames) and in Topics 8 and 9.

OPEN READING FRAMES

A **reading frame** is a sequence of consecutive, non-overlapping codons in DNA or RNA. There are three possible reading frames in any piece of DNA or RNA. The start codon determines which of these should be used. An **open reading frame** (ORF) is a length of DNA from a start codon to a stop codon that is long enough to code for a polypeptide. One hundred or more triplets of nucleotides are usually expected in an ORF. They can be on either strand of DNA but the start codon must be at the 5′ end of the DNA sense strand and therefore also at the 5′ end of mRNA. The start codon is AUG (in mRNA) and the three possible stop codons are UAG, UAA and UGA.

Example:

DNA 5′	T	G	C	G	A	T	G	A	C	T	T	A	A	C	G
DNA 3′	A	C	G	C	T	A	C	T	G	A	A	T	T	G	C
RF1 5′	U	G	C	G	A	U	G	A	C	U	U	A	A	C	G
RF2 5′		G	C	G	A	U	G	A	C	U	U	A	A	C	G
RF3 5′			C	G	A	U	G	A	C	U	U	A	A	C	G

Reading frame 1 (RF1) does not contain stop or start codons. RF2 contains a start codon but soon after a stop codon so is not an ORF. RF3 includes a stop codon but no start codon.

Genetic modification of crop plants

AIMS OF CROP GENETIC MODIFICATION

1. **Novel products** – a gene is inserted to allow crop plants to make something that they could not before.
2. **Overcoming environmental resistance** – genes are inserted to give greater tolerance to saline soils, frosts, or some other factor limiting crop growth.
3. **Pest resistance** – a gene is inserted for making a toxin that kills pests eating the crop – see Topic 3.
4. **Herbicide resistance** – a gene is inserted so a herbicide can be used without killing the crop plants.

TECHNIQUES OF GENETIC MODIFICATION

DNA containing the target gene, marker gene and sequences for controlling gene expression is prepared. This is known as **recombinant DNA**. For successful genetic modification it must be inserted into a cell of the crop plant and be taken up by a chromosome in the nucleus or by the DNA of the chloroplast.

In some cases the recombinant DNA is introduced into a **whole plant**. In other cases it is introduced into a leaf disc, or into a protoplast which is a single cell from which the cell wall has been removed.

The DNA can be introduced by different methods.

Direct physical methods:

- **electroporation** – electric fields cause pores to open briefly in membranes so DNA can enter cells
- **microinjection** – one micropipette holds the cell while another injects the DNA through a tiny needle
- **biolistics** (gunshot) – tiny metal balls (gold) with DNA on the surface are fired at the plant and penetrate cells

Direct chemical methods:

- **calcium chloride** – cells incubated in a cold $CaCl_2$ solution take up DNA when given a heat shock
- **liposomes** – artificial vesicles containing the DNA fuse with the cell membrane of protoplasts

Indirect methods using vectors:

- ***Agrobacterium tumefaciens*** – a bacterium that inserts a plasmid into plant cells with the target gene
- **tobacco mosaic virus** – a virus that inserts RNA into plant cells, with the RNA including the target gene.

AMFLORA POTATO

Potato starch is used as an adhesive and a coating for paper. It is the amylopectin form of starch that is useful for these purposes. Granule-bound starch synthase (GBSS) is an enzyme used to make amylose but not amylopectin.

A gene was inserted into potato cells with the same base sequence as the gene for GBSS, but in reverse. The mRNA transcribed from it therefore has a base sequence complementary to the base sequence of mRNA transcribed from the normal GBSS gene. It has the **antisense** sequence rather than the sense sequence. The antisense mRNA pairs with the **sense** mRNA to form double-stranded RNA. This cannot be translated by ribosomes and GBSS is not produced, so more than 99% of starch made by the potato plants is amylopectin. The new genetically modified variety of potato was named **Amflora**.

PRODUCTION OF HEPATITIS B VACCINE IN TOBACCO PLANTS

Tobacco mosaic virus (TMV) enters tobacco cells and then uses the metabolism of the cell to translate some of its genes into proteins and to replicate the RNA that is its genetic material. It can spread from cell to cell to infect a whole tobacco plant.

If TMV is genetically modified, the novel genes are expressed in infected tobacco cells.

Vaccination programmes in remote areas are difficult because of problems with access and refrigeration of vaccines. Hepatitis B vaccine contains Hepatitis B small surface antigen (HBsAG).

TMV has been genetically modified with the gene from the Hepatitis B virus for making HBsAG. Tobacco plants are then infected with the genetically modified TMV, to try to produce HBsAG in bulk. The tobacco plants are harvested and dried. If a person eats some of the dried material the HBsAG in it should stimulate production of antibodies against Hepatitis B and therefore induce immunity to the disease. This is an easier way to give a vaccine than sterile injection of a liquid vaccine that has to be refrigerated.

GLYPHOSATE RESISTANCE IN SOYBEANS

General herbicides (weedkillers) such as glyphosate kill all plants so cannot normally be sprayed onto growing crops, but a gene for glyphosate resistance has been transferred to soybeans and other crops, making this possible. The gene was transferred using a strain of the bacterium *Agrobacterium tumefaciens* that contains a tumour-inducing plasmid (Ti plasmid). The bacterium injects the Ti plasmid through a pilus into plant cells and DNA from it becomes incorporated into chromosomes in the nucleus. A glyphosate resistance gene was inserted into the Ti plasmid along with a kanamycin resistance gene. The recombinant Ti plasmids were reinserted into *A. tumefaciens* and sections of soybean leaf were exposed to these bacteria. Kanamycin was used to kill leaf cells that had not taken up the Ti plasmid and a glyphosate-resistant variety of soybeans was developed from surviving cells.

The introduction of glyphosate resistant varieties of crop plants has been controversial. Potential benefits include the reduced need for weed control by ploughing. Ploughing increases soil erosion and has harmful effects on the soil community. Lower concentrations of glyphosate may be sufficient to control weeds and other more harmful herbicides may not have to be applied. There also significant risks, especially the possible escape of glyphosate resistance genes into wild populations of plants. There have already been problems with glyphosate resistant strains of weeds that could previously have been controlled with this herbicide, but now cannot. This is an agricultural rather than an environmental problem. Careful evaluation of environmental risks and benefits is essential before decisions are made about the use of genetically modified crops.

Bioremediation

USING MICROBES IN BIOREMEDIATION

Bioremediation is the use of microbes to remove environmental contaminants from water or soil. The metabolic diversity of microbes allows a wide range of contaminants to be treated by bioremediation. The contaminant is absorbed by the microbe and used in its metabolism, with non-toxic waste products released back into the environment. Some contaminants are used as an energy source in cell respiration in the microbe. Bioremediation is sometimes combined with other procedures. **Physical** methods include removal of oil floating on water using skimmers. **Chemical** treatments include injecting oxidizing chemicals such as ozone or hydrogen peroxide into soils to destroy toxins.

EXAMPLES OF BIOREMEDIATION

1. **Benzene**

Offshore oil wells generate large volumes of saline wastewater contaminated with hydrocarbons. Benzene is of particular concern as it can persist in the environment for a long time, is moderately soluble in water and is carcinogenic. Bioremediation is difficult as the saltwater kills most bacteria. Some Archaea are adapted to live in highly saline water (halophiles). *Marinobacter hydrocarbonoclasticus* is a halophilic archaean that degrades benzene.

2. **Crude oil**

Where oil occurs in rock near the ocean floor, it can seep into the water through cracks and vents. Some members of the genus *Pseudomonas* can use the crude oil as an energy and carbon source. Clean-up at oil spills often involves seeding the spill with *Pseudomonas*. These microbes also need nutrients such as potassium and urea to metabolize the oil at a faster rate so they are sprayed on to the oil spill to aid the bacteria in their work.

3. **Methyl mercury**

Mercury ends up in garbage dumps as a component of certain paints and light bulbs. Elemental mercury is converted in dumps into the highly toxic compound methyl mercury by the bacterium *Desulfovibrio desulfuricans*. This form of mercury adheres to cell membranes and then dissolves in them. It is not easily removed, so builds up in food chains by biomagnification.

The bacterium *Pseudomonas putida* can convert the methyl mercury to methane and mercury ions. Other bacteria then use the soluble mercury ion as an electron acceptor resulting in insoluble elemental mercury being reformed. If this process is carried out in a bioreactor, the elemental mercury can be separated from waste water as it is insoluble and sinks due to its density.

BACTERIOPHAGES AND WATER TREATMENT

Biofilms can form inside pipes or other parts of water supply systems. Bacteria in the interior of these biofilms can be resistant to disinfectants. Viruses that kill bacteria (bacteriophages) are used increasingly to remove biofilms. Biofilms of *E. coli* are removed by the bacteriophage T4 for example.

BIOFILMS

Although bacteria can exist as single cells, some species also form **cooperative aggregates**. For example, layers of bacteria called **biofilms** can form on rocks or other surfaces. The cells jointly secrete an extracellular matrix of adhesive polysaccharides, sticking the cells to the surface and to each other. Single cells cannot produce enough polysaccharide for efficient adhesion.

This is an example of **quorum sensing** where the density of a population triggers particular types of behaviour. Each cell secretes signalling molecules that bind to receptors on other cells and if the population density is low, not enough of the signal is received to trigger secretion of extracellular polysaccharide.

Microbes in biofilms are sometimes very resistant to antimicrobial agents because the polysaccharide matrix acts as a physical barrier to penetration. Antibiotics that kill dividing cells sometimes fail to work due to reduced cell division rates in biofilms. This increased resistance to antimicrobials is an example of an **emergent property**. Biofilms have several emergent properties, but multicellular organisms have more.

The structure of biofilms is being investigated with **laser-scanning microscopes**. This type of microscope was developed in the latter part of the 20th century. It allows high-resolution images to be obtained at different depths in a specimen, so the three-dimensional structures can be determined. Using laser-scanning microscopes researchers have been able to obtain a deeper understanding of the structure of biofilms.

Biofilms can cause environmental problems such as clogging and corrosion of pipes, transfer of microbes in ballast water on ships and contamination of surfaces in food production areas. There have been many media reports about these problems recently and, as with all such reports, it is important to evaluate the scientific evidence for any claims that are made.

BIOFILMS IN SEWAGE TREATMENT

The diagram shows a trickle filter bed for sewage treatment.

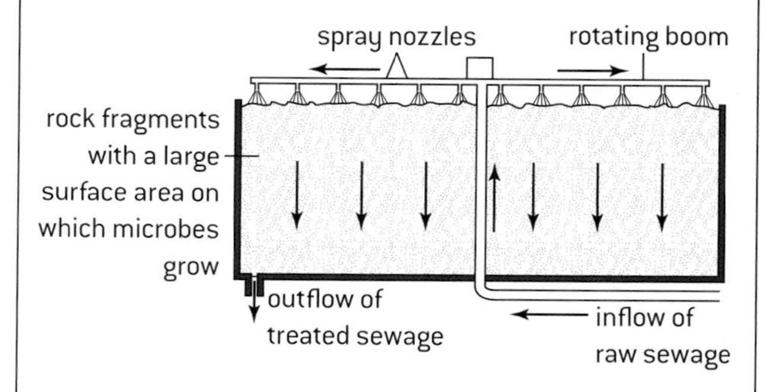

Biofilms form on the surface of the rock fragments. They contain decomposers, which digest organic matter in the sewage into inorganic compounds such as ammonia, and also nitrifying bacteria that convert ammonia to nitrates.

Biotechnology in diagnosis (HL only)

DETECTING DISEASE

1. **Metabolic diseases** are detected by the presence of specific metabolites in blood or urine. For example diabetes is detected by the presence of high concentrations of glucose and PKU is detected by elevated levels of phenylalanine.
2. **Predisposition to a genetic disease** can be detected by the presence of genetic markers. These are particular alleles that either contribute to the disease or are genetically linked to genes that influence the disease. The marker is detected by various methods such as **microarrays** or **PCR** combined with **gel electrophoresis**.
3. **Infectious diseases** are caused by pathogens. It is important to be able to identify the precise strain of a pathogen that is causing a disease. This is done by detecting the presence of the pathogen's genetic material, for example with **microarrays** or **PCR**. It can also be done by detecting a chemical produced by the pathogen that acts as an antigen, for example with the **ELISA** test.

   The diagnostic methods used for detecting genetic material and antigens have been improved greatly, but scientists continue to look for technological innovations that could be used to improve the diagnosis or treatment of diseases.
4. **Tumours** can be located using **tracking experiments** with fluorescent (luminescent) probes.

MICROARRAYS

A **microarray** is a small surface with a large range of DNA probe sequences adhering to it. Millions of probes per square centimetre may be present. Microarrays are designed to test for specific mRNA sequences in tissues. Reverse transcriptase is used to make a DNA copy (cDNA) of the base sequence of each type of mRNA in a tissue. Fluorescent dye is linked to each cDNA. The microarray is exposed to the cDNAs long enough for hybridization between fixed probes and cDNAs that have complementary base sequences. The microarray is then rinsed to remove cDNA that has not hybridized.

The microarray is exposed to laser light, which causes the fluorescent dye to give off light. This shows which probes on the microarray have hybridized with cDNA and thus which mRNA sequences there were in the tissue.

Patterns of gene expression in two tissues can be compared by preparing cDNA samples from their mRNA with different fluorescent dyes to mark them. If green and red dyes are used, these colours on the microarray indicate the presence of an mRNA in one tissue but not the other and yellow indicates that the mRNA was in both tissues.

Analysis of a microarray thus involves examining the pattern of dots of the different possible colours. The pattern changes as gene expression changes. There are marked changes in gene expression when cells become tumour cells, allowing detection and characterization of cancer. More subtle differences can also be used to test an individual for genetic predisposition to diseases or to test for a specific genetic disease.

PCR

The basic procedure for PCR is described in Topics 2 and 3. A modified version can be used to detect different strains of the influenza virus. This virus uses RNA as its genetic material. Cells are taken from an infected patient and the RNA is extracted from it. This will include both influenza RNA and the patient's own mRNA. Reverse transcriptase is used to produce cDNA copies of all the RNA in the sample. Primer sequences specific to the strain of influenza being tested for are then added and PCR is carried out. If this strain was present in the patient, many copies of double-stranded DNA are produced, with viral base sequences on one of the strands. Fluorescent dyes that bind to double-stranded DNA can be used to detect this DNA.

ELISA

This test can be used to detect antigens specific to a pathogen or antibodies indicating infection with the pathogen. It involves adsorption of antibodies to antigens or vice versa and also a colour change caused by an enzyme, so is called **enzyme-linked immunosorbent assay** (ELISA). If antigens are the **target molecule** of the test, antibodies that bind to the antigen are used as the **capture molecule**, and vice versa if antibodies are the target molecule. The test involves these steps:

1. Capture molecules are linked to the surface of a well on a plastic microtitre plate. The plate has many small wells, each of which is used for one test.
2. The sample being tested is placed in a well to expose it to the surface and let any target molecule bind to a capture molecule (immunosorbence).
3. Antibodies that have been linked to an enzyme are added. These antibodies bind to any target molecules that are adsorbed to capture molecules.
4. The surface is rinsed to remove all enzymes that are not bound to adsorbed target molecules.
5. A substrate is added that the enzyme changes to a different colour, indicating the presence of the target molecule.

Interpretation of the results of ELISA tests is quite straightforward. If the colour change has occurred in a well, the test result is positive and the target molecule (antigen or antibody) was present in the sample. The stronger the colour, the more of the target molecule was present.

TRACKING EXPERIMENTS

The movement of specific proteins in the body can be followed using tracking experiments in which a probe is attached to the protein. Radioactive probes can be located using a PET scan and fluorescent dyes used as probes are located with a microscope. For example, tumour cells have more receptors for the protein transferrin in their plasma membranes than normal body cells, so if fluorescent dyes are attached to transferrin, tumour cells in a sample are revealed by fluorescence on the cell surface. This method may help with difficult-to-diagnose cancers.

Biotechnology in therapy (HL only)

BIOPHARMING

Use of genetically modified bacteria for production of insulin is described in Topic 3. Some proteins that have therapeutic uses cannot easily be made using bacteria because post-translational modifications carried out by a Golgi apparatus do not occur. An example of a modification is addition of sugars to convert a protein to a glycoprotein.

Genetically modified eukaryotes are used to make these proteins. Animals and plants have both been used, either whole or in cell cultures. This branch of biotechnology is whimsically called **biopharming**.

Example of biopharming: **antithrombin**

Antithrombin is a glycoprotein that regulates blood coagulation. Antithrombin deficiency is a genetic disease that results in excessive blood clot formation. It can be treated using antithrombin from donated blood, but the supply has been increased hugely by production in goats. The gene for human antithrombin was inserted into goat embryos by microinjection, together with regulatory sequences that ensure the gene is only expressed in the mammary glands of lactating female goats. The antithrombin is therefore secreted in milk and can easily be purified from it.

CHROMOSOME 21 AND ENSEMBL

Chromosome 21 is of particular interest because a person with three copies of its genes has Down syndrome. Ensembl software can be used to explore information about the genes on this chromosome in databases. Sequence data stored in databases is increasing exponentially. Scientists throughout the world can get access to databases easily via the internet – an example of cooperation and collaboration.

Ensembl is a database and genome browser that collates information for 75 organisms and facilitates analysis of the coding and non-coding sequences of each of the chromosomes from these species. To explore the capabilities of Ensembl, open the Ensembl website and choose 'Human' then 'View karyotype' then 'chromosome 21' and 'Chromosome summary'. Ensembl shows the location of protein coding genes on the chromosome and other genes.

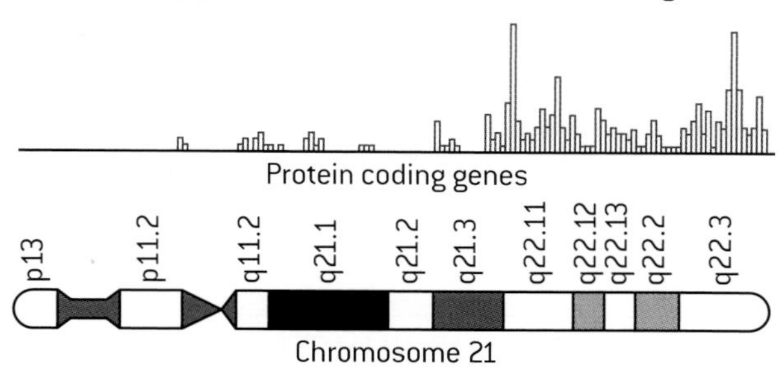

Ensembl allows the genes on human chromosome 21 to be compared with the equivalent chromosome in other species, helping investigate functions of genes and perhaps therapies for Down syndrome.

VIRAL VECTORS AND GENE THERAPY

In gene therapy, working copies of a defective gene are inserted into a person's genome. In **somatic therapy** the modified cells are somatic (body) cells. In **germ line therapy** therapeutic genes are introduced into egg or sperm cells so the missing gene is expressed in all cells of organisms derived from these gametes.

Viruses have had millions of years to evolve efficient mechanisms for entering mammalian cells and delivering genes to them. They sometimes also incorporate these genes into the host cell's chromosomes. Viruses are therefore obvious candidates for the gene delivery system, needed in gene therapy. Modified viruses must be produced containing the desired gene, which will infect target cells but which are not virulent because they will not replicate to form more virus particles. A modified virus that is used in this way is called a **viral vector**. Retroviruses are the most widely used viral vectors. Adenoviruses are also sometimes used; they do not insert DNA into the host cell's nucleus, which avoids some potential problems but means that the gene is not routinely passed on to the next generation of cells by mitosis so treatment has to be given more frequently. A challenge of using any virus as a vector is that the host may develop immunity to it.

One example of the use of viral vectors is in the treatment of **SCID** (severe combined immuno-deficiency), a genetic disease that is due to the lack of an enzyme called ADA. A famous early case involved a baby called Andrew (right).

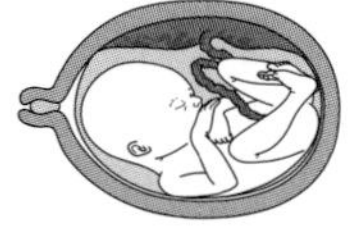

Genetic screening before birth shows that Andrew has SCID

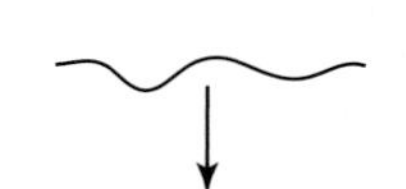

The allele that codes for ADA is obtained. This gene is inserted into a retrovirus

Blood removed from Andrew's placenta and umbilical cord immediately after birth contains stem cells. These are extracted from the blood

Retroviruses are mixed with the stem cells. They enter them and insert the gene into the stem cells' chromosomes

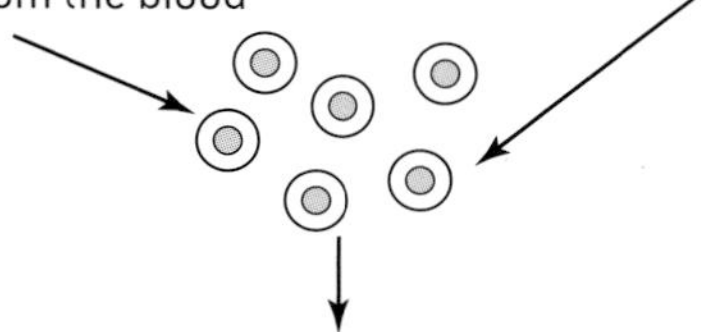

Stem cells containing the working ADA gene are injected into Andrew's blood system via a vein.

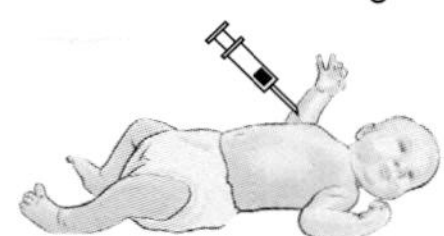

For four years T-cells (white blood cells), produced by the stem cells, made ADA enzymes, using the ADA gene. After four years more treatment was needed.

Bioinformatics (HL only)

BLAST SEARCHES

Because of common ancestry, similar base and amino acid sequences are often found in different organisms. As a result of polyploidy and gene duplication there are sometimes two or more similar sequences in one organism. The Basic Local Alignment Search Tool (**BLAST**) can be used to find similar sequences in databases. This software identifies the sequences and displays them alongside each other (**sequence alignment**), allowing analysis of similarities and differences. There are different versions: **BLASTn** is used to align nucleotide sequences in nucleic acids; **BLASTp** is used to align amino acid sequences in proteins. A typical use of BLAST is this:

- A researcher finds an open reading frame (ORF) in an organism's DNA but is unsure whether it is a protein coding gene.
- The amino acid sequence corresponding to the base sequence is deduced.
- If BLASTp identifies similar amino acid sequences in other organisms, the ORF is probably a protein-coding gene.
- If the similar sequences have known functions, the protein coded for by the new gene probably has the same or a similar function.

To compare two protein sequences, go to www.ncbi.nlm.nih.gov and explore the current versions of BLAST. Find the GI (sequence identification number) of two potentially similar proteins. Choose 'BLASTp' and then 'Align two (or more) sequences'. Enter the GI codes and then click on BLAST. The amino acid sequences will be shown alongside each other. For example, if you enter the codes for elephant insulin (69307) and ostrich insulin (69327) the sequence alignment is this:

```
Elephant insulin  FVNQHLCGSHLVEALYLVCGERGFFYTPKTGIVEQCCTGVCSLYQLENYCN
Ostrich insulin   AANQHLCGSHLVEALYLVCGERGFFYSPKAGIVEQCCHNTCSLYQLENYCN
```

The amino acids are shown using a standard single-letter code that can easily be downloaded. For example the first amino acid in elephant insulin is phenylalanine (F) but in ostrich insulin it is alanine (A). There are 51 amino acids in each version of insulin. The amino acid sequence is identical across much of the proteins but there are some differences that can be identified by careful comparison of the sequences.

EXPLORING GENE FUNCTIONS

Model organisms are species that have been intensively researched not only because of interest in that species but because related species are likely to be similar in many ways. A list of model organisms is given in Option A. The function of a gene in an organism can be predicted if the function of a gene with a similar sequence in a model organism is known. The mouse is the mammal that has been used as a model organism so gene functions in humans are predicted from those in mice.

One approach to determining the function of genes in mice is **knockout technology**. Mice are genetically modified so they only have non-functional versions of a specific gene. From the change to the phenotype of the mouse, researchers can deduce the gene function. For example, a strain of knockout mice were produced that only had non-functioning versions of the leptin gene. These mice became very obese, showing that leptin has a role in regulating fat deposition or energy metabolism.

DATA MINING WITH ESTS

ESTs are **expressed sequence tags**. If a gene is being expressed, mRNA transcribed from it can be extracted from a cell. To make ESTs, cDNA copies of the mRNA are made using reverse transcriptase and 200 to 500 nucleotides-long sequences are copied from both the 5′ end and the 3′ end of the cDNA. These are the ESTs. The 5′ end tends to have a sequence conserved across species and gene families. The 3′ end is more likely to be unique to the gene.

ESTs can be used to find the locus of a gene within the genome – its position on a particular chromosome. They can also be used to search for similar sequences in databases of ESTs, to try to match the gene to other similar genes of known function. This is an example of **data mining**.

CONSTRUCTING CLADOGRAMS

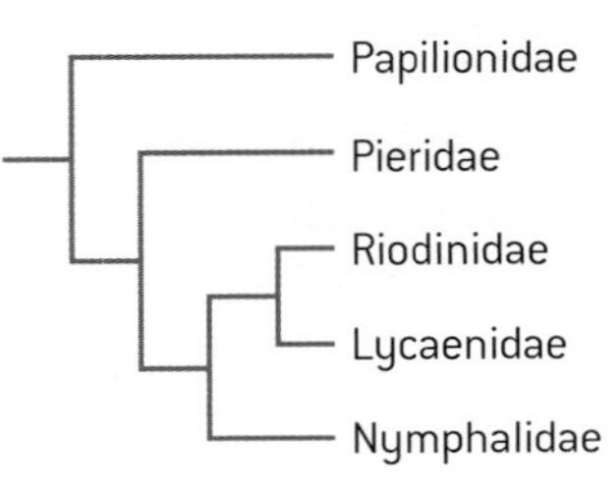

Phylogenetics is the study of the evolutionary history of groups of organisms. Bioinformatics has revolutionized phylogenetics. Sequence alignment software allows comparison of base or amino acid sequences from many organisms. Then computer software is used to construct **cladograms** (tree diagrams). The basic features of cladograms are described in Topic 5.

Some cladograms are also **phylograms**, because the length of each branch is proportional to the amount of difference. The example (above) is a cladogram for five families of butterfly but is not a phylogram.

The sequence differences between organisms in a group are the result of mutations. Computer software that produces cladograms uses the principle of maximum parsimony – the most likely origin of organisms in a group is the one involving the fewest mutations.

Cladograms do not show the evolutionary history of groups of organisms with certainty, but merely what is most probable based on the sequences used. If further cladograms based on sequences in different genes suggest the same evolutionary history, it is less likely to be falsified.

Simple cladograms of related organisms can be constructed with DNA sequences available on the NCBI website and with **ClustalX** and **PhyloWin** software. Detailed instructions are not given here as details of websites and software are evolving rapidly.

Questions – biotechnology and bioinformatics

1. The diagram shows a biogas fermenter used on farms.

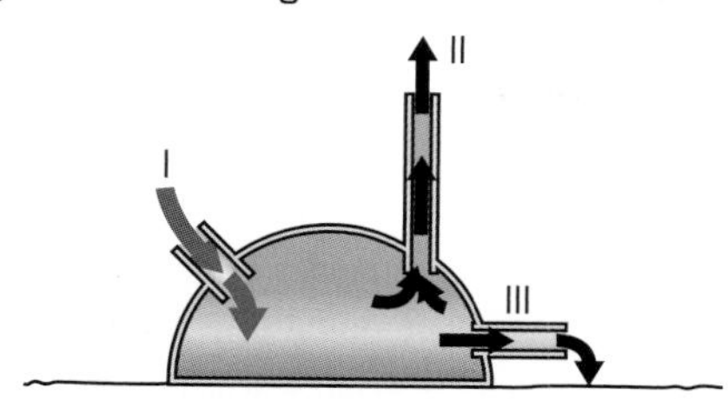

   a) Suggest two materials that might be loaded at I. [2]
   b) Outline the ideal conditions inside the fermenter. [3]
   c) Describe the substances that emerge from II and III and how they are used. [5]
   d) Outline an environmental benefit of using biogas fermenters. [2]

2. The electron micrograph shows a biofilm of *Staphylococcus aureus* bacteria inside a catheter used to drain urine continuously from a patient's bladder.

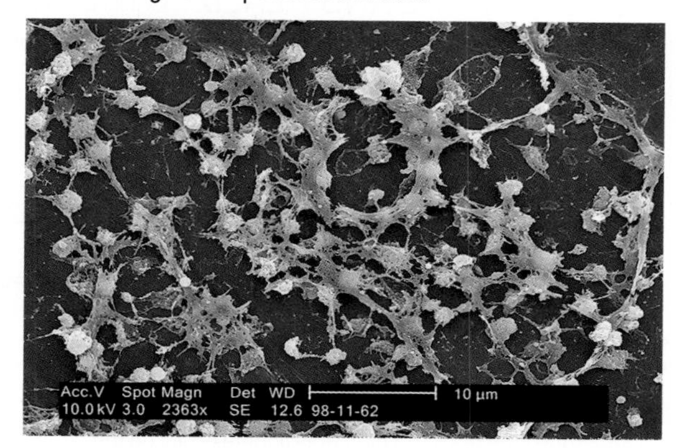

   a) (i) Calculate the magnification of the micrograph [2]
      (ii) Calculate the actual size of the bacteria. [2]
   b) Explain how the bacteria developed a biofilm. [4]
   c) Suggest two concerns about biofilms in catheters. [2]

   The graph below shows the mean depths of biofilms of *S. epidermis* with and without antibiotic treatment. The positive control received no antibiotic. The bacteria in the negative control were killed with ethanol.

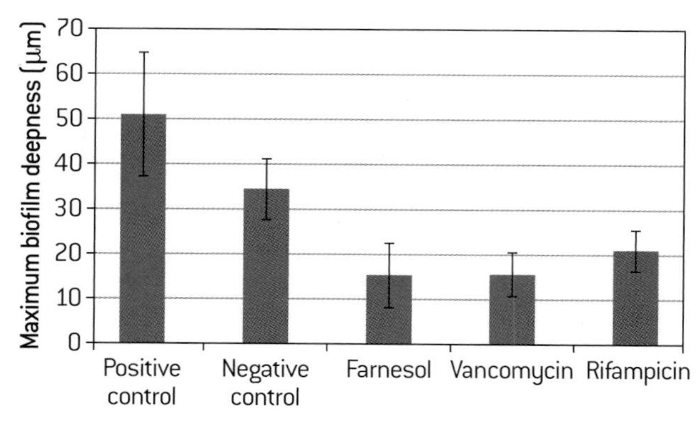

   d) Discuss the effectiveness of the three antibiotics in controlling the biofilms. [3]
   e) This research was carried out using confocal scanning laser microscopy. Suggest an advantage of this new technique for this research. [2]

3. The diagram below shows how a soy bean cell can be genetically modified using *Agrobacterium tumefaciens*.

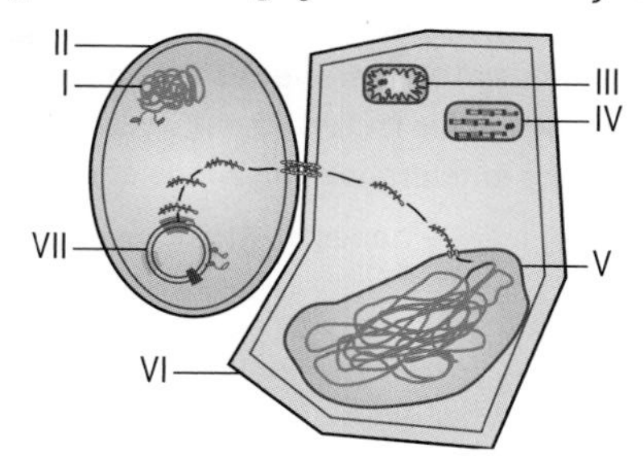

   a) Name structures I to VI. [6]
   b) Structure VII is the Ti plasmid. It contains genes needed for attachment to the host cell and for tumour induction.
      (i) State two other genes that are linked in the Ti plasmid to genetically modify soybeans. [2]
      (ii) Explain reasons for using each of these genes. [4]
   c) Outline the stages in producing GM crop plants after the stages shown in the diagram. [3]

4. (HL) The diagrams show four stages in a test for HIV antibodies that is carried out in wells on a plastic plate.

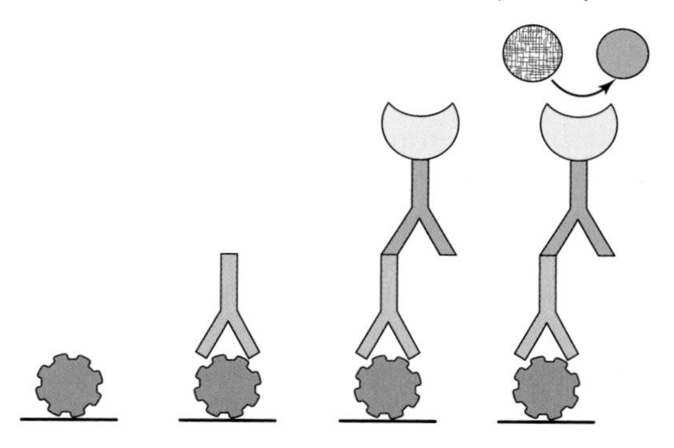

   a) State the name of this type of diagnostic test. [1]
   b) Explain what has happened in each of the four stages. [6]
   c) Distinguish between the events shown in the diagram and the outcome if the sample being tested was HIV-negative. [3]

5. (HL) (a) Compare and contrast BLASTn and BLASTp in bioinformatics. [4]
   b) Outline the use in bioinformatics of
      (i) databases [2]
      (ii) model organisms [2]
      (iii) multiple sequence alignment [2]
      (iv) knockout technology [2]
      (v) EST data mining. [2]
   c) Explain two conclusions that can be drawn from the phylogram showing three species of *Mycosphaerella* fungi, which grow on bananas. [2]

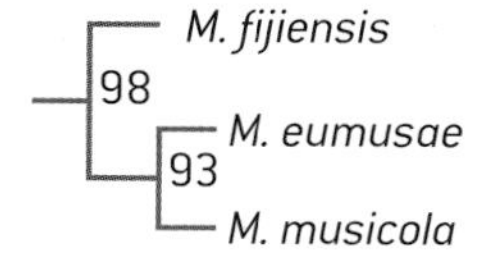

Community structure

PLANT AND ANIMAL DISTRIBUTIONS

The **distribution** of a species is the range of places that it inhabits. Plant and animal species are limited in their distributions by abiotic factors. These limits are shown in graphs as **limits of tolerance** and **zones of stress**.

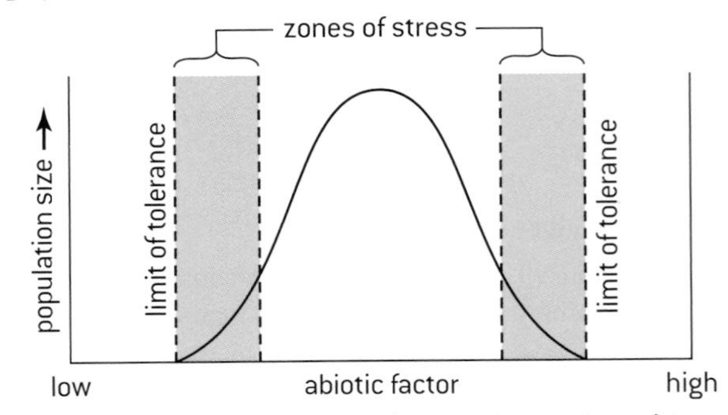

Plant example **Black mangrove** (*Avicennia germinans*) is a tree that inhabits areas with mean temperatures of 20 °C or more. It cannot survive cold and frost stress. It thrives in waterlogged soils that contain little or no oxygen with reducing conditions as low as −200 mV. It does not survive if soils periodically dry out. It tolerates a wide range of soil pH from 5.3 to 7.8 and also a wide range of soil salinity up to 90 parts per thousand (ppt), which is higher than seawater (35 ppt). Black mangrove is found in intertidal forests (mangroves) in tropical and subtropical areas with waterlogged, anaerobic soils and very variable soil salinity.

Example **Dog whelk** (*Nucellus lapillus*) is a sea snail that can survive out of water for a limited period but not for the lengths of time experienced above the high water neap line. It requires saline water and can tolerate limited increases in salinity above 35ppt when some water evaporates from rock pools, but not large increases. It requires a mean temperature between 0 and 20 °C and can survive some exposure to wave action by clinging onto fixed rocks, but not the battering on very exposed shores. Dog whelks are found on the lower to middle parts of rocky shores.

ECOLOGICAL NICHES

Every species plays a unique role in its community. The role of a species is its **ecological niche** and combines these elements:

- **spatial habitat** – where the species lives
- **interactions** – how the species affects and is affected by other species in the community, including nutrition.

If two species have a similar ecological niche, they will compete in the overlapping parts of the niche, for example for breeding sites or for food. Because they do not compete in other ways, they will usually be able to coexist. However, if two species have an identical niche they compete in all aspects of their life. One will inevitably prove to be the superior competitor and will eventually cause the other species to be lost from the ecosystem. The principle that only one species can occupy a niche in an ecosystem is called the **competitive exclusion principle**.

FUNDAMENTAL AND REALIZED NICHES

The niche that a species could occupy based on its limits of tolerance is often larger than the niche it actually occupies. The niche that a species could potentially occupy is its **fundamental niche**. The niche that it actually occupies is its **realized niche**. Differences between fundamental and realized niches are due to **competition**. Other species prevent a species from occupying part of its fundamental niche by out-competing or by excluding it in some other way. An example is used in a question at the end of this option.

KEYSTONE SPECIES

A **keystone species** has a disproportionate effect on the structure of an ecological community. Some keystone species are the direct or indirect food source for most other species in the community, for example a dominant tree species in a forest. Others are predators that have major effects on population sizes by limiting the numbers of their prey. The conservation of keystone species is essential for the overall conservation of an ecosystem.

TRANSECTS

A **transect** is a method of sampling at regular positions across an ecosystem, to investigate whether the distribution of a plant or animal species is correlated with an abiotic variable. Sampling usually involves recording numbers of individuals in quadrats positioned along the transect line. The data can be displayed in various types of chart with distance across the ecosystem as one of the axes.

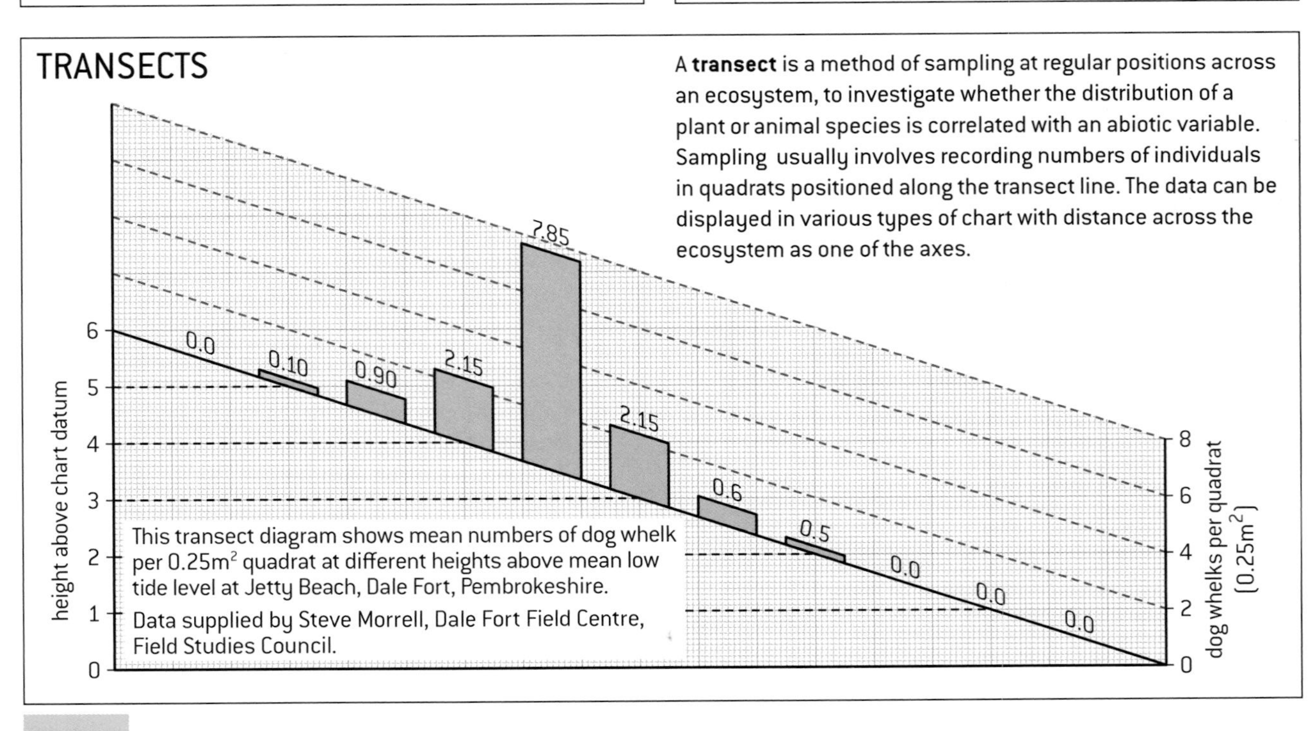

This transect diagram shows mean numbers of dog whelk per 0.25m² quadrat at different heights above mean low tide level at Jetty Beach, Dale Fort, Pembrokeshire.
Data supplied by Steve Morrell, Dale Fort Field Centre, Field Studies Council.

Interactions and energy flow

INTERACTIONS BETWEEN SPECIES

The types of interaction between species in a community can be classified according to their effects:

Herbivory – primary consumers feed on plants or other producers; this harms producers but reduces competition between producers.
Example: monarch butterfly caterpillars feed on milkweed.

Predation – predators benefit as they feed on prey; predation affects numbers and behaviour of prey.
Example: coyotes are predators of white-tailed deer.

Parasitism – a parasite that lives on or in a host, obtaining food from the host and harming it.
Example: *Ixodes* ticks use white-tailed deer as a host.

Competition – a species using a resource reduces the amount available to other species using it.
Example: red oak and sugar maple in mixed forests.

Mutualism – different species living together in a close relationship, from which they both benefit.
Example: zooxanthellae and corals – see below.

MUTUALISM IN REEF-BUILDING CORALS

Most corals that build reefs contain mutualistic photosynthetic algae called **zooxanthellae**.

The coral provides the alga with a protected environment and holds it in position close to the water surface where there is enough light for photosynthesis to occur. The zooxanthellae provide the coral with products of photosynthesis such as glucose, amino acids and also oxygen. The coral also feeds on organic particles and plankton suspended in the sea water, using its stinging tentacles. The coral's waste products are all used by the zooxanthellae: carbon dioxide, ammonia and phosphates.

The relationship between the two organisms is **symbiotic**, because they live together, and **mutualistic**, because they both benefit.

Zooxanthellae make coral reefs one of the most biologically productive ecosystems. They improve the nutrition of corals enough for the building of coral reefs by the deposition of their hard exoskeletons.

ENERGY CONVERSION RATES

Gross production is the total amount of energy in food assimilated by an animal or in food made by photosynthesis in producers.

Net production is the amount of energy converted to biomass in an organism. It is always less than gross production because some food is used in cell respiration and the energy released from it is lost from the organism and the ecosystem.

The efficiency with which a species uses food is assessed by calculating a **feed conversion ratio** (FCR):

$$\text{Conversion ratio} = \frac{\text{intake of food (g)}}{\text{net production of biomass (g)}}$$

The higher the ratio, the higher the respiration rate of the species and the lower the percentage of ingested energy that is converted to biomass.

Conversion ratios are sometimes used to assess the sustainability of food production practices. Typical ratios for meat production are shown below:

Production method	FCR
Salmon in fish farms	1.2
Chicken in broiler houses	1.9
Pork reared in housing	2.7
Beef reared on feed lot	8.8

Birds and mammals usually have high respiration rates because they maintain constant body temperatures so their FCRs are relatively high.

COMPARING PYRAMIDS OF ENERGY

Pyramids of energy can be used to model energy flow through an ecosystem. Topic 4 includes examples for a stream and for a salt marsh. They can be used to compare the two ecosystems:

- how many trophic levels are there?
- how much gross production by producers is there?
- how much energy reaches each trophic level?

The most productive ecosystems have high gross production by producers, so large amounts of energy flow to higher trophic levels and there can be relatively large numbers of trophic levels.

FOOD WEBS AND CHAINS

A **food chain** is a single sequence of organisms, each of which consumes the previous one in the chain. Most species of consumer eat a variety of other organisms, so are in many different food chains. Many species eat organisms from more than one trophic level so are themselves in different trophic levels. A **food web** diagram is used to show all the possible food chains in a community.

The example (right) is for the Arctic marine community. Polar bears feed as 3rd, 4th and 5th consumers.

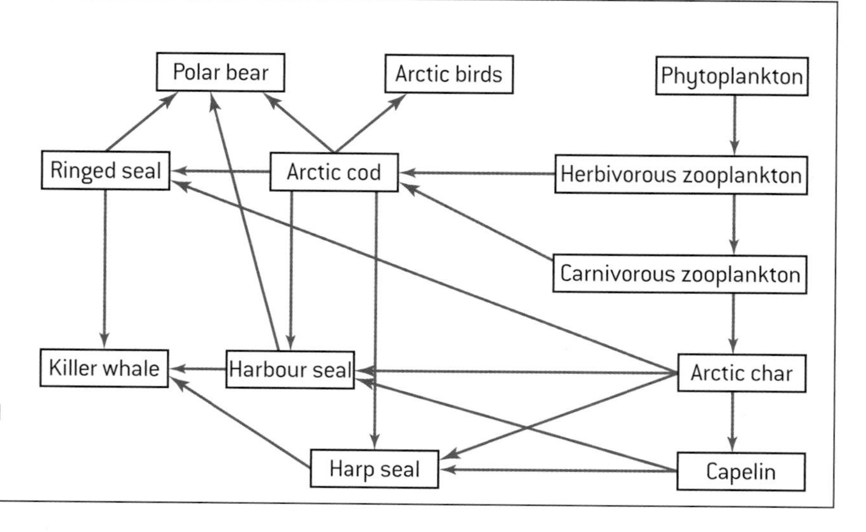

Nutrient cycles and change in ecosystems

NUTRIENTS IN ECOSYSTEMS

Energy enters ecosystems, flows through food chains and is then lost from the ecosystem. In contrast, nutrients can be retained in an ecosystem for an unlimited time. An ecosystem that does not exchange nutrients with its surroundings is a **closed ecosystem**. The carbon cycle is described in Topic 4; other nutrient cycles follow similar principles.

In **terrestrial ecosystems** there are three main **storage compartments: biomass** (living organisms), **litter** (dead organic matter) and the **soil**. Nutrients flow between these compartments and in an **open ecosystem** they also flow to or from the compartments and the surroundings. A Gersmehl diagram is a model of nutrient storage and flow for terrestrial ecosystems. The amount of nutrients in each compartment is indicated by the size of the circle and the flow rates are indicated by the size of the arrows. Gersmehl diagrams for taiga (boreal forest), desert and tropical rainforest (right) show considerable differences in the storage and flow of nutrients.

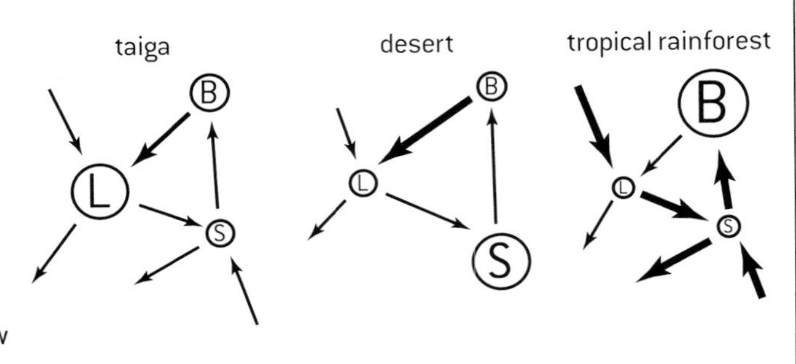

HUMANS AND NUTRIENT CYCLES

The nitrogen cycle is affected greatly by human activity. Fertilizers containing nitrates and ammonium are produced by the Haber process from gaseous nitrogen. Runoff from fields results in raised nitrogen concentrations in lakes and rivers. Nitrogen oxides from vehicle exhausts dissolve in water in the atmosphere to form nitrates, which are deposited in rainwater. These extra inputs to the nitrogen cycle cause **eutrophication** and **algal blooms**.

ECOLOGICAL SUCCESSION

An ecological succession is a series of changes to an ecosystem, caused by complex interactions between the community of living organisms and the abiotic environment. **Primary succession** starts in an area where living organisms have not previously existed, for example a new island, created by volcanic activity.

Analysis of examples of primary succession reveals some characteristic features:

- **species diversity** increases overall with some species dying out but more joining the community
- **plant density** increases as measured with the leaf area index (leaf area per unit of ground surface area)
- **organic matter** in the soil increases as more dead leaves, roots and other matter are released by plants
- **soil depth** increases as organic matter helps to bind mineral matter together
- **water-holding capacity** of soil increases due to the increased organic matter
- **water movement** speeds up due to soil structure changes that allow excess water to drain through
- **soil erosion** is reduced by the binding action of the roots of larger plants
- **nutrient recycling** increases due to increased storage in the soil and the biomass of organisms.

Example of primary succession: retreating glaciers in Iceland leave areas of sand, gravel and clay. Mosses and lichens colonize, then small non-woody plants (herbs), larger herbs, then shrubs and small trees, together with many animal species.

CLIMAX COMMUNUNITIES AND CLIMOGRAPHS

Ecological succession usually stops when a stable ecosystem develops with a group of organisms called the **climax community**.

The two main factors determining the type of stable ecosystem that develops in an area are temperature and rainfall. The figure below is a **climograph** that shows the relationship between temperature, rainfall and the type of stable ecosystem that is predicted to develop.

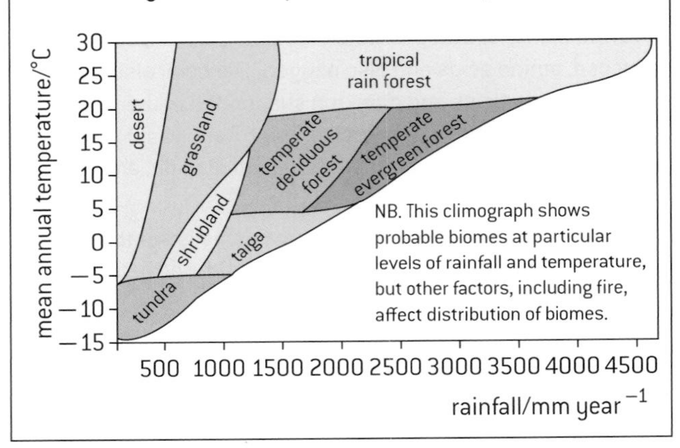

DISTURBANCE AND CHANGE IN ECOSYSTEMS

Communities sometimes change from those predicted by climographs to other communities as a result of **environmental disturbance**.

Fires, floods and storms are examples of natural disturbance, but humans are also sometimes responsible, as in the introduction of alien species, logging of forests and land drainage. Changes to the structure of ecosystems following disturbance can be rapid and profound.

In some ecosystems periodic disturbance is natural and contributes to biodiversity, by creating a patchwork of communities at different stages of development since the last disturbance.

Impacts of humans on ecosystems

INTRODUCTION OF ALIEN SPECIES

An **endemic species** naturally occurs in an area. An **alien species** is a type of organism that humans have introduced to an area where it does not naturally occur. Alien species that are released either accidentally or deliberately into local ecosystems often become invasive because predators from their natural community that would control their numbers have not also been introduced. Unless an alien species is adapted to an ecological niche not exploited in a community, it will compete with endemic species for resources and may cause them to become extinct by competitive exclusion. Two examples are given here.

Starlings have been introduced to North America and many other parts of the world and there are now hundreds of millions of these birds. They compete with endemic species for nest holes and food. They damage crops on farms and spread weed seeds in their feces. Trapping, netting and shooting have successfully reduced the population in Western Australia to a very low level, but destruction of nests and eggs and poisoning of millions of birds in the US have not reduced numbers significantly.

Cane toads are endemic in parts of Central and South America. They were introduced to Australia in the 1930s. Cane toads have toxins in their skin and also release toxins from glands when disturbed. Potential predators in Australia are killed by these toxins. A female can produce 30,000 eggs and adults naturally migrate to new areas so numbers of cane toads have risen exponentially. They are generalist predators so many endemic species, of both prey and predators, are threatened by them.

BIOLOGICAL CONTROL AND ALIEN SPECIES

Biological control is the use of a predator, parasite or pathogen to reduce or eliminate a pest. In some cases biological control methods have been introduced with great success, especially for pests of crops grown in greenhouses. There are also examples of biological control that were unsuccessful or harmful.

Cane toads were introduced to Queensland in Australia to control beetles that had become pests of sugar cane crops. The toads were unable to climb up into sugar cane plants to feed on the beetles and instead attacked native species of insect. They have become a far larger problem than the beetles. This type of example emphasizes the need for careful research into the effectiveness of proposed biological control methods and potential risks, before they are used.

BIOMAGNIFICATION

Some pollutants are absorbed into living organisms and accumulate because they are not efficiently excreted. When a predator consumes prey containing the pollutant, the level in the body of the predator rises and can reach levels much higher than those in the bodies of its prey. Concentration of pollutants in the tissues of organisms is called **biomagnification** and happens at each stage in food chains, with higher trophic levels reaching toxic doses.

To research the causes and consequences of an example of biomagnification, the level of toxin is measured in each organism in the food chain and also in the environment. Concentration factors can then be calculated, which are ratios between the level of toxin in two organisms or between one organism and the environment. Tissue from top carnivores (at the end of food chains) that are found dead is tested to see if it contains a toxic dose. The source of the pollutant in the environment is located.

Example: The radioactive isotope caesium-137 was released by the Fukushima nuclear disaster. Levels of it were measured in organisms from marine food chains near Fukushima. They rose more quickly in organisms at the start of food chains but reached much higher levels in organisms at the end of food chains. Killer whales (orca) reached caesium-137 levels that were 1000 times higher than Chinook salmon (their main prey) and 13,000 times higher than phytoplankton at the start of the food chain.

DDT POLLUTION AND MALARIA

Several species of mosquito transmit the disease malaria. When the insecticide DDT was sprayed onto water where the larvae of these mosquitoes were living, malaria became less common. As DDT killed other insects and, by biomagnification, had devastating effects on top carnivores such as ospreys, its use was mostly banned. This led to rises in malaria in some areas so was controversial.

PLASTIC POLLUTION IN OCEANS

Plastics dumped at sea or washed out from land are resistant to decomposition. Large pieces of plastic called **macroplastic debris** eventually degrade into many small fragments of **microplastic debris**. Huge amount of plastic have accumulated in marine environments, especially in five areas called **gyres**.

Examples of harm to marine organisms:

The **Laysan albatross** is a large marine bird that nests on the island of Midway Atoll in the Pacific Ocean. The North Pacific Gyre transports large volumes of macroplastic debris onto its beaches. Parent albatrosses confuse this with food and give it to their chicks resulting in gut blockages and high mortality.

The **lugworm** lives on muddy sea shores in Europe and North America, ingesting mud and digesting organic matter in it. Microplastic debris is ingested if it is contaminating the mud. The microplastic debris may contain toxic additives and tends to accumulate hydrophobic toxins from sea water, such as tributyl tin. The lugworms absorb and are harmed by some toxins, and organisms that feed on lugworms are at even greater risk, because of biomagnification.

Biodiversity and conservation

FACTORS AFFECTING BIODIVERSITY

Biological diversity, or biodiversity, has two components. **Richness** is the number of different species present. **Evenness** is how close in numbers the different species are. Sites with moderate populations are considered to be more biodiverse than if there are large numbers of some dominant species and much smaller numbers of others.

The number of species that live in an area is greatly affected by **biogeographic factors**. For example, there is a positive correlation between **island size** and the number of species on islands. Large unbroken areas of forest usually contain more species than a similar total area of fragmented forest. This is because of **edge effects**. Some species avoid the parts of forests close to an edge, so these species are absent from fragmented forest.

SIMPSON'S RECIPROCAL INDEX OF DIVERSITY

It is sometimes useful to have an overall measure of biodiversity in an ecosystem. Simpson's reciprocal index is suitable. Instructions for calculating it follow.

1. Use a random sampling technique to search for organisms in the ecosystem.
2. Identify each of the organisms found.
3. Count the total number of individuals of each species.
4. Calculate the index (D).

$$D = \frac{N(N-1)}{\Sigma n(n-1)}$$

N = total number of organisms

n = number of individuals per species

Example: Organisms were found and identified in the River Enningdalselva in a part of Sweden where some rivers have been affected by acid rain. Six sites in the river were chosen randomly and at each site organisms were collected by kick sampling along a 10 m transect. The results are shown below.

Group	Species	Name	
Ephemerida	*Dixa* species	Mayfly larva	8
Odonata	*Tipula* species	Dragonfly larva	5
Trichoptera	Species unidentified	Caddisfly larva	4
Plecoptera	*Nemoura variegata*	Stonefly larva	4
Hemiptera	*Gerris* species	Pond skater	3
Isopoda	*Asellus aquaticus*	Water louse	2
Acari	*Arrhenurus* species	Water mite	1
Platyhelminthes	*Dendocoelum lacteum*	Flatworm	4
Platyhelminthes	*Dugesia* species	Flatworm	3
Hirudinea	Species unidentified	Leech	1
Oligochaeta	*Lumbriculides*	Annelid worm	2
Gastropoda	*Lymnaea* species	Snail	4
Bivalvia	*Margaritifera*	Pearl mussel	1

$$D = \frac{42(42-1)}{140} = 12.3$$

The high diversity index suggests that the river has not been damaged by acid rain, or any other disturbance. This fits in with observations of a thriving salmon population in the river.

If Simpson's reciprocal index is calculated for two local communities using the same methods, an objective comparison of biodiversity can be made.

IN SITU AND EX SITU CONSERVATION

Conservation of species happens at a local, national and international level and often involves the cooperation of inter-governmental and non-governmental organizations.

The ideal place to conserve a species is in its own habitat. This is called **in situ** conservation. Many **national parks** and **nature reserves** have been established for this purpose. It may not be enough to designate an area for nature conservation. There is sometimes a need for **active management**, such as control of alien species.

Despite the advantages of in situ conservation, it does not always ensure the survival of species. Loss of natural habitat or catastrophic population declines sometimes force conservationists to transfer threatened populations from their natural habitats to zoos, botanic gardens or wild refuges. This is **ex situ** conservation.

Example: The **Mauritius kestrel** dropped to a population of four individuals due to loss of habitat, invasive alien species and DDT pollution. A captive breeding centre was established on a small island off the main island of Mauritius. Eggs were removed from the birds' nests, hatched in incubators, reared, then trained to catch prey and finally reintroduced. The population started to recover and is now close to the carrying capacity of Mauritius – about 1000.

INDICATOR SPECIES

Problems in natural ecosystems are detected quickly if environmental conditions are monitored. They can be measured directly or indicator species can be used. An indicator species needs particular **environmental conditions** and therefore shows what the conditions in an ecosystem are.

Example: lichen species vary in their tolerance of sulphur dioxide so can be used to assess the concentration of this pollutant in an area.

To obtain an overall environmental assessment of an ecosystem, a **biotic index** may be used. The number of individuals of each indicator species is multiplied by its pollution tolerance rating. These values are added together and then divided by the total number of organisms, to obtain the biotic index.

Examples of pollution tolerance ratings: stonefly nymphs need unpolluted, well-oxygenated water so have a rating of 10, whereas rat-tailed maggots and tubifex worms thrive even in low oxygen levels with much suspended organic matter so their rating is 0. All freshwater invertebrates have a rating between 1 and 10. The higher the biotic index calculated with these ratings, the less polluted the water.

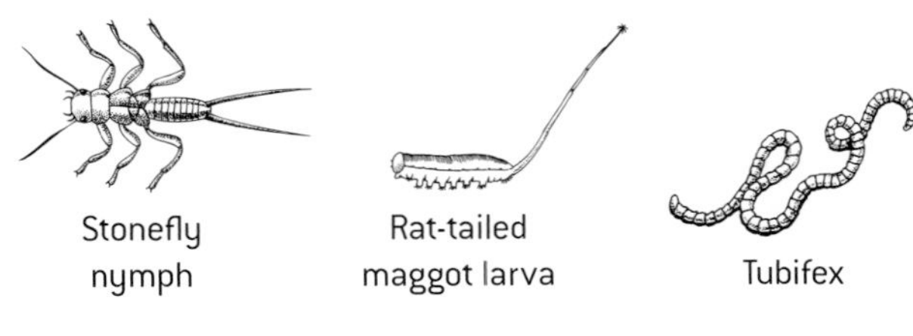

Populations (HL only)

ESTIMATING POPULATION SIZES

Populations are usually too large for every individual to be counted, so an estimate is made based on random sampling techniques. A **random number generator** helps ensure that sampling is free from bias. Quadrat sampling is described in Topic 4. It is not suitable for species of animal that move, so the **capture–mark–release–recapture** method is used.

1. Capture as many individuals as possible in the area occupied by the population.
2. Mark each individual, without making them more visible to predators.
3. Release all the marked individuals and allow them to settle back into their habitat.
4. Recapture as many individuals as possible and count how many are marked and unmarked.
5. Calculate the estimated population size

Estimated population size (Lincoln index) $= \frac{n_1 \times n_2}{n_3}$

n_1 = number caught and marked initially

n_2 = total number caught on the second occasion

n_3 = number of marked individuals recaptured

Example: 248 snails were caught in a pond and marked. 168 were recaptured, with 42 marked.

Estimated population size $= \frac{248 \times 168}{42} = 992$ snails

FISH POPULATIONS AND FISHERIES

The usual method of estimating fish stocks involves collecting data on catches. The numbers of each age are counted, allowing survivorship curves and spawning rates to be deduced, from which the total population is estimated. The problem with this approach is uncertainty about the proportion of the total population that was caught.

Capture–mark–release–recapture methods are inaccurate because the proportion of marked fish that can ever be recaptured is too small. Echo sounders can be used to measure the size of shoals of fish, but they must not be swimming too deeply and trawls must be used for calibration and to check which species of fish has been detected by the echolocation.

None of these methods estimate stocks with confidence.

Wild fish populations are an important food source for humans and with **sustainable fishing practices** they are a renewable resource. The maximum sustainable yield is the largest amount that can be harvested without a decline in stocks. It is essential to know the **age profile, reproductive status** and **size of the population**. If the population drops too low for effective breeding, there must be a ban on fishing. Sufficient larger fish that are mature enough to reproduce must be left for the population to replenish itself as least as fast as fish are caught.

POPULATION GROWTH

Population sizes change due to four factors:

natality – offspring produced and added to the population.
mortality – individuals die and are lost from the population.
immigration – individuals move into the area from elsewhere.
emigration – individuals move from the area to live elsewhere.

Populations are often affected by all four of these things and the overall change can be calculated using an equation:

Population change = (natality + immigration) − (mortality + emigration)

The graph (right) is a sigmoid (S-shaped) population growth curve. This growth curve can be **modelled** using organisms such as **yeast** grown in a nutrient solution in a fermenter, or ***Lemna*** (duckweed), a small floating plant that can be grown on water in beakers. Cell counts of samples of the yeast culture are done each day. Numbers of *Lemna* can easily be counted on the water.

1. Exponential phase
If a population is established in an ideal unlimited environment, it follows an exponential growth pattern, with the population increasing more and more rapidly. This is because the natality rate is higher than the mortality rate. The resources needed by the population such as food are abundant, and diseases and predators are rare. With abundant resources, immigration to the area is more likely than emigration.

2. Transitional phase
Population growth slows as the **carrying capacity** of the environment is reached –the maximum population size that can be supported by the environment. The natality rate starts to fall and/or the mortality rate starts to rise. Natality is still higher than mortality, but by a decreasing amount.

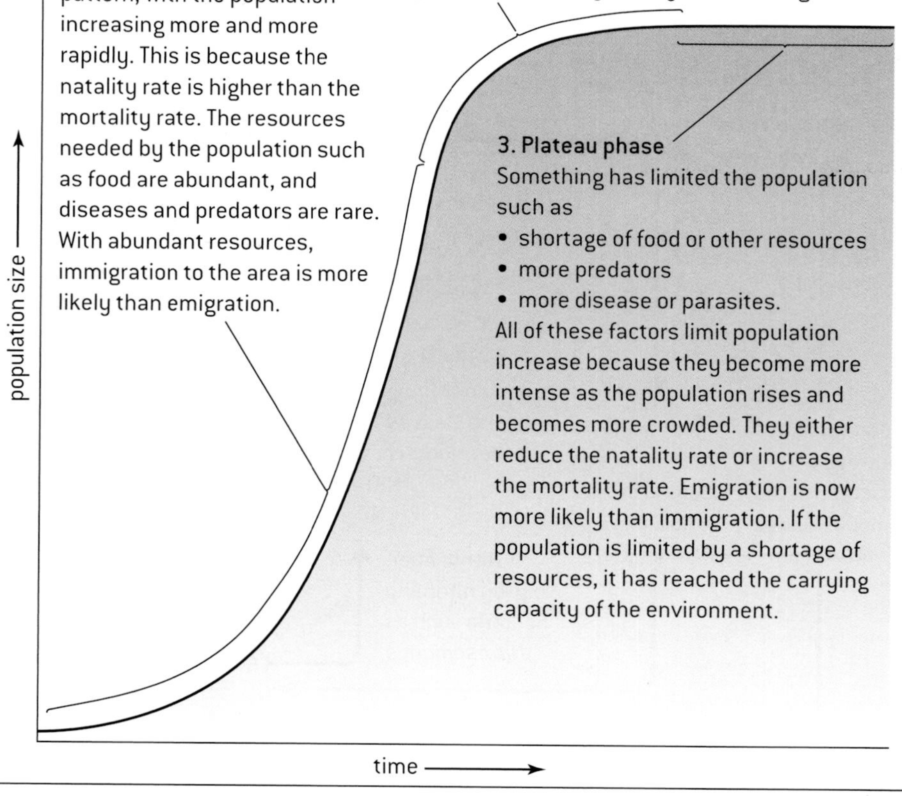

3. Plateau phase
Something has limited the population such as
- shortage of food or other resources
- more predators
- more disease or parasites.

All of these factors limit population increase because they become more intense as the population rises and becomes more crowded. They either reduce the natality rate or increase the mortality rate. Emigration is now more likely than immigration. If the population is limited by a shortage of resources, it has reached the carrying capacity of the environment.

Nitrogen and phosphorus cycles (HL only)

MICROBES AND THE NITROGEN CYCLE

Many microbes have roles in the nitrogen cycle, shown below.

1. **Nitrogen fixation** is conversion of atmospheric nitrogen into ammonia, using energy from ATP. Two nitrogen-fixing bacteria are *Azotobacter*, living free in soils, and *Rhizobium*, living mutualistically in roots.
2. **Nitrification** is conversion of ammonia to nitrate. It involves two types of soil bacteria. *Nitrosomonas* convert ammonia to nitrite and *Nitrobacter* convert nitrite to nitrate.
3. **Denitrification** is conversion of nitrate into nitrogen by denitrifying bacteria. This process only occurs in the absence of oxygen in the soil.

THE PHOSPHORUS CYCLE

Plants absorb phosphate from the soil for production by photosynthesis of compounds with phosphate groups. Phosphate is released back into the soil when decomposers break down organic matter. The rate of turnover in the phosphorus cycle is much lower than in the nitrogen cycle.

Phosphorus is added to the phosphorus cycle by application of fertilizer or removed by harvesting of agricultural crops. Phosphate fertilizer is obtained from rock deposits. These deposits are quite scarce and there are concerns that phosphate availability may limit agricultural crop production in the future.

WATERLOGGING AND NITROGEN CYCLE

Supplies of oxygen in waterlogged soils are rapidly used up. This prevents the production of nitrate by nitrifying bacteria and causes nitrate to be converted to nitrogen gas by denitrifying bacteria. Waterlogged soils are therefore deficient in nitrate. Plants show deficiency symptoms such as yellow leaves. Insectivorous plants overcome the low nitrogen availability in waterlogged soils by trapping insects and absorbing the ammonia released by digesting them.

SOIL NUTRIENT TESTS

Garden supply companies sell soil nutrient test kits that can be used for measuring concentrations of N, P and K in samples of soil. The samples are first dried and then ground to form a powder. To test for each nutrient, a measured volume of liquid reagent is added to a measured quantity of dry powdered soil. The colour that develops is either compared with a chart to deduce the concentration of nutrient or is assessed quantitatively with a colorimeter.

EUTROPHICATION AND ALGAL BLOOMS

Leaching of mineral ions from agricultural land can have harmful effects on aquatic habitats. Water with high concentrations of nitrate and phosphate ions is **eutrophic**.

Eutrophication causes algae to multiply excessively, resulting in an **algal bloom**. Some of the algae are deprived of light and die. Bacteria decompose the dead algae, using oxygen taken from the water. There is therefore an increased **biochemical oxygen demand (BOD)**. If oxygen levels drop very low fish and other aquatic animals die.

In natural ecosystems algal blooms are unusual due to two types of **limiting factor**, which are named according to their position in the food chain in relation to algae. If shortage of nutrients in the water limits the growth of algae, this is **bottom-up** control. If feeding on algae by large populations of herbivorous animals in the water limits populations of algae this is **top-down** control. When a population is limited by the carrying capacity it is bottom-up control, whereas control by means of predators or parasites is top-down.

THE NITROGEN CYCLE

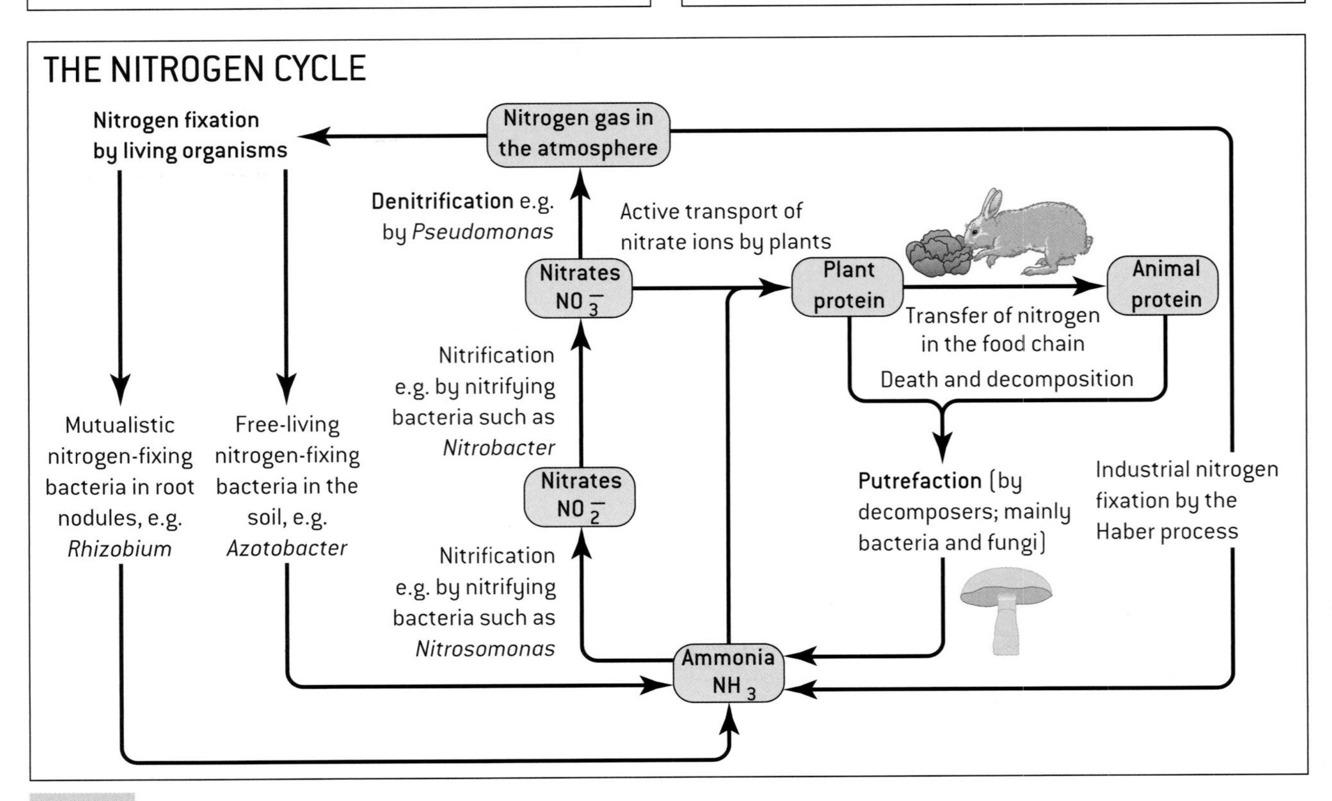

Questions – ecology and conservation

1. *Typha latifolia* and *Typha angustifolia* are plants that grow on the margins of ponds and lakes.

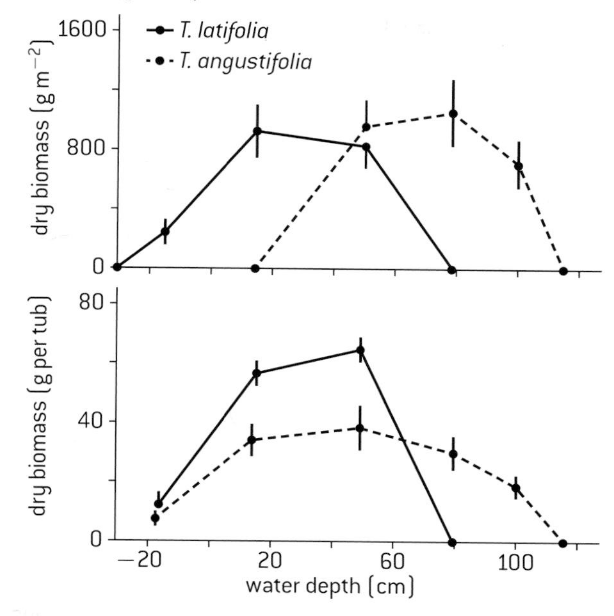

a) The upper graph shows the natural distribution of *T. latifolia* and *T. angustifolia* in a lake. Compare and contrast the two distributions. [3]

b) The lower graph shows the results of an experiment in which the species were planted separately in tubs, and placed at different depths in water to assess their growth. Deduce the depths that are within the limits of tolerance and zones of stress of each species. [3]

c) Explain the differences between the fundamental and realized niches of *T. angustifolia*. [4]

d) State one sampling technique that can be used to investigate the distribution of plants at increasing depths of water from the shore of a lake. [1]

2. The graph shows the concentrations of methyl mercury in feathers taken from museum specimens of the black-footed albatross (*Phoebastria nigripes*).

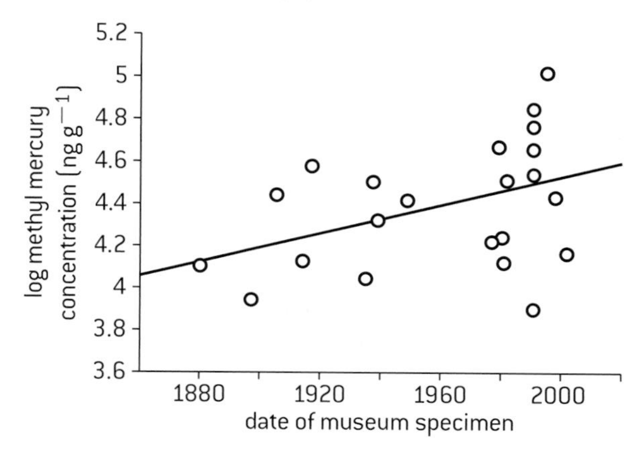

a) Explain how the methyl mercury concentrations found in the albatross feathers can be far higher than those in seawater. [4]

b) (i) Outline the trend shown in the graph. [2]

(ii) Suggest reasons for the trend. [2]

c) Outline one other threat to albatross populations due to pollution of the oceans. [2]

3. The biotic index used by Water Action Volunteers (WAV) in Wisconsin involves assigning aquatic macroinvertebrates to one of four pollution tolerance groups: *sensitive* (scoring 4), *semi-sensitive* (3), *semi-tolerant* (2) and *tolerant* (1). The WAV index is the mean score for a sample from a stream, river or other aquatic habitat: 3.6 to 4.0 indicates excellent water quality, 2.6 to 3.5 good, 2.1 to 2.5 fair and 1.0 to 2.0 poor.

a) (i) Outline features of aquatic macroinvertebrates. [3]

(ii) Explain advantages of using them for assessing river pollution rather than chemical tests. [3]

b) (i) Calculate the WAV biotic index for a sample that contained 12 stonefly larvae (sensitive) 8 alderfly larvae (sensitive), 8 mayfly larvae and 4 amphipods (both semi-tolerant). [4]

(ii) Explain what can be concluded. [2]

4. (HL) The graph below shows the growth of a population of ring-necked pheasants (*Phasianus colchicus*) on Protection Island off the north-west coast of the United States. The original population released by the scientists consisted of 2 male and 8 female birds. Two females died immediately after release.

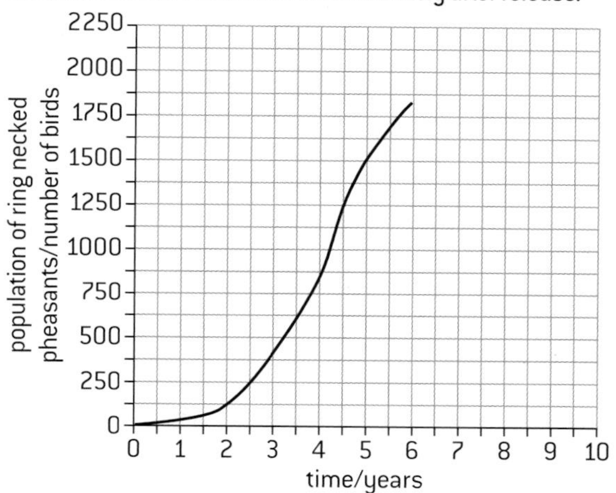

a) State the term used to describe the shape of a growth curve of this type. [1]

b) (i) Distinguish between the phases of the growth curve before and after 4.5 years. [3]

(ii) Explain the difference between these two phases in terms of the processes that can increase or decrease the size of a population. [4]

c) (i) The scientists predicted that the population would reach its carrying capacity of 2000 by year 8. Draw a line on the graph to show the population growth between years 6 and 10. [2]

(ii) Suggest factors that could cause the plateau. [3]

d) (i) Predict the results if all female birds in the original sample had survived. [1]

(ii) Predict the effect on the carrying capacity if all the female birds in the original sample had survived. [1]

5. (HL) (a) Explain the reasons for low concentrations of nitrate in the soils of wetlands. [3]

b) Describe the method used by *Dionaea muscipula* (Venus fly trap) to obtain nitrogen. [3]

c) *Dionaea muscipula* thrives in the years after fire in its habitats in subtropical wetlands but then is shaded out by taller growing plants. Discuss the importance of fire in this habitat. [4]

Human nutrition

NUTRITION AND MALNUTRITION

Nutrients are chemical substances in foods that are used in the human body. **Nutrition** is the supply of nutrients.

In humans there are **essential nutrients** that cannot be synthesized by the body so must be in the diet. They are divided into chemical groups:

- **minerals** – specific elements such as calcium and iron
- **vitamins** – chemically diverse carbon compounds needed in small amounts that cannot by synthesized by the body, such as ascorbic acid and calciferol
- some of the twenty **amino acids** are essential because they cannot be synthesized in humans and without them the production of proteins at ribosomes cannot continue
- specific **fatty acids** are essential for the same reason, for example omega-3 fatty acids.

Carbohydrates are almost always present in human diets, but specific carbohydrates are not essential.

Malnutrition is a deficiency, imbalance or excess of specific nutrients in the diet. There are many forms of malnutrition depending on which nutrient is present in excessive or insufficient amounts.

ENERGY IN THE DIET

Carbohydrates, lipids and amino acids can all be used in aerobic cell respiration as a source of energy. If the energy in the diet is insufficient, reserves of glycogen and fats are mobilized and used.

Starvation is a prolonged shortage of food. Once glycogen and fat reserves are used up, body tissues have to be broken down and used in respiration.

Anorexia is a condition in which an individual does not eat enough food to sustain the body even though it is available. As with starvation, body tissues are broken down. In advanced cases of anorexia even heart muscle is broken down.

Obesity is excessive storage of fat in adipose tissue, due to prolonged intake of more energy in the diet than is used in cell respiration. Obese or overweight individuals are more like to suffer from health issues, especially **hypertension** (excessively high blood pressure) and **Type II diabetes**. Most people do not become obese, because leptin produced by adipose tissue causes a reduction in appetite. A centre in the **hypothalamus** is responsible for feelings of **appetite** (wanting to eat food) or **satiety**.

MEASURING ENERGY CONTENT

A simple method for measuring the energy content of a food is by combustion. To heat one ml of water by one degree Celsius, 4.2 Joules of energy are needed so:

$$\text{energy content of a food } (\text{J g}^{-1}) = \frac{\text{temp rise } (^{\circ}\text{C}) \times \text{water volume (ml)} \times 4.2\text{J}}{\text{mass of food (g)}}$$

More accurate estimates of energy content can be obtained by burning the food in a food calorimeter which traps heat from the combustion more efficiently.

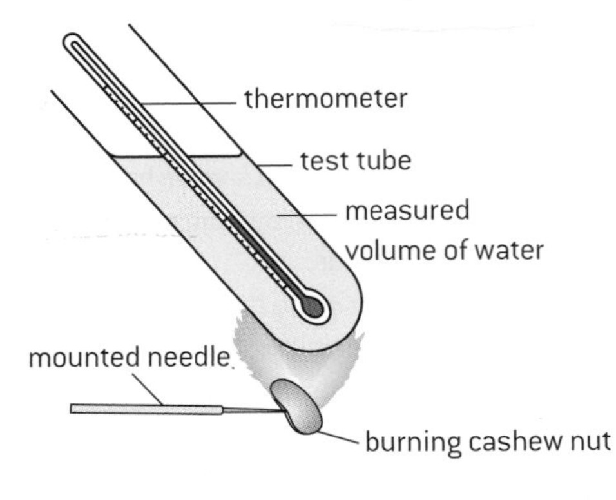

USE OF NUTRITION DATABASES

Databases are available on the internet with typical nutritional contents of foods. They can be used to estimate the overall content of a day's diet. The mass of each food eaten during the day is required. The nutritional analysis can be done very easily using free software also available on the internet such as at this site: http://www.myfoodrecord.com. The example below shows some of the nutrients in 50 g of salted cashew nuts, the recommended daily amount (RDA) of the nutrient for a 14–18 year-old boy and the percentage of this that the cashew nuts contain:

Nutrient	Total	RDA	RDA%
Protein (g)	293.5	3000	9.8%
Saturated fat (g)	4.88	33.3	14.6%
Cholesterol (mg)	0	300	0%
Iron (mg)	2.5	12	20.8%
Vitamin B_1 thiamine (mg)	0.16	1.2	13.3%

By carrying out this sort of analysis on a whole day's diet it is possible to determine whether sufficient quantities of essential nutrients have been eaten.

CHOLESTEROL AND HEART DISEASE

Research has shown a correlation between high levels of cholesterol in blood plasma and an increased risk of coronary heart disease (CHD), but it is not certain that lowering cholesterol intake reduces the risk of CHD, for these reasons:

- Much research has involved total blood cholesterol levels, but only cholesterol in LDL (low-density lipoprotein) is implicated in CHD.
- Reducing dietary cholesterol often has a very small effect on blood cholesterol levels and therefore presumably has little effect on CHD rates.
- The liver can synthesize cholesterol, so dietary cholesterol is not the only source.
- Genetic factors are more important than dietary intake and members of some families have high cholesterol levels even with a low dietary intake.
- There is a positive correlation between dietary intake of saturated fats and intake of cholesterol, so it is possible that saturated fats, not cholesterol, cause the increased risk of CHD in people with high cholesterol intakes.

Deficiency diseases and diseases of the gut

VITAMIN D DEFICIENCY IN HUMANS

If there is insufficient vitamin D in the body, calcium is not absorbed from food in the gut in large enough quantities. The symptoms of vitamin D deficiency are therefore the same as those of calcium deficiency including **osteomalacia**.

Osteomalacia is inadequate bone mineralization due to calcium salts not being deposited or being reabsorbed, so bones become softened. Osteomalacia in children is called **rickets**.

Vitamin D is contained in oily fish, eggs, milk, butter, cheese and liver. Unusually for a vitamin, it can be synthesized in the skin, but only in ultraviolet light (UV). The intensity of UV is too low in winter in high latitudes for much vitamin D to be synthesized, but the liver can store enough during the summer to avoid a deficiency in winter.

VITAMIN C DEFICIENCY IN MAMMALS

Ascorbic acid is needed for the synthesis of collagen fibres in many body tissues including skin and blood vessel walls. Humans cannot synthesize ascorbic acid in their cells so this substance is a vitamin in the human diet (vitamin C).

Scurvy is the deficiency disease caused by a lack of it. Attempts to induce the symptoms of **scurvy** in rats were unsuccessful because these and most other mammals have the enzymes needed for synthesis of ascorbic acid.

A theory that scurvy was specific to humans was falsified when scurvy was induced in guinea pigs by feeding them a diet lacking ascorbic acid. Apes and chimpanzees also require vitamin C in the diet.

PHENYLKETONURIA

Phenylalanine is an essential amino acid, but tyrosine is non-essential because it can be synthesized from phenylalanine.

$$\text{phenylalanine} \xrightarrow{\text{phenylalanine hydroxylase}} \text{tyrosine}$$

In the disease **phenylketonuria (PKU)** the level of phenylalanine in the blood becomes too high. The cause is an insufficiency or complete lack of phenylalanine hydroxylase, due to a mutation of the gene coding for the enzyme. PKU is therefore a genetic disease; the allele causing it is recessive.

The treatment for PKU is a diet with low levels of phenylalanine, so foods such as meat, fish, nuts, cheese and beans can only be eaten in small quantities. Tyrosine supplements may be needed if amounts in the diet are insufficient.

In a fetus the mother's body ensures appropriate concentrations of phenylalanine, so symptoms of PKU do not develop, but from birth onwards the level of phenylalanine can rise so high that there are significant health problems. Growth of the head and brain is reduced, causing mental retardation. Phenylalanine levels are now routinely tested soon after birth, allowing very early diagnosis of PKU and immediate treatment by means of diet that prevents most if not all harmful consequences.

CHOLERA

Cholera is a disease caused by infection of the gut with the bacterium *Vibrio cholerae*. The bacterium releases a toxin that binds to a receptor on intestinal cells. The toxin is then brought into the cell by endocytosis. Once inside the cell, the toxin triggers the release of Cl^- and HCO_3^- ions from the cell into the intestine. Water follows by osmosis leading to watery diarrhoea. Water is drawn from the blood into the cells to replace the fluid loss from the intestinal cells. Quite quickly severe dehydration can result in death if the patient does not receive rehydration.

EXCESSIVE STOMACH ACID SECRETION

The secretion of acid into the stomach is carried out by a **proton pump** called H^+/K^+-ATPase, in parietal cells in the stomach epithelium. These pumps exchange protons from the cytoplasm for potassium ions from the stomach contents. They can generate an H^+ gradient of 3 million to one making the stomach contents very acidic and potentially corrosive. A natural mucus barrier protects the stomach lining. In some people the mucus barrier breaks down, so the stomach lining is damaged and bleeds. This is known as an **ulcer** (see below). There can also be a problem with the circular muscle at the top of the stomach that normally prevents acid reflux, which is the entry of acid stomach contents to the esophagus, causing the pain known as heartburn. These diseases are often treated with a group of drugs called **proton-pump inhibitors** or PPIs, which bind irreversibly to H^+/K^+-ATPase, preventing proton pumping and making the stomach contents less acidic.

STOMACH ULCERS

Stomach ulcers are open sores, caused by partial digestion of the stomach lining by the enzyme pepsin and hydrochloric acid in gastric juice. Until recently, emotional stress and excessive acid secretion were regarded as the major contributory factors, but about 80 per cent of ulcers are now considered to be due to infection with the bacterium *Helicobacter pylori* (below).

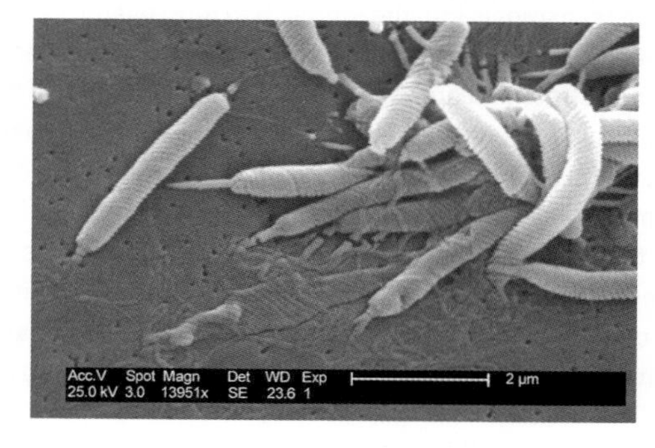

This theory was put forward in the early 1980s by Barry Marshall and Robin Warren. They cured ulcers using antibiotics that killed *H. pylori*, but it took some time for this treatment to become widely available. As so often in science, there was inertia due to existing beliefs. Doctors and drug companies had convinced themselves that they already knew the cause of ulcers and Marshall and Warren's infectious-agent theory did not immediately displace this mindset.

Digestion and absorption

SECRETION OF DIGESTIVE JUICES

There are two types of gland: **exocrine** and **endocrine**. Exocrine glands secrete through a duct onto to the surface of the body or into the lumen of the gut. The glands that secrete digestive juice are exocrine. Endocrine glands are ductless and secrete hormones directly into the blood.

EARLY RESEARCHES INTO GASTRIC JUICE

In 1822, Alexis St. Martin survived a gunshot injury, but the wound healed in such a way that there was access to his stomach from outside. William Beaumont, a surgeon who treated the wound, did experiments over an 11-year period. He tied food to a string and followed its digestion in the stomach. He digested samples of food in gastric juice extracted from the stomach. Beaumont showed that digestion in the stomach is a chemical as well as physical process. His research is an example of **serendipity**, as it only took place because of a fortuitous accident.

ACTIVITY OF GASTRIC JUICE

Gastric juice is secreted by cells in the epithelium that lines the stomach. Hydrogen ions are secreted by the **parietal cells**. This makes the contents of the stomach acidic (pH 1–3), which helps to **control pathogens** in ingested food that could cause food poisoning. Acid conditions also favour some **hydrolysis reactions**, for example hydrolysis by pepsin of peptide bonds in polypeptides. Pepsin is secreted by **chief cells** in the inactive form of pepsinogen; stomach acid converts it to pepsin.

CONTROL OF GASTRIC JUICE SECRETION

Secretion of digestive juices is controlled using both **nerves** and **hormones**. Control of the volume and content of gastric juice is described here as an example. The sight or smell of food stimulates the brain to send nerve impulses to parietal cells, which respond by secreting acid. This is a **reflex action**. Sodium and chloride ions are also secreted, causing water to move by osmosis into the stomach to form gastric juice. When food enters the stomach chemoreceptors detect amino acids and stretch receptors respond to the distension of the stomach wall. Impulses are sent from these receptors to the brain, which sends impulses via the **vagus nerve** to endocrine cells in the wall of the duodenum and stomach, stimulating them to secrete **gastrin**. The hormone gastrin stimulates further secretion of acid by parietal cells and pepsinogen by chief cells. Two other hormones, **secretin** and **somatostatin**, inhibit gastrin secretion if the pH in the stomach falls too low.

VILLUS EPITHELIUM CELLS

The structure of intestinal villi was described in Topic 6. Two recognizable features of epithelium cells on the villus surface adapt them to their role and are visible in the electron micrograph below:

Microvilli – protrusions of the apical plasma membrane (about 1μm by 0.1μm) that increase the surface area of plasma membrane exposed to the digested foods in the ileum and therefore food absorption.

Mitochondria – there are many scattered through the cytoplasm, which produce the ATP needed for absorption of digested foods by active transport.

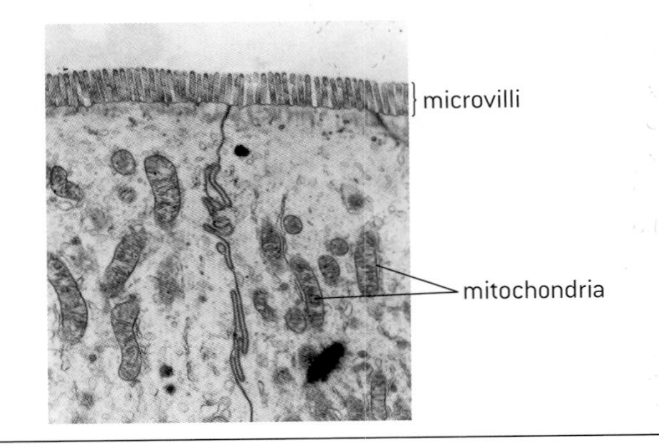

EXOCRINE GLAND CELLS

The exocrine gland cells that secrete digestive enzymes can be identified by the large amounts of rough endoplasmic reticulum, Golgi apparatus and secretory vesicles. The electron micrograph below shows several chief cells and one parietal cell.

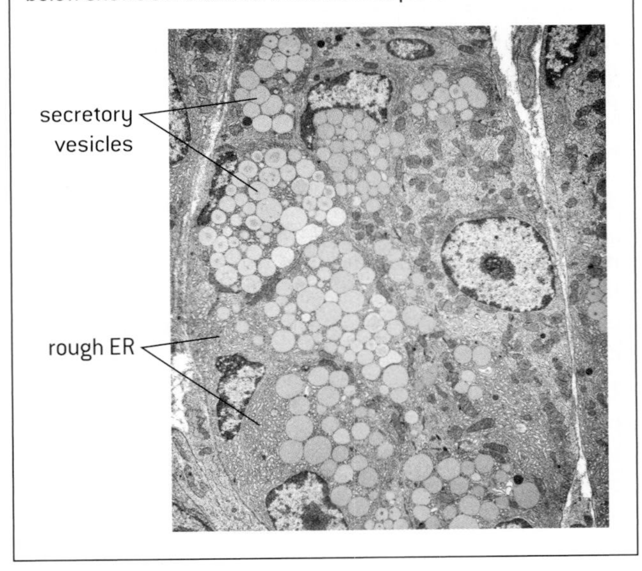

FIBRE AND FECES

Some materials, known as **dietary fibre**, are not digested or absorbed and therefore pass on through the small and large intestine and are egested. Cellulose, lignin, pectin and chitin are not readily digested in the human gut.

The average time that food remains in the gut is **mean residence time**. There is a positive correlation between mean residence time and the fibre content of the food that has been consumed. If the diet contains only low-fibre foods, the rate of transit of food through the gut becomes too slow (constipation), increasing the risk of bowel cancer, haemorrhoids and appendicitis.

Liver

FUNCTIONS OF THE LIVER

The liver is composed of hepatocytes that carry out many important functions:

Detoxification

Hepatocytes absorb toxic substances from blood and convert them by chemical reactions into non-toxic or less toxic substances.

Breakdown of erythrocytes

Erythrocytes (red blood cells) have a fairly short lifespan of about 120 days. **Kupffer** cells in the walls of sinusoids in the liver are specialized macrophages that absorb and break down damaged red blood cells by phagocytosis and recycle their components. The hemoglobin is split into heme groups and globins. The globins are hydrolysed to amino acids, which are released into the blood. Iron is removed from the heme groups, to leave a yellow coloured substance called **bile pigment** (bilirubin). The iron and the bile pigment are released into the blood. Much of the iron is carried to bone marrow, to be used in production of hemoglobin for new red blood cells. The bile pigment is used for bile production in the liver.

Conversion of cholesterol to bile salts

Hepatocytes convert cholesterol into bile salts which are part of the bile that is produced in the liver. When bile is secreted into the small intestine the bile salts emulsify droplets of lipid, greatly speeding up lipid digestion by lipase.

Hepatocytes can also synthesize cholesterol if amounts in the diet are insufficient.

Production of plasma proteins

The rough endoplasmic reticulum of hepatocytes produces 90% of the proteins in blood plasma, including all of the albumin and fibrinogen. Plasma proteins are processed by the Golgi apparatus in hepatocytes before being released into the blood.

Nutrient storage and regulation

Blood that has passed through the wall of the gut and has absorbed digested foods flows via the hepatic portal vein to the liver where it passes through sinusoids and comes into intimate contact with hepatocytes. This allows the levels of some nutrients to be regulated by the hepatocytes.

For example, when the blood glucose level is too high, insulin stimulates hepatocytes to absorb glucose and convert it to glycogen for storage. When the blood glucose level is too low, glucagon stimulates hepatocytes to break down glycogen and release glucose into the blood.

Iron, retinol (vitamin A) and calciferol (vitamin D) are also stored in the liver when they are in surplus and released when there is a deficit in the blood.

BLOOD FLOW THROUGH THE LIVER

The liver is supplied with blood by two vessels – the hepatic portal vein and the hepatic artery. Blood in the hepatic portal vein is deoxygenated, because it has already flowed through the wall of the stomach or the intestines. Inside the liver, the hepatic portal vein divides up into vessels called **sinusoids**. These vessels are wider than normal capillaries, with walls that consist of a single layer of very thin cells. There are many pores or gaps between the cells so blood flowing along the sinusoids is in close contact with the surrounding hepatocytes. The hepatic artery supplies the liver with oxygenated blood from the left side of the heart via the aorta. The hepatic artery branches to form capillaries that join the sinusoids at various points along their length, providing the hepatocytes with the oxygen that they need for aerobic cell respiration. The sinusoids drain into wider vessels that are branches of the hepatic vein. Blood from the liver is carried by the hepatic vein to the right side of the heart via the inferior vena cava.

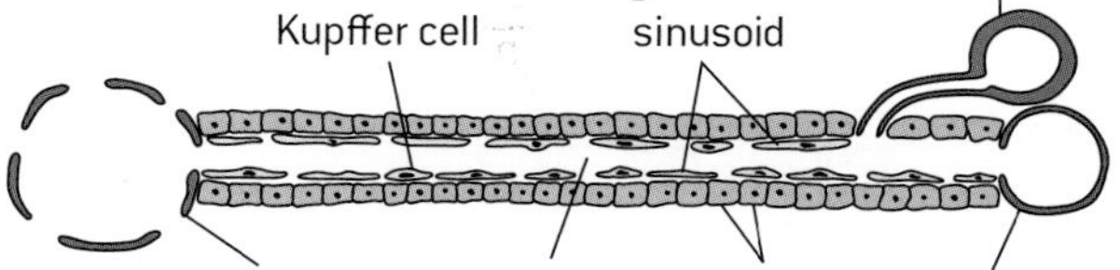

JAUNDICE

- **Jaundice** is a condition in which the skin and eyes become yellow due to an accumulation of bilirubin (bile pigment) in blood plasma.
- It is caused by various disorders of the liver, gall bladder or bile duct that prevent the excretion of bilirubin in bile, for example hepatitis, liver cancer and gallstones.
- There are serious consequences if bilirubin levels in blood plasma remain elevated for long periods in infants, including a form of brain damage that results in deafness and cerebralpalsy.
- Adult patients with jaundice normally just experience itchiness.

HIGH-DENSITY LIPOPROTEIN

Cholesterol is associated by many people with coronary heart disease and other health problems. This is not entirely justified as cholesterol is a normal component of plasma membranes and hepatocytes synthesize cholesterol for use in the body. High levels of blood cholesterol are not necessarily worrying – it depends on whether the cholesterol is being carried to or from body tissues. Cholesterol is transported in **lipoproteins**, which are small droplets coated in phospholipid. Health professionals are trying to educate the public to think of low-density lipoprotein (LDL) as 'bad cholesterol' because it carries cholesterol from the liver to body tissues. High-density lipoprotein (HDL) is 'good cholesterol' as it collects cholesterol from body tissues and carries it back to the liver for removal from the blood.

Cardiac cycle

EVENTS OF THE CARDIAC CYCLE

The main events of the cardiac cycle are described in Topic 6. The figure below shows pressure and volume changes in the left atrium, left ventricle and aorta during two cycles. It also shows electrical signals emitted by the heart and recorded by an ECG (electrocardiogram) and sounds (phonocardiogram) generated by the beating heart.

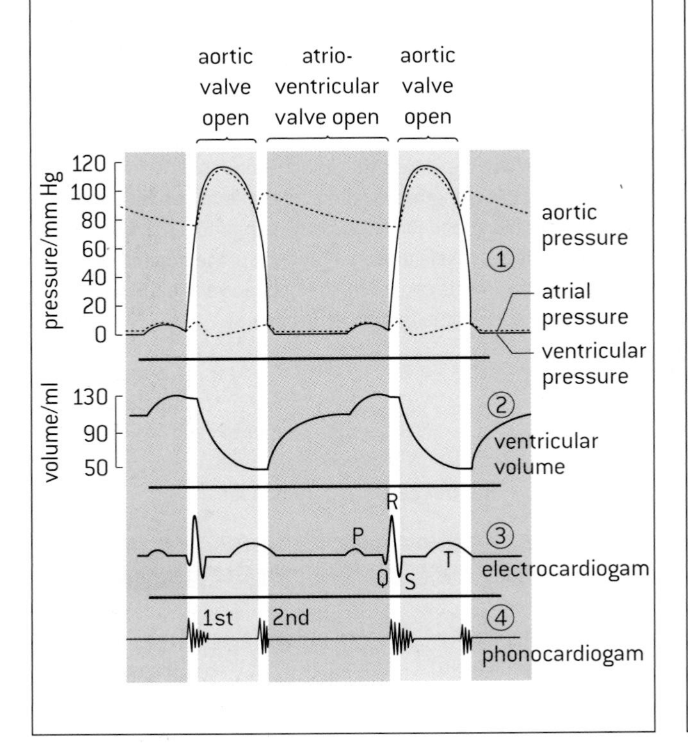

CARDIAC MUSCLE

The electron micrograph shows junctions between cardiac muscle cells. The junctions have a zigzag shape and are called **intercalated discs**. In these structures there are cytoplasmic connections between the cells that allow movement of ions and therefore rapid conduction of electrical signals from one cell to the next. Sarcomeres and mitochondria are also visible in the electron micrograph.

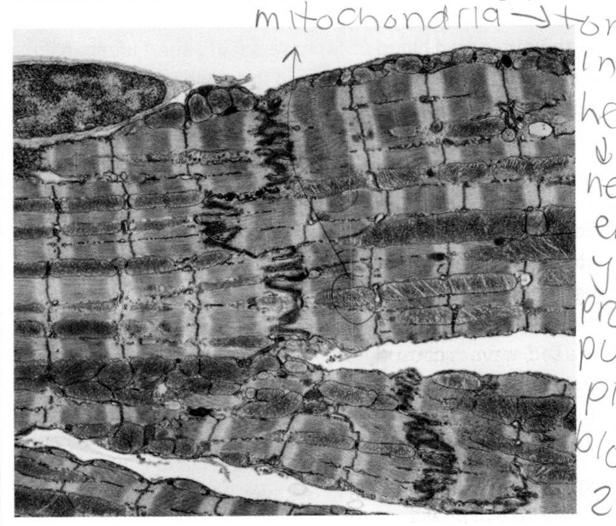

The cell on the left of the micrograph is connected to two cells on the right. This illustrates another property of cardiac muscle cells – they are branched. This helps electrical stimuli to be propagated rapidly through the cardiac muscle in the walls of the heart.

CONTROL OF THE CARDIAC CYCLE

Cardiac muscle cells have the special property of being able to stimulate each other to contract. Intercalated discs between adjacent cardiac muscle cells allow impulses to spread through the wall of the heart, stimulating contraction.

A small region in the wall of the right atrium called the **sinoatrial node** (SA node) initiates each impulse and so acts as the pacemaker of the heart. Impulses initiated by the SA node spread out in all directions through the walls of the atria, but are prevented from spreading directly into the walls of the ventricles by a layer of fibrous tissue. Instead, impulses have to travel to the ventricles via the **atrio-ventricular node** (AV node), which is positioned in the wall of the right atrium, close to the junction between the atria and ventricles.

Impulses reach the AV node 0.03 seconds after being emitted from the SA node. There is a delay of 0.09 seconds before impulses pass on from the AV node, which gives the atria time to pump blood into the ventricles before the ventricles contract. Impulses are sent from the AV node along **conducting fibres** that pass through the septum between the left and right ventricles, to the base of the heart. Narrower conducting fibres branch out from these bundles and carry impulses to all parts of the walls of the ventricles, coordinating an almost simultaneous contraction throughout the ventricles.

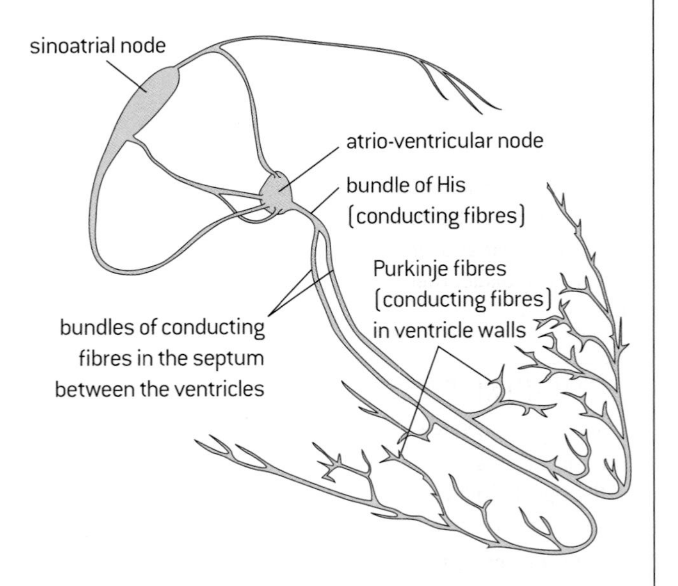

The diagram above shows the nodes and conducting fibres in the walls of the atria and ventricles that are used to coordinating contractions during the cardiac cycle.

Cardiology

STETHOSCOPES AND HEART SOUNDS

The stethoscope was invented in the early 19th century and has changed little since about 1850. It consists of a chestpiece with diaphragm to pick up sounds, and flexible tubes to convey the sounds to the listener's ears. Although a simple device, the introduction of the stethoscope led to greatly improved understanding of the workings of the heart and other internal organs. Normal heart sounds detected with a stethoscope are a 'lub' due to the closure of the atrio-ventricular valves (1st sound) and a 'dup' due to the closure of semilunar valves (2nd sound). Murmurs (other sounds) indicate problems such as leaking valves.

ELECTROCARDIOGRAMS

Electrical signals from the heart can be detected using an **electrocardiogram (ECG)**. Data-logging ECG sensors can be used to produce a pattern as shown in the figure below. The **P-wave** is caused by **atrial systole** (contraction of the atria) and the **QRS wave** is caused by **ventricular systole**. The T-wave occurs during **ventricular diastole**.

Specialists use changes to the size of peaks and lengths of intervals to detect heart problems.

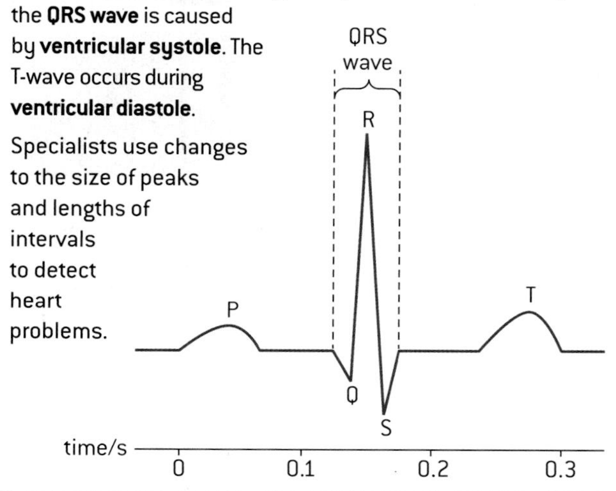

MEASURING THE HEART RATE

The heart rate can be measured easily using the radial pulse at the wrist or the carotid pulse in the neck. The rate is the number of beats per minute. Heart rate depends on the body's demand for oxygen, glucose and for removal of carbon dioxide. There is therefore a positive correlation between intensity of physical exercise and heart rate.

ARTIFICIAL PACEMAKERS

Artificial pacemakers are medical devices that are surgically fitted in patients with a malfunctioning sinoatrial node or a block in the signal conduction pathway within the heart. The device regulates heart rate and ensures that it follows a steady rhythm.

Pacemakers can either provide a regular impulse or only when a heartbeat is missed. They consist of a pulse generator and battery placed under the skin below the collar bone, with wires threaded through veins to deliver electrical stimuli to the right ventricle.

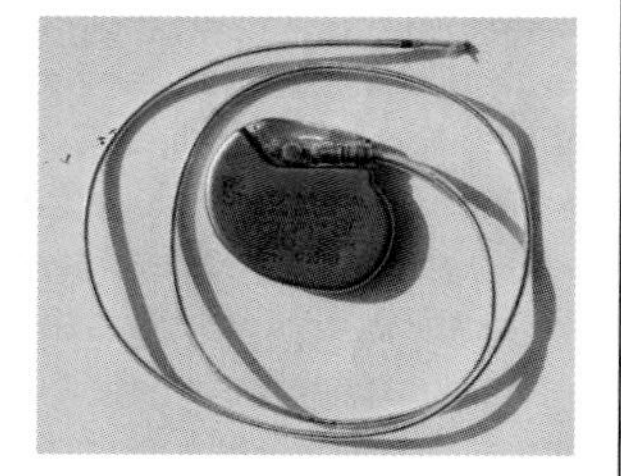

MEASURING BLOOD PRESSURE

To measure blood pressure, a cuff is placed around the upper arm and is inflated to constrict the arm and prevent blood in the arteries from entering the forearm. The cuff is slowly deflated and the doctor listens with a stethoscope for sounds of blood flow in the artery. This occurs when the cuff pressure drops below the systolic pressure. The cuff is further deflated until there are no more sounds, which happens when the cuff pressure drops below the diastolic pressure. The table indicates how blood pressures (such as 130 systolic over 90 diastolic) are interpreted.

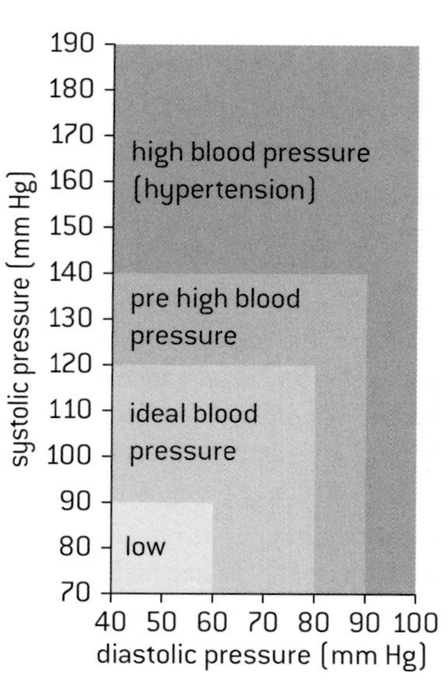

HYPERTENSION AND THROMBOSIS

The causes of **hypertension** are not clear, but there are various risk factors that are associated with this condition and may help to cause it: being obese, not taking exercise, eating too much salt, drinking large amounts of coffee or alcohol, and genetic factors (e.g. having relatives with hypertension). If left untreated, hypertension can damage the kidneys, or cause a heart attack or a stroke.

The causes of **thrombosis** (formation of blood clots inside blood vessels) are also unclear, but risk factors include high HDL (high-density lipoprotein) levels in blood, high levels of saturated fats and trans-fats in the diet, inactivity for example on air flights, smoking, hypertension and genetic factors. Thrombosis in coronary arteries causes a heart attack, and in the carotid arteries that carry blood to the brain it causes a stroke.

INCIDENCE OF CORONARY HEART DISEASE

Coronary heart disease (CHD) is damage to the heart due to blockages or interruptions to the supply of blood in coronary arteries. Investigation of CHD by experiment is unethical, so research is focused on analysis of epidemiological data. An example is included in the questions at the end of this option.

DEFIBRILLATORS

One of the features of a heart attack is ventricular fibrillation – this is essentially the twitching of the ventricles due to rapid and chaotic contraction of individual muscle cells. It is not effective in pumping blood. When 'first responders' reach a patient having a heart attack, they apply the two paddles of a defibrillator to the chest of the patient in a diagonal line with the heart in the middle. The device first detects whether the ventricles are fibrillating, and if they are it delivers an electrical discharge that often stops the fibrillation and restores a normal heart rhythm.

Endocrine glands and hormones (HL only)

STEROID AND PEPTIDE HORMONES

Hormones are chemical messengers, secreted by endocrine glands directly into the bloodstream. The blood carries them to **target cells**, where they elicit a response. A wide range of chemical substances work as hormones in humans, but most are in one of two chemical groups:

steroids e.g. estrogen, progesterone, testosterone

peptides (small proteins) e.g. insulin, ADH, FSH.

These two groups influence target cells differently.

Steroid hormones enter cells by passing through the plasma membrane. They bind to receptor proteins in the cytoplasm of target cells to form a hormone–receptor complex. This complex regulates the transcription of specific genes by binding to the promoter. Transcription of some genes is stimulated and other genes are inhibited. In this way steroid hormones control whether or not specific enzymes or other proteins are synthesized. They therefore help to control the activity and development of target cells.

Peptide hormones do not enter cells. Instead they bind to receptors in the plasma membrane of target cells. The binding of the hormone causes the release of a secondary messenger inside the cell, which triggers a cascade of reactions. This usually involves activating or inhibiting enzymes.

USE OF GROWTH HORMONE IN ATHLETICS

Growth hormone (**GH**) is a peptide secreted by the pituitary gland. It stimulates synthesis of protein and breakdown of fat, proliferation of cartilage cells, mineralization of bone, increases in muscle mass and growth of all organs apart from the brain. GH has been used by athletes since the 1960s to help to build their muscles. There is some evidence that it does enhance performance in events depending on muscle mass, but most sports ban GH and tests have been developed to catch illegal users.

IODINE DEFICIENCY DISORDER

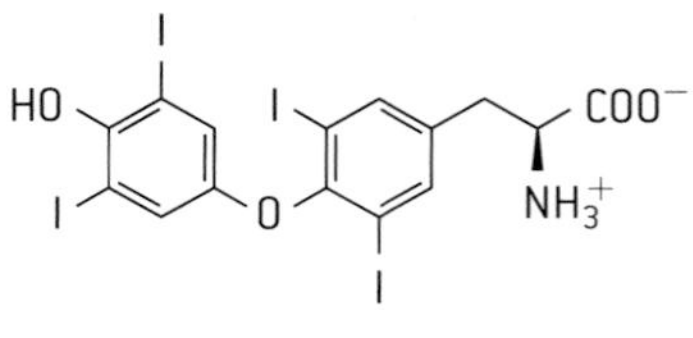

Iodine is needed for the synthesis of the hormone **thyroxin**, by the thyroid gland. An obvious symptom of iodine deficiency disorder (IDD) is swelling of the thyroid gland in the neck, called goitre. IDD also has some less obvious but very serious consequences. If women are affected during pregnancy, their children are born with permanent brain damage. If children suffer from IDD after birth, their mental development and intelligence are impaired. In 1998 UNICEF estimated that 43 million people worldwide had brain damage due to IDD and 11 million of these had a severe condition called cretinism. The International Council for the Control of Iodine Deficiency Disorders (ICCIDD) is a non-profit, non-governmental organization that is working to achieve sustainable elimination of iodine deficiency worldwide. It is a fine example of cooperation between scientists and many different other groups.

HORMONES AND THE HYPOTHALAMUS

The **hypothalamus** is a small part of the brain that links the nervous and endocrine systems. It controls hormone secretion by the pituitary gland located below it. Hormones secreted by the pituitary gland control growth, developmental changes, reproduction and homeostasis.

Some neurosecretory cells in the hypothalamus secrete releasing hormones into capillaries that join to form a portal blood vessel leading to capillaries in the **anterior lobe** of the **pituitary gland**. These releasing hormones trigger secretion of hormones synthesized in the anterior pituitary. FSH is released in this way. Other neurosecretory cells in the hypothalamus synthesize hormones and pass them via axons for storage by nerve endings in the **posterior pituitary**, and subsequent secretion that is under the control of the hypothalamus. ADH is a hormone that is released in this way.

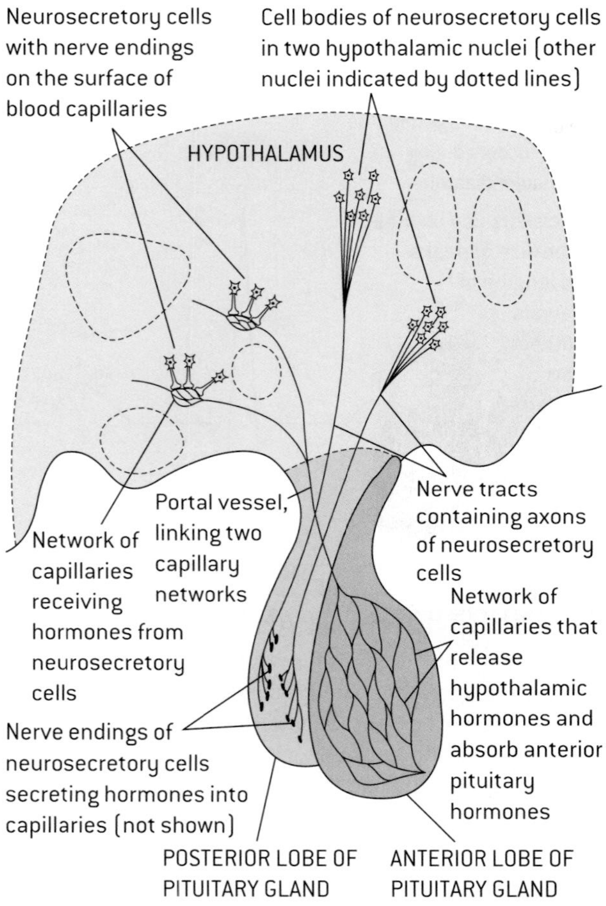

Neurosecretory cells with nerve endings on the surface of blood capillaries
Cell bodies of neurosecretory cells in two hypothalamic nuclei (other nuclei indicated by dotted lines)
HYPOTHALAMUS
Portal vessel, linking two capillary networks
Network of capillaries receiving hormones from neurosecretory cells
Nerve tracts containing axons of neurosecretory cells
Network of capillaries that release hypothalamic hormones and absorb anterior pituitary hormones
Nerve endings of neurosecretory cells secreting hormones into capillaries (not shown)
POSTERIOR LOBE OF PITUITARY GLAND
ANTERIOR LOBE OF PITUITARY GLAND

CONTROL OF MILK SECRETION

Milk secretion is regulated by pituitary hormones. **Prolactin** is secreted by the anterior pituitary. It stimulates mammary glands to grow, and to produce milk. During pregnancy, high levels of estrogen increase prolactin production but inhibit its effects. An abrupt decline in estrogen following birth ends this inhibition and milk production begins. The milk is produced and stored in small spherical chambers (alveoli) distributed through the mammary gland. **Oxytocin** stimulates the let-down of milk to a central chamber where it is accessible to the baby. The physical stimulus of suckling (nursing) by a baby stimulates oxytocin secretion by the posterior pituitary gland.

Carbon dioxide transport (HL only)

LUNG TISSUE IN MICROGRAPHS

The structure of alveoli in the light micrograph below can be interpreted using the diagram of an alveolus in Topic 6. The alveolus walls consist of one layer of pneumocytes. Capillaries between the walls of pairs of alveoli are only wide enough for red blood cells to pass in single file.

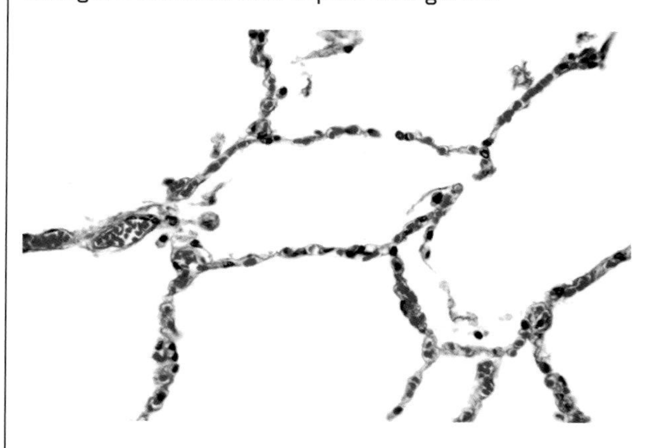

The electron micrograph below shows parts of two alveoli and a capillary with six red blood cells. Separating the air in the alveoli from the hemoglobin in the red blood cells are just two layers of cells: the **epithelium** and **endothelium** that form the walls of the alveolus and capillary respectively.

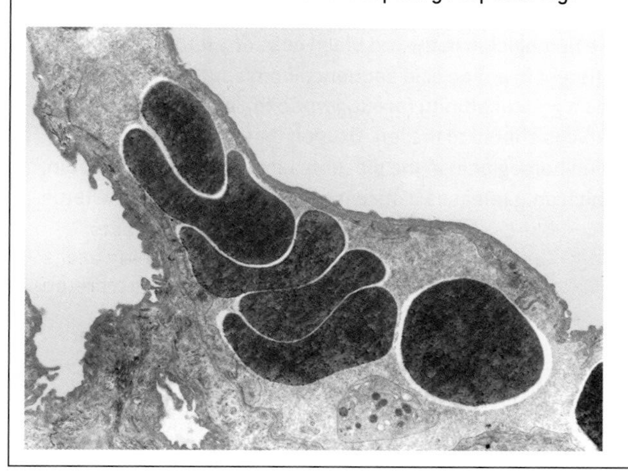

TREATMENT OF EMPHYSEMA

The causes and consequences of emphysema are described in Topic 6.

Treatment is by providing a supply of oxygen-enriched air, training in breathing techniques to reduce breathlessness, surgery to remove damaged lung tissue and less commonly lung transplants, and of course quitting smoking.

PUBLIC ATTITUDES TO SMOKING

Scientific research in the second half of the 20th century produced abundant evidence of the damage done to human health by smoking. Scientists have played a major role in informing the public about this, which has led to a change in public perception of smoking. As a result politicians have had enough support to allow them to raise taxes on tobacco and introduce increasingly extensive bans on smoking.

METHODS OF CARBON DIOXIDE TRANSPORT

Carbon dioxide is carried by the blood to the lungs in three different ways. A small amount is carried in solution (dissolved) in the plasma. More is carried **bound to hemoglobin**.

Even more still is **transformed into hydrogencarbonate** ions in red blood cells. After diffusing into red blood cells, the carbon dioxide combines with water to form carbonic acid. This reaction is catalysed by carbonic anhydrase. Carbonic acid rapidly dissociates into hydrogencarbonate and hydrogen ions. The hydrogencarbonate ions move out of the red blood cells by **facilitated diffusion**. A carrier protein is used that simultaneously moves a chloride ion into the red blood cell. This is called the chloride shift and prevents the balance of charges across the membrane from being altered.

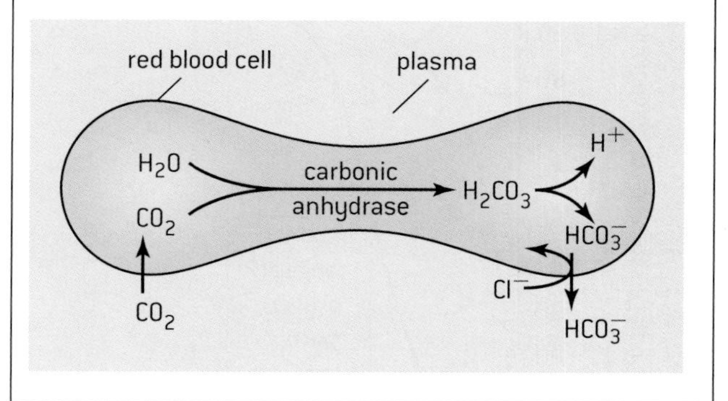

CONTROLLING THE VENTILATION RATE

In the walls of the aorta and carotid arteries there are **chemoreceptors** that are sensitive to changes in blood pH. The normal range is **7.35–7.45**. The usual cause of blood pH dropping to the lower end of this range is an increase in carbon dioxide entering the blood from respiring cells.

When a decrease in pH is detected signals are sent from the chemoreceptors to the **respiratory control centre** in the **medulla oblongata**. The respiratory control centre responds by sending nerve impulses to the diaphragm and intercostal muscles, causing them to increase the rate at which they contract and relax. This increase in **ventilation rate** speeds up the rate of carbon dioxide removal from blood as it passes through the lungs, so blood pH rises and remains within its normal range. The increase in ventilation rate also helps to increase the rate of oxygen uptake, which allows aerobic cell respiration to continue in muscles and helps to repay the oxygen debt after anaerobic cell respiration.

During vigorous exercise, the energy demands of the body can increase by over ten times. The rate of aerobic respiration in muscles rises considerably, so there is a significant increase in the amount of CO_2 entering the blood and the concentration rises. Blood pH therefore falls, but still usually remains within the normal range because of the large increase in ventilation rate. After exercise, the level of CO_2 in the blood falls, the pH rises and the breathing centres cause the ventilation rate to decrease.

Oxygen transport (HL only)

OXYGEN DISSOCIATION CURVES

Oxygen is transported from the lungs to respiring tissues by hemoglobin in red blood cells. The oxygen saturation of hemoglobin is 100% if all the hemoglobin molecules in blood are carrying four oxygen molecules, and is 0% if they are all carrying none.

Percentage saturation depends on oxygen concentration in the surroundings, which is usually measured as a partial pressure (pressure exerted by a gas in a mixture of gases).

The percentage saturation of hemoglobin with oxygen at each **partial pressure** of oxygen is an indication of hemoglobin's **affinity** (attractiveness) for oxygen. This can be shown on oxygen dissociation curves (below).

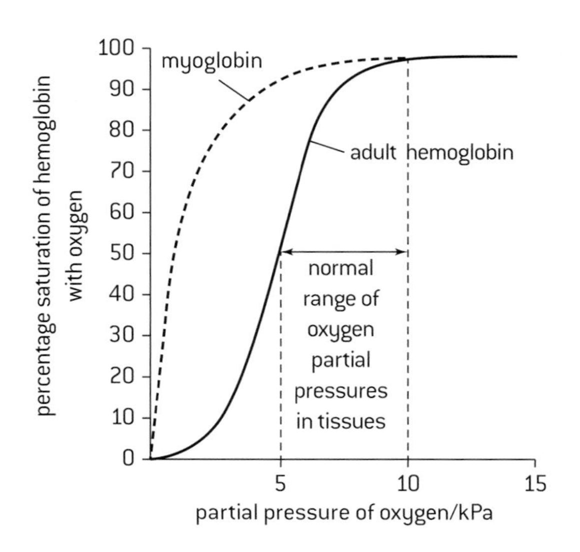

The curve for hemoglobin is S-shaped (sigmoid). This is because of interactions between the four subunits in hemoglobin that make it more stable when four oxygen molecules are bound or none. As a result, large amounts of oxygen are released over the range of oxygen partial pressures normally found in respiring tissues.

Myoglobin's curve is not sigmoid as it consists of only one globin and heme. The partial pressure of oxygen in alveoli is about 15 kPa. The dissociation curve shows that blood flowing through the lungs will therefore become almost 100% saturated. It also shows that the lower the oxygen concentration in a tissue through which oxygenated blood flows, the lower the saturation reached, so the greater the oxygen released.

Myoglobin consists of one globin and heme group, whereas hemoglobin has four. Myoglobin is used to store oxygen in muscles. The oxygen curve for myoglobin is to the left of the curve for adult hemoglobin, showing that myoglobin has a higher affinity for oxygen. At moderate partial pressures of oxygen, adult hemoglobin releases oxygen and myoglobin binds it. Myoglobin only releases its oxygen when the partial pressure of oxygen in the muscle is very low. The release of oxygen from myoglobin delays the onset of anaerobic respiration in muscles during vigorous exercise.

THE BOHR SHIFT

The release of oxygen by hemoglobin in respiring tissues is promoted by an effect called the **Bohr shift**. Hemoglobin's affinity for oxygen is reduced as the partial pressure of carbon dioxide increases, so the oxygen dissociation curve shifts to the right. The lungs have low partial pressures of carbon dioxide, so oxygen tends to bind to hemoglobin. Respiring tissues have high partial pressures of carbon dioxide so oxygen tends to dissociate, increasing the supply of oxygen to these tissues.

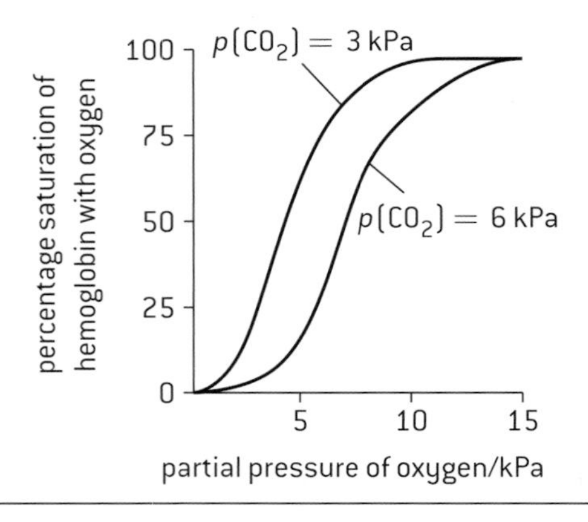

FETAL HEMOGLOBIN

The hemoglobin in the red blood cells of a fetus is slightly different in amino acid sequence from adult hemoglobin. It has a greater affinity for oxygen, so the oxygen dissociation curve is shifted to the left. Oxygen that dissociates from adult hemoglobin in the placenta binds to fetal hemoglobin, which only releases it once it enters the tissues of the fetus.

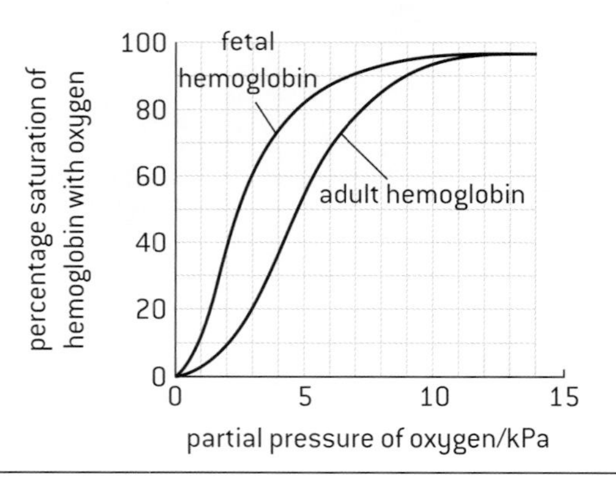

GAS EXCHANGE AT HIGH ALTITUDE

The partial pressure of oxygen at high altitude is lower than at sea level. Hemoglobin may not become fully saturated as it passes through the lungs, so tissues of the body may not be supplied with enough oxygen. A condition called **mountain sickness** can develop, with muscular weakness, rapid pulse, nausea and headaches. This can be avoided by acclimatization to high altitude during which time muscles produce more myoglobin and develop a denser capillary network, ventilation rate increases and extra red blood cells are produced. Some people who are native to high altitude show other adaptations, including a high lung capacity with a large surface area for gas exchange, larger tidal volumes and hemoglobin with an increased affinity for oxygen.

Questions – human physiology

1. A survey was done of patients who had complained of pain in their digestive system. The lining of their esophagus and stomach was examined using an endoscope and the patients' blood was tested for the presence of antibodies against *Helicobacter pylori*. The table below shows the results of the survey.

Endoscopy finding	Antibodies against *H. pylori* (number of cases)	
	Present	Absent
Normal	51	82
Esophagus inflamed	11	25
Stomach ulcer	15	2
Stomach cancer	5	0

   a) Explain why the researchers tested for antibodies against *H. pylori* in the blood of the patients. [2]
   b) Discuss the evidence from the survey results, for *H. pylori* as a cause of stomach ulcers and cancer. [3]
   c) Explain how *H. pylori* causes stomach ulcers. [3]
   d) Outline two reasons for acidic conditions being maintained in the stomach. [2]

2. a) Distinguish between essential and non-essential nutrients. [2]
   b) Explain the consequences of a deficiency in the diet of an essential amino acid. [3]
   c) Outline two conditions that might cause the breakdown of heart muscle tissue. [2]
   d) Outline two conditions caused by being overweight. [2]
   e) Outline the mechanism that can prevent the body from becoming overweight. [2]
   f) Explain how the content of energy and essential nutrients in a diet can be assessed. [3]

3. The electron micrograph shows tissue around a branch of the hepatic vein.

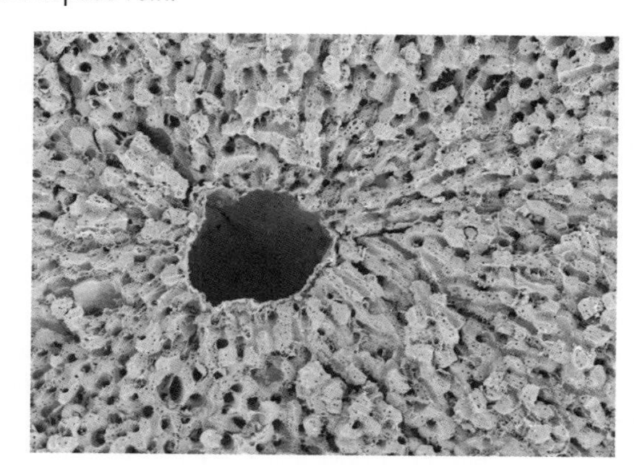

   a) Outline the structure of the liver around the vein. [3]
   b) Rough ER and Golgi apparatuses are prominent features in most liver cells. Outline their function. [2]
   c) State one example each of a vitamin, mineral and carbohydrate that is stored in liver cells. [3]
   d) Predict, with a reason, the difference between the concentration of ethanol in the hepatic portal vein and the hepatic vein. [2]

4. The figure is part of an ECG trace for a healthy person. The larger squares on the x axis are 0.1 seconds.

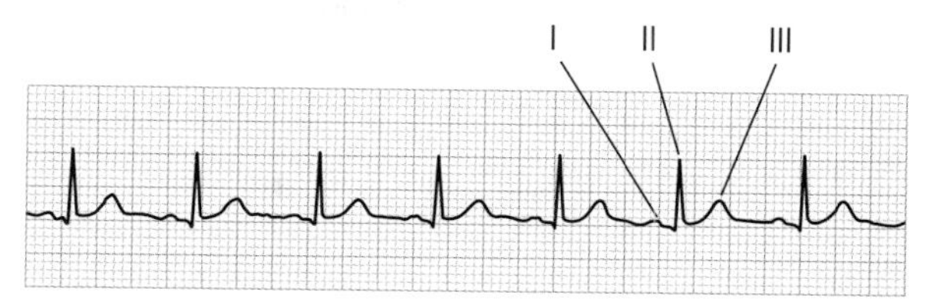

   a) Calculate the heart rate using data in the ECG. [3]
   b) (i) State the names given to I, II and III. [3]
   (ii) Deduce the events in the heart at I, II and III. [3]
   c) An ECG test is normally performed lying down, but it can also be done with the person on a treadmill or exercise bike. Predict how this will alter the results. [2]

5. (HL) The diagram shows the action of two types of hormone on a cell.

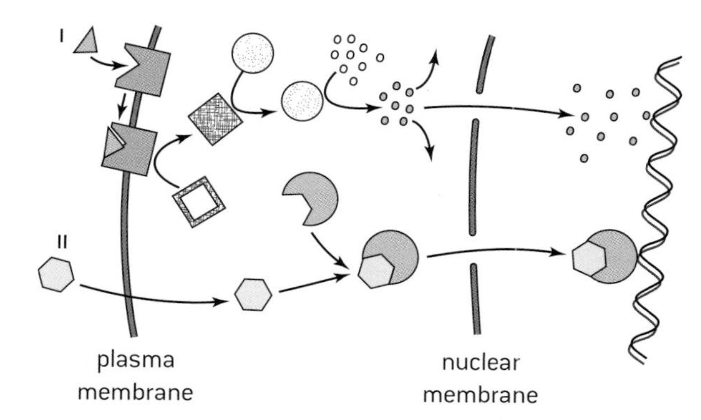

   a) Deduce the two types of hormone, I and II. [2]
   b) Suggest an example of each type of hormone. [2]
   c) Explain all the events shown in the diagram. [6]

6. (HL) V_E is the total volume of air expired from the lungs per minute. The graph below shows the relationship between V_E and the carbon dioxide content of the inspired air.

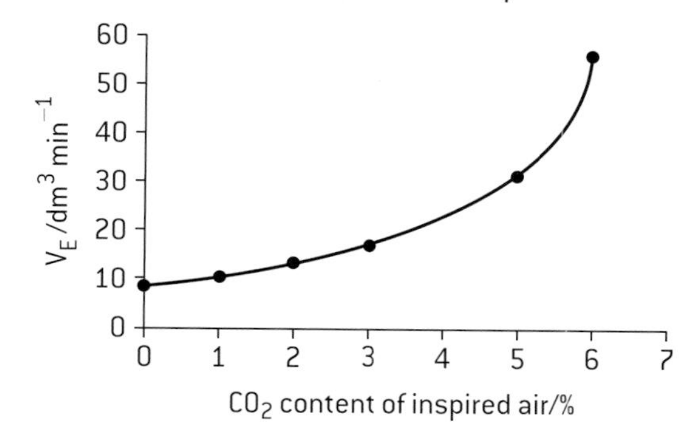

   a) Outline the relationship between the carbon dioxide content of inspired air and V_E. [2]
   b) Explain the effect of increasing CO_2 content of air on V_E. [3]
   c) Predict the effect on V_E of increasing the carbon dioxide concentration of inspired air above 7%. [4]
   d) Suggest one other factor that increases V_E. [1]
   e) State three ways in which carbon dioxide can be transported in blood. [3]
   f) Outline the effect of increasing carbon dioxide concentration on the affinity of hemoglobin for oxygen. [2]

Exam advice

There are three exam papers for both SL and HL Biology

	Standard Level (SL)			Higher Level (HL)		
	Time (min)	Marks	Types of question	Time (min)	Marks	Types of question
Paper 1	45	30 (20%)	Multiple choice based on the Core	60	40 (20%)	Multiple choice based on Core and AHL
Paper 2	75	50 (40%)	• Data-based question • Short answer questions • Extended response question (one from a choice of two) – all based on the Core	135	72 (36%)	• Data-based question • Short answer questions • Extended response question (two from a choice of three) – all based on Core and AHL
Paper 3	60	35 (20%)	Section A Short answer questions on experimental skills, techniques, analysis and evaluation of data based on the Core Section B Short answer and extended response questions from one Option	75	45 (24%)	Section A Short answer questions on experimental skills, techniques, analysis and evaluation of data based on Core and AHL Section B Short answer and extended response questions from one Option

If you want to do well in these final exams, you must prepare for them very carefully in the weeks beforehand. You will need to spend many hours on this task and find tactics that work for you.

You should practise answering exam questions using the questions at the end of topics in this book, after revising each topic. Your teacher should also give you some whole past exam papers to try.

There are four styles of question in IB Biology exams.

- **Multiple choice questions** – These are questions where you choose one of four possible answers. Read all the answers before choosing the best one. If you cannot decide on an answer, try to eliminate those that are obviously wrong to narrow down the possibilities. Leave difficult questions until you have answered the straightforward ones. Give an answer to every question – marks are not deducted for wrong answers. Calculators are not permitted as there are no multiple choice questions with difficult calculations.
- **Short answer questions** – These questions are broken up into small sections, each of which you answer in the space or on the lines provided. If you run out of space, you can continue your answer on extension pages but try not to do this – the best answers almost always fit into the space provided. You must indicate very clearly which questions you have continued on extension pages and ideally it should be none! The number of marks for each section is indicated and this tells you how detailed your answer needs to be.
- **Data-based questions** – These are a special type of short answer question. There is one main data-based question at the start of Paper 2, which will contain published research data that you are unlikely to have seen. There are also data-based questions in Section A of Paper 3, but they are based on practical work that you have done in the labs or on classic pieces of research.
  Look through the questions in this book to see some of the ways in which data can be presented. You should always study it very carefully before answering the questions, for example the scales and labelling on the axes of graphs.
  If there are calculations, remember to show your working and give units with your answer, for example grams or millimetres.
- **Extended response questions** – These questions require detailed answers on lined paper at the end of the exam booklet. You can decide what style of answer to give. Usually continuous prose is best, but sometimes ideas can be shown in a table or on a carefully annotated diagram. Read the whole of each question before choosing which to answer. As the question is divided up into sections (a), (b) and so on, you must answer it in these sections. If you plan out your answer it will be easier to ensure that you have arranged your ideas in a logical sequence. You can plan out your answer on the lined pages for answers to extended response questions, but remember to cross out the plan when you have written your full answer, so the examiner does not think that it is part of your answer. Do not include irrelevant material and express all your ideas clearly, without repetition. There is a mark for the quality of construction of your answer so it is worth paying attention to it. As with all questions, you must write legibly or the examiner may not be able to mark your work. This may mean that you have to write more slowly than normal.

COMMAND TERMS

The word at the start of each Paper 2 and Paper 3 exam question tells you what to do. These words are therefore called command terms. In IB exams each command term has a particular meaning. Your teacher can give you a complete list but some of the terms are obvious or are never used in biology exams. These are the most significant for biology:

Calculate: The answer will be a number and usually the SI units are needed as well, such as millimetres or seconds. It is best to include working, rather than just give the answer to the calculation. Sometimes there is a mark awarded for correct working, even if the final answer is wrong.

Compare/Compare and Contrast/Distinguish: In a 'compare' question you give only similarities, in a 'distinguish' question you give only differences and in a 'compare and contrast' you give both similarities and differences. In each of these types of question it is important not to describe things separately. Every sentence in your answer should be referring to both or all of the things that the question is asking about. When giving similarities the key word that might go at the start of your answer is 'Both' as in 'Both plant and animal cells have a nucleus'. With differences the key word is often 'whereas', for example 'Plant cells have a cell wall and plasma membrane whereas animal cells only have a membrane'. The best way to give differences is often in a table. Use the columns of the table for the things that you are comparing and the rows for the individual differences. Similarities can be included by merging cells in one row of the table.

Deduce: The answer is worked out logically from the evidence or information given in the question.

Describe/Outline: All details are expected in a 'describe' question but in 'outline' questions a briefer answer is expected.

Discuss: There won't usually be a simple straightforward answer to these questions, for example your answer should often include arguments for and against something. Try to give a balanced account. Sometimes your answer should consist of a series of alternative hypotheses – you could indicate how likely each one is but you don't need to make a final choice.

Evaluate: This usually involves assessing the value, importance or effects of something. You might have to assess the strengths and limitations of a technique, or of a model in helping to explain something. You might have to assess the expected impacts of something on the environment. Whatever it is that you are evaluating, you will probably have to use your judgement in composing an answer.

Explain: Sometimes this involves giving the mechanism behind something – often a logical chain of events, each one causing the next. This is a 'how' sort of explanation. A key word is often 'therefore'. Sometimes it involves giving the reasons or causes for something. This is a 'why' sort of explanation. A key word is often 'because'.

Label/Annotate: labels are simple names of structures on a diagram whereas annotations are notes added to a diagram giving more information than a label.

Suggest: Don't expect to have been taught the answer to these questions. Use your overall biological understanding to find answers – as long as they are possible, they should get a mark.

Nature of science – a summary

By studying IB biology you will develop an understanding of the nature of science, including the methods used by scientists to investigate the natural world. Nature of science is the overarching theme in all of the IB sciences. Some aspects of the nature of science are particularly important in biology and you should know examples of each – there may be questions on them in any of your biology exams. A summary of these aspects, with page numbers of examples in the Core or AHL is given below. (AHL examples are only studied by HL students.)

Aspect of nature of science	Example	Page
Some scientific discoveries are unplanned, surprising or accidental (serendipity)	• Discovery of cyclins	15
Scientists make careful observations	• Franklin's X-ray diffraction DNA research (HL) • Morgan's discovery of gene linkage (HL)	88 124
Scientists obtain evidence for theories	• Meselson and Stahl semi-conservative replication • Epidemiology and causes of lung cancer	29 79
Developments in scientific research follow improvements in apparatus	• Electron microscopes and ultrastructure • Use of gene sequencing machines • Autoradiography and DNA in chromosomes • Harvey's problems with reproduction research • Calvin and ^{14}C in photosynthesis research (HL) • Aphids and ^{14}C radioactive labels in phloem (HL) • Detecting traces of plant hormones (HL) • Radioactive calcium in muscle contraction (HL)	5 49 38 86 109 117 117 133
Developments in research have followed improvements in computers	• Advances in bioinformatics (HL) • Research into metabolic pathways (HL)	91 101
Scientists use theories to explain natural phenomena	• Hydrogen bonds and water's properties • Energy flow and the length of food chains • Natural selection and antibiotic resistance	18–19 57 66
Theories are regarded as uncertain and must be tested	• Cells only come from pre-existing cells	13
Theories that are found to be false are replaced with other theories	• Davson–Danielli model for membrane structure • Vitalism falsified by synthesis of urea • Reclassification of the figwort family • Harvey's discovery of the circulation of blood	8 17 70 74
Understanding of phenomena sometimes changes radically – this is a paradigm shift	• Chemiosmosis and ATP production (HL) • Pollinators and ecosystem conservation (HL)	104 119
Models are used as representations of the real world and help us to understand it or test theories about it	• Models of membrane structure • Crick and Watson's model of DNA structure • Dialysis tubing as a model of the gut wall • Simple models of xylem transport (HL)	8–9 28 73 115
Scientists look for patterns and trends and then look for discrepancies that don't fit	• Exceptions to the cell theory • Non-standard amino acids in proteins • Plants and algae that are not autotrophic • Pentadactyl limbs in vertebrates • Environmental effects on epigenetic factors (HL) • Non-Mendelian ratios and gene linkage (HL) • Chromosome numbers and polyploidy (HL)	1 24 54 63 93 123–124 127
Scientists make accurate quantitative measurements	• Monitoring CO_2 and methane concentrations • Measurements in osmosis experiments	60–61 11
Variables must be controlled in experiments	• Variables in photosynthesis experiments	35
Replicates are needed to ensure reliability	• Replicates in enzyme experiments • Mendel's genetic crosses with pea plants	27 42

Scientists cooperate and collaborate with each other	• Use of the binomial system for naming species • Multidisciplinary research – memory and learning	67 154
Research has ethical implications	• Ethics of stem cell research • Use of invertebrates in respirometers • Jenner's testing of smallpox vaccine (HL)	3 33 131
Scientists must balance the risks and benefits of their research	• Genetically modified crops and livestock • Florey and Chain's testing of penicillin • Contraceptive pills and male fertility (HL)	51, 160, 163 77 142
Claims made in the media should be tested scientifically	• Health claims about lipids in the diet • Climate change and human activities	23 60–61

Advice for Internal Assessment (IA)

Internal Assessment (IA) makes up 20% of the marks that decide your grade in IB Biology. Your teacher will guide you through the procedure and will assess your work according to five criteria. If they are too generous or too harsh, the scores may be changed by an external moderator, so you need to make it as clear as possible that you deserve high scores.

For Internal Assessment you carry out a scientific investigation and write a 6–12 page report on it. The first task is to choose a research question. You should choose a question that you are genuinely interested in, but also that is suitable for investigation by doing experiments. IB regulations allow database investigations and the use of simulations or models, but the guidance given here assumes that you are doing experimental work. Humans cause problems in experiments, especially with the control of variables, so it is best if your research question refers to other species. The IB has strict ethical rules for animal experiments and although they do not ban using animals, they prohibit procedures that cause suffering to animals. Experiments with plants, bacteria and fungi are less likely to raise ethical concerns.

Criterion 1 Personal engagement (8%)

- Show that you have specific reasons for being interested in the research question.
- Show that your investigation involved as many of your own ideas as possible and the research question, experimental design and presentation of results are not all copied from elsewhere without modification.

Criterion 2 Exploration (25%)

- Describe the biology relating to your investigation so that it is clear why you want to ask your research question.
- State the research question clearly. It should be focused rather than broad, so you can obtain enough experimental evidence to develop convincing conclusions.
- Explain how you designed and developed your experimental procedures, including preliminary trials, how you will ensure that the data is reliable and that there is enough to provide strong evidence relevant to the research question.
- Include a risk assessment of the experimental methods used and an audit of environmental and ethical considerations.

Criterion 3 Analysis (25%)

- Present all the raw data generated in your experiments in properly constructed results tables. Include a full title to explain what data is shown in each table. The left hand column should show the different levels of the independent variable. The other columns are for the results for the dependent variable. All rows and columns should have headings. Units should be with the headings, not with the numbers in the body of the table. Use correct SI units. The number of decimal places should be consistent in a column and appropriate for the precision of the apparatus used. A value for the measurement uncertainty should be given with the units in the column heading. This is usually plus or minus the smallest divisions on the measuring device, for example ± 0.1 cm^3 with a 1 cm^3 syringe. Mean results should be in the right hand column and should not include more decimal places than individual results.
- Ensure that you have enough raw data to provide evidence for any statistical tests and for detailed and valid conclusions to your research question.
- Analyse your data by calculating mean results and a measure of the variation in the repeats, such as a standard deviation. If possible, carry out a statistical hypothesis test, so you know the significance level of differences in the data. The test must be appropriate for your data so check carefully the conditions under which the test you intend to use is valid.
- Make it clear what the effect of measurement uncertainty is on your analysis.
- Plot graphs to display the data. Include an informative title. Plot mean results with your measure of variation. Join means with straight lines and do not extrapolate with a line beyond the first or last data points. Aim to include a few large graphs rather than many tiny ones.

Criterion 4 Evaluation (25%)

- Explain in detail your conclusions to the research question, based on your data and analysis of it.
- Compare your conclusions with published research or with the general scientific consensus among biologists about your research question. Do your conclusions conform to the consensus or are they unexpected?
- Assess the strengths and weaknesses of all aspects of your investigation, especially all the possible sources of error in your data and the reliability of the experimental methods used.
- Discuss suggestions for improving and extending your investigation, for example by seeing if a different species shows the same trends, or by investigating a different independent variable. The suggestions must be realistic and relevant.

Criterion 5 Communication (17%)

- Make sure that the report of your investigation that you hand in to your teacher for assessment is presented as clearly as possible.
- Divide the report up into sections with suitable headings. Consider structuring the report into these sections: 1. Developing the research question 2. Methods 3. Results and analysis 4. Conclusion 5. Evaluation.
- Keep the report concise by not including any material that is irrelevant to the research question. Your report should be between 6 and 12 pages long. If it is longer than this your mark for communication will be reduced.
- Check for errors and make sure that all biological terminology is correct.
- If you use ideas or data from another source in your report, you must make this clear. Include a numbered list of such sources at the end of your report and wherever one of these sources has been used in the report, put a numbered reference. The method of referencing does not matter, as long as you do it and thereby demonstrate your academic honesty.

Answers to questions

Marking points in answers with two or more marks are indicated with semi-colons. Alternative answers are indicated with a forward slash. Words in parentheses (brackets) are not essential but help to explain an answer.

Topic 1 Cell biology

1. (a) X = Rough ER, Y = mitochondrion

(b) magnification = size of image/size of specimen; 18,000 μm/2 μm = × 9,000;

(c) (eukaryotic) because nucleus present; because mitochondria/membrane-bound organelle present;

(d) ATP (by mitochondria); proteins (by rough ER);

(e) animal/not plant cell as glycogen/not starch stored; liver/muscle cell as these cells store glycogen;

2. (a) mass changes are +1.1g and −0.9g (πesp); % mass change = mass change/initial mass × 100%; % mass changes are −15%, +5% and −4%;

(b) x axis (horizontal) legend is Concentration of sucrose (mol dm^{-3}) and y axis legend is Percentage mass change; both scales are evenly spaced and use more than half of the graph paper; all points are plotted to within 0.5mm of the correct position; data points are joined by ruled straight lines not a line of best fit;

(c) (i) 0.31 milliOsmoles per litre (+/− 0.1) **(ii)** no mass change at this point so no osmosis; tissue must have same osmolarity as the sucrose solution;

3. (a) nucleus; vesicles; Golgi apparatus;

(b) phospholipid bilayer; proteins on either side;

(c) freeze-fracture split membranes in the middle of the phospholipid bilayer; lumps are visible in the middle of the bilayer; the lumps are transmembrane proteins;

(d) proteins in two cell types marked with different colours; when the cells fuse the colours mix rapidly; proteins in the membrane must be free to move; membrane proteins have hydrophobic parts so must be embedded in the phospholipid bilayer;

4. (a) I metaphase, II anaphase, III prophase, IV telophase

(b) mitotic index = number of cells in mitosis/total number of cells; $\frac{5}{43}$; = 0.12 (+/− 0.05);

(c) DNA replication; division of mitochondria/chloroplasts.

Topic 2 Molecular biology

1. C **2.** A **3.** D **4.** C **5.** B

6. (a) (i) DNA **(ii)** DNA **(iii)** RNA

(b) experimental error

(c) (i) DNA is double stranded; A pairs with T and C pairs with G; one base in each pair is therefore A or G, so A + G = 50%; **(ii)** any two of A = T; C = G; C + G = 50%; A + G/C + G = 1.00

(d) (i) influenza virus **(ii)** RNA contains uracil instead of thymine; single stranded so amounts of G and C not equal.

7. (a) (i) CO_2 concentration falls in the light and rises in the dark; **(ii)** CO_2 concentration falls when it is warmer and rises when it is cooler;

(b) CO_2 concentration is more closely related to light intensity; when there is a temporary dark period during the third day but it stays warm pH drops so CO_2 concentration rises;

(c) (i) respiration; producing CO_2; **(ii)** photosynthesis; causing CO_2 uptake;

8. (a) (i) temperature **(ii)** time taken for all starch to be digested

(b) pH; starch concentration/amylase concentration in saliva;

(c) best to check reliability by repeating each temperature; 20 °C to 60 °C is a better range than 20 °C to 40 °C; higher temperatures could be tested to show denaturation;

(d) starch

(e) starch to maltose; amylase shown over the arrow;

(f) (i) temperature on x axis and time taken on the y axis; U shaped curve; minimum time at 40/50 °C;
(ii) exponential rise from low temperature upwards to a maximum (at 40–50 °C); steeper drop to zero above the maximum.

Topic 3 Genetics

1. (a) T2 phage, *E. coli*, *D. melanogaster*, *H. sapiens*, *P. japonica* (4 marks minus 1 per incorrect placing);

(b) *P. equorum*, *O. sativa*, *H. sapiens*, *P. troglodytes*, *C. familiaris*;

(c) (i) diploid; two sets of chromosomes;
(ii) non-disjunction/trisomy/Down syndrome

(d) positive correlation between complexity and genome size; but exceptions such as *Paris* larger than *Homo*; no clear relationship between chromosome number and complexity; *Pan* and *Canis* not more complex than *Homo* but have more chromosomes.

2. (a) I first telophase; II first prophase; III first anaphase; IV second prophase/second metaphase; V first metaphase;

(b) first prophase with pairs of homologous chromosomes; inside nuclear membrane; first metaphase with chromosome pairs on the equator; with microtubules from each pole attached to opposite sides of the centromere; first anaphase with chromosomes of two chromatids being pulled to opposite poles by microtubules; with pairs of homologous chromosomes moving to opposite poles; first telophase with chromosomes inside a nuclear membrane; with two nuclei inside one cell; second prophase/metaphase with unpaired chromosomes of two chromatids; inside a nuclear membrane (prophase)/aligned on the equator (metaphase);

3. (a) O group individual must be genotype ii because it is due to a recessive allele; B group individual in generation 2 must be I^Bi because the parent that was blood group A could not have passed on I^B; B group individual in generation 3 must have been I^Bi because the O group parent must have passed on i;

(b) parents could have been group O; parents could have been group A with genotype I^Ai; parents could have been group B with genotype I^Bi (genotypes could have been ii × ii, $I^Ai \times I^Ai$, $I^Bi \times I^Bi$ and $I^Ai \times I^Bi$)

(c) parental genotypes both shown as $I^A I^B$; gametes from both parents shown as I^A and I^B; four genotypes and phenotypes shown correctly on a Punnett grid as I^AI^A group A, two of $I^A I^B$ group AB and as I^BI^B group B; ratio is 1 group A : 2 group AB : 1 group B;

4. (a) clone

(b) nucleus removed from a cell in an adult organism; nucleus removed from an egg cell and replaced with the nucleus from the adult animal

(c) (i) fragments had moved down; larger fragments are nearer the top and move more slowly; (ii) culture cells have the same profile as udder cells as they have the same pattern of bands; Dolly's blood cells have the same profile as the udder/culture cells as they have the same pattern of bands; Dolly was cloned from the udder cells; sheep 1–12 are genetically different;

(d) paternity tests; forensic investigations;

5. (a)(i) chromosomes from a bivalent should move to opposite poles in anaphase 1; both chromosomes in the bivalent move to the same pole if they fail to split; one cell from meiosis I has two of the pair of chromosomes and the other has none; (ii) one chromatid becomes separated from a bivalent; the separated chromosome may move to the same pole as the intact chromosome of two chromatids in the bivalent; centromeres should divide in meiosis II;

(b) total percentages are 16.4 for 25–34, 28% for 35–39 and 42.3% for 40–45; positive correlation/chance of non-disjunction increases as maternal age increases/ hypothesis supported; due to bivalents failing to split; no clear trend for premature centromere division; confidence levels/statistical significance /sample sizes /standard deviation /standard error need to be considered.

Topic 4 Ecology

1. D 2. A 3. B 4. C 5. B

6. (a) I = secondary consumers II = primary consumers III = producers

(b) chemical energy

(c) arrow from the sun to box III

(d) any two of: arrows represent energy losses; heat produced because energy transformations are never 100% efficient; energy not passed along the food chain to another organism; energy released by respiration;

7. (a) methane causes an increase in the Earth's temperature by the greenhouse effect; temperature increases as a result of an increase in atmospheric methane; methane emissions to the atmosphere must be greater than losses

(b) methane emission is a natural process, for example swamps and marshes; humans cause methane emission, for example coal burning/cattle and sheep/ rice paddies; most emissions are caused by humans/ humans have increased emissions considerably

(c) any three of: drain swamps and marshes; reduce cattle and sheep farming; stop growing rice in paddies; control releases of natural gas; reduce burning of coal; prevent forest fires/burning of biomass.

8. (a) 7ppm (+/− 1)

(b) CO_2 falls due to photosynthesis; photosynthesis exceeds respiration in summer; respiration exceeds photosynthesis in winter; May to October are summer in Northern hemisphere;

(c) (i) 395−316; = 79ppm; (ii) 370 to 390 = increase of 200 ppm; 200/10 = 2 ppm per year increase; (iii) concentration in 2100 will be 390 + (2 × 90) = 570ppm;

(d) CO_2 emissions due to human activities; combustion of fossil fuels; forest fires; drainage of wetlands and decomposition of peat;

9. (a) totals are 4.32 and 2.88; divide total by 5; means are 0.58 (control); and 0.86 (treated);

(b) mean is higher for treated mesocosms; lowest three results are controls /similar comparison of rank order; but highest result (1.89) is a control; considering that the open fjord result is 0.14 this result may be anomalous; results are very variable reducing the strength of the evidence;

(c) difference between the means is quite small; in relation to the variability within the control results and within the CO_2 treatment results; test statistic is not in the critical region; so we cannot reject the hypothesis that there is no difference between the means; at the 5% significance level; evidence for the treatment having an effect has not been shown; there may be an effect but the risk of there not being one is too great;

(d) calcium carbonate dissolves; reef-building corals cannot deposit calcium carbonate; coral reefs threatened/lost; loss of habitat for other species; loss of protective reefs close to land; molluscs unable to make shells; disruption to food chains.

Topic 5 Evolution and biodiversity

1. C 2. D 3. C 4. B 5. B

6. (a) mollusca

(b) cnidaria

(c) chordata

(d) porifera

(e) annelida

7. (a)(i) domain (ii) Archaea and Eubacteria;

(b) not living; do not carry out any of the functions of life/not similar enough to living organisms;

(c) kingdom; phylum; class; order; family; genus; species;

(d) adaptive radiation;

(e) fossils; of extinct species/of species with different characteristics;

8. (a) increase in all three species between 1999 and 2010; smaller increase in *Enterobacter*; largest increase in *Klebsiella*; no increases from 2009 to 2010;

(b) mutation; transfer of a gene from another type of bacterium;

(c) natural selection; non-resistant bacteria are killed if antibiotic is used; resistant bacteria divide and pass on their resistance gene;

(d) percentage of resistant infections will increase; fluoroquinolone will become ineffective so rates of use will eventually drop;

9. (a) (i) 2 (ii) 7 (iii) 9 (iv) 6 (v) 9 (vi) 7

(b) cladogram with four species; first split between rabbit and other three species; second split between lemur and other two species; final split between humans and orang-utans;

(c) rabbits in one group and the other three species in a different group; orang-utans and humans more closely related to each other than to lemurs.

Topic 6 Human physiology

1. (a) −90; mV;

(b) both rise steeply; both then drop; both end up level; action potential took much longer in cardiac myocyte;

(c) rise from −90mV to +30mV is depolarization; drop from +30mV to −90mV is repolarization;

(d) 5 ms (+/− 3 ms) in neuron; 340 ms (+/−30) in cardiac muscle;

(e) neuron repolarizes quickly to allow another impulse soon after; so effects of impulse are brief; cardiac muscle repolarizes slowly so contraction lasts longer; to allow time for pumping of blood;

2. (a) I is an artery; thick wall and narrow lumen; IV are veins; thin wall and wide lumen;

(b) dorsal nerve cord/spinal cord;

(c) liver

(d) intercostal muscle; located between ribs;

(e) longitudinal muscle; circular muscle; mucosa; epithelium;

(f) stomach; same layers as esophagus but larger:

3. (a) I = trachea; II = bronchioles/bronchial tree; III = bronchus;

(b) maintains concentration gradients of oxygen and CO_2 between air in alveoli and blood; ensures rapid diffusion/gaseous exchange

(c) alveolus wall consisting of single layer of very thin cells; Type I and Type II pneumocytes distinguished; blood capillaries adjacent to alveolus; capillary wall consisting of a thin layer of very thin cells; moist lining of alveolus; bronchiole connected to alveolus; diameter of alveolus indicated;

4. (a) Table with two rows/columns for treated and untreated; table with two rows/columns for alive and dead; four alive treated, zero dead treated; zero alive untreated; four dead untreated; all expected frequencies are 2;

(b) (observed frequency minus the expected frequency) squared and then divided by the expected frequency; sum of these values for each of the four groups; $(4-2)^2/2$; chi-squared $= 2 \times 4 = 8$;

(c) critical value with 5% significance is 3.84; critical value with 1% significance is 6.635;

(d) there is evidence for an association between treatment with penicillin and survival; at both 5% and 1% significance levels;

(e) 5 mice needed in each of four groups; total of 20 mice needed;

(f) not enough penicillin available to treat more mice; ethical concerns about killing more mice with *Streptococcus*;

5. (a) absorbed as part of a droplet of lipids; endocytosis/pinocytosis;

(b) facilitated diffusion; from a higher concentration in the lumen of the small intestine to a lower concentration in epithelium cells; active transport; from a lower to a higher concentration/against the concentration gradient; using energy from ATP;

(c) absorption is by active transport; rate of uptake limited by zinc concentration at lower concentrations; pump proteins work at a maximum rate at higher concentrations; rate of uptake would continue to rise with diffusion.

Topic 7 Nucleic acids

1. A 2. A 3. C 4. D

5. (a) (i) short (ii) short segments are replicated; on the lagging strand; (iii) Okazaki fragments;

(b) (i) 60 s has much larger peak at 0.5cm; 120 s has much larger peak at 2cm; 120 s has more DNA/radioactivity at all distances from 1 to 3.5;
(ii) in the second 60 s period of the 120 s results DNA polymerase continues to add more bases to the leading strand; DNA ligase links up Okazaki fragments on the lagging strand;

6. (a) globular

(b) number and sequence of amino acids

(c) (i) X is alpha helix and Y is a beta-pleated sheet
(ii) hydrogen bonding

(d) any two of: tertiary structure determines the enzyme's shape; determines the active site's shape; makes the enzyme substrate-specific; shape ensures that when the substrate binds it is distorted/induced fit

(e) mutation; most mutations are deleterious; cancer caused by mutation;

(f) methylation blocks transcription/gene expression; patterns of methylation can be passed on from parent cell to daughter cell;

7. (a) (i) anticodon; consists of sequence of three bases; different sequence coding for different amino acids;
(ii) position where amino acid is attached; opposite end from anticodon; same structure in all tRNAs;

(b) tRNA is produced by transcription; tRNA is used in translation;

(c) differences allow tRNA activating proteins to recognise a specific tRNA; and attach the correct amino acid; similarities allow tRNAs to bind to the same tRNA binding sites on ribosomes.

Topic 8 Metabolism, respiration and photosynthesis

1. **(a)** **(i)** higher than 40 °C; initial rate was faster; then reaction stopped due to denaturation **(ii)** lower temperature than 40 °C because the rate is slower; 30 °C because the rate is half that at 40 °C

   (b) **(i)** curve drawn below curve W; similar shape to curve W; **(ii)** curve drawn above curve W; not reaching as low a substrate concentration by the end;

2. **(a)** each volume of oxygen divided by 5; units shown as volume of oxygen per minute; rates not to more than one decimal place; no Cu rates are 0.0, 1.5, 2.1, 2.3, 2.4, 2.4; results with Cu are 0.0, 0.8, 1.1, 1.3, 1.3, 1.3;

   (b) suitable title for graph; concentration on x axis and rate on y (vertical) axis; legends on the axes are concentration of hydrogen peroxide and rate of production of oxygen; units on the axes shown as % and ml min^{-1}/ml per minute; points plotted not bars or other types of graph; all points correctly plotted; points joined with straight lines /lines of best fit drawn with no extrapolation;

   (c) copper ions are an inhibitor of the enzyme catalase; because the rate is lower when they are present; non-competitive inhibitor; because the rate is lower even at high substrate concentrations;

3. **(a)** oxidative phosphorylation and photophosphorylation

   (b) barrier to proton movement; allows a proton gradient to develop; location of ATP synthase;

   (c) plasma membrane;

4. **(a)** Any two of: double membrane; cristae/infoldings of inner membrane; ovoid shape;

   (b) double outer membrane shown; inner membrane shown folded in to form a crista

   (c) **(i)** label indicating the matrix **(ii)** label indicating the inner membrane/cristae **(iii)** label indicating the cytoplasm outside the mitochondria.

5. **(a)** peaks in the red and blue sections of the spectrum; minimum in the green section at about one third to half of maximal rate

   (b) action and absorption spectra are closely correlated; because pigments absorb the light energy used in photosynthesis; the more light absorbed at a wavelength the more photosynthesis.

Topic 9 Plant biology

1. B **2.** D

3. **(a)** thick waxy cuticle; reduces transpiration;

   (b) palisade mesophyll;

   (c) succulent; water storage; in case of periods of drought

   (d) low transpiration rates; less need for water transport;

   (e) stomatal aperture can be narrower; enough CO_2 will still diffuse into the leaf; less loss of water vapour through narrow stomata;

   (f) **(i)** the higher the salinity the lower the density of stomata; **(ii)** salinity makes water uptake more difficult; more need for water conservation at higher salinities;

4. **(a)** mitosis; cell division;

   (b) **(i)** auxin efflux pumps; in the plasma membrane of shoot apex cells; at one end of the cell; reduce concentration in the cell and increase it in an adjacent cell; **(ii)** causes more growth on one side of the stem than the other; causes the stem to bend; response to brighter light on one side; shoot bends towards that side; phototropism;

5. **(a)** $2.7/3 = 0.9$ g $hour^{-1}$; $2.61/3 = 0.87$ g $hour^{-1}$;

   (b) water passes through semi-permeable membrane into bag on left; by osmosis; increasing the pressure in the bag; pressure is then lower in the bag on the right; fluid moves through the tube from the left bag to the right bag;

   (c) **(i)** phloem **(ii)** high sucrose concentration causes osmosis in both; high pressure due to osmosis in both; flow of fluid due to a pressure gradient in both; left side is like the source in phloem and right side is like the sink;

6. **(a)** I anther/stamen; II stigma; III style; IV petal; V ovary; VI ovule;

   (b) animal pollinates the flower; transports pollen between anther and stigma; animal uses pollen as food; and/or plant supplies nectar;

   (c) pollen not deposited on stigma (from another flower); no male gamete; no fertilization;

   (d) change in gene expression in the shoot apex; different genes for making leaves/stem and flowers; some stimulus triggers this change; triggered by long nights in short-day plants; or by short nights in long-day plants;

   (e) radicle/embryo root; plumule/embryo shoot; cotyledon.

Topic 10 Genetics and evolution

1. **(a)** upper is anaphase II; lower is anaphase I;

   (b) upper produces haploid cells; lower also produces haploid cells;

2. **(a)** first; prophase;

   (b) **(i)** four chromatids **(ii)** five chiasmata;

   (c) breakage of chromatids; rejoining of non-sister chromatids; exchange of material between chromatids;

3. **(a)** polygenic

   (b) AaBb; blue-flowered

   (c) all gametes shown with one allele of each gene only; four homozygous genotypes shown AABB AAbb aaBB and aabb; four double heterozygous genotypes shown AaBb; eight other genotypes shown AABb AAbB aaBb aabB AaBB aABB Aabb and aAbb; all sixteen phenotypes indicated

   (d) 9 blue 3 red and 4 white

   (e) gene A converts white to red and gene B converts red to blue;

4. **(a)** dihybrid

   (b) black body long wing; grey body vestigial wing

   (c) **(i)** 1:1:1:1; grey-bodied long-winged flies: grey-bodied vestigial-winged flies: black-bodied vestigial-winged flies: black-bodied long-winged flies; **(ii)** G for grey body allele and g for black body allele; W for long wing and w for vestigial wing (or other suitable symbols); genes are linked/found on the same chromosome; parental combinations are kept together; unless there is a cross-over between the genes; grey-bodied long-winged parent genotype is G W g w; test crossed with g w g w; non-recombinants are G W g w and g w g w; recombinants are G w g w and g W g w;

5. (a) fewest i in Andamanese; most I in Navajo; fewest/no I^B in Navajo; most I^B in Kalmyk; I^A almost equal in Navajo and Kalmyk (27% and 28.5%); most I^A in Andamanese;
   (b) natural selection favours different blood groups in different environments; founding populations had different frequencies;
   (c) immigration of people with different frequencies; differential survival and reproduction of different blood groups;
   (d) gene pool;
6. (a) plant with 16 chromosomes is diploid; plant with 32 chromosomes is tetraploid/polyploidy; the plant with 24 chromosomes is a triploid hybrid; meiosis fails in triploids because three chromosomes cannot pair; triploids are infertile;
   (b) geographical; behavioural; temporal;
   (c) can interbreed; but do not produce fertile offspring; but have similar characteristics so hard to distinguish as separate species; have different chromosome numbers and a species is expected to have one characteristic number.

Topic 11 Animal physiology

1. (a) 490 (+/−5)
   (b) number of cases drops lower than it was before when vaccination starts; and remains lower; stays below 20 cases; steep drop with first mass vaccinations of children; drops lower each year;
   (c) antigens in the vaccine stimulate lymphocytes; T lymphocytes activate B lymphocytes; activated B cells multiply to form a clone of plasma cells; plasma cells secrete antibodies; activated B cells produce memory cells;
   (d) (i) booster shot stimulates production of memory cells; faster/greater production of antibodies; (ii) some children may miss the vaccination in one year; babies born that have not yet been vaccinated;
   (e) polio not eliminated globally yet; danger of spread of polio back to Brazil by travel;
2. (a) myosin
   (b) actin;
   (c) I stays the same length; II becomes shorter; (ii) I stays the same length; II becomes longer;
   (d) sarcomere shown from one labelled Z line to another; actin filaments attached to each Z line; gap between actin filaments stretching from one Z line and those stretching from the other; myosin filaments between actin filaments; actin and myosin filaments overlap as the myofibril is partly contracted;
3. (a) I humerus; II radius; III ulna
   (b) (i) triceps attached to humerus; and ulna; biceps attached to humerus; and radius; (ii) biceps contracted; triceps relaxed; because the arm is flexed at the elbow;
4. (a) (i) arthropoda; jointed legs/appendages; exoskeleton; segmented body; (ii) animal kingdom; arthropods are animals;
   (b) exoskeleton; joints between sections of leg; muscles attached to inside of exoskeleton; flexors and extensors/antagonistic muscles; muscles cross joints;
   (c) Malpighian tubules; absorb ammonia; convert ammonia to uric acid; uric acid discharged into hindgut; passes out with feces; ions reabsorbed from feces; water reabsorbed by osmosis; to achieve osmoregulation;
5. (a) I cortex; II medulla;
   (b) (i) V, VI, IV, VII, III (2 marks minus 1 per error); (ii) Bowman's capsule; collects filtrate from the glomerulus; proximal convoluted tubule; selectively reabsorbs useful substances; loop of Henlé; maintains hypertonic conditions in the medulla; distal convoluted tubule; regulates water/ion/pH concentration of blood; collecting duct; osmoregulation by reabsorbing variable amounts of water;
   (c) ADH;
6. (a) both involve mitosis; and cell growth; and two divisions of meiosis; and differentiation; both result in production of haploid gametes;
   (b) more male gametes produced than female gametes; four per meiosis versus one per meiosis; cytoplasm eliminated from male gametes but increases in female gametes;
   (c) cortical granules expel contents after one sperm has entered the oocyte; zona pellucida converted into impenetrable fertilization membrane;
   (d) early stage embryo feeds on egg cytoplasm/yolk; blastocyst implants in uterus wall; nutrients diffuse into embryo from mother's blood; placenta develops to increase rate of nutrition as embryo grows larger; nutrients pass from maternal to fetal blood as they flow close together;
   (e) hCG produced by embryo; stimulates ovary to secrete progesterone (for about 12 weeks); placenta takes over from ovary.

Option A Neurobiology and behaviour

1. (a) I cerebral hemisphere; II cerebellum; III medulla oblongata; IV vertebra; V spinal cord;
   (b) I for higher order functions; voluntary muscle control/other higher function; II for control of muscular coordination movement; balance; III for automatic/homeostatic functions; swallowing/other example;
   (c) neurulation; infolding of ectoderm; to form the neural tube;
   (d) photoreceptors; rods; cones;
   (e) lesions/autopsy; animal experiments; fMRI;
2. (a) more cones than rods in the fovea; more rods than cones everywhere else; rods more evenly distributed; maximum density of cones is higher than rods;
   (b) rods are more sensitive to light than cones; so are more useful in dim light than cones; one type of rod but three types of cones sensitive to different wavelengths; cones give colour vision but rods give only monochrome;

(c) both send impulses to bipolar cells; which pass impulses to ganglion cells; impulses from several rods passed through same ganglion cell but only from one cone cell;

3. (a) (i) vehicles are in the right side of the visual field; so are processed in the left visual cortex; (ii) sounds of vehicles are received by both ears but will be louder in the right ear; sounds from right ear processed by right cerebral hemisphere;

(b) detected by hair cells in semicircular canals; hairs move with canal/head but fluid tends to remain still so flows past hairs;

(c) olfactory receptors in nose; receptors detect smells from the garbage in odorant receptors in cell membranes/cilia;

(d) microphone picks up sounds; speech processor filters out frequencies above or below those of speech; external transmitter passes signals to internal receiver; stimulator converts signals from receiver into electrical impulses; electrodes pass impulses directly to auditory nerve;

4. (a) trial and error learning; operant conditioning;

(b) (i) reinforcement; encourages the rat to repeat a behaviour; (ii) operant response; carried out by the rat to get the food reward;

(c) give the rat a food reward; when it presses the lever; in response to a stimulus; for example a red light/green light/sound from the loudspeaker; no reward if the rat presses the lever at other times; rat learns to associate a stimulus with pressing the lever and getting a food reward; repeats to increase reliability;

5. (a) cannot be measured objectively by the experimenter; subjective effects may not be reported accurately; may be affected by factors other than the drug;

(b) positive mood; increased extroversion; increased self-confidence; altered perception; emotional excitation;

(c) citalopram reduced the subjective effects of MDMA; citalopram specifically inhibits serotonin re-uptake/serotonin release; ketanserin reduced MDMA-induced perceptual changes/emotional excitement ; ketanserin binds to/blocks serotonin receptors; haloperidol reduced MDMA-induced positive mood; haloperidol binds to dopamine receptors;

6. (a) animal behaviour; in natural conditions;

(b) (i) can be either; migration by blackcaps is genetic; feeding on cream in milk bottles in blue tits is learned; (ii) learned behaviour pattern spread more rapidly; feeding on cream spread very rapidly across Europe;

(c) females give birth at the same time; females lactate at the same time; females can suckle each other's cubs; male cubs are ready to leave the pride at the same time; can form a group of males; have more success in competing for dominance of another pride; can defend another pride more effectively as a group;

Option B Biotechnology and bioinformatics

1. (a) sewage/slurry; manure; crop wastes;

(b) anaerobic/no oxygen; warm; many methanogenic bacteria/archaeans;

(c) biogas emerges from II; contains methane; burned to generate heat/electricity; slurry/compost/decomposed organic matter/humus removed from III; used as a fertilizer/soil conditioner;

(d) prevents methane escaping into the atmosphere; reduces the need to burn fossil fuels; avoids manure entering rivers/watercourses;

2. (a) (i) Magnification = 15 mm/10mm; = x 1,500; (ii) diameter of bacteria in image = 2.0–2.5 mm; actual size = 1.33–1.66 mm;

(b) quorum sensing; if population density is high enough bacteria cooperate to form a biofilm; polysaccharide produced; glues cells together and to the surface;

(c) can cause infection of the patient; difficult to remove;

(d) all reduce the thickness of the biofilm compared to the positive control; none of them remove the biofilm completely; farnesol is the most effective; all reduce the biofilm more than when the bacteria are killed with ethanol;

(e) allows thickness of the biofilm to be measured; with great precision/quickly;

3. (a) I bacterial DNA; II *Agrobacterium (tumefaciens)* cell/wall; III soybean mitochondrion; IV soybean chloroplast; V nucleus/nuclear membrane; VI soybean cell/wall;

(b) (i) glyphosate resistance gene; kanamycin resistance gene; (ii) allows glyphosate to be sprayed onto growing soybean crops; kill weeds but not the crop; reduces need for plowing/other herbicides; kanamycin kills plant cells; resistance gene kills all cells that have not received DNA from the plasmid/have not been transformed;

(c) soybean cells grown in tissue culture/on nutrient agar gel; plantlets differentiate from the tissue; plantlets excised and grown on; seeds of GM variety produced;

4. (a) ELISA; HIV antigen attached to well in plastic plate; test sample placed in well; HIV antibody attached to antigen; secondary antibody attached to HIV antibody; with enzyme attached; enzyme has converted a colourless substrate into a coloured compound;

(c) no antibodies bind to the antigen; secondary antibodies do not attach so are washed away; coloured substance not produced;

5. (a) both are used to identify similar sequences; both help to assign functions to newly discovered sequences; BLASTp aligns amino acid sequences in proteins; BLASTn aligns nucleotides in DNA/RNA;

(b) (i) databases can store vast amounts of information/sequences; allow easy access; can be searched to find similar sequences; (ii) model organisms can be used to study gene function; similar genes in related organisms tend to have similar functions; avoids having to do experiments on humans; (iii) multiple sequence alignment allows base sequences in related organisms to be compared; allows relatedness to be assessed; used to construct cladograms/phylograms; (iv) used in mice; to determine gene function; (v) used for discovery of genes with similar sequences in databases; allows functions of genes to be deduced;

(c) *M. fijiensis* is more distantly related to the other two species than they are to each other; *M. fijiensis* split from the other two species longer ago than they split from each other.

Option C Ecology and conservation

1. (a) both occur over a range of depths but not all; *angustifolia* extending into deeper water; *latifolia* extending out of the water;

   (b) −20 to +80 are in limits of tolerance for *latifolia*; −20 to +115 for *angustifolia*; −20 is in zone of stress for both species;

   (c) fundamental niche is wider than realized; includes depths from −20 to +100 cm whereas realized niche is only from +75 to +100 cm; *angustifolia* is excluded from shallower depths; by competition from *latifolia*;

   (d) transect

2. (a) bioaccumulation; absorbed from food but not excreted; higher concentrations at each trophic level; albatrosses are top carnivores/later stage in food chain;

   (b) (i) positive correlation; higher concentration of methyl mercury over time; (ii) release of mercury into the oceans; from industry;

   (c) macroplastic debris; parents feed plastic wastes from the oceans to their young;

3. (a) (i) live in water; large enough to be seen without magnification; animals with no backbone; (ii) vary in their sensitivity; live continuously in the water; so indicate pollution levels over a longer period than a single chemical test;

   (b) (i) $(12 \times 4) + (8 \times 4) + (8 \times 2) + (4 \times 2); = 104$; divided by 32; = 3.25; (ii) water quality is excellent; very low pollution level;

4. (a) sigmoid/S-shaped

   (b) (i) exponential before; increases at a faster and faster rate; transitional after; increases slowing down; (ii) natality increasing before; mortality low; natality decreasing after; and/or mortality increasing;

   (c) (i) line reaching a plateau at 2000; by year 8 (ii) any three of: food supply; predation; breeding sites; disease

   (d) (i) population would have reached carrying capacity more quickly (ii) carrying capacity would have been the same;

5. (a) waterlogged soils are oxygen-deficient; denitrifying bacteria; convert nitrate to nitrogen gas; nitrifying bacteria cannot convert ammonia to nitrate;

   (b) catches insects; digests insects; absorbs nitrogen compounds from insects;

   (c) fire is a form of environmental disturbance; harms some organisms and helps others; changes the structure of the ecosystem; increases biodiversity; *Dionaea* might not survive without periodic fires in its habitat.

Option D Human physiology

1. (a) *H. pylori* is implicated as a cause of stomach ulcers/cancer; antibodies show that the patient has been infected with *H. pylori*;

   (b) incidence of stomach ulcers and cancer is higher in patients who had been infected with *H. pylori*; all patients with stomach cancer had been infected with *H. pylori*; some patients with stomach ulcers had not been infected so there must be alternative causes; correlation does not prove causation;

   (c) *H. pylori* infects the stomach; causes inflammation of the stomach wall; allows stomach acids/pepsin/protease to attack the stomach wall;

   (d) helps hydrolysis reactions; controls pathogens in ingested food;

2. (a) essential nutrients cannot be made in the body but non-essential nutrients can; essential nutrients must be included in the diet;

   (b) protein synthesis blocked; when translation reaches the amino acid;

   (c) anorexia; starvation;

   (d) type II diabetes; coronary heart disease;

   (e) leptin is secreted by adipose tissue; leptin causes a reduction in appetite;

   (f) record daily diet; types and amounts of each food; use a nutrition database to assess energy and nutrient content;

3. (a) sinusoids; with rows of adjacent hepatocytes; capillaries leading to sinusoids;

   (b) synthesis of plasma proteins; albumin/fibrinogen;

   (c) vitamin A/retinol/vitamin D/calciferol; iron; glucose/glycogen;

   (d) ethanol concentration lower in the hepatic vein; due to detoxification by hepatocytes;

4. (a) count number of beats using QRS wave/other repeating pattern and number of squares, one beat takes 0.71 seconds; 84 beats per minute;

   (b) (i) I is P-wave; II is QRS wave; III is T-wave; (ii) I atrial systole; II ventricular systole; III ventricular diastole;

   (c) shorter gaps between QRS waves; faster heart rate; shorter gaps between T-wave and P-wave;

5. I is a peptide; II is a steroid;

   (b) insulin/glucagon/ADH/leptin/oxytocin/prolactin/other peptide; testosterone/progesterone/estrogen;

   (c) peptide hormone binds to receptor; receptor converts secondary messenger to active form; triggers off cascade of reactions; steroid hormone enters cell; binds to receptor protein; activated receptor promotes/inhibits transcription of specific genes;

6. (a) V_E increases as CO_2 concentration increases; greater increases in V_E with successive increases in CO_2 concentration;

   (b) increase the blood concentration; detected by chemoreceptors in aorta/carotid artery; impulses sent to respiratory centre of medulla oblongata;

   (c) further increases in V_E; until maximal V_E is reached; fall in blood pH; fatal if blood pH drops below 6.8;

   (d) exercise/muscle contractions;

   (e) dissolved in plasma; bound to hemoglobin; converted to hydrogen carbonate ions;

   (f) reduce the affinity of hemoglobin for oxygen; Bohr shift.

Index